MOON MORPHOLOGY

One of the first detailed photographs (ca. 1851) of the Moon, taken by Professor W. C. Bond, of the Harvard College Observatory, and John Adams Whipple, a Boston photographer, using the newly completed 15-inch Harvard refractor, then the largest telescope in the world. Previous images were small and revealed only the highlands and maria. Because the photograph is a daguerreotype, the image is reversed.

This is one of a series of images of the Moon under different phases and represents the beginning of the photographic record of the lunar surface, culminating in the Lunar Orbiter images used in this book and the returned Apollo photographs. (From the collection and with the kind permission of Walter A. Johnson.)

MOON MORPHOLOGY

INTERPRETATIONS BASED ON LUNAR ORBITER PHOTOGRAPHY

By Peter H. Schultz

Foreword by William R. Muehlberger

UNIVERSITY OF TEXAS PRESS • AUSTIN AND LONDON

Library of Congress Cataloging in Publication Data

Schultz, Peter H   1944–
    Moon morphology.

    A revision of the author's thesis entitled A preliminary
morphologic study of the lunar surface, University
of Texas at Austin, 1972.
    Bibliography: p.
    Includes index.
    1. Lunar craters. 2. Moon—Surface. I. Title.
QB591.S38 1975   523.3   74-22176
ISBN 0-292-75036-6

Design/
Jo Alys Downs
Eje W. Wray
David Mortimore
Typesetting/G & S Typesetters
Printing/Meriden Gravure
Binding/Universal Bookbindery

To the memory of J. Hoover Mackin

CONTENTS

PUBLISHER'S NOTE

In 1965 Professor J. Hoover Mackin of The University of Texas at Austin submitted to NASA a proposal for support of research on ignimbrites (ash-flow tuffs) as a basis for interpretation of lunar plains. An initial grant for the study was awarded the following year, followed by supplements thereafter as Mackin and his graduate students intensified their investigation of terrestrial ash flows in western North America. Concurrently, Mackin began to assemble a collection of photographs of lunar topographic features, and he soon enlisted the assistance of Peter H. Schultz. In July 1968 he wrote NASA as follows: "With Peter Schultz as a research assistant, I am preparing a selection of Orbiter photographs of lunar features, with an explanatory text appropriate to students and to the public, for publication by NASA." He indicated that V. R. Wilmarth of the NASA staff was taking an active interest in the compilation.

Following Mackin's unexpected death on August 12, 1968, several University of Texas geology professors assisted Mackin's students with their research. Most active among them was W. R. Muehlberger, who shared Mackin's keen enthusiasm for lunar geology, and who asked to be designated principal investigator for continuation of the NASA grant for study of ignimbrites and compilation of the catalog of lunar photographs. Muehlberger himself soon became so involved in the geologic training of astronauts that he took temporary leave from the university and recommended that S. E. Clabaugh succeed him as principal investigator. Clabaugh subsequently supervised studies of ignimbrites in western Mexico and Texas, and in July 1972 he obtained NASA approval for a revised work plan under Grant NGL 44–012–045, which provided subvention funds to the University of Texas Press to assist in the publication of the photographic atlas of lunar surface features. NASA funds for this purpose have enabled the Press to produce a photographic atlas of high quality at a reasonable price, consistent with Mackin's plan for making it available to students of lunar surface features and to the public.

FOREWORD

The past several years have seen an all-encompassing revolution in the geological sciences. On Earth, through extensive oceanographic research, the unifying concept of plate tectonics evolved by which it can be shown that this planet is divided into a set of rigid plates that are moving apart along spreading lines, are sliding under one another at the site of ocean trenches, or are sliding past one another, as along the San Andreas fault zone in California. From space has come the study of samples from another planet, the culmination of many years of intense effort with the successful Apollo manned landings on the Moon. Each line of research is a spectacular collaboration of engineering and scientific talent: one, the exploration of the wet ocean of Earth; the other, the empty ocean of space.

Telescopic studies of the Moon, since the pioneer synthesis of G. K. Gilbert in 1893, have demonstrated that the dominant process that shaped the lunar surface was impact events of all sizes. It also seemed likely that the dark maria were covered with lava flows, but resolving the problem of what proportion of the lunar craters was volcanic had to await the Orbiter photographic missions and the later Apollo landings. It seemed likely, however, that impact processes dominated because, as F. E. Wright pointed out in the 1930's, the fact that at full moon the surface was equally bright at the limbs demonstrated that the entire surface was covered with dust. The depth of this dust (the crushed and pulverized surface layer of impact debris now known as regolith) could not be determined with telescopic resolution.

The three successful Ranger missions furnished an early close-up view of two mare regions and a highland crater floor, Alphonsus. These missions extended the determination of crater size versus frequency distribution down to diameters of 1 meter. Furthermore, these photos showed that the slopes on the maria were sufficiently gentle to be of no serious hazard to either manned or unmanned spacecraft (obvious exceptions being large, blocky craters). The identification of blocks lying on the surface indicated that the regolith had a bearing strength presumably firm enough to support a spacecraft without its sinking deeply into the regolith. Blocky rimmed craters, along with the derivation of a steady-state cratering theory, permitted estimates of

regolith thickness for these landing sites—less than 10 meters for the maria.

The five successful Surveyor missions furnished the next steps in lunar investigations by exploring four mare sites spread across the Moon's equatorial belt and a fifth on the rim of Tycho, a large fresh crater in the Southern Highlands. These crafts showed that the lunar surface did have the predicted strengths to support a spacecraft, returned thousands of television views, dug several trenches in the lunar surface, performed chemical analyses, measured mechanical properties, and carried out a number of other experiments.

Both the Ranger and Surveyor systems obtained data on small areas in great detail. With the advent of the Orbiter spacecraft, however, came the photographic coverage of the entire Moon—both nearside and farside. For the first time, it was possible to compare and analyze features over the entire planet. The stunningly beautiful photographs provide resolutions ranging from hundreds of meters down to meters for a few selected areas. These photographs have been and will continue to be the prime reservoir of data for comparison of morphologic features and interpretation of their origin for as long as the Moon and the samples obtained by the manned American Apollo and unmanned Soviet Luna missions are studied.

Dr. J. Hoover Mackin, William Stamps Farish Professor in the Department of Geological Sciences at The University of Texas at Austin and a member of the National Academy of Sciences, was for many years an active and effective member of the Project Apollo Field Geology Planning group. In addition, he had begun, with financial support from the National Aeronautics and Space Administration, a morphological study of lunar surface features from the Lunar Orbiter photography, selecting key examples of each and putting them into a coherent classification scheme. By one of those happy coincidences, Peter H. Schultz, then a graduate student in astronomy at The University of Texas at Austin, occupied the seat next to Mackin on a plane trip to Austin. As a result of their discussions, Mackin invited Schultz to work with him on the morphologic atlas. To the loss of all who knew or worked with him, Dr. Mackin died before the project was well under way and shortly before the first Apollo landing on the Moon. Peter Schultz continued the atlas to completion, largely independently, over the next several years. In my discussions with him during the compilation stages, it was obvious that he was not only compiling, but also interpreting and synthesizing—frequently better and more clearly than any studies in the literature of that period. In a field that is in such an active state of ferment as is the analysis of the Moon, it is inevitable that several workers will reach identical conclusions at about the same time. I am sorry that Pete did not publish his new interpretations as he made them; I believe he could well have been the first to offer many of them.

This atlas is different from any other one prepared from Orbiter photographs. NASA SP-200, *The Moon as Viewed by Lunar Orbiter*, contains a set of some of the best photographs displaying a variety of lunar features and it presents tentative interpretations. NASA SP-206, *Lunar Orbiter Photographic Atlas of the Moon*, is a selection of Orbiter photographs that gives nearly complete coverage of the entire Moon. Photographic support data and locations of principal lunar features are given as well. It contains no interpretation of the photographs. NASA SP-241, *Atlas and Gazetteer of the Near Side of the Moon*, uses Orbiter IV photographs as the base for presenting the locations of named features. The present volume groups photographs to illustrate and analyze a morphologic type, thus permitting the reader either to study and interpret on his own or to follow the text description and interpretation of the author. This book is, then, a ready and convenient reference to the best photographs for those interested in interpretation, analysis, and synthesis. It also furnishes a convenient source of photographs for teaching—especially because many of the morphologic forms are illustrated in stereoscopic pairs. And for the amateur lunar scientist, it places in his hands a distillation of raw data that enables him to focus on the processes that shaped the Moon's surface.

William R. Muehlberger
Department of Geological Sciences
The University of Texas at Austin

PREFACE

Moon Morphology is essentially my doctoral dissertation, which was completed in August 1972 ("A Preliminary Morphologic Study of the Lunar Surface," The University of Texas at Austin). It began in the spring of 1968 after I had compiled a preliminary outline of lunar surface features under the guidance of the late J. Hoover Mackin of the Department of Geological Sciences, The University of Texas at Austin. The death of Professor Mackin in the summer of 1968 was an immeasurable loss both to this effort and to the scientific community, but his continuous drive and creativity have remained as constant sources of stimulation. I deeply regret that the final product of this study has not had the benefit of his input and legendary critique.

Although six Apollo missions have supplied an invaluable record of the surface composition of the Moon, the sampled areas represent a miniscule portion of the total surface area. To attempt to interpret the entire geologic history of the Moon from these data alone would be clearly a precarious undertaking. The Apollo missions fortunately involved numerous experiments that will enable researchers to fit the returned samples into a larger and more complex puzzle. This puzzle finally must form an image of the lunar surface as we see it today—in particular, as the remarkable Lunar Orbiter and Apollo missions recorded it.

As the photographic record from the Apollo missions became available, a seemingly endless project of selecting the best examples of surface features was looming; consequently, I restricted myself to the illustration of features shown by the Lunar Orbiter missions. This restriction has resulted in lunar views that may look familiar, but it does not jeopardize completeness.

In recent years, several books using Lunar Orbiter photography have been released. Lowman (1969) and Kosofsky and El-Baz (1970) published selected Orbiter photographs with preliminary interpretations, and Mutch (1970) presented an excellent introductory text on the Moon as seen through the eyes of an experienced stratigrapher. In addition there are two NASA atlases based on a large collection of Orbiter images (Bowker and Hughes, 1971; Gutshewski et al., 1971).

I have taken a different tack from these efforts; in particular, the wide assortment of surface forms has been classified by their morphologic appearance. My intent was not to develop a rigid classification scheme as an end in itself; rather, it was to use the classification to describe and interpret both formative and modifying processes. Although occasional references are made to Earth analogs, most of the discussions are based on the premise that lunar surface features may not have correspondent analogs on Earth. A starting point is necessary, however, and, typically, I have chosen features that, because of the visible existence of small-scale surface detail, appear to be well preserved. The variety in their initial appearances permits discussions about their modified appearances. The broad scope of this study also makes intercomparisons of dissimilar features possible. For example, the inferred preservation of a small crater can be compared to the inferred preservation of a nearby flow terminus. Even where such intercomparisons are perhaps inappropriate for determining relative ages, they do provide a handle for understanding different types of modifying processes at various size scales and frequently raise questions about formative processes.

As space probes return images of other planets, new groups of researchers will be introduced to the field of planetology. The excellent preservation of lunar topography over eon upon eon makes it an ideal template with which to fit theories of processes acting on other planetary surfaces. Two dominant processes have sculpted the lunar surface: impact events and volcanism. A morphologic study of the nature of this one must incorporate the philosophy of multiple working hypotheses—or perhaps more accurately, schizophrenia: a Jekyll seeing impact-produced features and a Hyde seeing volcanically produced forms. This peculiar type of lunacy permeates the text, occasionally leading either to inconclusive alternatives or to interpretations not currently popular. But these loose ends are meant to be used as the bases for further research.

The nomenclature and discussions frequently assume a knowledge of basic geology. For those readers who desire an illustrated tour of surface features, the photographs are juxtaposed with the text and are placed in a meaningful sequence. As a result, the book should be useful both to the novice and to the advanced lunar interpreter.

After Professor Mackin's death, the encouragement from William R. Muehlberger, V. Richard Wilmarth, John Dietrich, and Ted H. Foss was directly responsible for the continuation of this research. From the outset, the Department of Geological Sciences at The University of Texas at Austin, under the chairmanships of Dr. Muehlberger and later Dr. Robert E. Boyer, provided unfailing support and generous facilities. In addition, the early part of this study was performed under a NASA Traineeship (1968–1970) and was put into its final form while I was a National Research Council Resident Research Associate at NASA Ames Research Center (1973–1974).

Several people deserve special acknowledgment. William R. Muehlberger, V. Richard Wilmarth, F. Earl Ingerson, and Harlan J. Smith have made invaluable reviews of the book as it mushroomed over the years. Other much-appreciated contributions both in labor and in discussions have been supplied by Walker D. Manley, Jr., and Ruth Fruland.

My special and loving gratitude goes to my wife, Barbara Williston Schultz, who has endured much more than our wedding vows had warned.

MOON MORPHOLOGY

Introduction

Planetologists are in a unique position relative to that of terrestrial geologists. Geologic views of Earth have evolved through detailed regional studies, and imaginative syntheses of such studies have yielded geologic histories of large provinces. Currently, aircraft and satellites are permitting a review from photographic overviews. Planetologists, on the other hand, must necessarily begin with the megaview and progressively narrow their studies to the microview as new techniques and space missions permit. Their initial questions often are much more general and usually are "answered" with relatively little data and crude assumptions. As a result, planetary bodies initially appear to have much simpler histories, and the causative processes seem to be more easily resolved. In many respects, the planetologist plays the role of the ultimate "arm-chair geologist."

Three general approaches are taken in the study of planetary surfaces: mapping; morphologic classification; and topical studies. Cartographic and topographic maps are invaluable aids and do not require assumptions concerning the origin or modification of the surface. Geologic maps also rely on careful observations, but ultimately informed and often controversial decisions concerning genesis are necessary. Pioneers in astrogeologic mapping are the geologists in the Astrogeology Branch of the United States Geological Survey (see Wilhelms [1970] for a history and summary of this work). Morphologic studies, which comprise the second approach, ideally should not depend on the genesis of features, but in practice such assumptions are used because of such problems as poor photographic resolution or excessively cumbersome descriptive lists. Typically, morphologic studies are required before geologic maps can be intelligently made. The third approach, topical studies, includes a wide variety of qualitative and quantitative analyses of more selected problems, such as spectrophotometry of the surface and crater statistics.

This volume is concerned with the second approach, that of morphologic classification. It is a broad survey of lunar surface features with interpretations, and it treats these features as products of an evolving planetary body rather than as data points. Since the surface is examined at resolutions provided by Lunar Orbiter photographs, a complex history emerges that has not been obvious through Earth-based telescopes. The very general approach taken here enables comparisons to be made between proposed surface processes acting on a variety of features over widely separated areas.

Various morphologic classifications have been made in the past. Notable studies include those by G. K. Gilbert (1893), J. E. Spurr (1944, 1945, 1948, 1949), R. P. Baldwin (1949, 1963), G. Fielder (1965), and G. J. H. McCall (1965). The present study considerably expands the list of typical surface forms, but some of the previously established nomenclature and general surface types are used. An important departure from these earlier works is the separation of craters according to the appearance of their floors, walls, and rims. My initial studies indicated that classification of characteristic crater types resulted in an enormous list that generally reflected different histories of these zones. Craters exhibiting appearances common to all three subdivisions exist, and these appearances presumably imply similar origins and histories.

Examples of characteristic forms were selected only from Lunar Orbiter photographs. This restriction results in a biased sample since the Lunar Orbiter series were designed to provide photographic coverage for future manned landings. Surface coverage of the first three missions was primarily restricted to low sun angles and to marelike terrains of the nearside equator. Orbiter IV returned very broad coverage, but the resolution was necessarily low, although much better than terrestrially based views. The coverage of Orbiter V included sites of scientific interest, predetermined by experienced lunar investigators.

Although selection of examples for the text was restricted to Lunar Orbiter photography, additional background was provided by inspecting photographs returned by Apollo Missions 8, 10, 11, 12, 14, 15, 16, and 17. I also made extensive reference to the *Consolidated Lunar Atlas* (Kuiper et al., 1967) to check the effects of different solar illuminations on topography and surface reflectivities.

Arranged in nine chapters, this volume devotes the first five to the subject of craters. The following outline provides a preview of the content and methodology:

I. Lunar craters
  A. Shapes in plan
  Characteristic geometries displayed by craters of different sizes are examined. In contrast to the practice in the following chapters, the largest examples are examined first in order to introduce the reader to features revealed by increasing resolutions.
  B. Floors
  Small craters are describable by their profiles: concave, flat, convex. Larger craters typically display flat floors, and, in many examples, the floor represents a volcanic basin formed by the initial cratering event or modified by subsequent multistage events. The floor is, in the latter case, a thermometer for lunar thermal history.
  C. Walls
  Craters with diameters less than 1 km have walls that may be a sensitive indicator of surface processes, but this typically requires assumptions concerning the initial crater form. For craters 1 km–15 km in diameter, the walls are screelike, whereas, in those exceeding 15 km, slumping is a dominant modifier. Walls provide a test for determining slope stability, which in turn is a function, at least in part, of crater age.
  D. Rims
  The rim zone is defined as the area beyond the rim crest. Its degree of preservation may be an indicator of surface processes, but this requires assumptions concerning the initial appearance of the ejecta blanket. The rim zone of some large craters is ordered into three broad areas that reflect relatively well defined depositional sequences.
  E. Crater groupings
  Chains and clusters of craters are found at all dimensions. Examples of both impact and internally produced groupings are examined.
II. Positive-relief features
  Positive-relief forms result from processes that destroy surrounding terrain or that construct them volcanically or structurally. The distinction in origin is not always obvious, but there is morphologic evidence that numerous volcanically constructed features are abundant over the lunar surface.
III. Negative-relief features
  Rilles dominate this class. They are described as curvilinear or rectilinear in plan. Curvilinear rilles are further subdivided into nonsinuous and sinuous. Sinuous rilles have meandering or nonmeandering plans. Both well-developed and incipient rilles are presented and discussed.
IV. Maria and other plains-forming units
  This discussion illustrates the wide range of implied flow viscosities responsible for major and local inundation.
V. Albedo contrast
  Contrasting surface reflectives commonly indicate separate geologic units where topographic discontinuities are either unresolvable or nonexistent.

In addition to the above discussions, Appendix A presents selected features and regions that have excellent photographic coverage. I consider it of more value to present these forms intact than to isolate and scatter their particular features throughout the text. The most compelling reason, however, for presenting these photographs as a concluding unit is that they provide a summary for the discussions in the chapters that precede them. The regions include the large multiringed Orientale basin; the craters Petavius, Copernicus, Aristarchus, Vitello, and Messier; the Harbinger Mountains; and the Marius Hills.

Each plate in this volume has accompanying text that records its selection and regional placement and provides descriptions and interpretations. This organization precludes reading the discussions strictly as a book; it also necessitates the repetition of ideas wherever clarity requires. To provide cohesiveness, general comments and summaries follow each chapter. Both typical and atypical features are included since the latter commonly add insight to the former. I do not intend for my interpretations to be taken as dogmatic conclusions; most often they offer plausible alternatives. More confident conclusions require more detailed and quantitative studies of specific features.

Lunar samples returned by Apollo missions place constraints on many of the early working hypotheses. I refer to the results and implications of these highly successful missions, but the intent is not to provide a detailed review of them here. Furthermore, the landings encompass a minuscule portion of the lunar globe, and a hypothesis that is found inapplicable at one of these sites may be applicable for other areas. The wide variety of lunar surface forms underscores this caveat.

Also included in the text are observations on other interesting or notable features visible in the adjacent photograph. Although occasionally such comments are seemingly unrelated to the theme of the chapter or section, they are introduced either to add information on regional processes or to provide an index for reference from other discussions elsewhere in the text.

The identifying data, which accompany each plate, were derived from Anderson and Miller (1971). The first identification is the Lunar Orbiter photograph number, in which the roman numeral denotes the particular Orbiter mission, and the following arabic number gives the particular photograph. The upper-case letter "M" indicates "medium resolution"; "H" indicates "high resolution." The high-resolution abbreviation is followed by a number that identifies which high-resolution strip is included (Langley notation).

The small figure below each photograph shows the approximate north direction. The photographic scale is given in kilometers and corresponds to the bar adjacent to the figure. It was calculated according to framelet widths in the photograph. Oblique views of the surface typically result in meaningless data and these examples are noted.

Additional information is given in Appendix B and includes the following:

1. North Deviation Angle (NDA). This is the north direction in degrees and corresponds to the angle drawn on the figure that accompanies most photographs.
2. Emission Angle (EA). This gives the angle between the idealized surface normal and the camera axis at the intersection of this axis with the surface (see figure). The emission angle is equivalent to the angle between the photographic image plane and the surface plane.
3. Incidence Angle (IA). This is the angle between the surface normal and the Sun.

These data refer to the intersection of the camera axis and the surface, whereas the selected photographic coverage is commonly off axis. Consequently, errors are present that are most apparent for distant views, such as oblique views and the medium-resolution Orbiter IV photographs.

The orientation of the photographs is typically with apparent illumination from the right, which corresponds to the east for most examples. Exceptions to this rule occur for oblique views of the surface (large "EA"), in which the feature is reoriented so that it appears right-side-up. Photographs of the lunar farside commonly have solar illumination from the west, and these are also reoriented so that the illumination is from the right, which results in the north direction pointed downward on the page. Additional exceptions are found in stereo coverage where Orbiter V and some Orbiter IV stereo pairs have the apparent solar illumination from the top of the page. These stereo pairs are designed for standard pocket stereoscopes.

Arrows placed on the plates are referenced in the text. It should be noted that surface features were identified and studied on the original 20" × 24" Orbiter photographs. If such features are not easily recognized in the reproductions included here, reference should be made to the originals.

Definition of Incidence Angle (IA) and Emission Angle (EA).

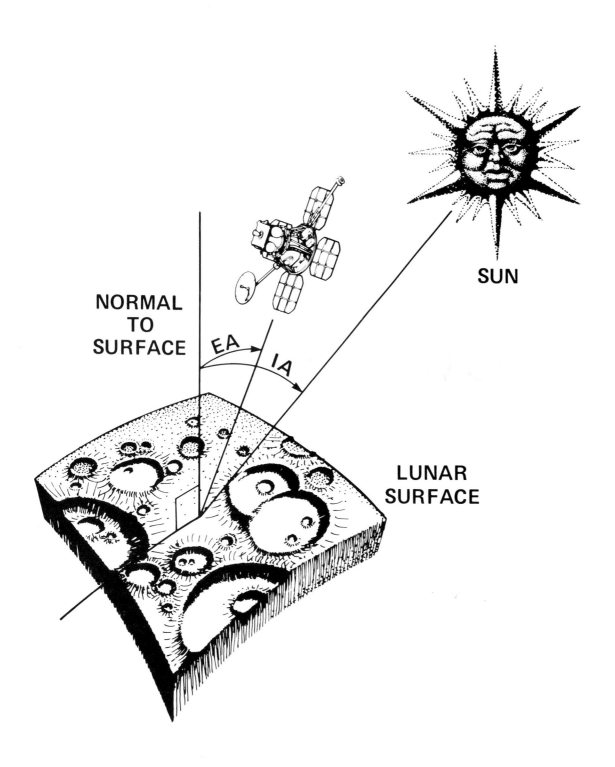

NORMAL TO SURFACE

EA

IA

SUN

LUNAR SURFACE

1. Craters: Shape in Plan

INTRODUCTION

Craters are the dominant surface form across the lunar surface. In the following sections their geometries are discussed under the following classes:

1. Concentric
2. Circular
3. Elongate
4. Scalloped
5. Polygonal

Each class is further subdivided into somewhat arbitrary divisions according to crater diameter:

1. Greater than 100 km
2. 15 km–100 km
3. 1 km–15 km
4. Less than 1 km

Each crater varies in its apparent degree of freshness or stage of degradation. These data, though subjective, provide insight into the relative ages of craters in each category and therefore are of value in the reconstruction of lunar events and history.

Not all craters resulted from impact events; some are volcanic in origin. Consequently, owing to the connotation of explosive formation, the term "crater" is misleading. A similar problem is faced in volcanology with the important distinction between craters (an explosive form) and several types of calderas (a volcanic collapse form). Attempts to avoid such confusion have led to more general terms, such as "cirques" (McCall, 1965) and "ring" structures (Fielder, 1965). A more specific label for each specific case depends on further studies and cannot be resolved through morphologic classifications alone. In the following descriptions, the term crater (unless stated otherwise) is used for approximately circular depressions without implications of genesis.

The classification of craters according to shape in plan could provide clues to differences in origin, but in most instances it represents differences in topography, subsurface structure, or later modifications. Within each class, a variety of processes may have occurred; hence, class members are not homogeneous samples with respect to such processes, much less with respect to origin. This is especially evident when contrasting craters of mammoth proportions with those less than a kilometer in diameter.

Several workers (Quaide and Oberbeck, 1968; Öpik 1968) have discussed the factors influencing initial shapes of impact craters. Öpik's discussion is largely theoretical and considers the effect of the projectile velocity, projectile density, angle of incidence, and character of the target (i.e., the lunar surface as granular, ductile, or brittle). Ideally, his analysis could be applied to craters of any diameter by adjusting the appropriate parameters, including changes in strength of the lunar crust with depth. However, Öpik's concepts cannot be used to predict the resulting shape of a crater if the crater is subsequently modified by slumping or volcanism. Furthermore, the theory requires additional attention for topographic influences. Thus, as is true in most geological problems, theory cannot be used without proper regard for regional peculiarities. Öpik introduces enough parameters so that potentially useful numbers could result. For instance, crater shapes might have been influenced early in lunar history by lower crustal viscosities as well as by low angles of incidence and low impact velocities for the projectiles. The eventual separation of these factors as well as such parameters as topographic control, gravity sliding, and later volcanism would seem virtually impossible without detailed "first-hand" inspection using geologic methods.

Quaide and Oberbeck (1968) also have examined the problem of crater shapes, but have limited themselves to small features that are influenced by the depth of the local regolith. Their analysis is strengthened by laboratory work and field simulation.

Once the initially stable crater shape has been determined, discussion then can turn to the influence of subsequent processes, whether catastrophic (such as volcanism, tectonics, or later impacts) or gradual (such as erosion by micrometeoroid impacts). The details of these shaping processes can be found in the chapters "Crater Walls" and "Crater Rims" of this volume.

As in any morphologic scheme, one can hope that fine enough subdivisions will reveal similarities or differences, which in turn will permit a narrowing of the possible hypotheses involved. It is the primary purpose of the following discussions to explore this rationale.

Concentric Plan (greater than 100 km)

Large basins typically display multiple concentric scarps as illustrated by the relatively unmodified prototype of this class—Mare Orientale (Plate 1a), and the subdued version, Mare Humboldtianum (Plate 1b). Orientale exhibits a complex geology and is considered in a separate section, Appendix A (see Plate 201). The origin of the Orientale-type geometry is not understood although hypotheses have been presented in which it represents major slump zones (Hartmann and Kuiper, 1962; Mackin, 1969), compressionally built mountain ranges (Hartmann and Kuiper, 1962), extrusive ring fissures (Fielder, 1965), standing waves (Chadderton et al., 1969), or frozen shock waves (Baldwin, 1963; Van Dorn, 1968, 1969). The Orientale-type structure is represented also by basins such as Imbrium, where only the outer scarp escaped complete inundation by mare material. Vestiges of the inner scarps might be indicated by wrinkle ridges and a few islands of positive relief.

Hartmann and Kuiper (1962) recognized several large-scale concentric basins on the lunar nearside and identified the concentric structure through annular mountain ranges, wrinkle ridges and/or concentric weaknesses revealed by grabens (see Plate 191a). Ratios of each ring diameter to the next interior ring diameter were found to be 1.4 ± 0.1 for basins Orientale, Imbrium, and Nectaris. Hartmann and Wood (1971) expanded this work and concluded that the typical ratio for Earthside and backside concentric forms is near $\sqrt{2}$. This value fits theories for shock-wave origin and for stresses developed by uniformly downward-directed hydrostatic pressure on a plate fastened at its edges (Lance-Onat features).

PLATE 1

b. IV-152-M   (oblique)

a. IV-182-M   (oblique)

Concentric Plan (greater than 100 km; 1 km–15 km)

Plate 2a reveals a single mountainous ring structure within a farside crater, Schrödinger, which is approximately 290 km in diameter. The inner ring is composed of isolated groups of connected hummocky peaks instead of blocklike massifs with gently outward-dipping slopes and steep inner scarps, as observed in Orientale (also see discussion accompanying Plate 49b).

Concentric and crescentic mountainous rings within craters smaller than 140 km exist, but the ratio of the outer ring to the interior ring commonly departs from $\sqrt{2}$ as described by Hartmann and Wood (1971). For example, the crater Sharanov (Lunar Orbiter II-034-MR) is 75 km in diameter, and the ratio between rings is approximately 7. Numerous craters smaller than 140 km exhibit an annular arrangement of central massifs. The relatively small diameter of such interior structures makes the label "concentric crater" inappropriate. The origin of these annular central peaks either may be similar to the origin of the concentric structure of enormous basins, or it represents subsidence of the central portion of the central-peak complex (see the discussion with Plate 49).

In contrast to Schrödinger and Orientale (Plate 1a), Posidonius (Plate 2b) exhibits an inner ring asymmetrically placed (or eccentric, following Fielder, 1965), which in general has a steep outer face and gently dipping inner slope. As is discussed in the chapter "Crater Floors" (Plates 34–39), this configuration might be the result of isostatic adjustment of the crater (Masursky, 1968). More likely, it indicates resurgent activity of the floor that has been raised significantly. The inner ring is interpreted as remnants of the base of the old crater wall, which has been dropped in an annular graben and buried by extrusions of mare material. Of additional interest in Plate 2b are:

1. The "tight meander" rille pattern (see Plate 177a) on the floor of Posidonius
2. A relatively well preserved ejecta blanket of Posidonius implies a recent event but extensively modified appearance

3. Mare flooding over portions of the ejecta
4. Slump block on the outer wall to the northeast

Interior ring structures similar to that in Posidonius occur in craters having diameters much less than 100 km. The craters shown in Plate 2c resemble Posidonius in that

1. Both have inner rings that meet the outer wall (generally on the wall adjacent to the lunar highlands).
2. Both are partly flooded.
3. Both have relatively symmetrical rims in profile.

For two adjacent craters to have two later impacts of the same relative sizes as the pre-existing craters is an unlikely event; furthermore, the lack of destruction of the outer wall implies that these examples are not statistical flukes. The inner wall of the larger crater is comparable to, if not higher than, the outer wall; therefore inward slumping is not plausible. Such features indicate endogenetic processes, which include weaknesses initiated by an impact. As in Posidonius, a rising floor plate may have triggered annular fracturing.

Plate 2d shows another interior ring structure of a crater in the 1 km–15 km size range. The encompassing crater (7 km in diameter) is subdued, with the interior ring exhibiting finer concentric detail. Clearly, if the inner structures were developed at nearly the same time as the crater, and if the subdued appearance of the latter is largely the result of an impinging flux of meteoroids, then there is a contradiction in the states of preservation. The floor probably was flooded at a considerably later date, with construction of the inner ring. As an alternative, the crater developed with the subdued appearance, such as a tuff ring, a theory that removes the necessity of imposing a long exposure age. Other regions of the Moon, however, show evidence for two or more stages of development separated by apparently long periods of quiescence; thus, the first suggestion is by no means unreasonable or unique.

PLATE 2

a. IV-094-M   (oblique)

b. IV-079-H1   13 km

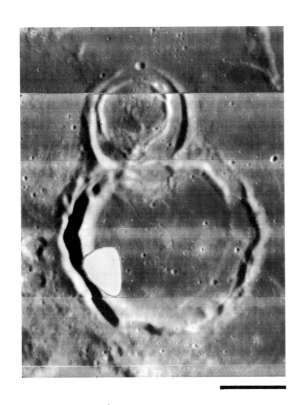

c. IV-183-H2,H1   12 km

d. III-120-M   (oblique)

11

Concentric Plan (1 km–15 km)

In contrast to the previous examples, whose development may be attributed to endogenetic processes, Plate 3a illustrates a probable statistical coincidence, as implied by the subdued nature of the outer wall compared with the fresh impact appearance of the central feature.

Plate 3b, however, reveals nested craters that are distinctly different from the pair shown in Plate 3a but are not atypical: concentric craters with doming between the inner and outer walls (see Plate 25). This feature is in a region characterized by endogenetic features, which include volcanolike domes, sinuous rilles, partially collapsed rilles, and irregular craters (also see Plates 130, a and b, and 181). Nested craters are concentric craters in the exact sense of the term in that either the interior or exterior crater resembles craters not having such a configuration. In contrast, the interior annular massifs of basins and the annular ridges within floor-fractured craters do not have good counterparts outside their domains.

The stereo pair (Plate 3, c and d) shows another crater with a concentric plan. It rests on a low-relief mound of mare material and its southwest rim is crossed by a grabenlike rille. Such an arrangement is comparable to nested volcanoes on the Earth and the inference is that it is endogenetic in origin.

PLATE 3

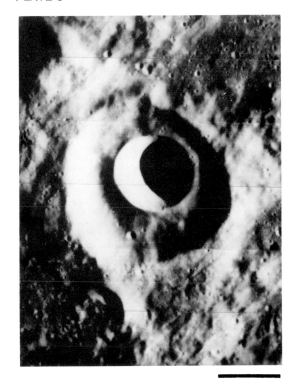

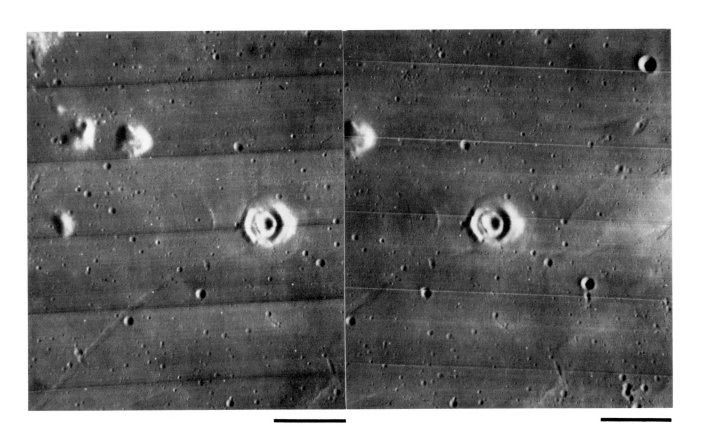

a. IV-169-H1   11 km

b. V-184-M   5.5 km

c. IV-136-H3   14 km

d. IV-131-H3   13 km

Concentric Plan (1 km–15 km)

The stereo pair seen in Plate 4, *a* and *b*, illustrates a concentric crater similar to that shown in Plate 3, *c* and *d*, but has, in contrast, a mounded floor within the inner ring. The inner ring may be the result of slumping around the entire wall, but more likely the configuration was produced by successive volcanic eruptions. Terrestrial analogs include nested tephra rings such as the Tagus Crater on Isabela Island, a member of the Galápagos Islands (see Green and Short, 1971: Plate 52B) and resurgent cones in calderas such as the Krasheninnikov Caldera (Steinberg, 1968). Nested craters on the Moon are typically found adjacent to or within highly fractured craters bordering the maria. Specifically, this example is adjacent to Pitatus (also see Plate 42, *a* and *b*), and similar examples are found on the floors of Lavoisier (Plate 35, arrow Z), and Humboldt (Plate 42c, arrow Z).

Large craters with fractured floors show abundant evidence for volcanic rejuvenation, and their association with the smaller nested craters suggests that they are also at least volcanically modified forms. The appearance of the outer rim of nested craters is typical of numerous circular craters in its size class, and this comparison includes a hummocky ejecta blanket. Therefore, either the multiring craters represent a later volcanic eruption within an impact crater or they were formed by volcanic processes. The latter alternative implies that numerous circular craters are the result of a single eruptive phase and are difficult to distinguish from impact craters.

Plate 4c shows a concentric crater in which the inner ring is closely associated with the wall of the "parent" crater and has relatively low relief. Based on its similarity with other craters, it is suggested that the inner ring represents termini of floor-flooding lava, which stand in relief. The lowering of the floor may be in response to lava devolatization or partial removal of the supporting magma through peripheral eruptions or migration of material in the magma chamber.

Plate 4d shows a concentric structure similar to that seen in Plate 2c in that the outer and inner rims are of comparable heights. One gets the impression that craters formed along the rim, and in several regions they coalesced to give the double-wall appearance. Such an arrangement was recognized by Fielder (1965) on much larger features, in which the encircling craters were termed "ring-wall craters." The resolution provided by the Lunar Orbiters reveals that numerous ring-wall craters within these larger craters are the effects of lighting conditions (but see Plate 110a).

PLATE 4

a. IV-120-H1   12 km

b. IV-119-H3   13 km

c. IV-156-H3   12 km

d. IV-107-H3   13 km

Circular Plan

Circular craters are numerous over the lunar surface yet are limited in their size distribution, rarely exceeding 15 km in diameter. Plate 5 shows selected examples in a sequence from fresh to subdued craters. Well-preserved rim sequences are illustrated by Mösting C (Plate 5a) and Copernicus H (Plate 5b), which rests on the ejecta blanket of Copernicus. Plate 5, c and d, shows unnamed craters in more subdued states. The craters seen in Plate 5 show that circular craters in this size range are not preferentially restricted to certain major lunar terrains, i.e., maria, ejecta blankets, and highlands. Thus their plans are independent of geologic setting. For this size range the plan remains unaltered through time by major mass movements, such as slumping, but may be affected by other processes, such as later impacts and surface creep. In time, the major change will be a blurring or smoothing effect, with only minor changes in plan since slumping is not likely, or possible. The restriction of smoothing to a relatively small size raises the question of limiting size before slumping will occur and whether or not this limiting size is at all determined by regional placement (see Plate 64).

PLATE 5

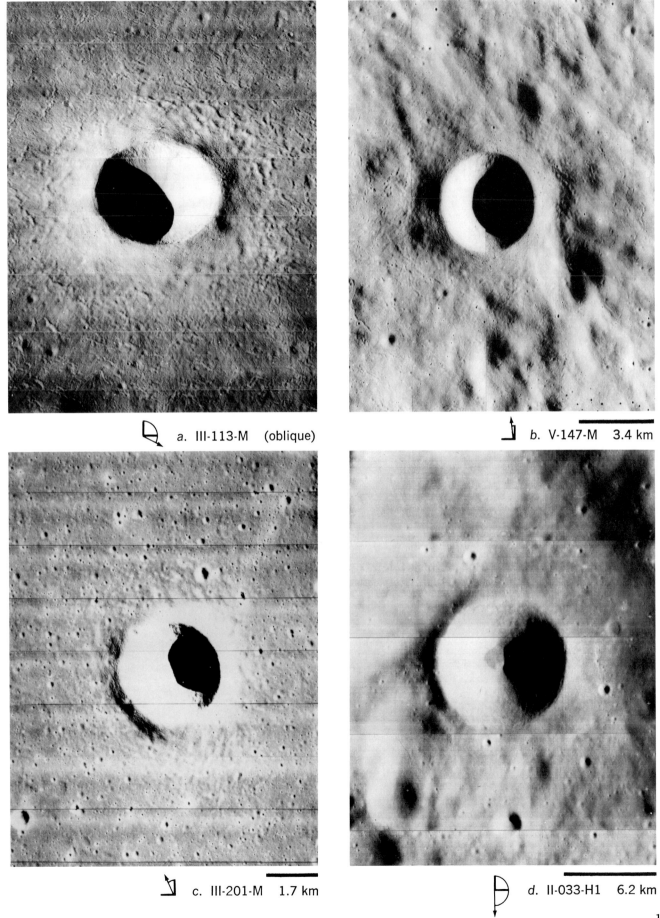

a. III-113-M (oblique)

b. V-147-M 3.4 km

c. III-201-M 1.7 km

d. II-033-H1 6.2 km

17

Small-Crater Geometry (less than 1 km)

Craters with diameters less than 1 km are circular, concentric, elongate, or irregular in plan. Their over-all form is extremely varied, depending on genetic and subsequent processes. Typical craters in this size class are shown in Plate 6. Plate 6a exhibits the circular plan, which is in contrast with the irregular plan in Plate 6b. The preservation of blocks along their rims indicates that both are of recent formation. The irregular crater may be evidence for an impact into interbedded competent and incompetent layers. It is also plausible that it is the result of non-impact processes (see Plate 76a). Plate 6, c and d, shows a stereo view of a crater with a concentric plan (also note the example seen in Plate 6b). It has been suggested that the bench along its wall, which produces the concentric appearance, indicates the underlying bedrock (Quaide and Oberbeck, 1968).

Included in these photographs are smaller craters with varying degrees of "freshness." Some may be degraded craters. Others could be in a pristine state and indicate an origin by endogenetic processes, such as collapse. Consequently, to derive a degradational sequence, the regional setting must be considered. This gives an appreciation for the original crater form that is commonly dependent on regional parameters, such as regolith depth. It also can give indications of catastrophic processes, such as extrusive activity, that may contribute not only to crater modification but also to crater density.

PLATE 6

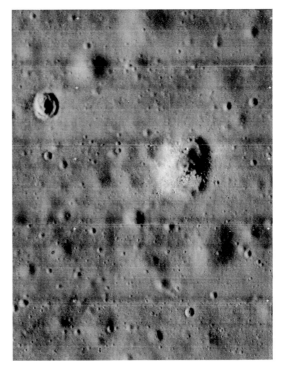

a. III-157-H1   0.21 km

b. II-207-H3   0.19 km

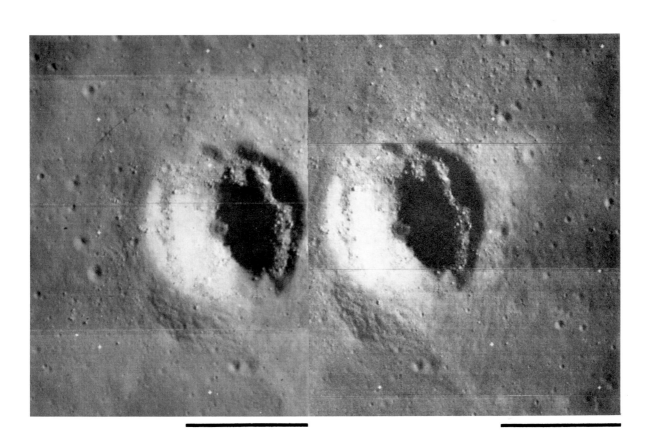

c. III-148-H3   0.20 km

d. III-149-H3   0.20 km

Elongate Plan (15 km–100 km; 1 km–15 km)

Elongate craters are not rare but are considerably fewer in number than other geometries. Schiller (Plate 7a) is an example of a large (100 km length) crater included in this category. The photograph reveals the raised rim, central ridge, extensive terracing, and flooded floor. Schiller may have been formed by one or several projectiles impacting at low angles from the surface. However, the largest volcano-tectonic depression or cauldron on Earth (the Toba depression in Sumatra) is elongate, and a similar mode of formation may be applicable in this example. The major axis of Schiller is located along the outer ring of an old multiringed basin (also recognized by Stuart-Alexander and Howard, 1970, and Hartmann and Wood, 1971), about 320 km in diameter. Also note the on-axis connection between Schiller and the smaller crater to the lower right. These coincidences suggest structural relations and add credibility to the interpretation that Schiller is a volcano-tectonic form.

Shown in Plate 7b, Messier (right) and Messier A (left) are approximately 15 km in length. The rim of Messier has a saddle-like appearance, with the highest portions parallel to the major axis. This configuration has been reproduced in the laboratory by low-angle impacts (Gault, personal communication). Plates 253–254 show more detailed views, and smooth lavalike units are identified along the rim and within subdued craters. The units likely represent impact melt splashed out of the crater. Messier A clearly is a more complex feature and could be the result of an oblique impact partially overlapping an older crater.

Plate 7c shows the elongate crater Torricelli to the north of Theophilus. The elongation in this example appears to be the result of collapse of the western wall. Owing to the absence of Theophilus ejecta on its floor, the event apparently postdates Theophilus. If Torricelli was impact produced, then the preservation of Theophilus ejecta on its rim implies pre-Theophilus formation. This contradiction can be resolved if secondary-impact features on the floor of Torricelli have been removed by subsequent and active mass movement in its interior. Alternatively, the crater was endogenetically formed after Theophilus.

Plate 7d illustrates convincing examples of endogenetically formed elongate craters. They lack rims and ejecta blankets—characteristic of impacts—and align parallel to a regional trend. These craters probably form part of an incipient rille, and their close association is a common phenomenon (see Plates 181–182).

PLATE 7

a. IV-154-H3   17 km

b. IV-060-H3   12 km

c. IV-077-H3   12 km

d. V-090-M   3.9 km

Elongate Plan (1 km–15 km; less than 1 km)

Plate 8, *a* and *b*, shows another variety of elongate craters that may be assigned either an endogenetic or exogenetic origin. These subdued features exhibit the "tree-bark" and grid patterns and are associated, at least spatially, with ridges. The crater plans do not appear to be produced by single oblique impacts. In Plate 8*a*, the plan is rectangular; in Plate 8*b*, the example seen in the center (arrow X) is very extended and slightly arcuate. Consequently, they are the result of multiple impacts, such as secondary impacts, or are endogenetic forms. I prefer an endogenetic origin for these examples because of their association with the ridges. This view also applies to the subdued elongate crater that overlies the wrinkle ridge (arrow Y). Secondary impacts, however, are thought to be responsible for the complex crater cluster shown in the upper right (arrow Z) of Plate 8*b*, owing to the herringbone pattern (see Plate 96).

At much smaller sizes, elongate craters may be produced by projectiles striking a sloped surface, such as the example seen in Plate 8*c*, where the wall boundary has been extended onto the floor of the larger crater. The block-strewn elongate crater shown in Plate 8*d* is an example of a recent low angle-of-incidence impact. Note that there are several other subdued elongate craters with parallel NNE-trending major axes (arrows X and Y). These may be very old secondaries or endogenetic features.

PLATE 8

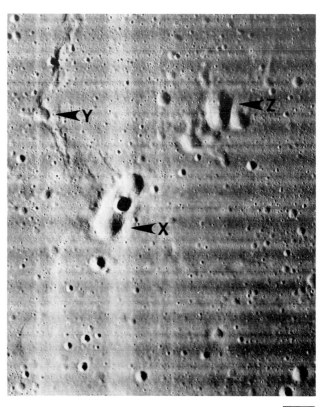

a. III-005-M   1.9 km

b. III-013-M   1.9 km

c. II-098-H3   0.17 km

d. II-187-H3   0.19 km

Scalloped Plan (greater than 100 km; 15 km–100 km)

The circular or polygonal plan of craters larger than 100 km in diameter becomes altered by considerable scalloping from slumping, collapse, or some other major mass movement; Clavius (Plate 9a) illustrates the results of such processes. For crater diameters 15 km–100 km, scallops affect the plan of smaller craters to a greater extent since the scallop size appears to remain approximately the same, whereas the circumference has been reduced. Newcomb (Plate 9b) and Bürg (Plate 9c) are excellent examples of crater walls extensively deformed by scalloping. In Newcomb, the smaller scallops are probably the result of mass movement either as massive slides or slumping, whereas the large scallop represents a much more extensive collapse into an underlying cavity.

PLATE 9

a. IV-118-H3   16 km

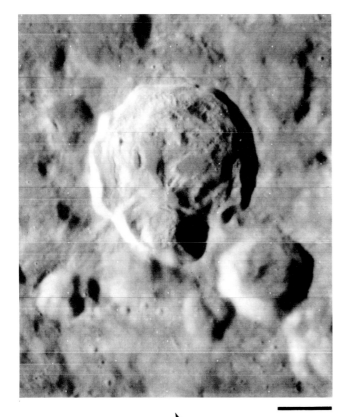

b. IV-067-H1   13 km

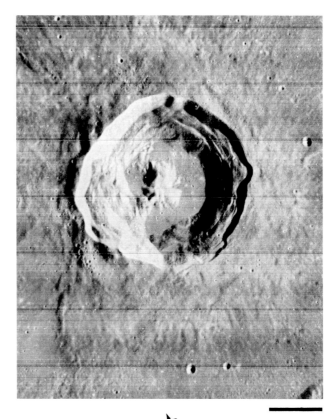

c. IV-091-H2   13 km

Scalloped Plan (15 km–100 km)

Triesnecker (Plate 10a) and Euler (Plate 10b), both about 27 km in diameter, illustrate relatively fresh craters modified by scalloping. In craters approximately 15 km in diameter, only single scallops typically are present (Plate 10, c and d). The absence of extensive scalloping at smaller scales indicates sufficient support at the base of the walls and/or insufficient load on the wall. The first alternative is dependent on the regional placement or depth of influence of the event. In some instances, the support appears to have been weakened by a nearby event or by later processes affecting the floor.

PLATE 10

a. IV-102-H1   12 km

b. IV-133-H3   11 km

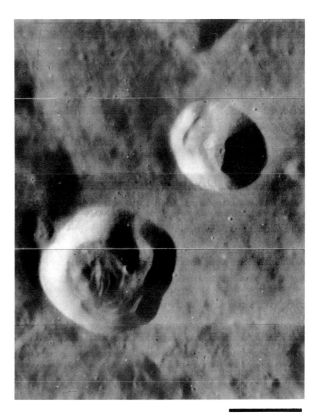

c. IV-085-H2   12 km

d. IV-076-H3   13 km

Polygonal Plan (greater than 100 km; 15 km–100 km)

Polygonal craters have portions of wall-rim contacts that depart from the curvilinear plan demonstrated for circular and scalloped craters. The alignment of the crater walls commonly parallels regional trends identified by crater chains, ridges, or rilles. These features are visible in Plate 11, a and b, which includes the large (greater than 100 km) polygonal craters Tsiolkovsky and Humboldt. Collapse, or some form of mass movement, preferentially followed pre-existing zones of weakness during or after crater formation. Guest and Murray (1969) believe that, although there is a slight structurally controlled polygonality, Tsiolkovsky is, in general, circular. Humboldt (Plate 11b) shows a similar alignment between the crater walls and regional trends revealed by ridges and crater chains. Lighting could play an important role in labeling a crater "polygonal" or "scalloped."

On a smaller but still impressive scale (15 km–100 km), linear wall segments are identified in Eudoxes (Plate 11c) and Barrow (Plate 11d). For the crater Eudoxes, only one portion of the wall is straight, whereas in the crater Barrow the polygonal plan is obvious around the entire wall except the northern portion. The linear trend on the northern wall of Barrow is approximately repeated on the eastern wall and reflects the general trend of regional lineations.

PLATE 11

a. III-121-M   49 km

b. IV-012-M   100 km

c. IV-103-H2   13 km

d. IV-116-H2   15 km

Polygonal Plan (15 km–100 km; 1 km–15 km)

The polygonal plan of craters having diameters 1 km–15 km is more easily identified than that of larger craters, owing to their smaller circumferences. Plate 12a shows a multifaceted crater that is on the wall of a large backside crater, Aitken, whose floor is partially flooded (see Plate 24a). The northern wall of the crater shown in Plate 12a reflects the trend of previous slumping in the larger crater. Its western wall, however, also has a straight segment that cuts across the old slump block. A preserved uniform structural trend in this latter area is not expected unless the stress developed after slumping of the large crater. It remains uncertain whether such a stress could be expressed within a crater that is presumed to be a jumbled mass. Consequently the western wall with its N-trending wall may not be structurally significant but may be merely completing the closed arc of the crater.

Plate 12b shows two craters having straight wall segments. The example at the top has a very prominent NW-trending NE wall, which is approximately paralleled by the NE wall of the smaller crater below it. The NE wall of the crater at the top cuts the wall and floor of a very subdued crater to the east. This straight segment probably is an effect of topography rather than slumping along a structural trend. This crater, as well as the crater seen in Plate 12a, indicates that crater enlargement occurred at a time considerably postdating crater formation. If these craters are impact produced, then smaller craters along their rims should not remain well preserved because of extensive ejecta blanketing. Existence in such a location indicates formation after the larger crater. Consequently, the small crater along the southwest rim of the large crater seen at the top of Plate 12b indicates considerable postformation expansion through slumping.

The crater shown in Plate 12c also displays a polygonal plan, in part, but in this case, its ejecta blanket is preserved (shown more clearly in the original photograph). Thus, the polygonal plan appears to have been established relatively soon after crater formation. The straight segment of the eastern wall aligns with a regional trend expressed by ridges radial to the Imbrium basin.

Plate 12d shows a subdued crater about 5 km in diameter on the floor of Hipparchus. Very little evidence of extensive mass transfer remains; yet the crater, owing to its straight southeastern wall, has a semipolygonal plan. If this segment reflects a structural weakness, then such a weakness occurs at relatively shallow depths and is expressed on the floor of a much larger crater—Hipparchus (see Plate 43).

a. II-033-H3   6.2 km

b. IV-082-H3   16 km

c. IV-163-H3   12 km

d. III-111-M   1.7 km

31

Polygonal Plan (1 km–15 km; less than 1 km)

Plate 13a shows an unusual example in the 1 km–15 km size range, in which the plan approaches a parallelogram. The feature is within the backside crater Barbier to the northwest of Mare Ingenii. The unusual plan and multiple-domed floor suggest an endogenetic origin.

At smaller dimensions, polygonality becomes increasingly difficult to recognize because of the rubbled nature of the walls and rims (Plate 13, b, c, and d). The samples presented are relatively subdued, and the linear walls may have been established by small-scale structurally produced grid patterns, inhomogeneities in the pre-existing surface, peculiarities of the crater event, nonuniform erosional processes, and/or uneven encroachment by later units of lava. Plate 13d shows a slightly rectangular crater, which is near the impact point of Ranger VIII and which was included in that photographic mission. Note the numerous "dimple" craters around this crater.

PLATE 13

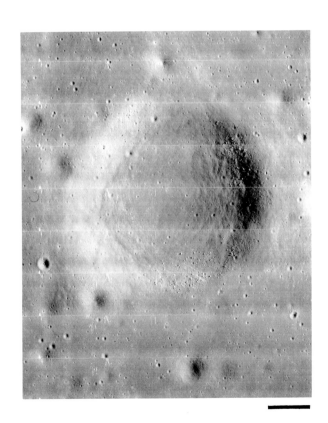

a. II-075-H2   6.5 km

b. II-187-H3   0.19 km

c. III-088-H3   0.20 km

d. II-070-H1   0.19 km

DISCUSSION AND SUMMARY

The foregoing sections of this chapter have provided a general survey that introduces the various forms of craters. The processes responsible for such variety are considered in greater detail in subsequent chapters dealing with crater walls, floors, and rims. The shape in plan of craters is a helpful criterion for determining origin but is usually insufficient as the only one.

Concentric craters are found in a wide range of sizes, from enormous multiringed basins down to small rubble-strewn craters. Large multiringed basins were shown by Hartmann and Wood (1971) to have a ratio √ 2 for one ring diameter to the next interior ring. They interpret this value as either the result of impact-induced microfractures (Lance-Onat features), along which major faults developed from later tectonism, or the result of impact-produced major fractures, which led to faulting. The expressions of this faulting could be as large slump features, horsts, or volcanic constructions. One of the smallest representatives of this type basin is Antoniadi (140 km in diameter), shown in Plate 49b. Several craters smaller than approximately 140 km in diameter exhibit interior massifs with annular and crescentric plans that yield ratios for the crater diameter to the annular massifs that are much greater than √ 2. These massifs are interpreted as remnants of the central-peak complex whose interior region has subsided.

Many concentric craters in the 150 km–30 km size range exhibit an inner ring that is ridgelike, in contrast to the massifs of multiringed basins. In further contrast, the inner ring is adjacent to the crater wall, although commonly separated from it by a moat of mare material. The location of this type of concentric crater along the borders of maria suggests a link to this epoch of volcanism, and the concentric plan is viewed as a product of isostatic adjustment or volcanic resurgence of the crater floor. More detailed discussion is deferred until specific examples are considered in the chapter "Crater Floors" (Plates 34–39).

Concentric craters smaller than 15 km in diameter are morphologically distinct from the larger forms. Many examples appear to be products of successive volcanic eruptions from the same vent. If these are solely volcanic, i.e., not impact triggered, then numerous circular craters in the same size range may be interpreted plausibly as volcanic forms having only a single eruptive phase preserved.

Still smaller concentric craters (less than 1 km) are distinct from larger volcanolike structures. These are considered to be the result of impacts penetrating competent and incompetent layers. Thus, they can be used as indicators of stratigraphy or depth of regolith (Quaide and Oberbeck, 1968).

Strictly circular craters are typically less than 15 km in diameter. Although the general plan of larger craters closely approaches circularity, they are almost always scalloped. Examples are shown in a variety of settings, apparently with different ages. Consequently, such craters illustrate an origin unaffected by location and a history of gradual smoothing without slumping. An impact origin is consistent with the former conclusion, but there are also numerous circular craters in this size range that appear to be volcanic because of their geologic setting (see Plate 159) or form (see Plate 116a).

Elongate "craters" are found in all size ranges. They are attributed to low-angle impacts (less than 5°) or to structurally controlled endogenetic formation, but the distinction between these origins on the basis of morphologic features is not always apparent. Endogenetic forms commonly lack raised rims or exhibit rims of about equal elevation around their circumferences and follow regional trends. In addition, elongate craters that are interpreted as endogenetically produced by virtue of their association with rilles exhibit beaklike extensions along the major axis. Low-angle impacts, however, typically have varying rim elevations that result in a saddle-like appearance, and they are without the beaklike extensions. Subsequent modifications can destroy these indicators.

Craters larger than 15 km in diameter are typically either scalloped or polygonal in plan. These departures from circularity may indicate, respectively, modifications by structurally controlled and nonstructurally controlled slumping. This mass transfer can occur immediately following crater formation or as later-induced events, and many craters are enlarged as much as 20 percent by this process (Quaide et al., 1965). Polygonal impactlike craters less than 15 km in diameter that lack slump blocks are interpreted to be structurally controlled during crater excavation. Terrestrial calderas typically are noncircular owing to pre-existing weaknesses, and analogous forms are present on the Moon. Scalloped or polygonal plans may not indicate slumping but may be the result of uniformly sloping walls that cut topographic irregularities. Apparent noncircularity can also result from low solar illuminations.

Adler and Salisbury (1969a) have made a study of the circularity of 487 craters larger than 25 km in diameter on the lunar Earthside. Their results showed:

1. There are at least two classes of craters, and in each the circularity decreases with apparent age.
2. Most polygonal craters occur along the borders of maria.

They concluded that the original circular outline was deformed as the result of stresses developed during mare flooding. A test for such a hypothesis, as they point out, would be to consider the proportion of noncircular craters on the farside.

2. Crater Floors

INTRODUCTION

Crater appearance is greatly dependent on size, and this is illustrated not only by crater floors but also by their wall and rim sequences. Craters having rim-to-rim diameters of 5 km or less are classified, in part, by their floor profiles: concave, convex, or flat. The concave appearance results from a gradual transition from wall to floor; detailed examples are examined in the next chapter, "Crater Walls." Convex and flat floors are illustrated by a variety of crater sizes, but such a description forms only a partial picture of the actual nature of the floor. For example, a flat floor may exhibit rubble, smooth marelike material, or a textured surface. In addition, a full description depends on the crispness of detail, which may indicate the extent of smoothing processes (hence relative age) or differences in initial crater-forming mechanisms.

The geometry and profile of small (less than 1 km in diameter) impact craters are determined by projectile and target parameters. Similarity of craters in a certain region and within a given size range provides evidence for the general structure of the upper surface layer, but subsequent events may significantly alter the resulting appearance.

The description of the appearance of a crater by its profile becomes much more ambiguous for larger craters since most craters with diameters greater than 5 km usually have generally flat or slightly domed floors. Crater floors may simply be rubble as a result of fall-back material or slumping from the walls (Plate 30, b and c). Others can be described loosely as volcanolike and may exhibit ropy, textured floors; marelike fill; or multiple doming. Within many larger and more recent craters, the floor becomes a volcanic basin that exhibits numerous constructional and collapse forms. This complexity can be appreciated by examining the detailed views of Copernicus (Plates 218–229) and Aristarchus (Plates 236–239).

Clearly, the complexity increases with size; therefore, the following discussion on crater floors first considers small craters with diameters less than 1 km (Plates 14–20). Craters with diameters greater than 1 km but less than 20 km are examined in Plates 21–29, and Plate 30 introduces characteristic floor features of larger craters.

A feature common to numerous crater floors is the central peaks or ridges. These are considered in Plates 44–49, which also provide further comparison of floor appearances.

Reference should be made to the selected regions of Petavius, Copernicus, Aristarchus, Vitello, and Messier, which provide detailed views of floor features of larger craters (see Appendix A).

Concave and Flat Floors

The lunar surface is covered with innumerable small craters (less than 1 km in diameter) having concave and flat floors. Plate 14 shows several examples of the former class, but discussion is deferred until the next chapter, "Crater Walls." Two types of flat-floored craters also are shown. The first type (Plate 14, a and b) exhibits rubbled floors of craters having diameters of less than 100 m. The crater shown in Plate 14a exhibits a well-defined ejecta blanket that includes a system of bright rays, whereas the nearby craters shown in Plate 14b generally do not. The difference in ejecta blankets probably indicates different ages of formation and subsequent degradation. Note that the rims and rubbled floors of the craters seen in Plate 14b persist, and consequently degradation was more efficient on the ejecta blanket.

The craters shown in Plate 14, a and b, fit into Oberbeck and Quaide's (1967) and Quaide and Oberbeck's later (1968) studies of impact cratering in which flat-floored craters were simulated if the regolith thickness was between $D_A/10$ and $D_A/3.8$, where D_A is the crater diameter. The floor is composed essentially of rubble from the impact. This is in contrast to craters with concave profiles, which resulted if the underlying competent layer was between $D_A/3.8$ and $D_A/4.2$ beneath the surface. They noted that such geometries (including the concentric plan shown in Plates 6 and 50 of this volume) were largely unaffected by impact variables, such as impact velocity, angle of impact, projectile properties, and surface strength. The substrate strength produced little or no effect.

The smallest flat-floored crater (upper center right in Plate 14b) is approximately 27 m in diameter, which suggests a local regolith depth on the order of 5 m. In addition, Plate 14b includes several large subdued concave craters. numerous small concave craters, and two dimplelike craters (i.e., small funnel-like craters, arrows X, Y).

Plate 14, c and d, shows the second type of flat floor that has a much smoother texture. The crater shown in Plate 14c is approximately 270 m in diameter and has a very small flat floor (arrow X); the crater seen in Plate 14d is about 530 m in diameter. Both craters appear to be relatively subdued, but numerous large blocks are preserved along their rims. Their floor may represent an accumulation of the finer fraction of fragmental debris through the sifting of the fines from the walls. Such a transport is suggested by the accumulation of material above the large blocks on the northern wall of the crater shown in Plate 14d. Seismic events and impacts by small meteoroids probably contribute to the relatively even distribution of such debris across the floor.

Alternatively, the floor debris may be derived from ejecta fallout during the initial crater-forming event, but this sorting requires a temporary atmosphere produced by the event. Such an atmosphere, or at least an interacting cloud of material, might accompany impacts by secondary projectiles. Therefore, it is significant that the crater seen in Plate 14d is within a group of secondary craters. Other possibilities:

1. The region was blanketed by particulate debris from either nearby impacts or volcanic events.

2. The floor material was much more friable than the wall exposures and was more efficiently pulverized by meteoroid bombardment.

3. The floor material represents the fused fraction created by an impact.

4. The crater and the floor are results of some endogenetic process, such as volcanism.

38

PLATE 14

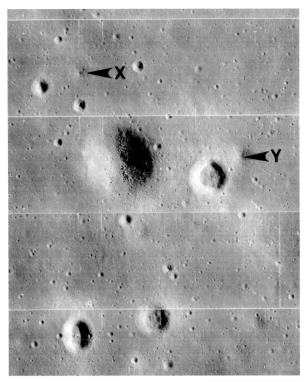

a. II-068-H2   0.19 km

b. II-068-H2   0.19 km

c. II-128-H2   0.21 km

d. II-171-H1   0.23 km

Flat Floors

Plate 15 illustrates the tendency for flat-floored craters to be clustered in a particular region. The craters shown in Plate 15a are similar to the examples seen in Plate 14, a and b, and fit into Quaide and Oberbeck's (1968) scheme, which describes the dependence of floor type on regolith depth. The clustering of such craters thus indicates a regolith of the same average thickness over the area covered in the photograph. The craters display a range of apparent ages; hence, they were not produced by members of a single meteoroid swarm. Arrow X identifies an irregular surface pattern believed to be produced by the impact of ejecta from a large impact crater (see discussion with Plate 100).

Plate 15, b and c, introduces a third flat-floored appearance, distinct from the previous examples, and a similar form is shown in detail in Plate 16a. The region shown is south of Maskelyne F in southern Mare Tranquilitatis, and, at Orbiter IV resolution, it appears to be a marelike surface with several insular positive-relief features. The craters (noted by arrow X) are very subdued and have large floor areas relative to the total crater areas. These and similar craters elsewhere very typically are found on surfaces that are relatively devoid of craters 100 m–300 m in diameter and that exhibit a high degree of small-scale texturing. Patches of this type of surface seen in Plate 15, b and c, are mixed with areas having a large number of small craters (arrow Y).

The formation of the flat floors shown in Plate 15, b and c, is likely related to the origin of the relatively craterless patches. Such patches probably do not represent simply products of long-term erosion by meteoroid bombardment. This conclusion is based on the following considerations:

1. There are nearby regions that are similar to the zone marked by arrow X but are devoid of craters larger than 50 m in diameter, relative to a typical mare surface. These regions are apparently on level ground, and consequently the absence is not thought to be the result of mass movement.

2. Craters in these zones do not display a smooth transition of appearances from fresh impactlike craters to subdued craters with concave floors.

3. If a flat-floor profile is the expected degradational product of secular erosion, then the typically subdued concave floors on other mare surfaces indicate an anomalous process.

In addition, the area in and around that shown in Plate 15, b and c, exhibits several elongate depressions (see Plate 8, a and b), unusual crater-rim features (see Plate 86, c and d), small-scale flowlike forms, platforms with irregular plans (see Plate 124), and hogbacklike ridges (Plate 15, b and c, arrows Z and XX). Note that one of these hogbacklike ridges is exposed across the wall of a subdued crater shown at arrow YY.

The foregoing observations suggest several alternatives for the local geologic history:

1. The smooth patches represent very old surfaces with a thick regolith, whereas the areas with numerous small craters represent a cluster of secondary impacts.

2. The smooth patches represent very young surfaces with endogenetically produced subdued craters, whereas the areas with numerous small craters represent exposed old surfaces subjected to long-term cratering by primary impacts.

3. The smooth patches represent regions heavily blanketed by pyroclastic eruptions, whereas the areas with numerous small craters represent unblanketed regions.

If the smooth patches correspond to old surfaces, then areas with highly degraded flat-floored craters and those with relatively few pristine craters larger than 50 m may reflect rapid degradation, owing to creep within a deep regolith. Such a regolith may represent an early catastrophism, such as blanketing by volcanic ash or by ejecta from a large basin. However, the small dome surrounded by a moat (arrow ZZ) suggests that endogenetic processes have been operative relatively recently or that these surfaces have been well preserved since the time of mare emplacement (see Plate 125e and Discussion and Summary at the end of Chapter 6). Either alternative is very significant with respect to our understanding of surface processes.

PLATE 15

a. II-197-H3   0.20 km

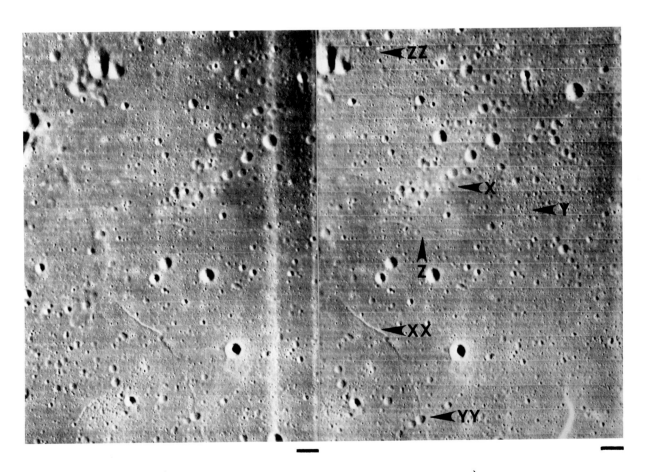

b. III-017-M   1.9 km

c. III-019-M   1.9 km

Flat Floors

Plate 16, *a* and *b*, permits a direct contrast between the flat-floor types shown in Plate 15, *b* and *c*, and those shown in Plate 14, *c* and *d*. In addition, Plate 16, *c* and *d*, introduces another floor type that is flat but surrounded by a doughnutlike ring adjacent to the wall.

The texturing shown in Plate 16a is commonly called a "tree-bark" pattern and may be an indicator of processes responsible for the flat floor. Crittenden (1967) compared the pattern to terrestrial textures produced by slow mass movement, such as creep, and the lunar tree-bark pattern commonly occurs along sloped surfaces both on maria and on highlands. The initiation of movement could be from seismic jostling induced by large or nearby impact events. There are regions that display much more extensive patterning than others but are not obviously more hilly. Such regions may have a thicker regolith produced by local blanketing, as discussed with plate 15b and c. The flat-floored craters seen in Plate 15, *b* and *c*, are found in this type area, which appears smooth at the resolution shown.

It is also possible that some flat-floored craters, such as the large one shown in Plate 16a, are endogenetically formed. The subdued craters seen in Plate 15, *b* and *c*, have approximately the same diameters and are within the smooth patches. The tree-bark surface, which characterizes the smooth areas at higher resolutions, may be remnant flow or fracture patterns of a relatively recently formed unit (Kuiper, 1966a), and the craters may be vents or collapse structures. This interpretation is supported by similar-appearing craters that are associated with wrinkle ridges (Plate 142, *a* and *b*), but the interpretation requires extremely well preserved surfaces. Although the tree-bark pattern probably does not represent flow patterns, the endogenetic origin of the subdued craters remains a reasonable hypothesis.

Plate 16b shows a smooth flat-floored crater, similar to the example seen in 14d, that has a low-relief textured rim and a concentration of blocks around the rim. The subdued appearance, preservation of blocks on the rim, and smooth floor suggest a nonsecular degradational process, such as blanketing by fragmental debris, either ash or ejecta. Secondary craters formed by projectiles ejected from a much larger crater typically appear subdued in their profile yet are surrounded by large blocks (see Plate 115b). Consequently, the crater shown in Plate 16b may be a relatively recent secondary crater rather than an old primary one. The paradoxical association of large blocks around a very subdued crater are thought to reflect properties of the secondary-impact event (see the discussion with Plate 100).

In contrast to the foregoing examples, the subdued flat-floored craters shown in Plate 16, *c* and *d*, exhibit doughnutlike rings adjacent to their walls. These might be erosional remnants of concentric craters similar to those shown in Plate 50, but this region is otherwise devoid of such benches. In addition, a process that subdued the craters to the extent shown also would be expected to erase the moatlike depression between the wall and floor, unless this depression is the result of continued subsidence.

The feature shown in Plate 16c can be examined in stereo in Plate 130, *a* and *b*, which also includes the regional setting. It is one of several such features and is immediately to the south of a volcanolike dome. This particular example is at the crest of a low-relief rise. Although not obvious, it is surrounded by an irregular platformlike rim, which resembles the terminus of an extrusion. Thus an endogenetic origin for such a feature is suggested. The doughnutlike bulge on the floor could have resulted from extrusions or from different degrees of settling of the crater floor. It also may be analogous to the aprons found at the base of many positive-relief features (see Plate 126).

The region shown in Plate 16d is in eastern Mare Tranquilitatis (west of Secchi A) near a ridgelike relief displaying a well-developed basal terrace (lower left, Plate 84a). It is similar to the previous example with the interior moat and ring structure but, in contrast, displays a lower wall slope. This might be an endogenetic structure or a pre-existing crater, which was flooded by mare material and subjected to differential settling. Also see the ring structures shown in Plates 151–152, which could be interpreted as flooded craters.

PLATE 16

a. III-010-H2   0.24 km

c. V-184-H1   0.73 km

b. II-207-H1   0.20 km

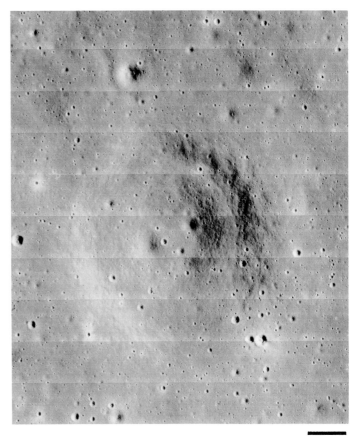

d. II-021-H2   0.21 km

Convex Floors

Small craters (less than 200 m in diameter) may have convex or mounded floors, as Plate 17 illustrates. The first two examples (*a* and *b*) permit comparison between a relatively recent impact and a more subdued and presumably older feature. The two craters, less than 1 km apart, are to the southeast of Lansberg (near the Apollo 12 landing site). Quaide and Oberbeck (1968) have reproduced similar floor appearance, and they suggest that such craters might indicate regolith thicknesses on the order of 1/7.5 to 1/4 the diameter of the crater (same range as the flat-floored geometry). Plate 17c shows several similar craters in the same region.

PLATE 17

a. III-155-H3   0.21 km

b. III-155-H3   0.21 km

c. III-153-H3   0.21 km

Convex Floors

The origin of convex floors in craters 0.5 km–1 km in diameter might be explained by Quaide and Oberbeck's (1968) thesis. But craters of this size would have penetrated the upper regolith layer attributed to long-term bombardment by meteoroids; therefore, the competent substratum probably would represent a transition of a distinctly different origin. An alternative is that presented by Roddy (1968) for the central uplifts produced in 100-ton and 500-ton (TNT) explosion craters. These mounds are thought to have been produced by elastic rebound through stress waves reflected by competent strata or discontinuities beneath.

Plate 18, *a* and *b*, shows a vertical and oblique view (inverted with respect to the north direction), respectively, of two craters with convex floors. In Plate 18a, it appears that the smaller-sized particles (presumably revealed by smoother surfaces) accumulated between the mound and the wall. Plate 18b shows a more subdued crater in an area north of Fra Mauro. These craters are 0.8 km–0.9 km in diameter, with a present floor at least 100 m below the rim. From Quaide and Oberbeck's work (1968), this depth indicates the depth of a competent layer, which perhaps is the pre-Imbrium surface for the crater shown in Plate 18b, owing to its location in ejecta from that basin. Roddy's (1968) thesis, however, suggests that this competent layer may be deeper than the crater floor.

On the floor of Flammarion (Plate 18c) are numerous craters that display the convex floor profile. They appear to be more subdued than the foregoing examples and are observed in craters up to 1.5 km in diameter (not shown in Plate 18c). The preservation of the central mounds could indicate that the material is competent and not composed solely of rubble. If they are largely rubble, then the processes that destroyed the ejecta blanket and smoothed the walls might be expected to erase the moat between the mound and the wall as well. Their preservation might indicate mound formation that continued or that was initiated at a time considerably after crater formation. A possible mechanism is seismically induced debris slides from the crater wall that converge on the floor to produce a low-relief mound. The paradox in preservation state also indicates the relatively rapid degradation of the ejecta blanket, which may have lacked small-scale structuring at the outset (see Plate 85a). The apparent restriction of such craters to certain regions suggests a locally thick layer of incompetent material. There are also similarly subdued craters in the same size range on the floor of Flammarion that do not exhibit the mounded floor geometry.

46

PLATE 18

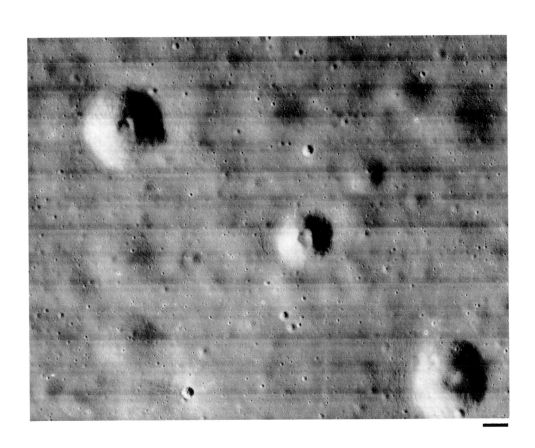

a. III-152-H3   0.21 km

b. III-132-H2   (oblique)

c. III-118-H2   0.20 km

Convex Floors

The stereo pairs of Plate 19 permit more detailed examination of convex crater floors. Plate 19, a and b, shows a crater of 330 m diameter on Sinus Medii. In contrast, Plate 19, c and d, shows a crater of 2 km diameter on the floor of Hipparchus. The jumbled floor and narrow floor border (arrow X) of the latter crater seem out of character within such a subdued crater and are different in profile and appearance from the crater floor seen in Plate 19, a and b. Perhaps the two floors, and even the craters, had different origins. The crater shown in Plate 19, a and b, is probably impact, whereas the one seen in Plate 19, c and d, might be either an impact crater or a volcanic structure that has had subsequent resurgence. The resurgence could have been an adjustment due to a plastic layer beneath or a viscous protrusion similar to that found in Mono Craters, California (Russell, 1889). Convincing examples of tholoid development within small craters in the Marius Hills are shown in Plates 264a and 265, a and b. Although there are nearby endogenetic forms, such as wrinkle ridges, the geologic setting of the example shown in Plate 19, c and d, is not comparable to the volcanic terrain of the Marius Hills. If this crater is not volcanic in origin or modification, then the contrasting appearance of the floor and ejecta blanket further illustrates either the very rapid erasure of the ejecta blanket or the absence of a hummocky blanket, owing to its formation in a deep layer of incompetent material.

PLATE 19

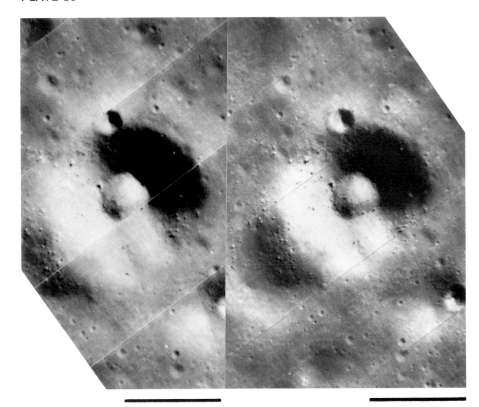

a. II-121-H1   0.20 km

b. II-129-H3   0.22 km

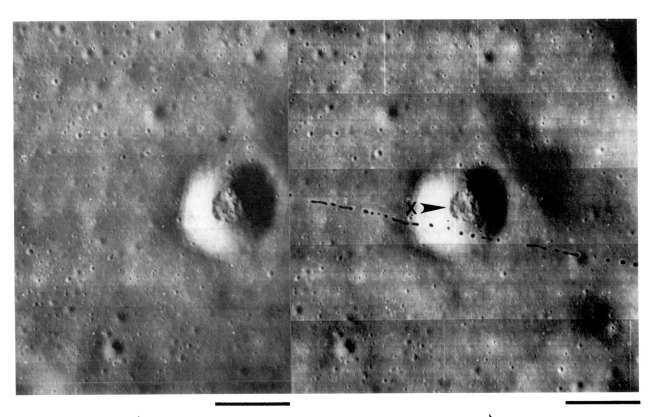

c. III-110-M   1.7 km

d. III-111-M   1.7 km

Dimpled Floors

Lunar craters in which the floor is merely the apex of an inverted cone are descriptively termed "funnel" or "dimple" craters. As shown in Plate 20, a (an oblique view) and b, variations in appearance occur as a result of variations in the profile. In Plate 20a, the wall slopes at approximately the same angle from the rim to the floor, whereas in Plate 20b it decreases in slope to an inflection point (arrow X), then increases. Neither profile is unique. Typically, the latter profile can be found in regions that appear to have undergone extrusive activity (Plates 122–123, a and b).

Additional variety may result from the presence or absence of a rim. Both craters shown in Plate 20, a and b, have subdued but identifiable rims, but some craters exist that are essentially rimless. Rimmed and rimless dimple craters can be divided into those having the following:

1. An approximately uniform increase in slope of the wall with depth, which produces a convex wall in profile
2. An approximately constant wall slope, which produces a funnellike crater
3. A combined profile that yields the appearance of a dimple crater within a larger subdued crater

In many craters the floor is not a true cone apex but a concave surface. There are also examples in which the floor is not pointlike but appears as a crease, i.e., it follows a linear trend. The tree-bark pattern typically is associated with such craters. In many examples, the initial break in slope at the rim is revealed by an enhancement of the pattern, which might appear more like peripheral fractures (see Plate 194).

As Plate 20c illustrates, small dimple craters are abundant in some regions. The area shown is on the ejecta blanket of Copernicus. Also note the craters along the inner rim of Vitello (Plate 252b).

Kuiper (1966a), Oberbeck (1970), and Greeley (1970) have discussed the possible mechanisms that result in dimple craters. Greeley noted two general classes of volcanic craters that correspond to the examples shown here. The first involves plastic deformation of the partially cooled crust of a lava flow that is still moving locally. This results in a uniform increase of wall slope or a convex wall. The second type forms after solidification of the lava surface when overlying ash or regolith drains—"funnels"—into openings, such as fractures, or into subsurface cavities (including lava tubes) that were created during the cooling stage. Kuiper discussed the first type and showed that some terrestrial dimple craters in basaltic flows are accompanied by considerable peripheral fracturing, which he suggests is analogous to the tree-bark pattern. Oberbeck suggested three possible origins, which include those formed by erosion of concentric craters, collapse on young surfaces, and drainage of the regolith into subsurface cavities. He noted that the last mode of formation must be small because of the relatively thin overlay represented by the regolith. It should be noted that the regolith could be very thick in regions of heavy blanketing by ash or by ejecta. Consequently, some dimple craters shown in Plate 20c may be produced endogenetically by particulate drainage.

Two other mechanisms likely will produce dimple craters. The first is by impact into a thick layer of incompetent material, followed by relatively rapid degradation by mass wasting, such as creep. This may apply to some of the craters seen in Plate 20c. The second mechanism is fortuitous impact of a small crater within a shallow subdued crater. If the subdued crater was produced by blanketing or continuous degradation, the floor presumably represents a thick and mobile bowl of particulate debris. A small crater formed within this debris would be subject to rapid degradation relative to craters on a level surface. Plate 55, a and b, shows two large subdued craters with small dimple craters on their lower walls. Formation of a small crater nearer the center of these debris traps could produce the profile shown in Plate 20b.

The unusual texturing around the northern rim of the crater shown in Plate 20a (arrow X) or the lobate boundaries of the tree-bark pattern seen in Plate 20b (arrows Y and Z) are possible indications of extrusive flow units. These are not unique examples.

PLATE 20

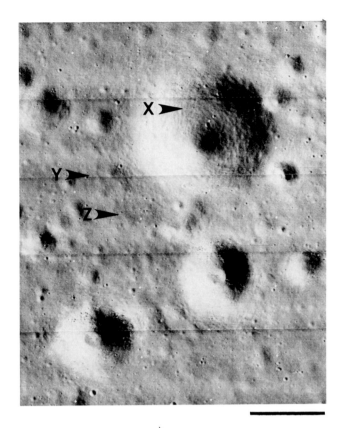

a. III-205-H3   (oblique)

b. III-208-H1   0.24 km

c. V-142-H3   0.44 km

Flat and Concave Floors

Lunar Orbiter photography was planned for typically low sun angles. Consequently, many craters in the size range 1 km–5 km were photographed with their floors in shadow. The examples discussed here are limited to those craters with sufficient diameter-to-depth ratios to reveal a sunlit floor. The stereo pair in Plate 21, a and b, shows examples of concave and flat-floored craters. The concave crater near the top is shown in high resolution in Plate 21c. The region included in Plate 21 is within the interior of the Flamsteed Ring (Plate 154a).

The closer view afforded by Plate 21c of the concave crater (1.5 km in diameter) reveals the highly textured rim area. The continuous merging of the wall region with the floor gives the concave appearance. Numerous blocks are strewn across the wall and a few are found at the crater bottom. The rounded rim also displays an abundance of blocks, even on the highly textured region beyond the wall-rim contact.

The over-all subdued appearance and concave profile of the crater floor shown in Plate 21c may imply a long degradational history. The absence of a hummocky ejecta blanket comparable to that around the crater seen near the bottom of Plate 21, a and b, supports this interpretation. However, the region shown in Plate 21 is within the Flamsteed Ring (Plate 154) and exhibits numerous subdued irregular depressions (Plate 21, a

and b, arrow X), irregular platforms (upper left, Plate 21b), and small craters that either are endogenetic in origin or represent craters inundated by mare units (Lunar Orbiter III-193-H2, framelet 157; see Plate 151 for analogous features). These small-scale features are interpreted as remnants of structures formed during or soon after the emplacement of mare units. Their survival, yet the almost complete destruction of the visible hummocky ejecta blanket, suggests four alternatives:

1. Degradational processes were more rapid on the ejecta blanket and crater interior than on the features associated with mare flooding.
2. The crater was formed by a primary impact prior to the last local emplacement of mare units, which buried the outer hummocky ejecta blanket.
3. The crater represents a subdued volcanic vent associated with mare emplacement.
4. Since secondary impacts commonly produce very subdued craters at the outset (see Plate 115b), this crater may be a secondary impact formed before or after local mare flooding.

Thus, the development of the concave crater profile should be considered in conjunction with the regional setting.

PLATE 21

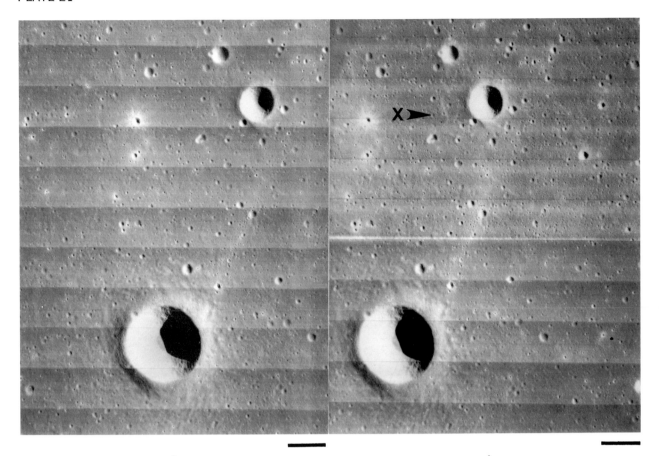

↘ *a.* III-201-M   1.7 km

↘ *b.* III-203-M   1.7 km

↘ *c.* III-194-H1   0.22 km

Flat Floors—Marelike and Hummocky

Concave floors commonly are displayed by craters having sub-dued features such as those shown in Plate 21c. Craters with screelike walls typically have small, smooth, and dark floors, such as that exhibited in Secchi X (Plate 57b) or the crater shown in Plate 22a. The existence or absence of such floors reflects the nature of either—or both—the crater wall or floor. The physical structure of the crater wall can determine its slope at the outset. One crater having coarse angular wall debris will exhibit a greater wall slope and expose a greater floor area than another crater of the same diameter having small spheroidal wall debris, provided that the walls do not meet in a funnellike profile. In addition, the slope of many crater walls appears to decrease with age, presumably owing to the redistribution and comminution of wall debris by meteoroid bombardment, and this decrease in slope would decrease the exposure of a flat floor. If the physical structure of walls of different impact craters is the same at the outset, then the nature of the floor material and the crater depth will determine the existence of a flat floor. For example, if the flat floor reflects influence by sub-surface structure or properties at a particular depth, then larger craters penetrating this horizon will exhibit greater floor areas. Perhaps more likely, a pristine crater having a limiting diameter, below which a flat floor does not occur, simply indicates that the diameter of different craters increases more rapidly than the depth, owing to a greater fraction of retained impact melt or fall-back debris.

Later crater history also will alter the exposed floor area. Regional blanketing by volcanic or impact events will increase the floor area since the crater acts as a debris trap. However, in some regions of the Moon, craters clearly have been subject to an epoch of filling by marelike material, which has resulted in almost complete inundation of several craters, and thus they appear to be rejuvenated vents for extrusions unrelated to the crater-forming process (see Plate 197). In addition, Apollo photographs reveal marelike floors that advanced onto the screelike walls apparently from internal processes. The origin of the floors shown in Plates 22a and 57b, however, is not obvious. Such floors can be found in craters at widely different states of preservation.

Plate 22b illustrates a flat but hummocky floor with relatively smooth dark material (arrow X) filling the low areas. This type of floor might represent debris floating on a layer of subse-quently extruded low-albedo material or debris with a lower albedo fine fraction filling topographic lows. It also could have developed from solidifying impact melt that produced the ropy units in larger craters (see Plates 26b and 30a). Note that the rim of this crater generally appears subdued without a hum-mocky ejecta blanket, yet, in contrast, blocks and radial grooves (arrow Y) reveal its well-preserved state. Such a paradox re-flects the subdued appearance of impact craters in the highlands.

The crater (3.1 km in diameter) shown in Plate 22c is one end of a crater chain near the Littrow Rille and is shown in stereo in Plate 118, a and b. Despite the relatively subdued ap-pearance of its rim, the wall appears to be rugged, and the floor to be recently constructed or modified. A smooth unit (arrow X) composes the central floor region and is surrounded by a hummocky ring adjacent to the wall (arrow Y). The latter unit could be debris derived from the wall or the product of differen-tial settling of the floor (also see Dionysius, Plates 26a and 89, a and b). The association of this crater with the chain and its location in a once endogenetically active area perhaps suggests a volcanic origin for the crater itself. However, the herringbone pattern between chain members and the higher albedo of the chain relative to the surrounding terrain are evidence for the origin by multiple impacts, such as secondary impacts (see Plate 118, a and b).

PLATE 22

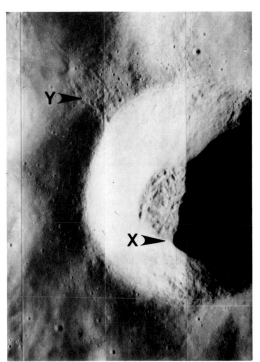

a. I-011-M   6.9 km

b. V-104-M   3.9 km

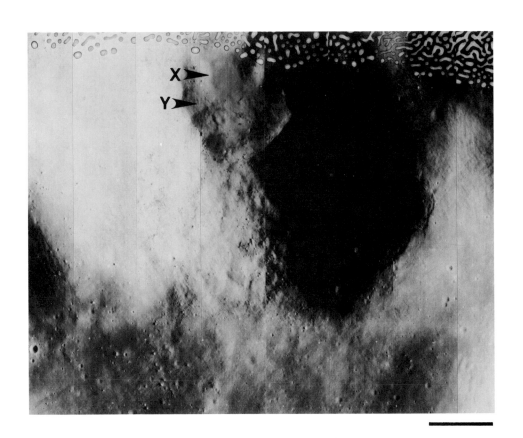

c. V-068-H2   0.51 km

Flat Floors—Hummocky

The general description "flat floored" for larger craters is inadequate, for most craters display such a profile. Plate 23, a and b, reveals in stereo the crater Bode E (about 6 km in diameter), adjacent to Rima Bode. The shallow hilly floor is about 300 m below the rim and does not have marelike material as a major unit. The crater wall has a coarse texture at low resolutions that appears as a bright ring under full illumination. The narrow moat (Plate 23, a and b, arrow X) that encompasses the floor of Bode E indicates a fracture zone developed by differential floor movement or by cooling and contraction of an originally molten floor.

Bode E is a very plausible candidate for a caldera. The surrounding area exhibits numerous indications of endogenetic activity, such as collapse features and a regional low albedo shown in Plate 119, a and b. In addition, nearby, and possibly related, flow termini of a highly cratered plains-forming unit can be identified in the lower left corners of Plate 23, a and b (arrow Y). The lobate unit (arrow Z) may represent results of a parasitic eruption.

The high-resolution view of Bode E (Plate 23c) reveals the very subdued appearance of the floor, similar to the floor of calderalike Hyginus (Plate 160). In the original print of Bode E, several isolated boulder fields can be located.

PLATE 23

a. V-123-M   3.5 km

b. V-121-M   3.5 km

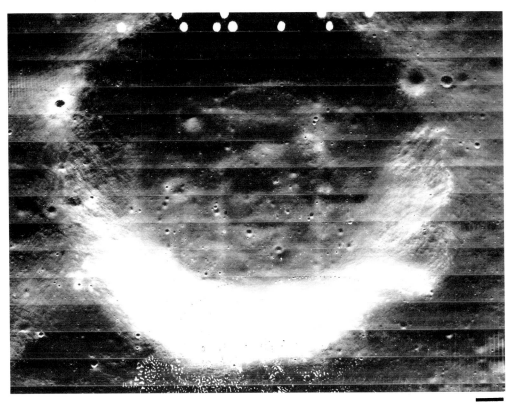

c. V-120-H1   0.44 km

Domed Floors

Multiple doming is another characteristic feature found on the floors of numerous craters that are generally smaller than 15 km. Plate 24a shows the location of four adjacent craters within the larger backside crater, Aitken. Plate 24b reveals the details of the domes and their restriction to the interior of the craters. It is apparent that the processes responsible for the domes are related to the craters and not to regional processes. The original appearance of the craters may have been similar to that of the mare-filled ringed plains (lower right center in the cluster) or the deeper concave crater (left center). As is common, multiple-domed craters may be subdued, with breached walls, but the doming nevertheless is restricted to the circular wall-floor interface. The crater shown on the floor near the top of Plate 24a has a flat floor with a single dome and possibly represents the beginning of multiple-domed formation. Note that the domes have a relatively narrow range of sizes in contrast to those in Dionysius (Plate 26).

Plate 24, c and d, provides a stereo view of two craters in Oceanus Procellarum (adjacent to the rille seen in Plate 176) that have similar multiple domes. The domes in the larger crater are concentric to the wall and surround a central platform. The resolution and lighting unfortunately prevent a definite statement concerning the age of the floor relative to the adjacent "fresh" crater. Numerous other craters in this region display similar doming.

By analogy with terrestrial calderas, dome construction represents late-stage or resurgent volcanic activity. A possible terrestrial analog is the Valles Caldera in New Mexico, which exhibits numerous resurgent rhyolitic domes (Smith et al., 1961).

PLATE 24

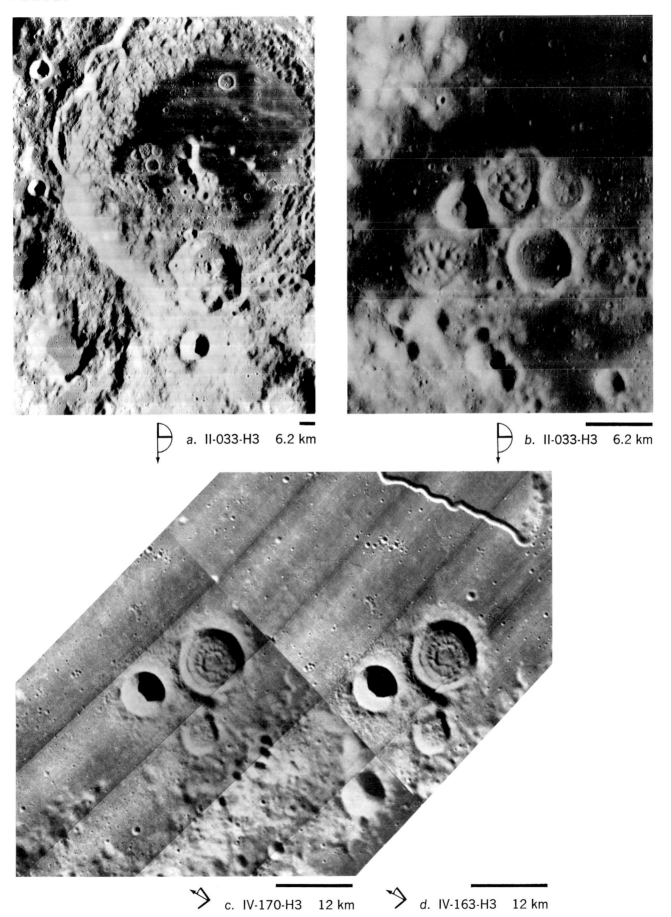

a. II-033-H3   6.2 km

b. II-033-H3   6.2 km

c. IV-170-H3   12 km

d. IV-163-H3   12 km

Domed Floors

Plate 25 shows several domes between an inner and outer crater, both of which appear relatively subdued. The main crater, similar to Bode E (Plate 23), displays a raised rim, small irregular scallops, textured rim, and patterned walls. The interior crater, however, is similar to the tuff-ringlike feature adjacent to the Alpine Valley (Plate 180d). Surrounding this region are the Gruithuisen domes (Plate 130, a and b) and a wrinkle ridge with elongate craterforms (Plate 181) that suggest endogenetic activity was a formative process.

The irregular depression to the northeast also displays multiple domes on its floor. Not far from these two examples, but not included in the photographs shown here, is a small circular depression (5 km in diameter) with numerous hummocks, producing a corrugated appearance. This description applies to the floors of other small subdued depressions (see Plate 13a).

At higher resolution (Plate 25c), the fine texturing of the floor and rim, as well as the grid pattern of the wall, becomes apparent. The domes seen on the north side have lobate boundaries that overlap the floor and may indicate lava termini, debris slides, or differential floor movement.

PLATE 25

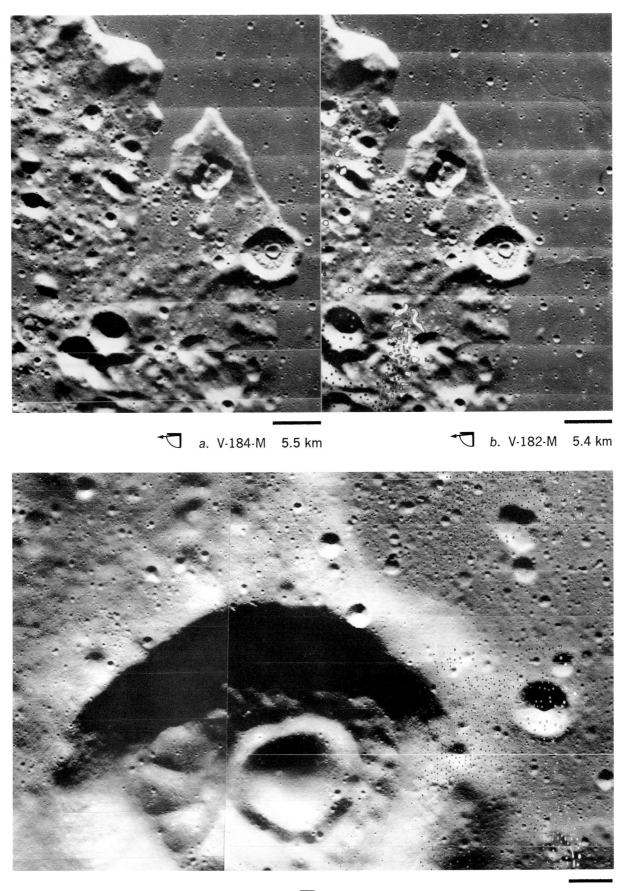

a. V-184-M   5.5 km   *b.* V-182-M   5.4 km

c. V-183-H3   0.69 km   V-182-H3   0.69 km

Complex Meltlike Floors

Several crater floor types are shown in Plate 26a, the area of which includes a small region on the western edge of Mare Tranquilitatis. The crater at left is Dionysius (approximately 19 km in diameter) which illustrates the complex floor discussed below. Several small (less than 1 km) craters to the east and south of Dionysius typify the concave profile of their size class, whereas several subdued craters of the same approximate diameter on the mare surface north of Dionysius have flat floors. To the lower right, the crater Ritter (33 km in diameter) represents a typical flat-floored crater with interior rings and fractures. This type of floor is commonly displayed by craters of similar size, and larger, that occur along the boundaries of the maria (see Plates 34–39). Ritter C, B, and D (in order northward from Ritter) illustrate relatively large subdued concave craters whose floors merge continuously with the walls.

The enlargement of Dionysius, shown in Plate 26b, reveals the small complex floor. Surrounding the floor is a doughnutlike ring (arrow X) that may represent

1. Extrusions along the weak wall-floor contact
2. Debris or molten material derived from sources on the wall or rim (possibly fall-back debris)
3. Differential floor movement resulting in a scarp along the wall contact

The ring appears to be buried beneath a slide (arrow Y) along the southern wall. Within the ring is a moatlike region of small hummocks and ropy relief, and within this zone are numerous domes of various sizes. The domes typically are studded with blocks, in contrast to the moatlike ropy zone.

This structured floor appears to be a combination of rubble in a matrix of molten material and is interpreted as a small-scale version of the floors of Tycho (Plates 31–33) or Copernicus (Plates 219–230). Such floors characterize many craters having extensive ejecta blankets, but it is not certain whether there is a limiting diameter below which these complex floors are buried or are not formed.

62

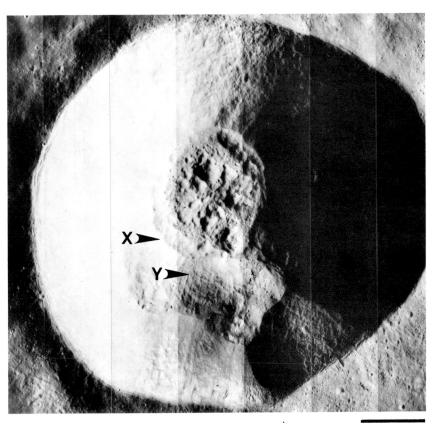

PLATE 26

a. V-081-M   2.9 km (*left*)   3.6 km (*right*)

b. V-081-M   3.0 km

Debrislike Floors

Dionysius (Plate 26) and Dawes (Plate 27) are similar in size (about 19 km in diameter), plan (slightly scalloped), and section (flat floors). At medium resolution (Plate 27b) and comparable enlargement, the major difference appears to be the greater floor coverage for Dawes. Plate 27a shows Dawes from Orbiter IV photography and is included for comparison with other craters at similar resolution.

The enlargement of Dionysius (Plate 26b) and the high resolution of Dawes (Plate 27c) reveal that in detail the floors are distinctly different. The floor of Dawes is hummocky at wavelengths of 0.3 km but appears to be relatively smooth at larger and smaller scales. The floor patterns generally reflect the scallops in the crater wall and are studded with numerous blocks. The arcuate ridges seen to the left in Plate 27c (arrow X) are common near the base of wall slumps in larger craters (see Plate 30c and Plate 65, a and b). In contrast, the hummocks of Dionysius (Plate 26), Petavius B (Plate 30a), and Tycho (Plate 31) consist of numerous small domes that are surrounded by a plains-forming ropy unit.

I suggest that the floor pattern in Dawes is debris resulting from collapse and slides, an inference drawn from its scalloped plan. If Dawes and Dionysius are impact craters, then either the evolution of their floors was different or the development in Dionysius was arrested. Perhaps an initial or rejuvenated lavalike floor triggered collapse and slides from the walls of Dawes. The plastic nature of the material beneath this debris permitted rapid adjustment and yielded a relatively level floor. The differences between Dawes and Dionysius could reflect differences in their locations and/or times of formation. Dionysius is on the edge of Mare Tranquilitatis, whereas Dawes is in Mare Tranquilitatis adjacent to Mare Serenitatis. If Dawes formed relatively soon after mare flooding, but prior to complete cooling, then it may have tapped still-molten material. Alternatively, its floor

was rejuvenated by volcanic processes, which were related to late stages of volcanism within nearby Mare Serenitatis.

A third possibility is that Dionysius and Dawes represent craters of different origins or mechanisms of eruption. Dionysius displays a highly reflective inner rim zone with dark rays, whereas Dawes is a dark-haloed crater. This might indicate differences in target materials or excavation depths, but it also might indicate a caldera formation for Dawes. Donaldson (1969) suggested this third interpretation and compares the floor patterns with terrestrial pahoehoe flows. At low (Plate 27a) and medium (Plate 27b) resolutions, the floor patterns resemble those in Halemaumau, the inner pit of Kilauea Caldera on Hawaii, at a much larger scale (see Jaggar, 1947, Plates 14–15), which are formed by successive and overlapping lava units. I believe, however, that this comparison is not supported by the high-resolution view in Plate 27c. Well-developed pressure ridges, flow termini, and ramparts are absent. Plates 33a and 105b show more plausible examples of pahoehoelike flows. In addition, the floor of Dawes at 0.3 km wavelengths is hummocky rather than flat. The absence of small craters indicates either a well-preserved or a mobile surface. The former alternative suggests that the lack of detail cannot be attributed to smoothing processes; the latter alternative is consistent with a debris-laden floor. It is possible, however, that Dawes is a caldera without extensive volcanic activity on its floor after its formation.

The difference between Dionysius and Dawes might simply be the result of their geologic environment, a determinant independent of subsequent volcanic processes. Both craters are near the limiting diameters at which slumping becomes a major factor in postformation modification. The location of Dawes on a mare seems to favor slumping at smaller diameters. One then could couple this observation to the possibility that the physical strength of the mare or submare material was not sufficient to permit formation of consolidated slump blocks.

PLATE 27

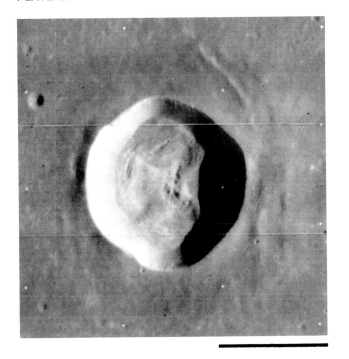

a. IV-085-H2   12 km

b. V-070-M   3.7 km

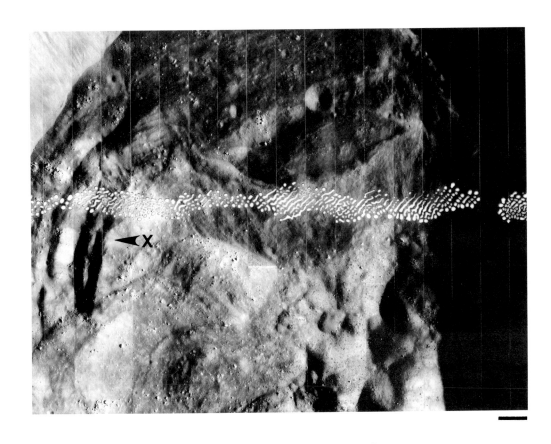

c. V-070-H2   0.46 km

Rejuvenated Floors

Plate 28a shows the crater Bessel (15 km in diameter), which has portions of the floor (arrow X) similar to that of Dawes (Plate 27). In addition, there is a smooth plains-forming unit near the floor center (arrow Y). The hummocky unit I interpret as debris from wall slumps that is overlain by the marelike unit. This is based on the observation that the contact between these two units is relatively well defined, which would not be expected had the slumping postdated the smoother unit.

The crater floor shown in Plate 28b appears to be more inundated than the floor of Bessel (Plate 28a). Vestiges of multiple slump blocks remain near the western wall, but the remainder of the floor exhibits a plains-forming unit that generally covers debris derived from the walls.

Plate 28c shows the crater Plato B, about 12 km in diameter. In contrast to the two foregoing examples, the floor is without obvious scallop patterns but contains numerous small hummocks surrounded by a darker floor unit. This floor type is not unique, but further comment requires photographs of better resolution. On the southern wall is a narrow ledge (arrow X) that perhaps represents a previous floor level.

The crater Calippus (Plate 28, d and e) is in the Caucasus Mountains on the northwestern edge of Mare Serenitatis. The floor is distinctly different from that of Dawes (Plate 27b) in that it is without the numerous small-scale scallops and has several relatively large hummocks. In general, the floor resembles the smaller crater Bode E (Plate 23). Its western wall has a large slump block (not shown) that may have been partly responsible for these hummocks. A narrow furrow (arrow X) separates the southwestern third of the floor, possibly indicating differential floor movement. A plains-forming unit comprises a portion of the southeastern floor, and a terrace (arrow Y) can be identified at the base of hummocks to the west, which suggests settling of the marelike unit.

I interpret the difference between the floor detail in Calippus and Dawes as the result of differences in regional rock type as well as subsequent floor evolution. Relative age might be another important factor, but this assumes identical initial appearances, an assumption that may not be warranted.

Rocca G (Plate 28f) displays another floor type that has numerous small (0.5 km–0.75 km in diameter) hummocks roughly arranged along ridges, which reflect the crater plan (arrow X). This corrugated appearance in Rocca G is generally within 5 km from the wall, but other similar-appearing craters display it across the entire floor (see Plate 13a).

Rocca G is in a region with numerous flow features and irregular depressions, interpreted as subsided lava lakes outside and within nearby craters. Consequently, this region exhibits results of recent volcanic activity, and the floor of Rocca G probably developed through differential movement related to this activity. The origin of Rocca G itself is probably volcanic.

The craters seen in Plate 28, g (arrow X) and h (arrow X), are very subdued but have raised floors separated from the walls by a moat. Such floors typically display fractures (see Plate 29, a and b). The contrast between the degree of preservation of the floors and that of the rims indicates either that the floors are rejuvenated or that the entire structure was formed volcanically. Note in Plate 28g that the larger crater at right overlaps the smaller crater with the moat.

Contrasting floor units, fracture systems, and narrow wall terraces suggest that floors of some craters in the size range of 1 km–20 km in diameter have undergone adjustment. Floor features commonly appear to be much more recent than the craters in which they occur. Consequently, floor adjustment is thought to represent rejuvenation of an old crater, whether it was initially impact or volcanic in origin. Such a process appears to be typical of many craters larger than 20 km in diameter.

PLATE 28

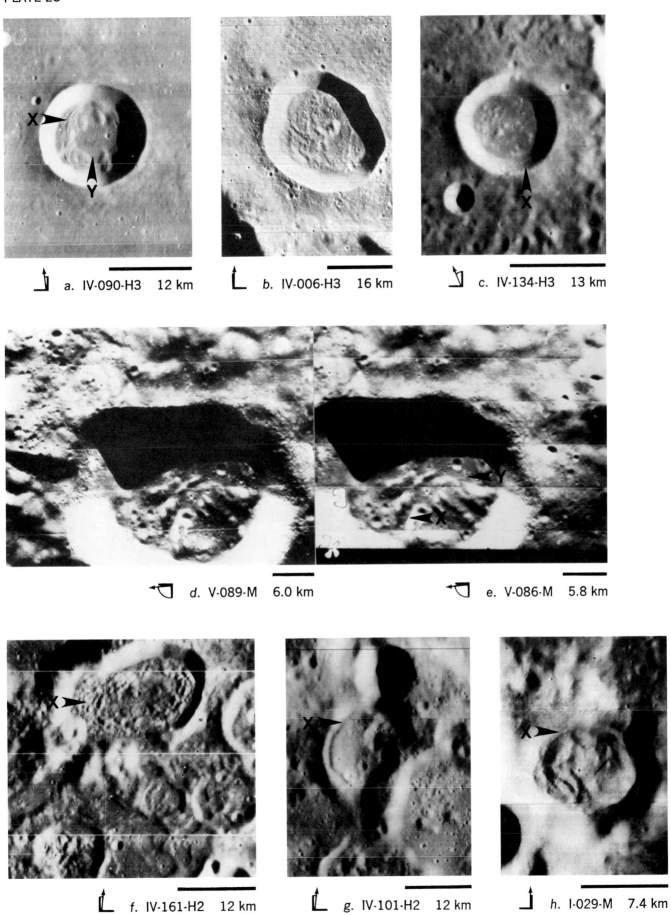

a. IV-090-H3 12 km

b. IV-006-H3 16 km

c. IV-134-H3 13 km

d. V-089-M 6.0 km

e. V-086-M 5.8 km

f. IV-161-H2 12 km

g. IV-101-H2 12 km

h. I-029-M 7.4 km

Fractured Floors

Plate 29 displays moderate-size craters with well-developed floor fractures. Fracture arrangements may be annular, radial, polygonal, or irregular. In Plate 29, a and b, the lobate fronts and discontinuous contacts with the wall suggest that the floor material once was molten. This is a floor type commonly associated with very subdued craters in the highlands. The fractures shown in Plate 29a (oblique view) are somewhat irregular, and it is difficult to distinguish between successive fronts and actual fractures. The top crater shown in Plate 29b, however, definitely displays annular and polygonal fracturing, and the floor appears to bulge as if subjected to upwelling. The floor of the upper crater joins with that of the lower crater, which also has an annular fracture arrangement. Small (12 km in diameter) craters seen in Plates 35 (center right) and 198d show radial fractures that are clearly the result of floor rejuvenation.

The floors shown in Plate 29, a and b, are new relative to the original crater and, after solidification of the fluidlike fill, underwent dynamic processes that produced fractures. The craters seen in Plate 29b are near the crater Tsiolkovsky (Plate 11a), and the proposed rejuvenation may have been related to the Tsiolkovsky event or a later stage of rejuvenation indicated by mare units on the floor of Tsiolkovsky.

In contrast to the two foregoing examples, the crater shown in Plate 29c is better preserved. The fractures cut a debris and lavalike floor in a semiannular arrangement.

Features of floor rejuvenation shown by Plate 29 are local in that they are not present in other nearby craters with similar size, depth, or appearance. These parameters are indicative of volcanic activity. Floor rejuvenation on a regional scale is illustrated by Plate 35, but even in these instances, not all the craters were subjected to such activity.

PLATE 29

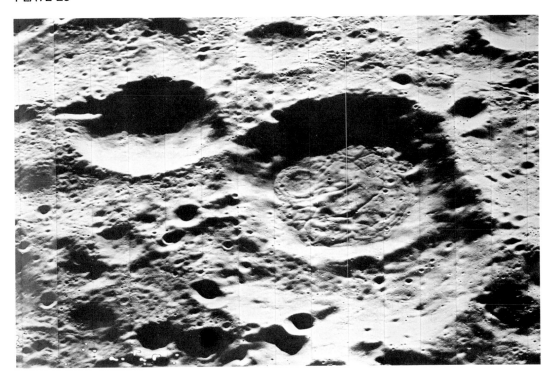

a. V-163-H1 (oblique)

b. I-115-H1,H2   5.8 km

c. IV-168-H1   12 km

69

Volcanolike and Debrislike Floors

Petavius B (Plate 30a) is about 35 km in diameter and illustrates a floor type that was introduced with the crater Dionysius (Plate 26) and is even better displayed in larger craters, such as Tycho (Plate 31). The floor is characterized by numerous small hillocks surrounded by a unit having ribbonlike patterns. A narrow and shallow trough generally marks the contact between the floor and wall. On the southeastern floor there is a flow feature (arrow X) that originated from a depression on the wall; in addition, note the smooth plains-forming units on the floor near the southern wall (arrow Y).

The general impression is that the floor of Petavius B was once molten, or at least fluidlike. Since the floor of Dionysius can also be described in these terms, perhaps it represents an earlier stage than Petavius B. Alternatively, the Dionysius event did not generate or trigger as much fluidlike material as a result of its location, size, depth, and/or energy.

In contrast to Petavius B, the 25-km-diameter crater Mösting (Plate 30, b and c) has a debrislike floor similar to that of Dawes (Plate 27). In particular, Mösting and Dawes both display arcuate slumplike features, multiple central hummocks, and debrislike floors. Close inspection (Plate 30c) reveals flow termini of pitted plains-forming units (arrow X) but nothing comparable to the extent and type displayed by Petavius B. Mösting's floor could be described loosely as concave, but the amount of slumping and other mass movement forces such a label to be gross simplification. The crater Römer (Plate 45, c and d) exhibits a similar floor.

PLATE 30

a. V-037-M   3.9 km

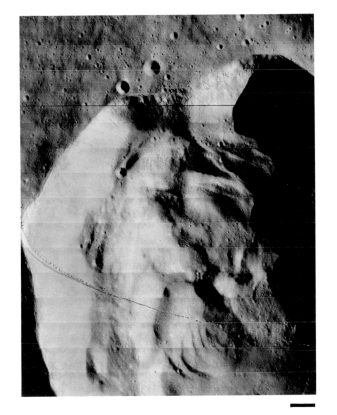

b. III-107-M   1.4 km

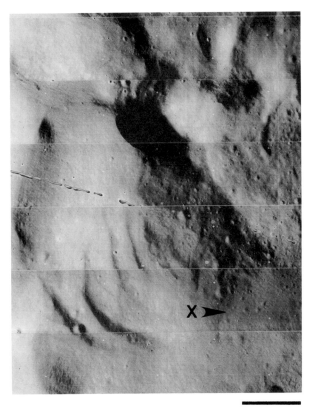

c. III-107-M   1.4 km

Volcanolike Floors

The floor of Tycho (a conspicuous Earthside crater 98 km in diameter) rests about 4.5 km beneath the surrounding rim and is designated as flat with central peaks. The stereo pair (Plate 31, a and b) reveals numerous hills and complex textures, shown in greater detail in Plates 31c, 32, and 33. Comparison of similar features should be made with those on the floors of Copernicus (Plates 219–229) and Aristarchus (Plate 237).

The textured pattern is due to both extensive fracturing, clearly identified in Plate 31c, and to a ribbonlike pattern resembling (at a much larger scale) pahoehoe lava flows (see Plates 32–33). Within the crater, the fractures tend to be preferentially arranged and appear to have been modified by the larger domes. In particular, fractures appear to be interrupted by and, in some cases, to converge toward the dome rather than being purely radial or concentric to the relief. In several instances, fractures extend onto the relief and terminate at a lip shown clearly in Plate 33.

The domes or mounds are interpreted as either active or passive features. If active, they might be volcanic domes or buoyed masses that protruded through the solidifying floor. If passive, the features may have been pre-existing relief subsequently surrounded by a lavalike unit. Both general interpretations apply to different features across the floor and are consistent with the modification of the stress fields indicated by the fracture patterns. Also note the craterlike features with broad rims to the northwest of the relief in the southeastern corner (arrow X) and that in the northwestern corner (arrow Y) of Plate 31c; these are probably analogous to terrestrial tephra rings.

The presence of volcanolike features does not necessarily imply that Tycho is a lunar volcano; it may be that volcanism was active in the later stages of crater formation or modification. The extensive ray pattern and ejecta sculpturing are indications of an impact, which must have generated a large volume of melt.

PLATE 31

a. V-126-M   7.0 km   b. V-125-M   7.1 km

c. V-125-H1   0.92 km

Volcanolike Floors

The northwestern part of the floor and lower wall of Tycho shown at high resolution in Plate 32 is much rougher than the area shown in Plate 31c. The ribbonlike pattern weaves around the relief and extends up to, and apparently onto, the crater wall as well as onto numerous domes. The "ropy" appearance resembles pahoehoe flows, but note the giant dimensions of these features. That the floor is quite obviously devoid of impact-type craters further confirms the youthfulness of the formation, but possible volcanic craters can be identified. Details of Plate 32 are shown in Plate 33.

PLATE 32

V-125-H2   0.92 km

Volcanolike Floors

Plate 33a is an enlargement of the western floor-wall contact in Tycho. The floor unit exhibits an extremely complex pattern composed of ropelike ridges that roughly parallel the wall boundary. Arrows X, Y, and Z identify the terminus of a more recent flow unit derived from sources on the wall. Close examination of the original photographs reveals numerous blocks on the floor unit and fractures that cross both the floor unit and the hummocky base of the wall.

Plate 33b shows a mound (arrow X) to the south of the area shown in Plate 33a and included in Plate 32 (bottom edge). The lip surrounding this positive-relief feature is clearly defined, but relatively few mounds exhibit such a contact. The lip also is found along the floor-wall contact (Plate 33a, arrow XX). Although these mounds may have protruded through the cooling floor melt, it is more likely that the melt subsided as the result of cooling or local subsurface removal and produced the raised-lip borders. It is also possible that such boundaries represent ramparts from the thrusting of the floor crust.

The stereo pair (Plate 33, c and d) is an enlargement of the northeastern corner of the area seen in Plate 32. The fracture pattern is generally parallel to the floor-wall contact and is believed to indicate peripheral extensional fracturing produced during floor subsidence. Note that this fracturing has been deflected by a lobate front at the upper left (arrow X). A subdued semiconcentric (ripplelike) pattern of ridges centered around the lobate front (arrow Y) can be traced onto the floor without alteration of the fractures. Possibly, less viscous material originally flowed onto a plasticlike floor and produced this low-relief ripple pattern. The deflected fracture pattern reflects the local stress field, which was altered by the flow unit during subsequent floor subsidence or shrinkage.

PLATE 33

a. V-125-H2   0.92 km

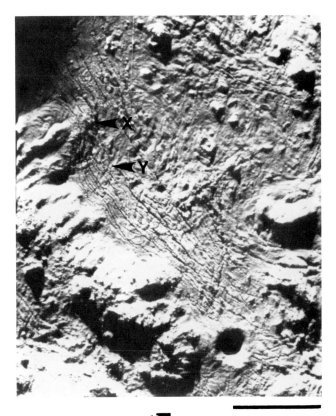

c. V-126-H2   0.92 km

d. V-125-H2   0.92 km

Shallow Fractured Floors

The craters Damoiseau (Plate 34, a and b) and Taruntius (Plate 34c) exhibit a floor type that typically is found along the borders of maria and that generally is characterized by shallow flat floors with numerous large (0.5 km–1 km wide) fractures.

Plate 34a is an oblique view of Damoiseau (center right, arrow X), a large crater in the highlands along the western edge of Oceanus Procellarum. It is approximately 35 km in diameter and is nested within a larger subdued crater that has been modified by the Orientale ejecta (see the vertical view, Plate 34b). The floor is cut by several fractures that surround the central ridge in a polygonal and concentric plan. Floors of this type commonly have been partly flooded by mare material and display a concentric or semiconcentric plan as in Posidonius (Plate 2b), Gassendi (Plate 36), and Doppelmayer (Plate 191b). Also characteristic is the absence of well-defined slump terraces along the walls.

The floor of Taruntius (Plate 34c) is similarly described except that here the fracture system is more annular. Note, however, that Taruntius exhibits a well-defined ejecta blanket, which is similar to that of Copernicus (see Plate 94). The ejecta generally overlie Mare Fecunditatis, although portions appear to have been blanketed by later mare material. The crater Atlas (Orbiter IV-74-H2), among numerous other examples, also has a preserved hummocky ejecta blanket but a highly fractured floor. If the preservation of an ejecta blanket can be used as an age criterion, then processes responsible for this floor type have affected a wide range of apparent crater ages.

Note the wrinkle ridges that appear to be radial to Taruntius. A similar radial pattern is exhibited around the ghost ring, Lamont, and is evidence for upward-directed stress associated with some craters or craterforms.

Plate 34d shows two overlapping craters (each about 25 km in diameter) with fractured floors. These are the only craters in the region that have this floor appearance, and they are found within an extremely subdued depression that is partly inundated by plains-forming units. If floor fracturing had occurred first in the overlapped crater at right, then later formation of the adjacent crater by an impact would have blanketed this previous history. Since both floors remain well preserved, either they were altered at the same time or the crater at left formed through caldera collapse without an extensive ejecta blanket. The first interpretation does not seem reasonable, because processes responsible for floor rejuvenation have not affected the rim of the more-recent crater. Thus, caldera collapse is the favored hypothesis, and the crater pair represents offset sites of eruption, formation that is typical for terrestrial calderas.

The 25-km-diameter crater seen in Plate 34e is in Mare Smythii and illustrates floor fracturing and a concentric moat. The multiring plan was introduced in Plate 2b, which shows Posidonius. The interior platform bordered by the inner ridge is interpreted as the original crater floor that has been raised to nearly the same level as the crater rim.

PLATE 34

a. III-213-M   (oblique)

b. IV-161-H3   12 km

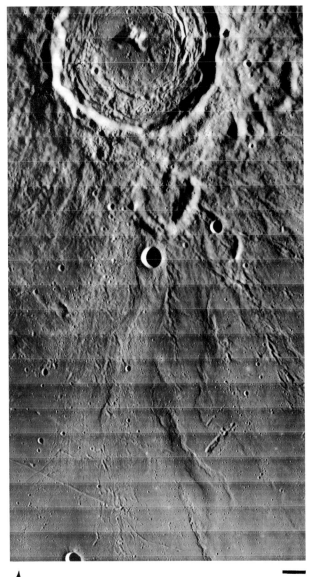

c. I-031-M   7.7 km (top)   8.6 km (bottom)

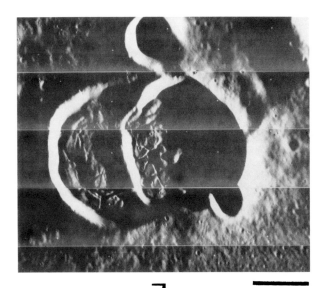

d. V-025-H1   13 km

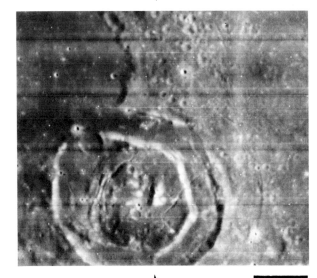

e. IV-017-H1   12 km

Shallow Fractured Floors

Floor fractures are described by four patterns: annular, radial, polygonal, irregular. Plate 35 displays each type in craters bordering western Oceanus Procellarum. The restriction of fractures to within the confines of the crater wall (and where more than one wall, within the inner wall) implies that their origin is the result of local processes. This supposition is substantiated further by the fact that not all crater floors in the region exhibit the fracture system. Consequently, spatial, temporal, and/or size dependencies are inferred.

Also note the following features that add to the complexity of the region:

1. The craters near the mare (for example, arrow X), whose cracked floors are encircled by a ridge as well as a moat of mare material (recall Posidonius, Plate 2b)
2. The two small craters (far right center, arrow Y) in the mare whose floors have multiple domes (Plates 24–25 show other examples)
3. The concentric crater on the large annular fractured floor of Lavoisier (arrow Z; see also Plate 4, a and b)
4. The numerous low-albedo patches along the fractures (arrow XX)
5. The absence of large intact slump blocks within the fractured craters
6. The hummocky relief (perhaps the remnants of slump blocks?) that is adjacent to the wall and is crossed by the peripheral fractures (arrow YY)
7. The range of crater "ages" affected by fracturing

Possible origins for floor fractures, as well as the general floor type, are discussed following the more detailed examination of Gassendi in Plates 36–37.

PLATE 35

IV-189-H2   12 km

Shallow Fractured Floors

Gassendi (Plate 36) is 110 km in diameter and is situated along the northern border of Mare Humorum (see Plate 191). In contrast to the examples shown in Plate 35, its factures are not so well ordered. A major NE-trending scarp (Plate 36a, arrow X) cuts off the southeast sector of the floor and is partly engulfed by mare material. Several fracture systems intersect to the east and west of the acentral ridgelike peaks.

Partial inundation of the floor of Gassendi has occurred on the side adjacent to the mare, a situation also recognized in the crater Posidonius (Plate 2b). As a consequence, the plan resembles a biringed structure with the inner ring eccentrically situated. The entire region appears to have been tilted toward the basin. Thus, at first glance, it is reasonable to envision the partial flooding of the southeastern floor as "spillover" from Mare Humorum. This assumption is not realistic, however, because the only possible breach in the wall is a very narrow gap (Plate 36a, arrow Y) that must have been closed by subsequent degradation of the walls (see Plate 36, b and c). If that gap was the gate for encroachment into the interior, then the mare material must have been extremely fluid. More likely, there were also multiple feeders within Gassendi that supplied much marelike material and that were interconnected with sources responsible for the inundation of the Humorum basin.

Plate 36, b and c, is a stereo view of part of the floor of Gassendi at higher resolution than that of Plate 36a. The fracture parallel to the outer rim appears to have a raised border (arrow X). I interpret this rim as the remnants of the base of the wall, the rest of which subsided along interior annular grabens (compare the fracture in the upper left with that in the upper right). Support for this interpretation is given by annular fracturing in the crater Taruntius (Plate 34c). Although this configura-

tion is not convincingly illustrated in Plate 34c, Apollo 11 photographs (see AS-11-42-6226) show that the interior floor of Taruntius clearly has been raised (or the wall region lowered), with an inner rim apparently derived from the base of wall slumps. Other fractures clearly intersect previously existing features, such as the central peaks and smaller mounds.

Plate 36, b and c, shows several volcanolike features:

1. Craterforms with wide subdued rims (arrow X; also shown in Plate 86a)
2. A craterform with ropelike features on its rim (arrow Y; also shown in Plate 86b)
3. Conelike feature (arrow Z)

In addition, note the chain of summit pits on a ridgelike segment of the central-peak complex (Plate 36a, arrow Z). These may represent extrusive centers. Also note the two wrinkle ridges (Plate 36a, arrow XX), that are shown at higher resolution in Plate 36, b and c. That these ridges can be traced into the interior of Gassendi indicates that they are the result of regional events independent of the floor.

Local areas within Gassendi display marked differences in their crater densities. The region identified by arrow XX (Plate 36, b and c) clearly exhibits a deficiency of craters relative to the region (arrow YY), which is nested within the annularly arranged central peaks. Such differences can reflect different cratering histories (endogenetic or exogenetic) or different crater-removal rates (determined by rock type or localized extrusive activity). Both processes appear to have been operative within Gassendi.

PLATE 36

a. V-179; 178; 177-M   4.2 km (top)   3.9 km (bottom)

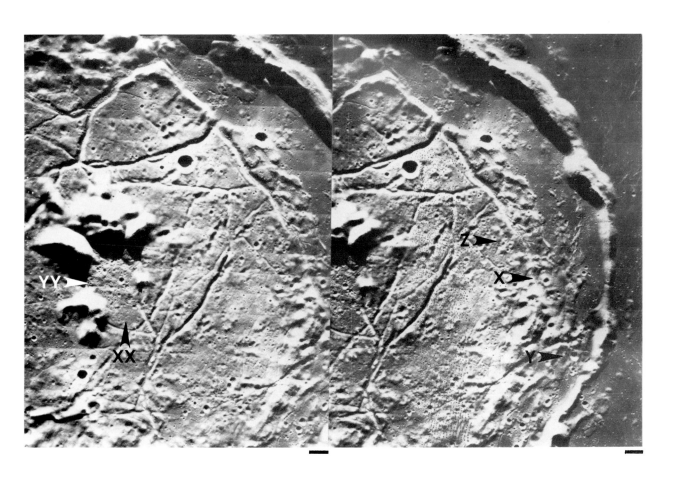

b. V-178-M   4.2 km (top)   3.8 km (bottom)   c. V-177-M   4.4 km (top)   4.0 km (bottom)

Shallow Fractured Floors

Plate 37 shows selected stereo views of the floor of Gassendi. Plate 37, a and b, covers an area of the southwest rim and floor (see Plate 36, b and c). Features of interest are the following:

1. Crater-topped cone on the lower slopes of the wall (arrow X)
2. Low-albedo unit (arrow Y)
3. Ring feature inside a crater (arrow Z)
4. Moatlike boundary at base of wall (arrow XX)
5. Subdued craterform that has an inner shelf along its upper walls, which may represent a previous level of the floor (arrow YY; shown clearly in the original photograph)

Plate 37, c and d, is a selected region of the eastern floor and reveals the different floor units that are crossed by fractures. The plains-forming unit (arrow X) has very few craters on its relatively flat surface. A more hummocky and textured unit identified by arrow Y also is without a large number of craters. In contrast, the plains-forming units identified by arrow Z and on the floor of the adjacent fracture are highly cratered. If crater counts alone are used as an indicator of surface age, then arrow Y marks one of the youngest surfaces. This may, however, indicate either rapid degradation by a relatively mobile unconsolidated surface or a large endogenetic component in the crater density. Note the irregular north-facing scarp (arrow XX) that apparently corresponds to terraces surrounding other relief (arrows YY and ZZ). This may indicate a previous local level of floor material, i.e., a lava terrace. Probable volcanic cones are identified by arrows XY and XZ.

PLATE 37

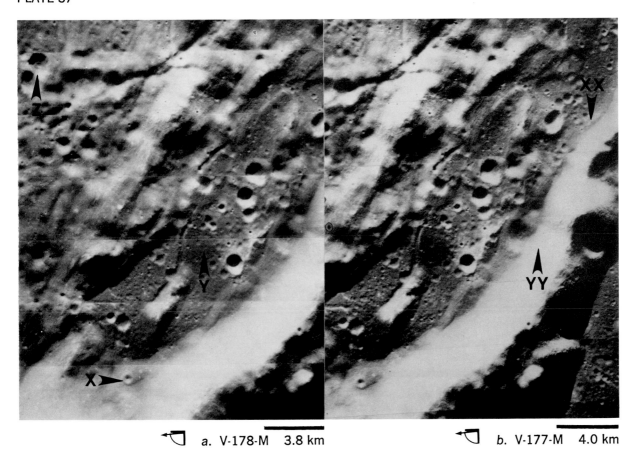

a. V-178-M   3.8 km   b. V-177-M   4.0 km

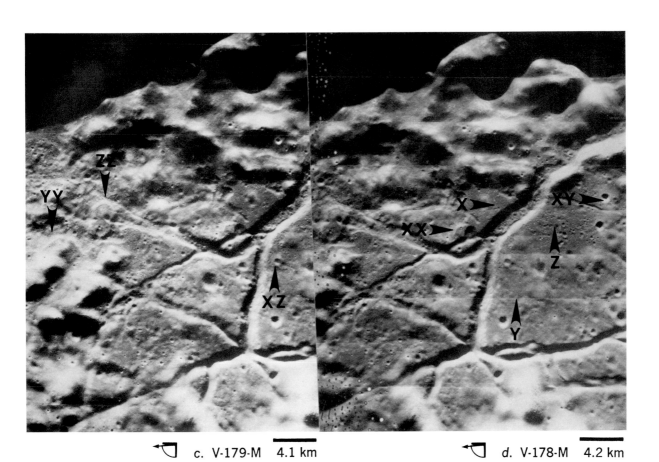

c. V-179-M   4.1 km   d. V-178-M   4.2 km

Shallow Fractured Floors

Plate 38, a and b, is a stereo view of the western floor of Gassendi. The prominent N-trending fracture at the bottom of the photographs separates the topographically higher floor to the east from wall debris to the west. The crater with the convex floor (arrow X) is crossed by this fracture, which has a ropelike ridge on its floor at this segment. Similar ropelike forms are found on the bottom of other segments of this (arrow Y) and other fractures (arrow Z; also refer to Plate 38c, arrow Y). Layering on the wall of one fracture (arrow XX) may represent either previous levels of flow material or outcrops of strata. If the latter alternative is accepted, then the hummocky units are not simply floor rubble, for such debris should not be expected to display horizontal stratification. Similar layering can be identified on the northern wall of the crater marked by arrow X.

Endogenetic modification of the crater floor is indicated by an irregular scarp (arrow YY) that borders an irregular depression. A similar depression was identified in Plate 37, c and d. Within the depression shown in Plate 38, a and b, is a dome surrounded by a narrow moat (arrow ZZ), which resembles features occurring on the floors of pristine craters, such as Tycho (Plate 33b, arrow X). Similar domes occur near the marelike moat of Gassendi as well as within the floor-fractured crater Vitello (Plate 250, arrows ZX, ZY). Such features within Tycho were suggested to be the result of encroachment of pre-existing relief by molten floor material. A similar explanation for the moat-surrounded domes within floor-fractured craters has important implications for the geologic history of such floors. This stereo pair (Plate 38, a and b) also includes a crater having an interior ring (arrow XY). Similar but larger concentric craters within or near floor-fractured craters (for example, Plate 35, arrow Z) are interpreted as resulting from multiple eruptions.

The western member of the central peaks is shown at the top of Plate 38, a and b, and its crescent-shape plan suggests that it is a breached crater. The small elongate crater (arrow XZ) on its southern summit is also breached, and close inspection of the original photograph reveals a possible flow feature extending from its floor. In addition, note the possible steeply dipping strata on the southwestern flanks (arrow YX).

Plate 38c is a high-resolution photograph of the fracture and central peak shown in Plate 38, a and b (note reorientation). The relief is surprisingly subdued and textured. The smooth merging of the floor and wall of the fracture with the surrounding terrain might suggest considerable age, but the crater count is much less than that of the plains-forming unit (Plate 38c, arrow X). As noted in the discussion accompanying Plate 37, c and d, this reflects differences in degradational rates, degradational processes, or a large number of endogenetic craters on the plains-forming units. Plate 38c also shows a high-resolution view of the ropelike ridge on the bottom of the fracture (arrow Y).

With the examples of Damoiseau, Taruntius, the craters along western Oceanus Procellarum, and Gassendi as background, consideration can now be given to the origin of this floor type. In summary, any hypothesis should provide an explanation of the following observations:

1. Craters similar to Gassendi typically have shallow fractured floors with central peaks (or ridges) and relatively symmetrical rims. In a few examples (IV-17-H1), the central-peak complex is replaced by a circular mare-filled depression; in others, it occurs on a relatively unfractured central floor plate (Plate 249).
2. Such floors typically are found along the borders of maria but do not occur in all craters at such locations.
3. Several craters with these floors are postmare in origin and have preserved ejecta blankets, whereas similar sequences around other craters have been destroyed, covered, or were nonexistent.

If all the craters here considered were once similar to Copernicus, then processes affecting the floor clearly have not been the same as those degrading the ejecta blankets; that is, they are not simple erosional remnants. The existence of Taruntius precludes the assumption that all such craters must have formed prior to the emplacement of the maria. Most likely their floors represent volcanic rejuvenations related to the mare-flooding epoch. In several examples, such as Damoiseau, the origin of the crater itself might be from caldera collapse. This has been suggested by De Hon (1971) for the floor-fractured craters Ritter and Sabine.

Masursky (1964) postulated that the shallow fractured floors resulted from rapid isostatic adjustment. Their apparent link to the maria, however, suggests that this adjustment was related to mare flooding and that perhaps a plastic layer extended beyond and beneath the borders defined by the lunar terrae (see discussions accompanying Plates 191 and 197). In addition, if the adjustment is viewed as a regional process, then either some craters were already in equilibrium or the effects of isostasy have been masked, except for the shallow floors.

Volcanic rejuvenation would affect only those pre-existing craters (or calderas) that have roots connected to one of the magma chambers related to mare flooding. The old floor floats as the level of tapped magma rises. The intrusion would produce a variety of geometries for the fracture patterns in response to the rigidity and thickness of the old floor, multiple intrusions, and multiple periods of elevation and subsidence.

The end product could be viewed as Gassendi or Plato (Plate 40a). In the case of Plato, the old floor foundered and was completely engulfed by mare material.

PLATE 38

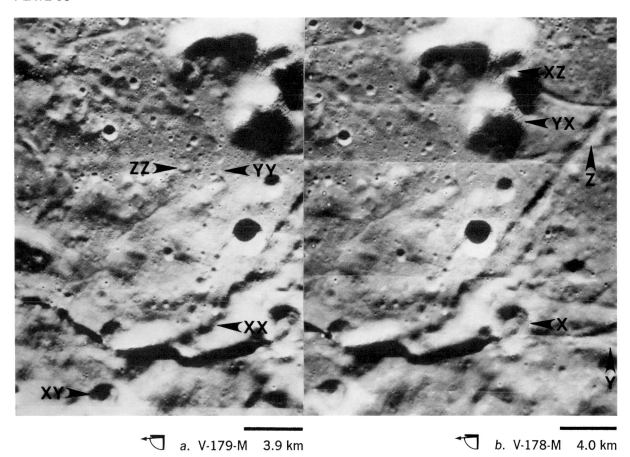

a. V-179-M   3.9 km

b. V-178-M   4.0 km

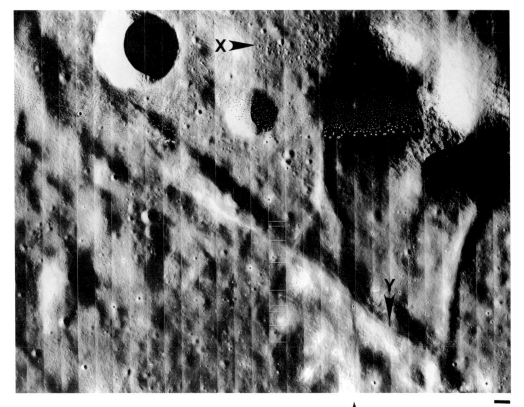

c. V-178-H3   0.53 km

Inundated Marelike Floors

Plate 39a illustrates partial inundation by marelike units of a crater 90 km in diameter, which is on the Orientale ejecta blanket. In contrast to the examples shown in Plates 34–38, inundation occurs without the development of a moat between the floor and wall and without the removal of old wall slumps (also see Plate 46a). Consequently, the inundation sequence does not always parallel that inferred from Gassendi. This crater exhibits, however, floor fracturing and islandlike separation of the central peaks. In addition, mare flooding generally occurs near the floor-wall contact.

Plate 39, b, c, and d, shows recent inundation of highly modified craters. Plate 39b includes a crater 70 km in diameter that is 50 km south of Tsiolkovsky on the lunar farside (see Plate 108a). It is well within the extensive ejecta blanket of Tsiolkovsky, which was responsible for the mutilation of the crater's walls and rims. The floor of this crater has been rejuvenated volcanically, and rejuvenation was probably related to the emplacement of marelike units on the floor of Tsiolkovsky (Plate 40b). Lobate-bordered flow units, which typically have highly cratered surfaces, are identified in the upper left (arrow X), and possible lava-lake depressions are to the right (arrow Y). Low-albedo units with ill-defined borders (arrow Z) overlie the units having lobate termini and are interpreted to be deposits of ash. Detailed examination of the original photograph reveals terracelike borders along the base of the walls and domes. These features imply a complex and well-preserved eruptive history.

Plate 39, c and d, shows the crater Murchison, which has been highly modified by ejecta from the Imbrium basin. The oblique view (Plate 39d) reveals highly fractured islandlike platforms (arrow X). These represent remnants of flow units from the Imbrium basin or remnants of an early stage of floor rejuvenation. The lip adjacent to the far wall of Murchison (arrow Y) probably resulted from settling of the marelike unit. Recent volcanic activity along this zone is indicated by the resurgent dome at left center (arrow Z) and the lava pool with well-defined termini at lower left (arrow XX).

The examples shown in Plate 39 demonstrate that floor rejuvenation with marelike inundation affects a variety of apparent crater ages and that typically this inundation originates and continues along the margins of the crater floor.

88

PLATE 39

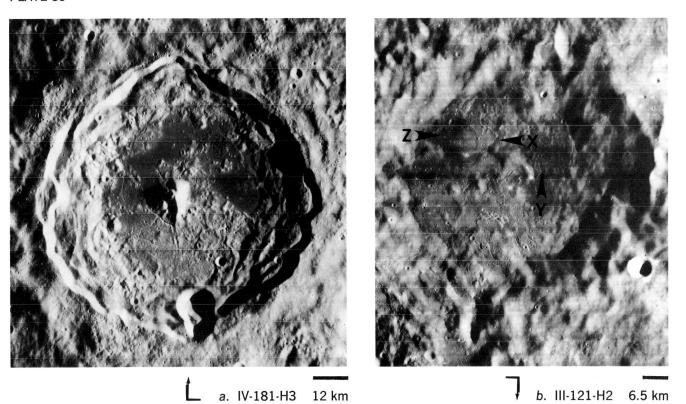

a. IV-181-H3   12 km

b. III-121-H2   6.5 km

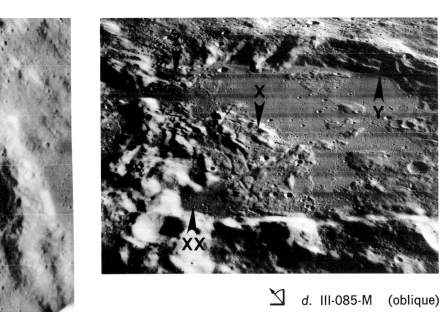

d. III-085-M   (oblique)

c. IV-102-H1   12 km

Inundated Marelike Floors

Plato (Plate 40a) and Tsiolkovsky (Plate 40b) have floors inundated by mare units. Plato is an Earthside crater on the northern border of Mare Imbrium and is approximately 95 km in diameter. In contrast to many flat-floored craters of this class, it displays remnants of an ejecta sequence similar to that of Taruntius or Copernicus. The floor material is darker than Mare Imbrium to the south and is mottled in appearance, which suggests multiple stages of flooding. The low-relief dome with a fracture near its crest (arrow X) might be a surface expression of underlying intrusions or a buildup of extrusions resembling a small-scale shield volcano (also see Plate 206, a and b).

Tsiolkovsky (about 180 km in diameter) is a striking feature on the lunar farside and can be studied in medium resolution in Plates 11a and 108. Plate 40b has been oriented so that the solar illumination is from the right, but north is directed downward. The floor is characterized by its very low albedo marelike unit, central peak, hummocky relief, rilles, and isolated domes. In contrast to Plato, Tsiolkovsky is not a mare-filled basin. The hummocky floor unit along the northern wall (arrow X) might be interpreted at first glance as unflooded floor debris, but closer examination shows that it is related to material that meets the wall as a flow front (arrow Y). In addition, arrow Z identifies another plains-forming unit that exhibits numerous craters and a higher albedo than that of the mare unit. I interpret these units as a separate stage of floor volcanism, which was modified by floor fracturing and later engulfed by marelike units. This stage may be related to the cooling of an originally impact generated or tapped molten floor analogous to Tycho (Plate 31) or Copernicus (Plate 218). Guest and Murray (1969) reached similar conclusions.

Whatever the composition of the marelike units, their viscosity must have been low, or there were many sources acting simultaneously, to fill such craters as Tsiolkovsky and Plato with such uniformity. Plate 39b shows a crater adjacent to Tsiolkovsky whose floor flooding was not completed.

Several hummocks on the floor of Tsiolkovsky are surrounded by aprons that are interpreted as flow fronts associated with the relief, slow mass movement extending onto the floor, or features resulting from flooding by the marelike floor material. In regard to these possibilities, the following observations are pertinent:

1. The extension of the apron is not dependent on the size of the relief; in particular, compare the small "island" (arrow XX) to the hummocky relief (arrow YY).
2. The apron is small around the central peak complex but can be recognized by a change in texture or albedo, rather than as an extension onto the marelike floor.
3. The apron can be easily identified only where positive relief is in contact with the mare material.
4. The apron is small or nonexistent at the base of precipitous walls.

The highly pitted platform on the north side of the peak (arrow ZZ) provides additional clues to the geologic history of this crater floor. One of the larger craters on this platform exhibits a wide rim and resembles terrestrial volcanic cones. The platform unit is thought to be volcanic in origin and may be related genetically to the central peak complex. Guest and Murray (1969) suggested that the peaks and platform may represent a denuded caldera. Alternatively, the highly pitted platform unit is an isolated portion of the hummocky floor unit that borders the northern wall. Oblique views of Tsiolkovsky from Apollo photographs (AS15-91-2383) show that the platform is at a significantly higher elevation than that of the hummocky floor unit to the northwest, and they show that the summit of the central peaks is comparable to the elevation of the crater rim. Consequently, I suggest that the peak complex was elevated during or prior to mare emplacement. Such a structural history is supported by floor-fractured craters whose central peak complex has been elevated with a central floor plate. The branching radial rilles adjacent to the central peaks and the concentric rilles on the mare units near the wall may indicate continued but relatively minor differential movement of the contrasting floor units. The crescent-shaped plan of the central peaks and the partially contained pitted platform, however, suggest volcanic modification of the central peak complex. Further discussion of this and other central peaks is found with Plates 44–49.

PLATE 40

a. IV-127-H3   12 km

b. III-121-H1   6.1 km

Inundated Floors

Plate 41 shows craters that have been inundated by a plains-forming unit, characteristic of the lunar highlands, having a higher albedo and greater crater densities than the mare units examined in Plate 40. Detailed views of such floors are shown in Plate 43. The region shown in Plate 41a is in the highly stippled and cratered terrain of the Southern Highlands and includes the elongate crater Clairaut, 65 km by 100 km, which has been inundated in part by the highland plains-forming unit. This unit also occurs at several different elevations within both the interior doublet crater (arrow X) and the highly subdued rim crater (arrow Y). The rim region of the interior crater (arrow X) exhibits outward-facing scarps and is embayed by the plains-forming unit (see also Plate 92b). This relation suggests that emplacement of the unit postdates the formation of the smaller craters. Arrow Z identifies a topographic discontinuity associated with this unit in a southern rim crater. Such discontinuities are common, and in several examples they resemble flow termini (see Plate 200). Whereas other small depressions in the rim region do not exhibit this unit, small pockets occur on the floor of Clairaut.

Plate 41b shows an area approximately 280 km south of the region shown in Plate 41a. Numerous craters contain the highland plains-forming unit, but of particular interest is the fractured floor of the crater Tannerus (arrow X). Such fracturing suggests differential floor movement or differential cooling. The former mechanism raises the possibility of floor emplacement by endogenetic processes. owing to an inferred (but not necessarily correct) relation between such movement and a subsurface magma chamber. Fracturing by differential cooling of the floor unit raises the question of why other similar-appearing floors have not been fractured. Note that the subdued-appearing wall is in distinct contrast to the preservation of the narrow floor fractures. If floor emplacement or modification and crater formation were not separate events, then the crater must have been highly subdued at the outset. As in the examples seen in Plate 41a, the plains-forming units do not occur in all topographic lows. In particular, the irregular depression (arrow Y) contains this unit, whereas the subdued crater (arrow Z) does not.

A viable hypothesis for the origin of the plains-forming units must account for the following observations:

1. The small amount of inundation within the pre-existing craters relative to that on the floors of Clairaut and its rim craters (Plate 41a)
2. Localized pockets of inundation commonly occurring in local topographic lows but not necessarily in regional lows associated with subdued craters and noncraterlike depressions
3. The relatively sharp contact between the subdued walls and the plains-forming unit of several craters, including the existence of "bathtub" rings on the crater walls
4. The existence of fracturing of several crater floors containing this unit

Several plausible interpretations have been proposed. Researchers at the U.S. Geological Survey, Astrogeology Branch, originally offered the alternatives of early lunar extrusions or ejecta from large impact basins (see Wilhelms [1970], for a historical summary). Eggleton and Schaber (1972) recently have added greater weight to interpretations involving fluidized ejecta flows and Moon-wide "dusting" from large impact basins. Beales (1971), however, suggested that such units in highland craters represent molten subcrustal material tapped by an impact. Finally, these units may be associated with secondary cratering associated with basin formation (see the discussion with Plates 112 and 199). Observational constraints on these suggestions are considered with Plate 43 and in Chapter 9 (Plates 197–200).

Wargentin is the large crater seen in the upper center of Plate 41, c and d, and it may be evidence that spillover of the high-albedo plains-forming unit from the crater floor has occurred. Wargentin is on the outer fringe of the Orientale ejecta sequence, and the lineated appearance of its northern rim was probably produced by such ejecta or faulting related to the formation of this basin. Identification of actual flow units from this crater is generally inconclusive, although a possible flow may have occurred near the breach on the southern wall (arrow X). Massive flow units related to Orientale are found west of Wargentin (Plate 199) and resemble the plains units. This suggests that Wargentin may not have been inundated by internal processes but by massive ejecta flows from Orientale.

PLATE 41

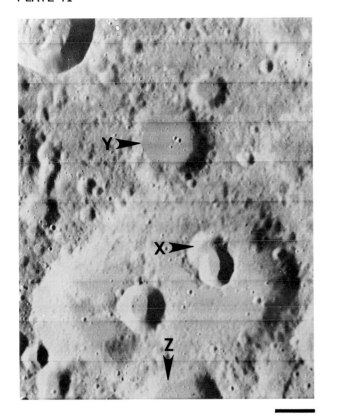

a. IV-100-H2   13 km

b. IV-094-H3   16 km

c. IV-172-H2,H1   14 km   d. IV-167-H2,H1   14 km

Marelike Floors with Fracturing

Pitatus (Plate 42, a and b) is an example of a mare-filled crater with annular fractures. The crater is more than 110 km in diameter and is on the southern border of Mare Nubium. As is true of most of the foregoing examples, the fractures are restricted to within the crater rim but cross both the lower rubbled walls and mare-filled floor. Although the over-all fracture plan is annular, the system is composed of numerous secondary fractures that commonly meet at 30° or 90°. Several fractures extend to the rounded peaks, north of the geometric center. In addition, note the following features:

1. The grabenlike extension to the northwest of Pitatus into the crater Hesiodus (arrow X)
2. The change in width and apparent depth of the fractures
3. The platformlike appearance of the eastern and western rims that have summit craters whose floors seem to extend to a similar depth as the mare floor (see Regiomontanus, Plate 47, a and b)
4. The complete blocklike separation of several portions of the floor by fracturing
5. The platform (arrow Y) that interrupts the fracture extending across the floor from the northwestern and southeastern walls; also, the domes (arrow Z) that align with this fracture to the east of the platform
6. The wide rillelike feature (arrow XX) that parallels the western wall-rim contact
7. The concentric crater seen at the bottom of the plate (see Plate 4, a and b)

Pitatus clearly has been modified by a variety of endogenetic processes, of which floor fracturing represents one of the last events. Floor subsidence, rather than cooling of the emplaced mare units, probably produced this fracturing, for other mare-floored craters, such as Plato (Plate 40a), are commonly unfractured. Floor movement included portions of the wall zone, and the very wide southwestern and southeastern wall areas suggest that crater expansion may have been triggered by floor rejuvenation prior to mare emplacement. Pitatus probably resembled Gassendi (Plate 36a) at an earlier stage in its history. This interpretation implies that the floor of Pitatus has been raised significantly. The islandlike ridges in the northeastern and southeastern portions of the floor may be remnants of the concentric ridge that typically borders the elevated floor plate of Gassendi-like craters.

The crater Humboldt (almost 200 km in diameter), shown in Plate 42c, can barely be seen through terrestrial telescopes. The extensive radial and concentric fractures of its floor are clearly visible. The radial fracture system converges toward an area between the central peaks, whereas the concentric system is ap-proximately symmetric about a point north of these peaks. Both systems have axes of symmetry north of the geometric center of Humboldt. The separation of several peaks and portions of the wall indicates that the fractures were formed after the general crater plan had been established. Whereas the upward-directed stresses within Gassendi (Plate 36a) appear to have been distributed somewhat evenly across the floor, the upward stresses in Humboldt apparently had a pronounced maximum near the central peak complex, perhaps analogous to a batholithic intrusion. The noncoincidence between the axis of symmetry of the fractures and the geometric center of the crater may reflect asymmetric crater enlargement by extensive slumping of the southern wall region. The crater outline indicates relatively surficial effects, whereas the fracture pattern is symmetric about the intrusive body, which was controlled by the point of deepest impact penetration.

Along several fractures, Plate 42c, are crater chains (arrow X) similar to those interrupting the Hyginus Rille (Plate 159). Two low-albedo units can be identified in the NW and NE portions of the crater, and additional examples can be located under higher angles of illumination (Plate 11b). Such zones typically are found near the floor-wall contact where fractures intersect. From the spatial distribution of fractures and pools, three alternatives are plausible:

1. The sources of the low-albedo units are localized beneath the floor-wall border.
2. The sources are closer to the surface near the floor-wall boundary.
3. The fractures are more deep-seated near the wall.

Other interesting features are:

1. The NE-trending structural weakness indicated by the central peaks, the southeast straight-wall segment, the apparent expansion of the crater to the northeast (near arrow YY), and the long furrow that extends from the rim (arrow Y; also see Plate 11b)
2. The concentric crater near the eastern edge of the floor (arrow Z)
3. The highly dissected lower wall (perhaps from flow of material to the floor [arrow XX])
4. The plains-forming units on the walls and rim (arrow YY)

Humboldt is near the northwestern edge of Mare Australe. Thus, the doming (revealed by the fracturing) and the rejuvenated volcanic activity (revealed by the dark-haloed craters and marelike pools) may be related to the regional mare-filling epoch.

PLATE 42

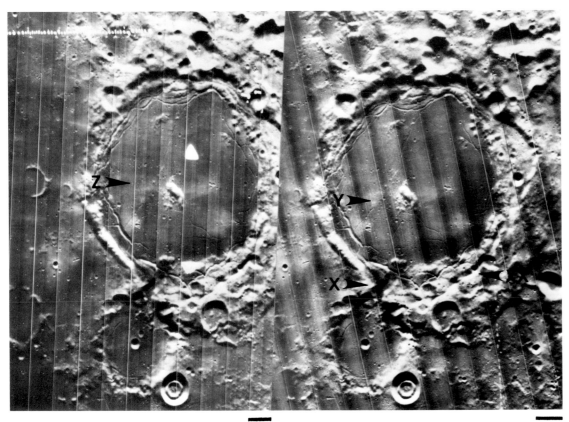

a. IV-120-H1   12 km   b. IV-119-H3   13 km

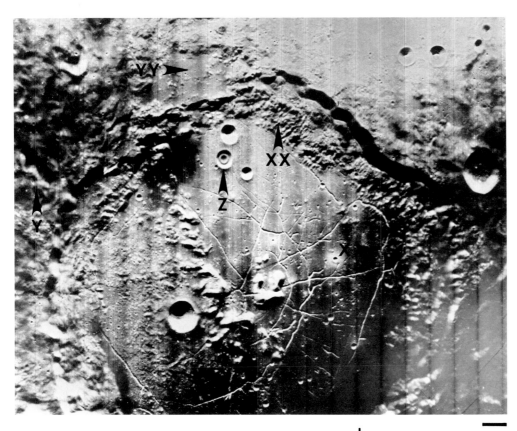

c. IV-027-H1   12 km

Highland-Type Floors

The craters Flammarion and Hipparchus display a class of floors characterized by a relatively high albedo and highly cratered plains-forming unit. This unit may represent an impact-battered version of a Plato-like floor, or a different extrusive or depositional type that is characteristic of the lunar highlands. Plate 43a encompasses the region south of Sinus Medii and includes Flammarion (arrow X, upper left), Ptolemaeus (arrow Y, bottom left), and Hipparchus (arrow Z, right). The prominent NW-trending lineations are part of the sculpture associated with the Imbrium basin.

Flammarion exhibits these large-scale features (see Plate 74d):

1. En échelon walls paralleling the Imbrium basin sculpture
2. Flat and mottled floor having a higher albedo than that of the maria
3. Superposition of several separate ejecta units

Plate 43a shows the southeastern portion of Flammarion at resolution considerably increased over that shown in Plate 74d and Plate 65, a and b, and displays several additional features:

1. Numerous subdued concave craters in the 1 km–5 km range and myriads of smaller subdued craters
2. Relatively smooth floor-wall contacts
3. A rolling floor topography considerably rougher than a typical marelike surface
4. Local regions having relatively low small-crater densities (see Plate 43c)

Craters with concave, dimple, and convex (also see Plate 18c) profiles are present, but all appear subdued. Several dimple craters appear to have material extending to their floors.

Because the sculpture from the Imbrium basin is not expressed across the floor of Flammarion, emplacement of the floor unit postdates the Imbrium basin-forming event. Plate 43b illustrates the greater abundance of craters—especially those approximately 1 km in diameter—on the floor of Flammarion than that on many mare surfaces. Such a difference has been recognized by Morris and Wilhelms (1967) and can result from:

1. A high influx of primary impact craters between the times of floor emplacement and mare flooding
2. A locally increased flux of secondary impacts
3. A large number of endogenetically produced craters

The generally subdued state of most kilometer-size craters can fit into any of these three possibilities.

Plate 43c presents a transition between two different small-scale topographies on the floor of Flammarion: one characterized by a textured terrain (top half, arrow X) and the other by an abundant display of small (50 m) subdued craters (bottom half, arrow Y). These represent either different units that responded differently to surface processes or the same unit, one area of which (bottom) has been bombarded by secondaries.

The floor of Hipparchus (Plate 43a, arrow Y) is shown at high resolution in Plate 43d. Although relatively smooth on the scale of hundreds of meters, the surface is dotted by small depressions that may have resulted from secondary impacts from the nearby crater Horrocks (see Plate 45, a and b). Such a surface, however, is not unique to Hipparchus; it also occurs on the floor of Alphonsus (V-119-H1).

The morphologic features of such craters as Flammarion, Hipparchus, and Alphonsus probably cannot be formed simply by progressive degradation of a Copernicus-type crater. The influence of the Imbrium basin event can be best appreciated by a glance at the Orientale ejecta sequence (Plates 209–213). Such a catastrophic event, combined with periods of volcanic modifications, isostatic adjustments, and blanketing by nearby sizable impacts, has probably been much more effective than the billions of years of exposure to the large influx of smaller meteoroids. The relative roles of these processes depend both on their locations on the Moon and on time. The latter dependence arises not only from lunar thermal history but also from an early epoch of intense bombardment, as suggested by Hartmann (1966, 1970). As a result, theories concerning the nature and origin of the high-albedo plains-forming units, which characterize the floors of highland craters, must allow for catastrophic as well as secular processes.

Preliminary results from the Apollo 16 mission indicate that at a site similar to the floor of Flammarion the surface and subsurface rocks are almost entirely breccia (Muehlberger et al., 1972). Data from the active seismic experiment are consistent with a model in which such breccia and impact-derived debris extend to more than 70 m in depth (Latham et al., 1972). In addition, the gravimetric data from the S-Band Transponder Experiment reveal that the crater Ptolemaeus exhibits a strong negative gravity anomaly, in contrast to a crater of comparable size filled with mare material (Sjogren et al., 1972). Such results have been interpreted as evidence for the origin of the highland plains-forming units by fluidized ejecta produced during the formation of the enormous multiringed basins (Eggleton and Schaber, 1972).

A deep layer of relatively unconsolidated or friable debris may account for the typically subdued state of small craters occurring on highland plains-forming units. As shown in Plate 19, c and d, some small craters on these units exhibit floor detail without surface detail in their ejecta blanket. Thus, it appears that a recent impact crater may be subdued at the outset as it is rapidly degraded owing to the relatively mobile and thick incompetent layer in which it was formed. The pebbled appearance shown in Plate 43d could be explained as a cluster of subdued secondaries, or perhaps collapse pits triggered by Moonquakes.

Although the data from the Apollo 16 landing site are consistent with Eggleton and Schaber's theory, there is evidence that this is not the entire (or correct) explanation for all highland plains-forming units (see the discussion with Plates 197–200).

PLATE 43

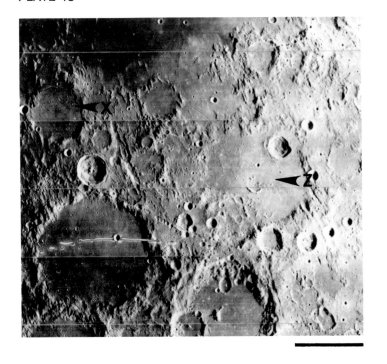

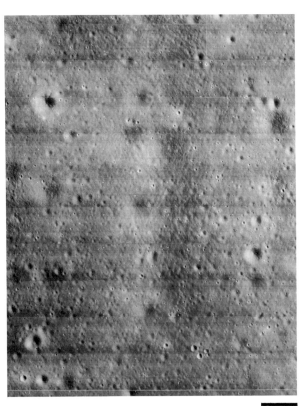

a. IV-101-M   100 km

b. III-119-M   1.5 km

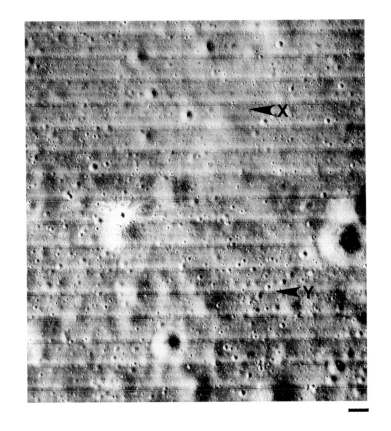

c. III-119-H1   0.20 km

d. III-111-H1   0.22 km

Central Peaks

The origin of central peaks has been the focus of considerable debate. Some researchers suggest that they represent nonvolcanic products of the initial cratering event, such as rebounded blocks (Roddy, 1968) or uplifted toes resulting from deep lateral mass transfer from the walls and floor of the crater (Dence, 1968). Others propose that these features are the result of subsequent volcanic processes, analogous to the last cycle of extrusions in terrestrial calderas (Fielder, 1965).

Early workers (Goodacre, 1931; Young, 1940) estimated that 20 percent of the lunar craters displayed such peaks. Baldwin (1963) indicated that 68 percent of 342 craters larger than 8 km in diameter have peaks. Recently, Wood (1968) reanalyzed the problem, using the *Photographic Lunar Atlas* of Kuiper et al. (1960). With due regard to possible selection effects, he concluded that only 6.8 percent of the Earthside craters with diameters greater than 10 km contain such peaks. If all craters equal to or larger than 3.5 km were considered, this would result in only 3 percent of the craters containing central peaks. Wood explained the difference between his results and those of other workers by differences in their sampling techniques, which resulted in highly biased samples for the earlier analyses.

Wood also considered a rough classification scheme for peak types: "weak" or "strong" and "single" or "multiple." These were then compared to the apparent "age" of the crater according to *The System of Lunar Craters*, the catalog of Arthur et al. (1963, 1964, 1965, 1966). In brief he found that

1. With increasing crater "age" (class), the ratio of "weak" to "strong" peaks increases.
2. Larger craters are more likely to display a central peak.
3. In each crater class a diameter is reached beyond which the percentage of craters having central peaks remains relatively constant.
4. "Fresher" craters are also more likely to display such peaks.
5. The percentage of multiple central peaks increases with size but is about the same for the various "age" classes.

His results seem to imply that the central peaks have, in general, undergone the same degradational processes as the crater as a whole; that is, they were formed at approximately the same time as the crater. As he pointed out, however, the analysis does not resolve the debate with respect to origin.

The Orbiter photographs provide numerous examples of craters with central peaks. The following discussions do not include the convex-floored craters found at small scales, although their

formation may, in some instances, be comparable. In general the peaks can be described by the following scheme:

1. Location
   Central
   Acentral
2. Plan
   Ridgelike
   Peaklike
   Annular
   Crescent shaped
3. Multiplicity
   Single (mono)
   Multiple
4. Morphologic appearance
   Blocklike
   Domelike
   Conelike
   Sharp crested
   Platformlike
   Crater topped

Plate 44 includes several peak types. Alphonsus (arrow X), Alpetragius (arrow Y), and Parrot C (arrow Z) display single conelike peaks, whereas Arzachel (arrow XX) contains an acentral ridge. The Alphonsus peak rises 1.1 km above the floor, that of Alpetragius is 1.8 km, and the ridge of Arzachel, 1.5 km. These elevation differences might be compared to their respective maximum rim-to-floor distances noted in type on LAC 77 and 95: 3.2 km, 2.9 km, and 4.0 km; however, contours from the charts reveal typical differences of 1.5 km, 2.4 km, and 2.4 km, respectively.

The ridge in Arzachel is oriented with the low linear relief that crosses the floor of Alphonsus (arrow YY) and a crater chain that continues into Ptolemaeus (arrow ZZ). This system aligns with the general lineations from the Imbrium basin. Alphonsus clearly predates the Imbrium event, whereas Arzachel postdates it. Hence, the low-relief feature across the former might be a structural or volcanic rejuvenation, but the ridge in Arzachel appears to have been formed along the pre-existing regional weakness developed by the Imbrium basin-forming event.

Central peaks also can be found in small-scale craters in the Marius Hills (Plates 264b and 265, a and b). These generally are large with respect to the craters and are interpreted as tholoids.

PLATE 44

IV-108-H2   12 km

Central Peaks

Horrocks (Plate 45, a and b) and Römer (Plate 45, c and d) are approximately the same size but have peaks of markedly different proportions. The small central hillocks within Horrocks appear to have resulted from the meeting of toes of slumps along the wall, possibly combined with rebounded material. Peaks with similar appearance can be examined more closely in Mösting (Plate 30, b and c; Plate 65, a and b) or Dawes (Plate 27).

In contrast, Römer (Plate 45, c and d) displays a single cone-like peak that may have a summit pit. The peak comprises the entire crater floor and crests 1.8 km above its base to the east but is about 2 km below the eastern crater rim. Thus the three-dimensional view has been greatly enhanced. Adjacent to its west slope is a ringed depression (arrow X) that might represent a collapse form.

Plate 45e shows a very subdued crater, Airy, which is about 120 km east of Parrot C (Plate 44, arrow Z). Airy, Parrot C, Alpetragius, Donati, Faye, Capella, and Behaim have similar large but subdued conelike peaks, and all interrupt, are in line with, or are adjacent to linear furrows, which typically are 5 km wide. The furrows associated with the first five craters are in the Southern Highlands and are radial to the Imbrium basin. All seven craters display very subdued appearances; i.e., hummocky ejecta blankets and well-defined wall slumps are absent. If such craters postdate the formation of the basin-related furrows, then the craters have been highly degraded from the typical impact-produced craters, and, despite this extensive degradation, their central peaks have remained prominent. If the furrows postdate the craters, then the furrows should cross the crater floors; furthermore, in the examples of Parrot C (Plate 44, arrow Z) and Capella, the furrows should have destroyed or highly modified the central peaks. Parrot C and Airy (Plate 45e, arrow X) exhibit lobate floor boundaries that suggest floor rejuvenation, and Airy displays floor fracturing. Consequently, it is possible that these peaks may be related genetically to floor rejuvenation, through either volcanic construction or uplift postdating crater formation. It is also possible that these craters are calderas related to the structurally formed furrows.

Plate 45f is the farside crater Icarus, approximately 100 km in diameter. Shadow lengths of the wall and conelike peak indicate that the peak protrudes to almost rim level. This is vividly illustrated in the Apollo 11 photograph AS11-44-6606, in which the entire floor and practically all of the sunward wall are in shadow, whereas the central peak is sunlit.

PLATE 45

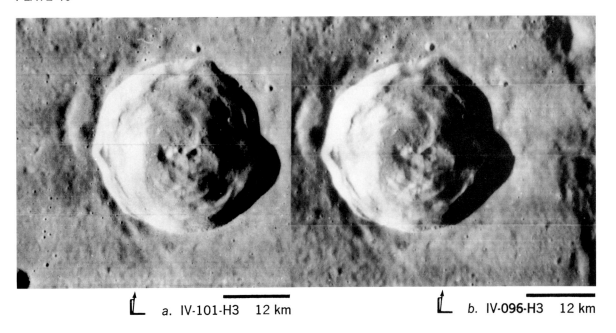

a. IV-101-H3   12 km

b. IV-**096**-H3   12 km

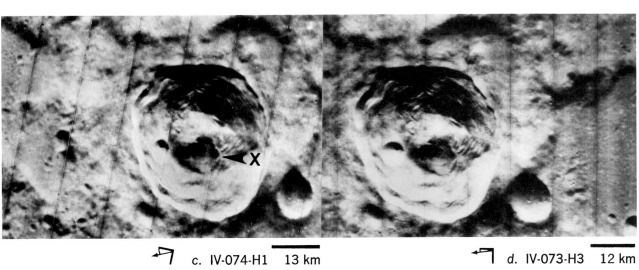

c. IV-074-H1   13 km

d. IV-073-H3   12 km

e. IV-101-H2   12 km

f. II-033-M   (oblique)

Central Peaks

The multiple peaks of the farside crater Jenner (Plate 46a) and Theophilus (Plate 46, b and c) contrast the subdued domal type versus the blocky type. The two examples are of further interest in that both craters have similar relative ages, as suggested by their preserved ejecta sequence.

The peaks of Theophilus (also see Plate 93) appear to be influenced by two structural trends: a major NW trend, probably associated with pre-existing weaknesses developed during the formation of the Imbrium basin, and a minor NE trend, which may represent secondary weaknesses related to the NW trend. Several flow termini can be identified at the base of the southern peak of Theophilus (arrows X and Y), and numerous other flows can be located along the base of the walls. In contrast, the peaks in Jenner (Plate 46a) are lower in relief and subdued in surface detail. The peaks are multiple but not separated and resemble the domal components of the Flamsteed Ring. In further contrast, the floor is marelike and exhibits wrinkle ridges, one of which is spatially related to one of the peaks (arrow X).

Several possible interpretations are applicable:

1. The domal structures are extrusive and have an origin different from that of the blocky peaks. This also includes pluglike protrusions—i.e., spines—that rapidly crumbled through exfoliation and unloading. This view allows a similar interpretation for numerous analogs on the maria, which commonly have been interpreted as pre-existing relief.
2. The domal structures represent blocky peaks that were originally similar to the blocky peaks in Theophilus but have undergone extensive mechanical degradation associated with mare flooding.
3. Both peak types are rebounded or deep-seated slide material associated with crater formation. The blocky peaks represent competent subfloor material, whereas the domelike peaks represent incompetent subfloor material.
4. The blocky peaks are structurally modified volcanic forms of a different type or are of a degradational state different from that of the domelike peaks.
5. The blocky peaks are rebounded or deep-seated slide material associated with crater formation and have been modified by minor volcanic activity.

Not all examples provide the contrast in type as do the two presented here. Several component peaks of the Copernicus central ridge system appear domelike, but the ridges also are de-

scribable as dissected blocks. To appreciate the variety and detailed morphology of such peaks further, one should examine:

1. Dawes (Plate 27, a and b): 17-km-diameter crater with multiple domelike peaks
2. Mösting (Plate 30, b and c; and Plate 65, a and b): 25-km-diameter crater with small multiple central domes or hillocks
3. Vitello (Plates 249–250): multiple domelike peaks arranged in an annular plan within a 42-km-diameter fractured-floor crater
4. Aristarchus (Plates 236–237): central ridge composed of domelike forms in a recently formed 36-km-diameter crater
5. Eudoxes (Plate 11c): a 70-km-diameter Copernicus-like crater that contains no prominent central peak but numerous small domes
6. Tycho (Plate 31): multiple domelike and blocky peaks in a recently formed 98-km-diameter crater
7. Copernicus (Plates 219–223): multiple domelike, conelike, and blocky peaks arranged as ridges within a recently formed 90-km-diameter crater
8. Alphonsus (Plate 186, b and c): about 105 km in diameter with a central conelike peak (also discussed with Plate 44)
9. Gassendi (Plates 36–38): shallow fractured-floor 108-km-diameter crater with multiple crested peaks arranged in a crescent-shaped plan
10. Petavius (Plates 215–217): large 175-km-diameter crater whose central peak complex contains an interior basin, probably produced by collapse
11. Tsiolkovsky (Plates 11a and 40b): 180-km-diameter crater with a large acentral crested relief having a highly pitted plains-forming unit
12. Humboldt (Plates 11b and 42c): 200-km-diameter crater with both a blocky ridge and an adjacent crested peak

Plate 46d shows the subdued crater Joliot-Curie (160 km in diameter) with both conelike (arrow X) and platformlike (arrow Y) central peaks. The conelike peak appears to have an offset pit that is breached on the northern slope. The platform may be a constructional feature or large crags (see Jaggar, 1947) formed during floor inundation. Floor inundation and the formation of the platform clearly postdate the pimpled terrain that characterizes the walls and rim and that is interpreted as modifications from antipodal Moonquakes produced by the Orientale impact (also see Plate 75a).

PLATE 46

a. IV-012-H2 13 km

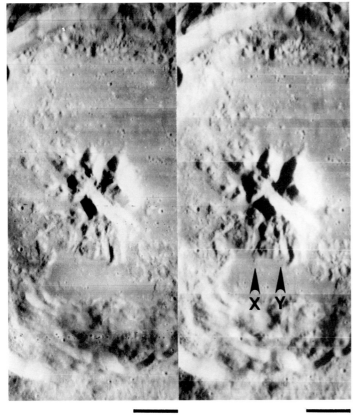

b. IV-084-H2 12 km c. IV-077-H2 12 km

d. IV-017-H3 12 km

Central Peaks

Considerable debate persists concerning the origin and significance of central peaks with summit craters. Such occurrences could be statistical flukes or volcanic peaks with summit pits. Regiomontanus (Plate 47, *a* and *b*) has a ridge to the northeast of center that aligns with the Alphonsus-Arzachel trend. At resolutions provided by terrestrially based telescopes, the crater on the acentral ridge appears to be unique and perhaps significant, but at the resolution shown in Plate 47, *a* and *b*, it is one of many secondary and primary craters on the floor of Regiomontanus. As a result, a statement of genesis from morphology alone is probably premature. Note that the outline of Regiomontanus appears to be composed of interconnected irregular platforms, some of which have large summit pits (arrows X, Y, and Z). The discussion accompanying Plates 110 and 113 examines in greater detail the possible origin of such rim platforms.

The crater Plinius, shown in Plate 47c, is near the northwestern border of Mare Tranquilitatis. In contrast to Regiomontanus, Plinius displays extensive wall slumps, a smooth floor, and a well-defined ejecta blanket. The central peak is approximately conelike and presents a relatively large breached summit pit. If the summit pit resulted from an impact, then its ejecta blanket should be preserved. Its absence suggests that the central peak is either a volcanically constructed cone with a summit pit or a structurally produced relief with a central collapse feature. The endogenetic origin of the central pit is supported by the existence of several other craters having central pits very similar to that of Plinius, such as Timocharis (IV-121-H3) and Lansberg (IV-126-H1). Craters displaying an annular or crescent-shaped arrangement of central peaks (Posidonius, Plate 2b; Gassendi, Plate 36a; Philolaus, IV-164-H2; Einstein A, Plate 198a, arrow YY; and Kunowsky, Plate 92a) are believed to represent more extensive subsidence of the central portion of the peak complex. Collapse was perhaps accompanied by ejected debris that blanketed the crater floor. In several cases, however, collapse appears to have occurred prior to the solidification of the impact-generated floor melt; consequently, this inferred layer may have been buried by later floor eruptions.

PLATE 47

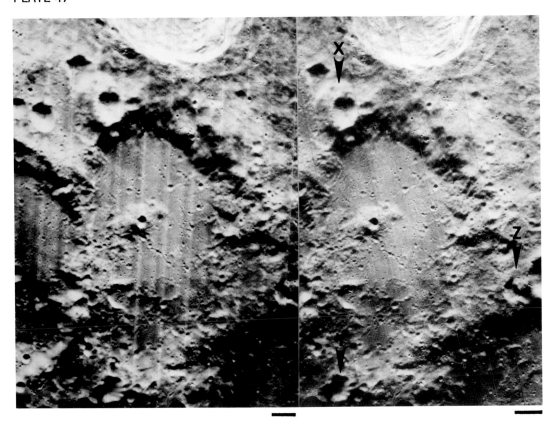

⌐┘ *a.* IV-108-H1 12 km   ⌐┘ *b.* IV-107-H3 13 km

⌐┘ *c.* IV-085-H2 12 km

Central Peaks

Plate 48, *a* and *b*, includes another example of a central peak with a summit crater (arrow X), which is in the crater Poisson U in the Southern Highlands. The summit crater is much larger, relative to the peak, than the examples shown in Plate 47, and it could be viewed simply as a crater with a large rim. The stereo pair reveals that the floor of this crater appears to be above the surrounding terrain. As in the example of Regiomontanus (Plate 47, *a* and *b*), the abundance of nearby craters of similar size reduces the significance of this centrally located crater. It may represent, however, a summit crater or caldera within a large irregular crater (caldera?).

The large elongate (approximately 60 km by 80 km in diameter) and subdued crater Poisson displays a relatively large central peak (arrow Y). LAC 95 gives its relief at 1.1 km above the plains to the east. This peak also exhibits numerous, but relatively small, summit pits that may be related genetically. More significant, however, is its location in a highly subdued crater. If the central peak was formed at the same time as Poisson, then its preservation relative to other features of the crater is clearly inconsistent. Consequently, it is proposed that this peak was formed volcanically at a considerably later date. It is also plausible that Poisson represents a caldera, but the morphologic distinction between this origin and an impact that has been modified extensively by later epochs of volcanism may be very slight. Note the subdued furrows that extend down the flanks of the peak, which may be a mass-wasting phenomenon. Also note the highly cratered plains-forming units that fill many local depressions inside and outside Poisson.

Plate 48, *c* and *d*, includes another highly modified crater, Censorinus D, which is in the complex highland terrain between Maria Fecunditatis and Nectaris. The floor exhibits numerous domelike and conelike peaks, some of which have summit pits. Many of the domes are confined to an interior craterlike depression (arrow X) about 12 km in diameter.

Censorinus D and the surrounding terrain indicate a complex history. The grabenlike rille (arrow Y) crosses this crater and therefore is one of the later events. In addition, note the following features:

1. The unusual southern rim that apparently has been altered by faulting (arrow Z)
2. The linear depression (arrow XX) that parallels the crater rim
3. The breach of the northern rim (arrow YY)

Censorinus D is proposed as a caldera that has undergone multiple phases of eruptions and collapse. The domelike and conelike peaks are suggested as the last or late-stage volcanic constructions within a smaller collapse form.

PLATE 48

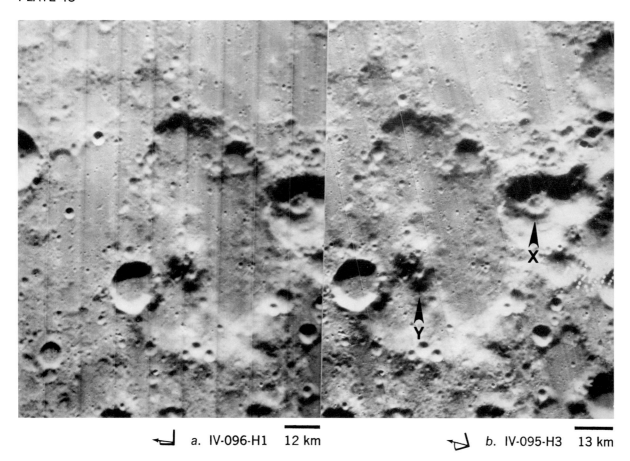

a. IV-096-H1   12 km   b. IV-095-H3   13 km

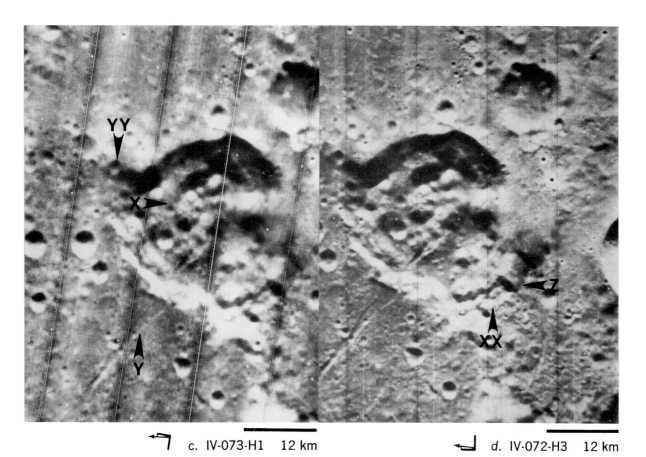

c. IV-073-H1   12 km   d. IV-072-H3   12 km

Central Peaks

Plate 49a gives a further example of the relation of regional structures to central ridges. This oblique view of the backside crater Fabry (about 150 km in diameter) reveals a prominent ridge (arrow X) aligned with a subdued, but obvious, grabenlike trough (arrows Y and Z). The crater Fabricius (Orbiter IV-71-H2) shows a similar relationship. The crater's wall and rim are degraded, but the plains-forming floor unit appears to be relatively unscathed. The ridge has several "fingers" extending northward, perpendicular to the trend of the ridge, which probably resulted from volcanic or debris flows (also see Plate 46, b and c). Arrow XX identifies the crater Giordano Bruno, mentioned in the discussion with Plate 185.

Fabry is at the end of an open chain of mare-filled craters that trend perpendicular to a belt encircling the Moon and containing a significant proportion of the lunar maria (see the discussion and summary accompanying the chapter "Maria and Plains-forming Units"). Although the plains-forming unit within Fabry has a higher albedo than that of the maria, it exhibits a similar appearance, such as embayment of pre-existing relief. This unit is interpreted as a relatively high albedo marelike basalt that was emplaced in association with the maria and considerably after crater formation. Volcanic eruptions may have been localized along the central ridge during this emplacement, although the ridge probably reflects structural control of rebounded material or deep-seated slides during crater formation.

The annular arrangement of peaks results in the concentric plan of several enormous craters. Hartmann and Wood (1971) suggest that there is a sequence from central peaks to mountainous rings. This can be taken only as a highly generalized statement, however, owing to numerous gaps in the sequence, which are perhaps the result of later processes. Plate 49b shows the craters Antoniadi (140 km in diameter, arrow X) and Schrödinger (300 km in diameter, arrow Y); the latter is also shown in Plate 2a under different illumination. As discussed with Plates 1 and 2a, such interior rings could represent the crests of large concentric slump blocks, zones of reinforced shock waves, frozen "tsunamis" in the crust, or volcanic peaks.

The outer wall of Schrödinger is similar to the walls of Copernicus, despite the crater's enormous size; that is, it is not broken into blocks or mountainlike peaks as is common around the Orientale basin or the northern edge of the Imbrium basin. In contrast to the extensive and thick ejecta blanket of Orientale, Schrödinger exhibits a relatively limited and thin blanket (generally within 1.5 crater radii from the outer rim crest) as indicated by the preservation of the crater pair to the north (lower left in Plate 2a). At least three large leveed furrows extend from the rim, probably analogous to the Alpine Valley (Plate 179) from Imbrium or Bouvard (Plate 209) from Orientale. The fractured floor appears to be composed of high-albedo plains-forming units and hummocky zones, with late activity indicated by the dark-haloed crater along one of the fractures. The annular peaks appear to be platforms, and, if there is a major scarp, it is to the interior.

In contrast, Antoniadi (arrow X) is smaller than Petavius (Plate 215) yet displays the remnants or partial construction of an inner ring of peaks as well as a small central ridge. Its ejecta overlie the floor of Schrödinger, thus indicating a younger relative age.

As previously discussed (Plates 1 and 2a), Mackin (1969) viewed the interior rings of large basins as the result of slumping that had been lubricated by a subsurface plastic (or potentially plastic) layer. The layer could be dependent both on time and location. His theory could allow for regional and temporal dependences of inner-ring formation as well as differences in the ejecta blanket. However, the Copernicus-like outer rims of Schrödinger and Antoniadi suggest an origin more analogous to the central peaks of small craters. A depth dependence could indicate a plastic layer or other conditions that reflect and reinforce the induced shock waves. The ring also could be envisioned as the toes of the wall slumps that no longer meet to form a central peak, but, owing to the relatively large height of the ring, this does not seem to be realistic. In the view of Dence (1968), slumping might be accompanied by lateral and vertical plasticlike transport of the country rock into the newly formed void. Therefore, the inner ring represents, in a sense, the toes of slumps that could reach appreciable heights. The fractures on the western floor of Schrödinger are directed radially from the annular peaks. This pattern suggests uplift associated with the peaks that postdate floor emplacement.

Another mechanism is indicated by smaller craters having a large pit within the central peak complex (Plate 47b) or an annular arrangement of central peaks (Plate 2b). In several examples, such as Sharanov (II-34-MR), the central peak complex resembles that of Antoniadi, but the ratio between the inner ring and the crater rim departs from $\sqrt{2}$. Following Baldwin (1963), I interpret this type of ring as the result of dropping of the central portion of the peaks. Another crater 120 km to the southeast of Sharanov (Orbiter II-34-MR) is approximately 50 km in diameter and exhibits a central peak complex that appears to be composed of a partial ring within which are multiple domes. This may represent an early or arrested stage of central peak destruction where the interior blocks have been partially dropped. The Sudbury structure is perhaps a terrestrial analogy, according to French's (1970) interpretation of that complex structure. Hence, the inner ring seen in Plate 49b I suggest is composed of remnants of the peak complex that remained after its central portion collapsed. This collapse resulted either from insufficient support given by the underlying shattered zone or from subsequent volcanism unrelated to crater formation. A remnant arc from this inner ring can result in a single acentral peak or ridge.

PLATE 49

a. V-181-M   (oblique)

b. IV-008-M   (oblique)

DISCUSSION AND SUMMARY

Crater floors are modified by erosion, deposition, and volcanic rejuvenation. Consequently, they are important indicators of regional histories. Craters smaller than 1 km in diameter have floor profiles that are generally concave, flat, convex, or dimpled. The term "concave floors" is only an approximate description for impact craters with rubbled floors but is appropriate for more subdued forms. Small subdued craters may represent degraded impact craters, but some, owing to their association with other types of endogenetic forms, appear to have been endogenetically formed. The distinction in origin is not always obvious from morphologic studies. A third alternative is that subdued concave craters were formed by secondary projectiles, which characteristically do not exhibit surface detail at the outset. Their subdued appearance probably results from the interaction between the ejecta of the secondary crater and either ejecta from an adjacent secondary or a cloud of debris accompanying the projectile.

Flat-floored craters less than 1 km in diameter typically have rubbled, smooth, or wrinkled surfaces. Those with flat rubbled floors commonly are thought to be impact produced and can be found at varying degrees of apparent degradation, as implied by the preservation of their ejecta blankets. Quaide and Oberbeck (1968) have produced impact craters with flat floors where underlying competent layers exist at depths between $D_A/10$ and $D_A/3.8$ where D_A is the crater diameter.

Craters (less than 1 km in diameter) with smooth flat floors commonly have subdued ejecta blankets but exhibit numerous blocks on the bulbous rim and interior walls. Such floors are here interpreted as the result of blanketing by a nearby event (exogenetic or endogenetic), fallback of sorted fine debris during crater formation, or migration of the finer debris from the walls.

Flat-floored craters (less than 1 km in diameter) with wrinkled surface patterns, the "tree-bark" pattern, typically lack extensive ejecta blankets but may exhibit blocks on the rim and walls. They might be interpreted as erosional products; however, they commonly are grouped on the maria in regions displaying considerable surface texturing and having few craters with 100-m diameters, evidence which may indicate a relatively recent surface. Consequently, these craters are interpreted as having been blanketed by a nearby event, rather than being the result of progressive smoothing, or of having been formed endogenetically.

Small craters (less than 0.2 km in diameter) with rubbled convex-floor profiles also are believed to be indicators of subsurface stratification. Convex floors occur within small-scale impacts produced in granular layers over a competent substrate at typical depths of $D_A/6.25$ (Quaide and Oberbeck, 1968). Larger (0.6 km in diameter) and highly subdued craters with convex floors also may indicate subsurface alterations of the crater-forming shock front. The preservation of the moat between the central mound and wall in several subdued craters indicates either that mass transfer was insignificant relative to *in situ* degradation or that floor mounding continues, perhaps through the convergence of seismically triggered debris slides from incompetent crater walls. The existence of pristine-appearing floors in craters having subdued rims and walls suggests either floor rejuvenation or the formation of subdued rims and walls at the outset, owing to a deep layer of incompetent material.

Dimple craters are classified according to their wall and floor profiles. Funnellike craters have walls with constant slopes or progressively increasing slopes with depth. Complex dimple craters have an interior shelf and appear to be concave or flat floored with a central dimple. Dimple craters typically are found in areas displaying a blanket of low-albedo material, a circumstance interpreted here as the result of either pyroclastic eruptions or an impact into a deep layer of pyroclastic material. The interior wall occasionally has a "silver lining" due to exposed rock surfaces on the upper wall (see Plate 166, *a* and *b*). Dimple craters can be interpreted as

1. Pre-existing craters with later blanketing and compaction of the fill material
2. Fortuitous impact within a larger debris-filled crater
3. Impact into a thick blanket of unconsolidated material with subsequent development of screelike walls
4. Collapse craters analogous to those found on terrestrial lava flows

5. Craters produced by funneling of unconsolidated overlay into a subsurface cavity
6. Maarlike eruptions

Each suggestion has merit dependent on the particular features of the crater and its regional placement. Some examples are found on small marelike pools within the highlands that show little evidence for regional blanketing (see Orbiter I-25M). These and occurrences in other settings are strong support for the last suggestion in the foregoing list, and, as maarlike eruptions, such craters possibly indicate frequent and relatively recent volcanism. Some small and poorly developed dimple craters appear to be erosional products whose dimple profile is enhanced by lighting.

Craters having diameters 1 km–5 km are also classified according to their floor profiles or particular morphologic features. Concave floors display tree-bark and/or grid patterns, and the craters typically do not have extensive ejecta blankets, though numerous blocks on the rim and walls are common. Concave craters commonly are associated with wrinkle ridges and have highly textured rims. This floor type appears to be the result of degradational processes acting on either impact or volcanic craters.

Flat-floored craters (1 km–5 km diameter) are described as smooth, hummocky, or composite. Refinements of these classes are expected as better photography under different illuminations becomes available. Smooth floors within deep craters are marelike in albedo but may represent an entirely different type of unit, such as accumulations of fine debris. Unbreached and shallow mare-filled craters, however, indicate either floor rejuvenation of impact craters or caldera formation. Hummocky floor types appear to be composed of rubble or small domes and may represent impact melt, fall-back material, or tapped extrusive units. Composite floors exhibit both hummocky and smooth units, the latter typically filling local topographic lows. Several craters in this size range show evidence for floor development after the screelike slopes of their walls have stabilized. Thus, they indicate upwelling of material from within.

Convex-floored craters 1 km–5 km in diameter are

sparsely represented, but this is probably a selection effect due to low solar illumination in most Lunar Orbiter photography.

Craters larger than 5 km in diameter typically have flat floors, and consequently the profile is of minor importance. Floors of craters 5 km–20 km in diameter are classified according to their particular morphologic features, which include the following descriptions:

1. Hummocky
2. Multiple domed
3. Debrislike
4. Complex meltlike
5. Fractured (of nonmarelike and marelike floors)
6. Inundated

The hummocky floor type includes a variety of specific floor appearances. In general it includes shallow crater floors having a disordered arrangement of small hills and undulations. Bode E (Plate 23) illustrated one example in which individual low-relief hummocks have a range of sizes, plans, and profiles. The evidence for differential floor movement, the general crater appearance, and the regional setting support the interpretation that Bode E is a caldera. The crater Hyginus (Plate 160) exhibits a similar floor, and its location at the intersection of two grabenlike rilles strongly suggests caldera formation by analogy with terrestrial calderas. The assortment of other hummocky-appearing floors, such as those shown in Plate 28, b, c, d, and h precludes assigning all hummocky-floored craters to the same origin as that of Bode E and Hyginus. Hummocky floors may represent slide debris from the wall, remnants of caldera collapse, products of differential floor movement, or intrusive and extrusive doming. Ordered arrangements of hummocks produce multiple-domed (Plates 24–25) and debris (Plate 27) floors. Small hummocks surrounded by distinctive plains-forming units produce complex meltlike floors. These three floor types are reviewed below.

If the individual hummocks are more symmetrical and have greater relief, the floor is termed multiple domed. These domes typically have approximately the same size and are arranged in a cell-like pattern. Multiple-domed crater floors occur along the margins of the

111

maria or within large craters and are often varied in the over-all crater appearance. Several examples near Sinus Iridum appear to be secondary craters, and chains of these craters are not uncommon. Many craters having such floors are similar to Gambart (Plate 91), which has a relatively symmetric ridgelike rim. The plans of these craters may be circular, scalloped, or irregular and may be breached. In several examples, domes are clustered within very ill defined ring boundaries or within small noncraterlike depressions in the highlands. Craters with such domes typically do not have ejecta blankets but otherwise appear similar to supposed impact craters. A preliminary survey by W. Manley, Jr. (personal communication) confirms these observations.

The more-pronounced domes are interpreted, by terrestrial analogy, as tholoids from late-stage viscous (silicic) eruptions or from eruptions accompanying floor resurgence within cauldrons (see Smith and Bailey, 1968). Multiple phases of activity are indicated by the presence of an inner tephra-ringlike structure in several lunar examples. In a few craters, less pronounced domes in symmetric clusters might be degraded products of floor fracturing due to the cooling or differential settling of a floor melt. This origin is inferred from clusters and chains of such craters that resemble secondary impacts associated with major multiringed basins. However, clusters and chains of craters are also characteristic of terrestrial volcanic fields. The location of multiple-domed crater floors near the borders of the maria and in the highlands may indicate geochemical evolution of the magma chamber associated with mare flooding, or they may indicate chambers having initially silicic magma. The significance of these domes is referenced further in the discussion of mamelonlike domes (Plates 126–129).

The remaining morphologic features of crater floors characterize not only craters in the 5 km–20 km range but also those with diameters greater than 20 km. Debrislike floors are illustrated by Dawes (19 km diameter) and Mösting (25 km). Such floors without wall slumps are difficult to recognize without adequate resolution; therefore, they can be assigned to the more general hummocky class of crater floors. From high-resolu-

tion photographs, the floor of Dawes is characterized by a relatively smooth yet chaotic appearance, with numerous boulders and boulder fields. At lower resolution, arcuate floor patterns emerge and generally reflect the crater plan. At even lower resolutions, such as Orbiter IV high-resolution photography, these patterns are subtle, and the floor appears mottled but relatively smooth. The floor of Mösting appears similar to that of Dawes at low resolution, but its walls have well-defined slump blocks. The toes of these slumps are clearly the dominant floor feature, with minor plains-forming units.

The absence of more complex floor units in Mösting-like craters similar to the floor in Petavius B results from burial by slumping or from the absence of subsequent floor extrusions. Petavius B is scalloped to a lesser degree than Mösting, and the slumping in Petavius B generally predates final emplacement of its floor. Consequently, burial of possible volcanolike floors by slumping in Mösting-like craters might be due to a later crater expansion. Numerous highly slumped Mösting-like craters have subsequent lavalike plains-forming units on their floors (see Plates 66, c and d, and 45, a and b). These units may be related to floor rejuvenation that initiated slumping or may represent squeeze-ups of the still-molten floor soon after crater formation.

Volcanolike floors are examined within Dionysius (19 km in diameter), Petavius B (35 km), and Tycho (almost 100 km). The label ''volcanolike'' is very ambiguous because of the wide variety of floor types that might be attributed to volcanism. It is used here in reference to floors similar to that of Tycho. In general, these floors exhibit numerous hillocks that are surrounded by complex and textured plains-forming units. The hillocks include symmetric mamelons, some of which have summit pits; composite domes and massifs; and low-relief, broad domes. Many hillocks apparently control local fracture patterns and some are split by fractures. The plains-forming units surrounding these domes are typically highly fractured, and, in areas where there are numerous hillocks, they appear to be ropy and once mobile. In several examples, this unit appears lapped onto positive-relief forms. This plains-forming unit becomes more extensive as crater size increases. Units with smooth surfaces occur, but they commonly are derived from small pools of plains-forming units on the crater walls or rims.

These complex floors are essentially large volcanic basins exhibiting a wide variety of extrusive forms developed over extended time periods. In most of these craters, the development of the floor postdates slumping of the walls, and the entire floor typically shows evidence of later settling. The floor of Copernicus (Plates 218–229) exhibits morphologic forms with more variety than that of Tycho, such as smooth-surfaced units around fractures, rimless pits, irregular depressions, volcanolike cones, moats surrounding domes, and a variety of crater types. Craters with such volcanolike floors have extensive ejecta blankets and are believed to be impact produced.

The general similarity of these floors to those of craters found in widely differing geologic settings indicates subsequent volcanism generated by melting of parent material or by tapping of a panlunar magma layer at depth. The specific differences between the floors of Tycho and Copernicus, however, suggest that the eruptive history in Tycho was limited relative to that in Copernicus. This may reflect differences in the impact-generated melt. For example, the impact melt in Copernicus may have been less viscous than the melt in Tycho, owing to the impact of the former into a mare, which is thought to be metal rich with respect to the highlands where Tycho was formed. A less viscous melt is more likely to exhibit a longer eruptive history.

Alternatively, the differences between the floors of Tycho and Copernicus reflect differences in the internal thermal gradient of the lunar crust at the time and location of crater formation. Copernicus appears to be older than Tycho and perhaps formed when (or where) the thermal gradient was greater. From Carr (1964), this occurrence could influence the formation of an impact-generated melt and its cooling history. If Copernicus actually tapped a large magma chamber, then I would suspect a shallower floor and an even greater display of related volcanism inside and outside the crater than is in evidence today. Unless such a magma chamber migrated at depth, extensive peripheral eruptions are required in order to cause significant subsidence of the

floor from an earlier level of eruptive activity. There are, however, other craters, such as King crater (AS11-43-6448) and Ölbers A (Plate 105c), that exhibit large areas of plains-forming units on their rims. Such units might be evidence for significant peripheral eruptions (see the discussion with Plate 105).

Fractured floors comprise another broad class of floor types. The fracturing process affects craters having both marelike and nonmarelike floors. It produces radial, annular, polygonal, or irregular fracture patterns that reflect the uniformity of the applied stress, the direction of this stress, and the homogeneity of the floor. Radial fracture patterns are commonly exhibited by enormous craters, such as Humboldt (Plate 42c) and Petavius (Plate 215), that have preserved ejecta blankets. This pattern is the result of upward-directed stress near the center of the crater, perhaps in response to rapid isostatic adjustment. Radial patterns also occur within several medium-size craters (15 km in diameter, Plate 198d) and probably reflect volcanic resurgence of the floor. Annular patterns are thought to represent two possible floor histories. Annular patterns related to a ring graben encircling the floor are interpreted as the result of a rising floor plate. Other annular patterns form a system of concentric fractures apparently unrelated to such a ring graben and are attributed to floor subsidence. The second process may be preceded by the first, and it is likely that a crater floor has undergone multiple periods of elevation and subsidence comparable to those of terrestrial calderas. Consequently, separation of the two processes is not always possible. Polygonal patterns can be produced by a variety of processes: multiple sites of upward-directed stresses, an inhomogeneous floor plate, cooling of a once-molten floor, intersecting radial and concentric fracture patterns. Irregular fracture patterns probably reflect stress fields altered by a nonuniform floor thickness, multiple sources of the applied stress, or a complex floor history.

Floor-fractured craters range from less than 10 km to 200 km in diameter, not including mare-filled basins, such as Mare Humorum. I have recognized 192 floor-fractured craters on the lunar nearside and farside. The mode of the frequency-diameter distribution is approximately 20 km, whereas the mean crater diameter is approximately 40 km. Such craters show a marked concentration near the borders of mare plains; for example, western Oceanus Procellarum exhibits 45 floor-fractured craters. However, not all craters adjacent to the maria have fractured floors, and there is not an obvious relation between the smallest size of an affected crater and the distance from the mare shore. Floor-fractured craters are most common around the irregular maria, such as Oceanus Procellarum, or partly inundated basins, such as the farside Mare Ingenii. Craters with and without ejecta blankets display fracturing. In numerous examples, emplacement of mare or marelike units occur in a moat between the old floor boundary and wall, which produces a concentric or eccentric plan.

Morphologic features of a nonmarelike fractured floor were illustrated in Gassendi (Plates 36–38). The floor displays a subdued hummocky terrain, which has extensive texturing in high-resolution photographs. Portions of this terrain are overlain by relatively pristine features that resemble volcanic cones, tephra rings, and lava flows. There are also irregular patches that exhibit low crater density and that are interpreted as either recently emplaced extrusive units or rapidly degraded surfaces owing to a deep layer of unconsolidated material. Layering within several fractures and craters possibly suggests relatively ordered stages of floor emplacement rather than chaotically developed rubble. More likely, these layers indicate previous levels of presumably molten material that drained into the fractures or to the moat adjacent to Mare Humorum. Terraces bordering irregular depressions also may reflect previous high-level marks. These morphologic features are not consonant with interpretations that such craters are simply ancient versions of Copernicus subjected only to degradation by meteorites.

Gassendi-like craters are interpreted as the result of floor rejuvenation associated with processes responsible for mare flooding. Some of these craters represent impact scars whose brecciated and fractured zones became conduits for the mare-producing magmas; others represent calderas. The floor floats on injected magma, and the moat develops as a result of the elevated floor and the down dropping of wall slumps. Further eruptions

of mare material result in foundering of the old crater floor and in the development of inundated craters, such as Plato. As discussed in Chapter 9, such floors indicate that the inundation of the irregular maria and old basins is probably related to the volcanic activity within these craters. Renewed floor rejuvenation produces fractured marelike floors, such as Pitatus. Detailed views of Gassendi-like floors are illustrated by Vitello (Plates 249–252) and Petavius (Plates 215–217).

Floor modifications also occur at considerable distances from the maria. Many highland craters, such as Flammarion and Hipparchus, indicate widespread inundation by plains-forming units that are more highly cratered and higher in albedo than mare material. The small-crater density appears to be dominated by subdued forms. It was recognized by Wilhelms (1964) and was termed the Cayley formation. The floor material was emplaced in the Southern Highlands after the formation of the Imbrium basin. Consequently, the highland plains-forming units must have been emplaced at least 3.6×10^9 years ago, which is the typical age of the sampled maria, but less than 3.9×10^9 years, which is the proposed age for the formation of the Imbrium basin (Sutter et al., 1971; Husain et al., 1971; Turner, 1971). The greater density of craters on the highland plains suggests that the cratering rate was much higher during this interim, a condition that has been suggested by Hartmann (1966). The highland floor units may represent an extrusive or nonextrusive unit altogether different from the lunar maria.

Preliminary results of surface samples from the Apollo 16 mission to the lunar highlands north of the crater Descartes suggest that the Cayley unit in that area is largely composed of breccia. This has revived interpretations of the Cayley unit in general as panlunar deposits of ejecta from basin-forming events, such as Imbrium and Orientale (Eggleton and Schaber, 1972). I suspect, however, that this interpretation is not appropriate for all similar high-albedo plains-forming units or the units responsible for filling highland craters. Evidence for and discussions of alternative interpretations are found in Chapter 9. Most likely the "Cayley unit" is a catch-all term for a variety of plains-forming units having different origins.

Peaks within craters are shown to have a variety of characteristic forms. Although most result from rebound or deep-seated lateral mass transfer (a form of slumping), a significant number appear to be modified volcanically, if not formed that way. This is indicated by extrusive units associated with the peaks, enormous peaks relative to subdued parent craters, alignments with structural trends, and summit pits. The transition from central peak complexes to interior mountainous rings is open to several interpretations. Craters less than 100 km in diameter commonly have relatively small interior rings that are suggested to be remnants of centrally dropped portions of peak complexes. Such a process may not be applicable for the interior rings of multi-ringed basins.

Thus, crater floors indicate numerous and varied processes, some of which have significance for interpreting early lunar history. In addition, small features recognized on the floors of pristine craters can be used to recognize the possible range of pristine morphologic forms. This is important for understanding surface processes as well as for placing constraints on the initial appearances of features outside the crater floor but related to crater formation. Refinements in this simple classification of crater floors are anticipated from high-resolution Apollo photography. Quantitative studies, it is hoped, will provide a means to decipher the importance of this variety.

3. Crater Walls

INTRODUCTION

The gross appearance of crater walls is dependent on the size range under discussion. The following presentation emphasizes this dependence by its arrangement of craters from the smallest to the largest, but omits the enormous multiringed basins that are discussed in the chapters "Craters: Shapes in Plan" (Plate 1), "Crater Floors" (Plate 49), "Crater Rims" (Plate 112), and Mare Orientale, the opening section of Appendix A (Plates 201–208).

The appearance of crater walls is affected by surface processes that smooth and degrade the over-all character. As is intuitively reasonable, the efficiency of different types of smoothing mechanisms should be dependent on crater size and the physical strength of the material. Thus, walls of a small crater are expected to degrade faster than those of a large one; furthermore, a small crater formed in unconsolidated material is expected to be obliterated more quickly than one formed in solid material. The walls of large craters commonly appear to be subdued in detail, but the gross structure is preserved (see Eratosthenes, Plate 106). In most craters, this is a problem of dimensions, in which smoothing processes first affect small-scale features. Several craters, however, appear to be subdued in low-resolution views but in high-resolution photographs they display considerable texturing and other detail attributed to small-scale mass movement.

The walls of craters might be used in determining surface processes, inasmuch as one crater in a given size range formed by an impact should appear like another if the effect of crustal materials is disregarded. This approach might be reasonable for medium-size craters (1 km–20 km) because small-scale inhomogeneities in the lunar crust assume a secondary importance. It becomes unreasonable, however, for smaller craters (less than 1 km in diameter) that are clearly affected by subsurface structure. Hence, small craters should be selected in the same region where they presumably have penetrated similar stratigraphies and have undergone similar degradational histories. The approach also may be unwarranted for tremendous impact events, where large-scale differences in stratigraphy and crustal structure are encountered. Consequently, the size range between 1 km and 20 km seems to be the most useful sample for examining surface processes; this, however, is the range in which volcanic structures typically are indistinguishable, at first glance, from impact structures. In fact, contamination of selected examples by volcanic features might render useless any inferences about gross changes in appearance for almost any size range. In addition, craters smaller than 20 km include both primary (high-velocity) and secondary (low-velocity) impacts, which may be morphologically different. Thus, the evaluation of surface processes through comparisons of crater walls is not so straightforward as might be anticipated initially.

Concentric Walls

The walls of small (about 0.2 km) "fresh" craters typically are strewn with rubble and boulders. Plate 50 illustrates an apparent sequence in degradation from a "fresh" crater (Plate 50a) to one primarily characterized by degradation of its ejecta blanket and rubbled wall (Plate 50d). These craters occur in the same region (southwest of Maestlin R) and display the terraced wall believed to have resulted from the impact with a surface having a competent layer beneath the regolith. The terraced wall produces a concentric plan that is introduced in Plate 6, c and d. Quaide and Oberbeck (1968) have simulated such features and found that this geometry typically results if the thickness of the regolith is less than some value between $D_A/8$ and $D_A/10$, where D_A is the apparent crater diameter. Hence, the regolith depth and, perhaps, the relative age of the local emplacement of the mare can be estimated.

Examination of selected regions of the Moon reveals that clusters of small concentric craters, such as the region shown in Plate 50, occur where there are other morphologic indications that the surface is relatively young (see Plates 195–196). Consequently, a presentation of a possible degradational sequence of crater walls requires that craters be selected from the same region; otherwise, the initial appearance of small craters may be quite different. Even this restriction of sample area, however, may not be sufficient, for differences in crater appearance can also result from local variations in regolith depth and regolith structure, which may be very important on youthful surfaces. In addition, such factors as projectile velocity, projectile density, and angle of incidence will alter the initial crater appearance, thereby affecting the reconstruction of a degradational sequence. For example, Plate 50b may not be an appropriate sequel to Plate 50a, because fresh craters are found locally that exhibit a bulbous rather than sharp rim.

Several plausible mechanisms may be responsible for the degradation of the series from the sharply defined crater shown in Plate 50a to the subdued example seen in Plate 50d. It is assumed that the craters were produced initially by the same mechanism (presumably impact), with similar initial appearances, and that subsequently the following processes were active:

1. *In situ* pulverization of detail by micrometeoroid bombardment
2. Statistical overlap of meteoroid impacts
3. "Sandblasting" and blanketing by ejecta from nearby impacts
4. One or a combination of the foregoing processes augmented by slow mass movement, such as thermally or seismically induced creep

The crater wall maintains a relatively blocky appearance in examples shown through Plate 50c. It is difficult to identify very subdued terraced craters in the surrounding region. This could be the result of degradation to a simple concave profile, which is illustrated by a number of the small craters, or degradation to an irregular profile (Plate 52c). If, on the other hand, concentric craters at this scale indicate a relatively recent surface, then perhaps degradation has not proceeded far enough to produce subdued concentric craters. If the terraces indicate the local base of a regolith that was constructed by meteoroid bombardment, then, in general, the older craters should exhibit terraces at shallower depths, and the oldest craters simply may exhibit concave profiles reflecting both the state of degradation and the shallowness of the wall terrace.

A block-strewn inner wall could persist as a result of the relatively steep slopes if the finer fraction between blocks slides to the bottom as a result of seismic events or thermal fluctuations. As degradation proceeds, more blocks probably would be exposed. Eventually, the slope becomes small enough to maintain any particulate overlay, whether constructed *in situ* or deposited, and thus to allow the eventual burial of the blocks.

PLATE 51

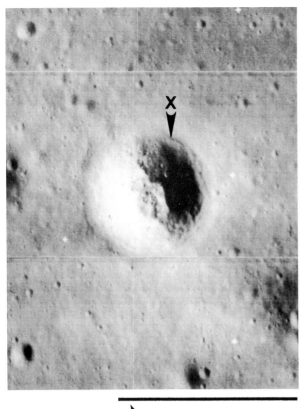

a. III-117-H3   0.19 km

b. III-118-H2   0.19 km

c. III-117-H2   0.19 km

d. III-119-H1   0.19 km

Blocky Walls

Plate 52 permits examination of a sequence of craters on the mare that have slightly larger diameters (0.4 km) and geometries different from those considered in Plate 50. The craters are relatively irregular in plan, possibly the result of interbedded competent and incompetent layers, and they generally have flatter floors throughout the sequence. All the craters shown occur in a small region within Sinus Medii. It is assumed initially that these craters were once similar to the crater shown in Plate 52a, but, through degradational processes, have been reduced to a form similar to the crater in Plate 52d.

The first two members of the sequence in Plate 52, a and b, reveal an abrupt transition in which surface expression of the bulk ejecta (essentially a continuous overlay of debris) is erased, but large blocks remain on the walls and rim. In addition to the degradation of the ejecta blanket, the walls and rim of the crater seen in Plate 52b apparently have been subdued. It is not clear which proposed processes (see discussion accompanying Plate 50) are major causes of degradation. Meteoroid bombardment with its associated gardening could be expected to destroy initially the smallest blocks and subdue the terrain. The numerous smaller craters that surround the main crater shown in Plate 52b support this interpretation. Erasure of the continuous ejecta blanket, degradation of the crater profile, and preservation of the largest rim and wall blocks might be produced by a regional blanket of fine debris, but a history of sub-

sequent secular or catastrophic meteoroid bombardment remains necessary for the observed crater density on the rim. Large blocks on the rim and wall of the crater shown in Plate 52b do not exhibit extensive aprons, which should be produced by in situ degradation. Such an absence suggests a sorting process compatible with debris creep, possibly induced by seismic events.

Most of the blocks around the crater seen in Plate 52c have been either destroyed or buried. If they were originally as large as those shown in either Plate 52a or 52b and were destroyed by only meteoroid bombardment, then erosion rates given by Shoemaker et al. (1970) would suggest that the ejecta, and therefore the crater, are older than the surface on which they rest. This obvious paradox may indicate that this crater did not penetrate to a sufficient depth to excavate such blocks, owing to a locally thick regolith or to the particular parameters of this impact. Therefore, the initial assumption that the craters seen in Plate 52 represent a consistent degradational sequence may not be entirely appropriate. See Plate 78 for further illustrations and discussion on the initial appearances and degradation of ejecta blankets.

Plate 52d reveals the tree-bark pattern on the crater floor, with very subdued walls having relatively small boulders. The crater walls exhibit inner slopes less than 12°, and presumably the processes acting to destroy the blocks also exposed a few.

PLATE 52

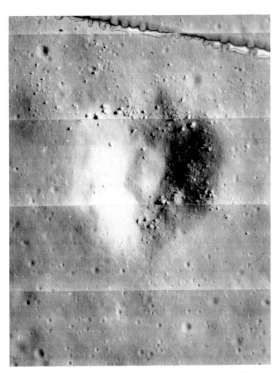

a. II-131-H2   0.22 km

b. II-130-H2   0.22 km

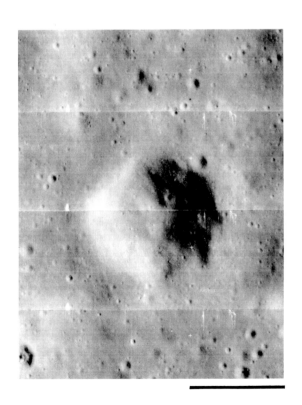

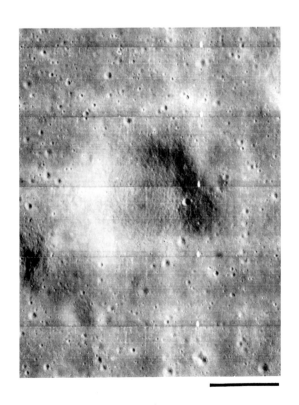

c. II-132-H1   0.22 km

d. II-130-H3   0.22 km

Blocky Walls

Although the craters shown in Plate 53 are almost equal in size to those shown in Plate 52, they disclose several differences in appearances that might lead to a better understanding of surface processes. Their plan is less scalloped than those of the other examples, and the ejecta of the pristine craters (Plate 53, a and b) contain a greater fraction of large blocks (greater than 7 m). These craters are within the Flamsteed Ring, and the difference in subsurface structure could be the major parameter responsible for the general difference in appearance relative to the examples on Sinus Medii (Plate 52).

If a degradational sequence is present, then note the transition in appearance illustrated in Plate 53:

1. The plan becomes more circular with age.
2. There appears to be a broadening of the rim profile.
3. The outer ejecta blanket disappears more rapidly than the inner, but several large blocks remain on the outer slopes of even the very subdued rims.
4. The tree-bark pattern increases in extent but becomes finer in texture with apparent age.
5. There is no significant accumulation of large blocks on the floor, and the total number of blocks exposed on the wall appears to decrease.
6. The largest number of exposed blocks occurs near the upper wall.
7. The "end product" (Plate 53d) displays a small number of craters on its relatively flat floor with respect to an equivalent area on the outer rim.

These transitions suggest the following degradational sequence. The outer extensive ejecta blanket illustrated by the crater in Plate 53a is rapidly subdued, whereas the wall maintains its blocky appearance. Eventually, the blocky inner ejecta blanket is replaced by a corrugated surface (Plate 53c) and finally a tree-bark pattern (Plate 53d). The disappearance of the ejected blocks presumably reflects both the *in situ* comminution of blocks by meteoroid impact and subsequent burial by ejecta from nearby small impacts and more distant large impacts. The appearance of the surface patterns is thought to represent creeplike movements of the rim deposits and the built-up regolith. In contrast, the steeper wall slopes maintain a blocky appearance through rock slides induced by impacts on the upper wall as well as seismically induced slope failure. The blocky rim crest evolves into a rounded and smooth-appearing (at the resolutions shown) rim zone that can maintain a regolith because of the decreased slopes. Blocky outcrops remain below this zone, where the slope is still great enough for creep of the finer fraction of surface material. Below the outcrop, relatively fine debris accumulates on the lower wall and floor from both surface creep and the trapping of ejecta due to small craters inside and outside the larger aging crater. The apparent increase in the circularity of the crater plan suggests that the dominant modifiers are local impacts, which are much smaller than the crater, and much larger and distant impacts, which produce extensive ejecta blankets.

Again, interpretations must be made cautiously because of the initial assumption. For instance, this region has a large number of subdued craters similar to that shown in Plate 53d. Nearby chains of such craters indicate that they were the result of either secondary impacts or internal processes. Several internally formed irregular depressions are similar in form to the supposed end product (Plate 53d). If such structures resulted during or soon after emplacement of the last local mare unit, then their preservation seems to place a constraint on the degree of degradation of later impact craters. The plausibility of non-impact-produced craters at these scales is discussed with Plates 78–79.

PLATE 53

a. III-193-H1   0.23 km

b. III-191-H3   0.21 km

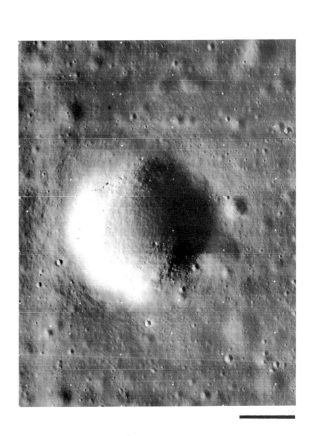

c. III-195-H1   0.23 km

d. III-191-H3   0.21 km

Blocky Walls

Larger craters are typically more circular in plan, as illustrated in Plate 54. The craters seen in Plate 54, *a*, *b*, and *d*, are approximately 650 m in diameter, whereas the crater in Plate 54c is almost 800 m. Despite this slight variation in size, roughly the same sequence noted in Plate 53 is observable for these craters southwest of Maestlin R. The transition in the appearance of ejecta blankets from Plate 54a to 54b is not so continuous as in the set shown in Plate 53, although numerous blocks remain identifiable along the crater rim seen in Plate 54b (arrows X and Y). Also note in Plate 54b the dimple crater adjacent to the northwest rim (arrow Z). The walls of the craters shown in Plate 54 exhibit the same sequence as that seen in Plate 53, in which the number of exposed blocks on the wall decreases with inferred crater age. The last example, Plate 54d, retains exposures in a zone downslope from the rounded rim crest but above the textured lower wall.

There is evidence that the regional history was complex and that the craters shown in Plate 54 cannot be described as products of continuous degradation. Specifically, numerous flat-floored craters similar to the one shown in Plate 54d are present in the area. This example, in particular, has an inner dimple crater (arrow X) and is similar to craters found in once endogenetically active regions. Plate 195 shows two larger craters (3 km and 2 km in diameter) in the same region that were surrounded by mare material during a late stage of local mare flooding. Features that I interpret to be related to this stage remain relatively well preserved. Smaller craters (less than 0.2 km in diameter) with unusual but not unique rim sequences also are scattered across the region (see Plates 76a and 196, *d* and *e*).

It is suggested that either the subdued craters were endogenetically formed or that they were subdued through a catastrophic event, such as deposition of ash or ashlike ejecta.

PLATE 55

a. III-132-H1 (oblique)

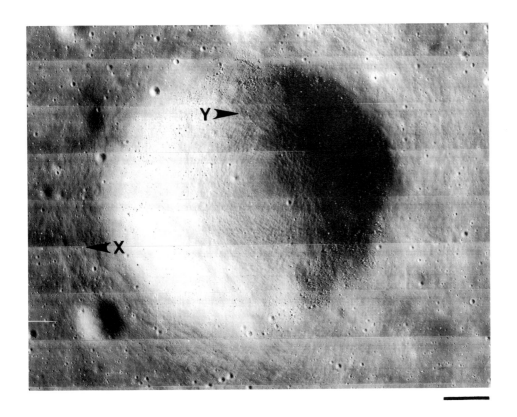

b. III-151-H1 0.21 km

Textured Walls

The tree-bark pattern can be identified easily on the walls of craters as well as in the surrounding region shown in Plate 56a. Craters of various depths are present, the deepest one having a block-strewn wall (arrow X). The area shown is to the south of Arago and it can be viewed in stereo in Plate 194, a and b. The region is very complex, with branching rilles that extend from Arago and cross very well preserved mare ridges.

The crater shown at top right (arrow X) displays numerous blocks across the outer rim area but lacks other indications of an ejecta blanket, which probably never exhibited a blanket comparable in appearance to that of Mösting C (Plate 80). The tree-bark pattern on its rim is connected to the surrounding textured relief that extends northward (arrow Y). A similar relation can be noted along wrinkle ridges (see Plates 79b and 141). Surrounding it are numerous dimple craters and linear depressions.

The tree-bark pattern displayed on the rims, walls, and floors of the other craters does not always coincide with changes in slope. Note the N-trending corrugated patterns that cross the floors of the two subdued craters, which are 0.6 km in diameter (arrows Z and XX). Whereas the tree-bark pattern reflects the contours, these coarse patterns do not.

Plate 56b shows a portion of the subdued crater identified by arrow XX. It reveals depressed zones containing blocks on the crater wall (arrows X and Y) that extend to the floor and are separated by smoother material. This is not the pattern expected from continuous smoothing by meteoroid bombardment. Similar blocky patches are shown more impressively on the walls of small craters within Hyginus (Plate 161b), near a volcanic ridge (Plate 132), and in other regions apparently subject to recent endogenetic processes. Near the example seen in Plate 56b (not shown here) is a small irregular depression that has well-defined borders and a flat blocky floor (see Plate 194d).

These patches having well-defined borders are not unique, therefore, and provide clues for interpreting surface processes. The patches might represent zones of avoidance (kipukas) from lavalike flows that cascaded into the crater. Their well-preserved borders, however, require recent formation or very slow degradational processes since emplacement of the mare units.

It is not likely that such patches are associated with slow mass movement of the surrounding regolith. Their well-defined lobate borders require a unitlike flow, which is unlike the movement anticipated for creep induced by micrometeoroid bombardment. Mass movement is also inapplicable for exposing blocky patches on the floors of dimple craters (Plate 161b) or within closed depressions (Plate 194d).

A plausible alternative is that subdued craters having such wall patches are formed by collapse and, therefore, are relatively recent yet subdued forms. The locally extensive corrugated and tree-bark patterns are proposed as expressions of surface creep of a regolith initiated by this collapse. The blocky zones might represent a blocky substratum exposed by the relatively rapid removal of the regolith. This theory is not satisfactory, however, for such zones at the bottom of depressions. It is possible that subsurface cavities with trapped volatiles were preserved beneath the surface until seismic events and slow volatile loss triggered collapse. The remaining volatiles may have escaped rapidly through extensional fractures along the crater wall and jetted away the overlying regolith. The relatively high albedo of these patches perhaps indicates a change in surface roughness or it may indicate sublimates associated with the escaping volatiles. Consequently, numerous large subdued craters may be recent formations on a possibly old surface; furthermore, preservation of the typically well defined borders of the blocky patches can be rationalized.

PLATE 56

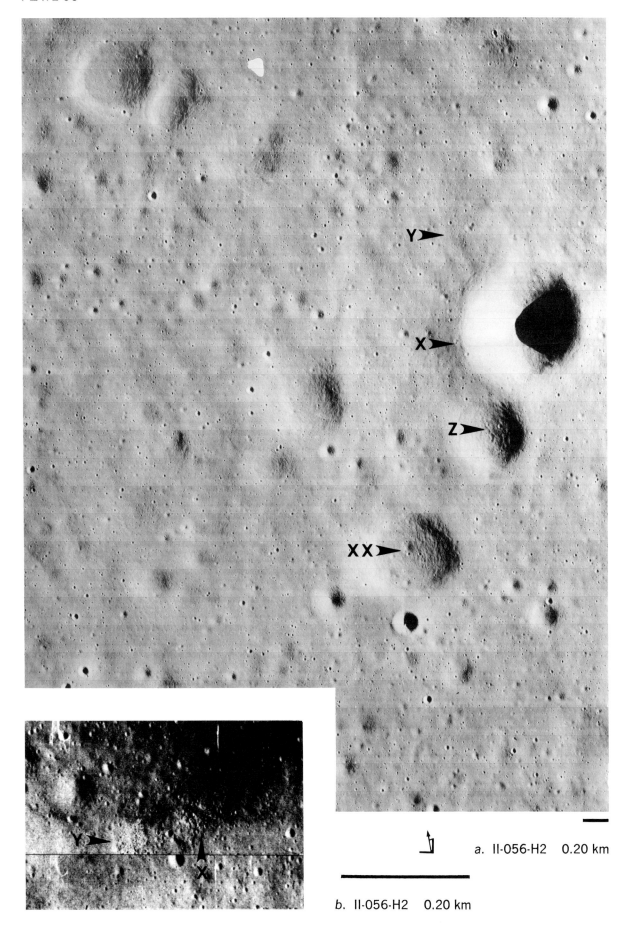

$\downarrow$

a. II-056-H2   0.20 km

b. II-056-H2   0.20 km

Screelike Walls (1 km–10 km)

Secchi X (5.6 km in diameter) is shown in Plate 57a in an oblique view at medium resolution (arrow X). It is on Mare Fecunditatis and is surrounded by small flat-floored and subdued craters. The grabenlike extension to the south is apparently part of a larger "ghost" crater (arrow Y), which Secchi X intersects along its northeastern rim (see Plate 150b).

As revealed in a high-resolution photograph, Plate 57b (for its corresponding medium resolution, see Plate 150b), Secchi X exhibits a truncated-cone or funnel-like profile with circular plan. The crater wall is divided into three zones according to texture and slope. The upper wall beneath the rim (arrow X) is the steepest and is characterized by outcrops, above which are accumulations of fragmental material. Below this zone, the decrease in slope is relatively abrupt, and low-albedo blocky deposits have accumulated on ledges (arrow Y), with strings of the material (arrow Z) extending toward the floor. The third and lower zone has a finer texture (arrow XX) and is believed to represent an accumulation of relatively low albedo talus. Numerous blocks are visible along the abrupt contact with the floor.

The development of these wall zones is at least in part the result of sorting of wall debris by mass transfer. Photographs brought back from the Apollo missions show numerous craters on the maria that have streaks of low-albedo material extending to the floor from strata on the walls. Most craters in the area around Secchi X display a low-albedo component covering the lower wall, and in some cases the floor (Plate 57a, arrow Z). Perhaps some of this material in Secchi X and nearby craters was derived from regional blanketing rather than "weathering" of outcropping strata.

If Secchi X was the result of an impact, then its ejecta blanket has been smoothed to a rolling topography that is in marked contrast to its wall. The particulate nature of the ejecta blanket responds to seismic events by compaction and creep and to repeated meteoroid bombardment by redistribution of the blanket material. The crater wall, however, responds to both processes by sheetlike slides of its scree slope, and the sharp floor-wall contact is maintained by the advancing wall at its base. Thus, the ejecta blanket loses its hummocky character, whereas the rugged appearance of the wall is constantly renewed. The zonal character of the wall probably develops from the combined action of parallel retreat of the upper wall and reduction in slope of the lower wall.

The possibility of volcanic formation of Secchi X cannot be ruled out. The placement of this crater on the ringed ridge or ghost crater may indicate parasitic volcanism if this ridge reflects control of subsurface weaknesses and vents. Note indications of similar activity adjacent to the ghost crater Lambert (Plate 150a). Endogenetic activity in the region is indicated by the low-albedo volcanic ridges (Plate 57a, arrows XX and YY; see also Plate 132a); the rectilinear rille in the upper left corner (arrow ZZ) that has a low-albedo maarlike head crater (Plate 178c; see also AS11-42-6309); and the numerous dimple craters in the region. In addition, the grabenlike rille, which extends to the lower left, appears to continue in a series of craters similar to, but smaller than, Secchi X. Thus, Secchi X could be a relatively recent or preserved volcanic structure, which might not require a rationale for a subdued ejecta blanket. Regardless of the validity of such an interpretation, the proximity of volcanolike forms—particularly the maarlike feature—suggests a source for the low-albedo wall and floor debris.

PLATE 57

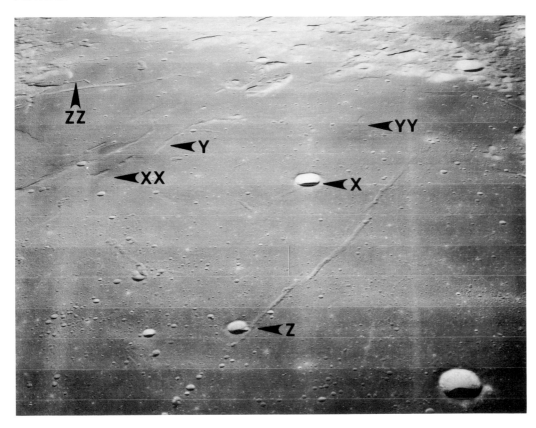

a. V-042-M (oblique)

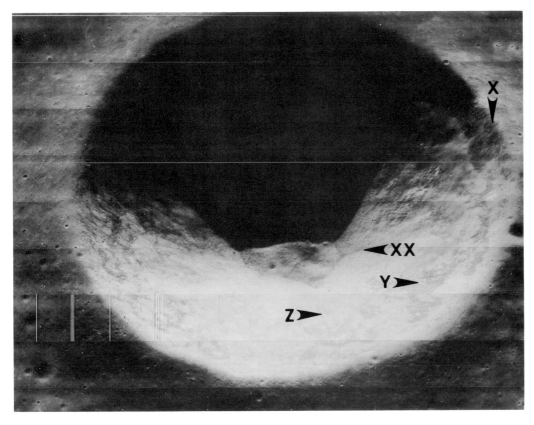

b. V-050-H1 (oblique)

Textured Walls (1 km–10 km)

The crater Hipparchus N (Plate 58, *a* and *b*) is on the floor of Hipparchus, which is a large (150 km in diameter) flat-floored crater shown to the right in Plate 43*a*. Hipparchus N is similar to Secchi X in that both have truncated-cone profiles and are about 6 km in diameter. In detail, however, these craters are different: the floor-wall contact of Hipparchus N is not well defined and its wall is largely covered by a grid pattern.

The rim-wall contact is relatively abrupt along portions of the rim (see V-98-MR). Numerous patches of blocky outcrops are found (Plate 58*b*) beneath the rim crest (arrow X). In addition, the upper wall is mottled with smooth areas (arrow Y) that generally are devoid of patterns and blocks. Beneath this upper third to one-half portion of the wall, blocks are sparsely distributed, and the grid pattern is apparent (general region of arrow Z). The lower wall exhibits numerous but widely separated blocks (arrow XX), some of which have left trails on the wall. Small craters on the wall indicate relatively stable conditions or recent bombardment by small secondary ejecta, but their density is clearly lower than that on the crater floor or rim.

The floor has a subdued doughnutlike ring that is most apparent at the base of its eastern wall (Plate 58*a*, arrow X; Plate 58*b*, arrow YY). The ring displays the tree-bark pattern and has few craters relative to the interior floor (arrow ZZ). Both of these observations suggest a mobile surface. Detailed examination of the original photograph reveals flowlike features on the ring that may be debris slides from the wall. Alternatively, the ring is analogous to a similar feature on the floor of Dionysius (Plate 26) that is interpreted as the previous level of a molten floor.

The walls of Hipparchus N apparently have attained a relatively stable slope, which is approximately 30°. Further slope reduction is probably in response to creep induced by seismic jostling or locally rapid mass transfer from meteoroid impact. Since the wall of Hipparchus N shows no evidence for debris cones and slides, the rate of mass transfer is apparently slower than that in Secchi X (Plate 57), Mösting C (Plate 80), and Gambart C (Plate 59). Furthermore, the walls of Hipparchus N appear to be composed of small blocks or fine debris, in contrast to those of the other craters.

Hipparchus N is, as already stated, on the floor of Hipparchus, which contains the highland plains-forming Cayley unit, and on the ejecta blanket of the crater Horrocks (30 km in diameter; see Plate 45, *a* and *b*), 18 km to the northeast. Both of these important regional units suggest impact excavation of possibly incompetent material. Consequently, the difference between the wall of Hipparchus N and those of the craters shown in Plates 57, 59, and 80 may be an expression of the competency of the subsurface as well as of relative age. Despite the subdued appearance of the ejecta blanket of Hipparchus N, small-scale lineations (0.2 km wide, not shown) associated with Horrocks remain. Therefore, it is likely that Hipparchus N did not display the complex hummocky ejecta blanket found around Mösting C.

PLATE 58

a. III-111-M   1.7 km

b. III-111-H3   0.22 km

Screelike Walls (10 km–15 km)

Gambart C (Plate 59, *a* and *b*) is approximately 13 km in diameter and is almost 300 km southeast of the crater Copernicus. It is also shown in Plate 90, where the discussion is directed to its rim sequence. The NW-trending lineations are associated with a loop of Copernicus secondary craters, some of which are shown in Plate 59*a*, north of Gambart C. Despite regional blanketing and scouring by Copernicus secondaries, the rim of Gambart C appears to be relatively unaffected (see Plate 59*b*). The rim exhibits numerous narrow (10 m wide) furrows (Plate 59*a*, arrow X) that trend toward Copernicus, but a similar NW-trending set can be identified (visible on the southeastern rim in original photograph).

Plate 59*b* shows the detail of the southern wall and rim. The wall displays intricate braided patterns (arrow X) that probably resulted from rapid downslope movement of material at differ-ent velocities. This movement was not initiated by the Copernicus or secondary events, because such detail would have been rapidly destroyed since that time. More likely, it is the result of continued but catastrophic backwasting of the upper wall because of Moonquakes from nearby impacts. The elongate depression (0.2 km long) on the rim (arrow Y) was apparently the source of part of the wall debris.

The contrast between the inner rim detail and wall detail of Gambart C should be compared to that of Mösting C (Plate 80c). Gambart C may represent a degraded version of Mösting C, and the screelike wall illustrates its continued rejuvenation, which is due to the steep slope. It is possible, however, that a detailed comparison is unwarranted because of differences in the initial appearances of these two craters.

PLATE 59

a. II-112-M   1.3 km

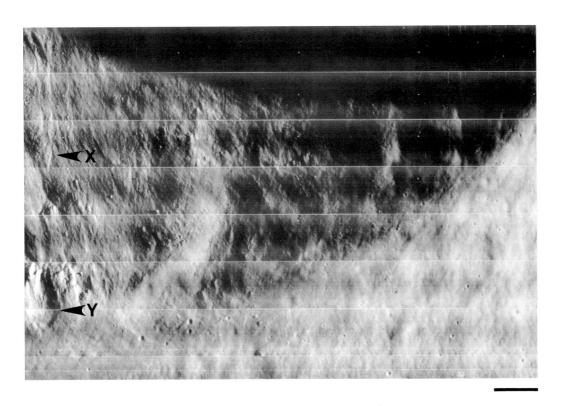

b. II-112-H1   0.18 km

Basal Aprons and Terraces

Plate 60 illustrates miscellaneous but important wall features. The crater Ritter D, about 9 km in diameter, is shown in Plate 60a (also see Plate 26a). The plate is oriented with north to the left because of the obliqueness of view. The wall of Ritter D makes a relatively abrupt contact with the rim. The surface textures of its wall, the rim profile, and the absence of a sculptured ejecta blanket are features similar to those of Secchi X (Plate 57). The numerous secondary craters west of Ritter D are from Dionysius.

Of particular interest are the two craters (arrows X and Y) adjacent to Ritter D. The rim-wall contact crosses the floor of the smaller crater identified by arrow X and is not displaced markedly by the smaller crater's boundary, although evidence for its rim is recognized on the wall of Ritter D. Hence, it is suggested that these craters are being destroyed at the expense of Ritter D through mass wasting, which has been more rapid than the smoothing of the wall-rim contact. During this time, either an extensive ejecta blanket comparable to that of Mösting C (Plate 80) has been destroyed, or Ritter D was formed by one or several large secondary impacts that did not produce such a structured ejecta blanket at the outset. An endogenetic origin for Ritter D also does not require the extensive degradation of a sculptured impact-produced ejecta blanket because such ejecta probably had slower velocities. Note the low-albedo cone (arrow Z), which is evidence for local volcanism.

Plate 60, b (arrow X) and c (arrow X), shows aprons along the base of crater walls. These aprons are characterized by lower slopes, convex profiles, and finer surface textures. They typically are found at the contact between positive-relief features and mare units. The insular features seen in Plate 60b are part of the Flamsteed Ring (Plate 154), whereas the crater shown in Plate 60c is within Mare Smythii (Plate 109b).

A detailed discussion on the origin of these aprons is found with Plates 126–128 and 154. They are believed to be either transferred debris from the wall or the results of mare emplacement, such as a previous floor level or extrusions along the wall-floor contact. The apron seen in Plate 60b (arrow Y), very large relative to the feature it surrounds, suggests that mass wasting may not be an adequate or complete interpretation.

Plate 60d includes a crater within the Orientale basin (see Plate 202, arrow XX). The crater exhibits a terrace (arrow X) that extends continuously along a considerable portion of the lower wall. This terrace is interpreted as a floor level prior to subsidence.

PLATE 60

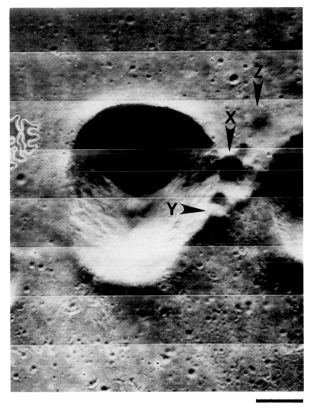

a. V-083-M   3.6 km

b. III-205-M   (oblique)

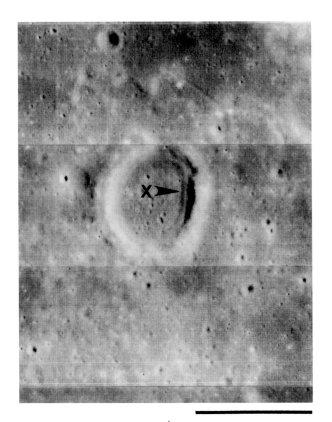

c. I-017-M   6.8 km

d. IV-195-H2   12 km

Terraced Walls (10km–15 km)

Plate 61 reveals unusual but not unique wall formations of narrow ledges on the upper walls. The crater shown in Plate 61, *a* and *b*, is on the Orientale ejecta blanket and is approximately 14 km in diameter (see Plate 207, arrow XY). In stereo, the shelf is clearly seen on the wall and is not a floor phenomenon. Plate 61c shows a similar crater of approximately the same size on the floor of Repsold, a large crater that is adjacent to northwestern Oceanus Procellarum.

The largest crater seen in Plate 61d—Damoiseau D—is 18 km in diameter. It is east of Grimaldi along a rille system that parallels the Sirsalis rille complex. LAC 74 shows that the crater rim rests approximately 1.2 km above the regional topography to the west and 1.5 km above the crater floor. The inner shelf appears to be a ridge rather than the upper terrace of a slump block.

Several interpretations for the shelves are suggested:

1. They are competent outcrops that are exposed by slides of less-consolidated material.
2. They are the terraces of narrow slump blocks.
3. They indicate multistage eruptions.
4. They are high-level marks of molten floors that have since subsided.
5. They are exposures of ring dikes on degraded walls.

The first suggestion is plausible for the crater shown in Plate 61, *a* and *b*, in which the outcrops might represent a pre-ejecta surface. The other examples shown in Plate 61, *c* and *d*, however, cross the walls of larger craters, where such a surface should be very complex. The second suggestion requires very unusual slumping in that the upper terrace is continuous. The third alternative was applied also to concentric craters seen in Plate 3, *b*, *c*, and *d*, and Plate 4. Those examples, however, have a well-developed inner symmetrical ring that remains craterlike even without the outer component, whereas the examples shown here are wall phenomena. The fourth suggestion requires considerable floor subsidence, plausible only if these craters are linked to a larger magma chamber. The last proposal implies caldera collapse along nearly vertically dipping faults that were intruded by magma. Subsequent mass wasting of the walls has exposed these intrusive features. I prefer the first and fourth suggestions as the most reasonable working hypotheses.

PLATE 61

a. IV-188-H1   12 km

b. IV-187-H3   12 km

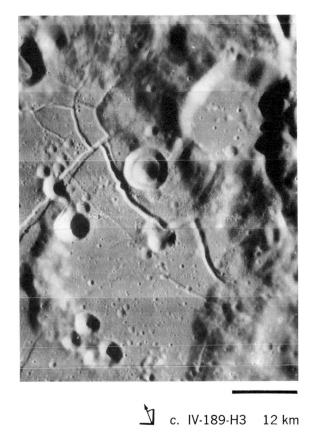

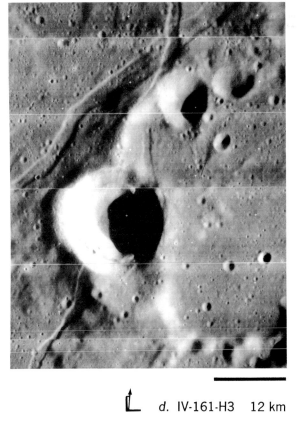

c. IV-189-H3   12 km

d. IV-161-H3   12 km

141

Screelike Walls (15 km–20 km)

The crater Dawes (Plate 27, *b* and *c*) is re-examined here with respect to the nature and origin of wall features. Plate 62*a* establishes the regional setting; Plate 62*b* shows wall features that are examined more closely in plate 63. Dawes is on the border of Mare Tranquilitatis, adjacent to Mare Serenitatis, and is slightly scalloped in plan and flat floored in profile. It is approximately 17 km in diameter and 1.5 km in depth from rim to floor. Its ejecta blanket is well preserved and similar to that surrounding Mösting C (Plate 80). Earth-based photographs show Dawes to be a dark-haloed crater. The nearby Plinius rille system (left corner and upper left of Plate 62*a*) reflects a pre-Dawes regional trend that is arcuate with respect to Mare Serenitatis.

Plate 62*b* shows the well-defined outcrops on the upper slopes of the northern wall. Beneath this stratum, the wall has apparently a constant screelike slope (see Plate 63*a*) that is slightly lower in over-all albedo than the wall above the outcrops. Although there are no slump blocks on the wall, the floor patterns may be remnants of these slides, which probably produced the scallops of the north and south rim. Numerous concentric fractures along the rim suggest crater expansion. Arrows X, Y, Z, and XX identify selected regions shown in Plate 63, *a*, *b*, *c*, and *d*, respectively.

PLATE 62

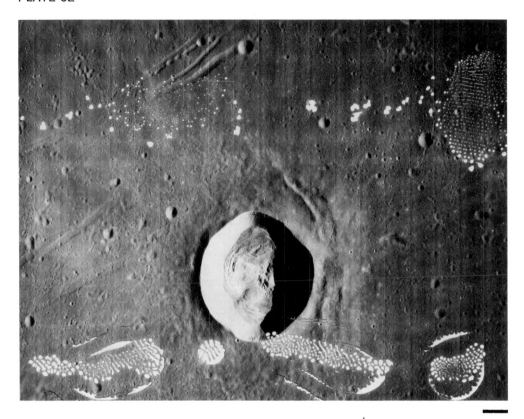

a. V-070-M   3.7 km

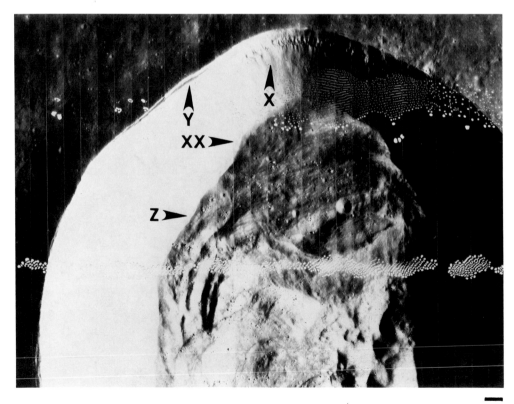

b. V-070-H2   0.46 km

Screelike Walls (15 km–20 km)

Plate 63 shows enlargements of Plate 62b; their locations are identified in that plate. Plate 63a (see Plate 62b, arrow X) shows the triangular-faced outcrops—pehaps upturned bedrock—that line the upper wall at approximately the same elevation below the rim. Notice that upslope from each exposure is an accumulation of material, whereas downslope are numerous debrislike cones and fans, some of which have well-defined multiple termini. The block size composing this debris is less than 3 m—probably less than 1 m. Coarser debris appears to follow the furrows between debris cones (arrow X) and apparently corresponds to low-albedo stripes that extend down the wall and that contribute to the slightly lower albedo of the wall beneath the outcrops.

Plate 63b permits examination of the wall beneath the rim crest on the northwest side of Dawes (Plate 62b, arrow Y). The smooth surface resembles the original rim and is a sliverlike terrace of a slump block.

Plate 63, c and d, shows the base of the northwestern (Plate 62b, arrow Z) and northern walls (Plate 62b, arrow XX), respectively. The conelike pattern is absent, but lobate termini are visible in the upper right portion of Plate 63d (arrow X). The absence of a generally extensive accumulation of blocks at the wall base probably reflects screelike advancement of the wall over such talus. The base of the wall seen in Plate 63d, however, is littered with 3 m–5 m blocks, and the grayish apron probably represents accumulations of smaller blocks (arrow Y). It is likely that they were derived from recent slides originating above the outcrops.

In general, a few large blocks are on the wall, and those on the rim typically occur in clusters. Several blocks on the floor (Plate 63c, arrows X and Y) are greater than 40 m across, whereas those on the rim are commonly less than 20 m. The blocks on the floor probably were originally within wall slumps that slid onto the floor.

Thus, the wall of Dawes is divided into two major sections, separated by a prominent outcrop. Above this stratum, material is being removed and redistributed across the lower wall. The incompetent units above the stratum may be ejecta derived from Dawes. The outcrops appear to have been cut and partly removed by transported debris. This suggests that the stratum either was or has become relatively friable owing to its recent exposure to conditions of space or to weaknesses developed during the formation of Dawes. Continued rapid mass transport of the wall will maintain the relatively pristine appearance of this crater despite the smoothing of its outer ejecta blanket and floor. Although the lower wall exhibits numerous well-preserved indications of mass movement, several small young craters are maintained on its slope. If these craters are from primary or secondary impacts, then they illustrate the infrequency of major slope failure.

PLATE 63

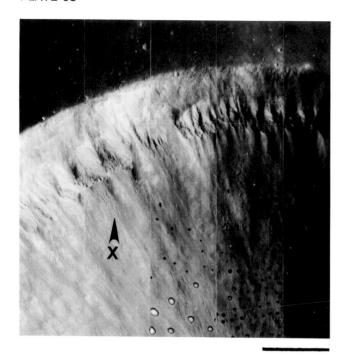

a. V-070-H2   0.46 km

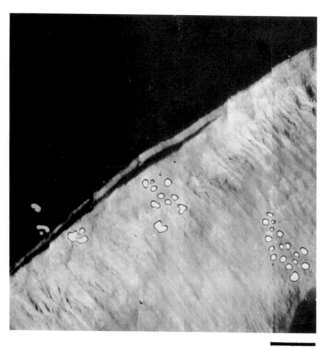

b. V-070-H2   0.46 km

c. V-070-H2   0.46 km

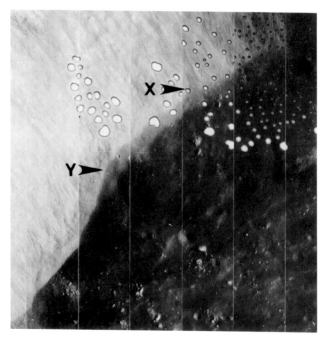

d. V-070-H2   0.46 km

Slumped Walls

Large-scale mass movement is common on the Moon and, as on Earth, exhibits a variety of forms including falls, slides, and flows. Plate 64, *a* and *b* (arrow X in both oblique views), illustrates rotational slumps within two relatively small craters, 30 km and 25 km in diameter, respectively. Slump features typically are restricted to craters greater than 10 km–20 km in diameter. The same size range corresponds to an abrupt change in slope of the depth to diameter distribution (Pike, 1967). Preliminary examination of Orbiter photographs (Manley, 1969) shows that the smallest craters affected by slumping in the maria are smaller, on the average, than those in the highlands. Mackin (1969) suggested that a limiting depth existed where a plastic zone acted as a lubricant after the release of support at the toe. Specifically, a cooling mare would have altered the lunar thermal gradient so that the peak temperature and the potential zone of flowage (after the release of overburden) would be at shallower depths; hence, smaller cratering events could reach this zone more readily on the maria.

Alternatively, the size dependence simply reflects differences in physical strength of the mare and highland materials. This might be revealed by smaller craters in which excavation exceeds a critical depth. Quaide et al. (1965) estimated that this depth is approximately 3 km. They also suggested that the diameter of a crater greater than about 10 km could be enlarged more than 20 percent by slumping. Consequently, a note of caution is in order regarding indiscriminate application of diameter-to-depth ratios to substantiate a favored thesis. Slumping is initiated at the time of crater formation, during later seismic events, or by withdrawal of subsurface magma in volcanically modified impact craters.

The two examples shown in Plate 64, *a* and *b*, display the undisturbed rotated head of the slump (arrow X in each), but in both cases the sequence toward the toe of the slide is difficult to identify. Of additional interest are the low-albedo striations down the wall and the concentration of this material at the base of the wall seen in Plate 64*b* (also see Plate 57*b*). This was also shown in Dawes (Plate 62) and indicates material of different lithology or size that has been transported down the wall as rock slides.

PLATE 64

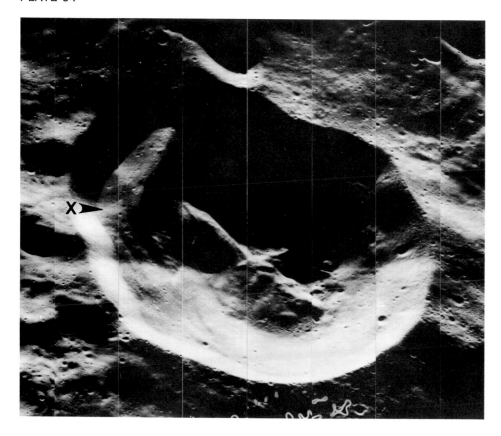

a. V-181-H3 (oblique)

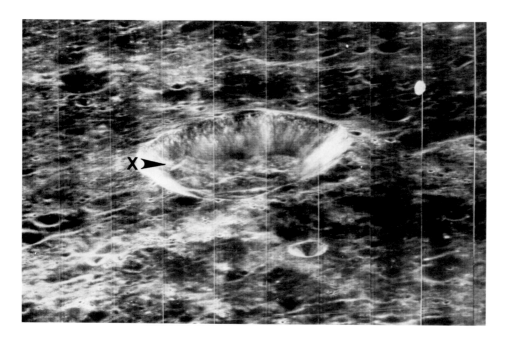

b. I-102-H1 (oblique)

Slumped Walls

The crater Mösting (about 25 km in diameter), Plate 65, displays the results of a more complex series of mass movements than those shown in Plate 64, a and b, and the scalloped crater plan is clearly the result of these movements. Mösting (the stereo pair, Plate 65, a and b) interrupts a ridge (arrow X) but for the most part was formed on a mare unit. At low resolution the ejecta blanket appears to be well preserved but is less well defined at higher resolutions (Plate 65, c and d). Debris fall and successive slumping have expanded the crater. Plate 65c shows more clearly the arcuate pattern in the southwestern portion of the crater (arrow X). This feature is very common in craters of this approximate size and probably represents successive slumping; Dawes (Plate 62b) has an analogous pattern on its floor. The toe with its associated flow can be identified at the wall base of Mösting (arrow Y). The numerous minor scarps that line the wall with cracks forebode further slumping on the northeast rim (arrow Z).

Detail of the northern upper wall is shown in Plate 65d. Outcrops of strata are visible in the medium-resolution photograph (Plate 65c, arrow XX), but in Plate 65d they are not so apparent as in Dawes. The rim-wall transition displays smooth rim material (Plate 65d, arrow X; this corresponds to arrow YY in Plate 65c) that extends down the wall, commonly as an overhang with abrupt and jagged borders (Plate 65d, arrow Y). Therefore, the break in slope does not coincide with the transition to the roughly textured wall. A similar overhang is shown in Plate 235c. The wall displays myriad separate blocks composing rock slides (for example, arrow Z), several of which have well-defined termini. In addition, several small boulders with trails can be identified on the original photograph.

Details of this part of the upper wall of Mösting are distinctly different from those of the walls of Dawes. Large debris cones are not abundant, and the walls appear much more chaotic. Although both Mösting and Dawes were formed on a mare unit, Mösting occurs in a region where this unit does not appear to be as thick as that near Dawes, owing to the numerous exposures of pre-existing ridges and hills. Mare units near Mösting also are underlain by thick ejecta from the Imbrium basin that probably exhibit a stratigraphy more complex than that near Dawes. Consequently, the wall exposures in Mösting would be expected to be more complex than those in Dawes.

The identification of blocks in the wall debris of Mösting is probably due to a factor of 2.7 higher resolution in these photographs. Although there are very few blocks on the rim of Mösting, it is probable that the inner rim zone, which is characterized by such blocks, has slumped toward the lower wall and floor.

PLATE 65

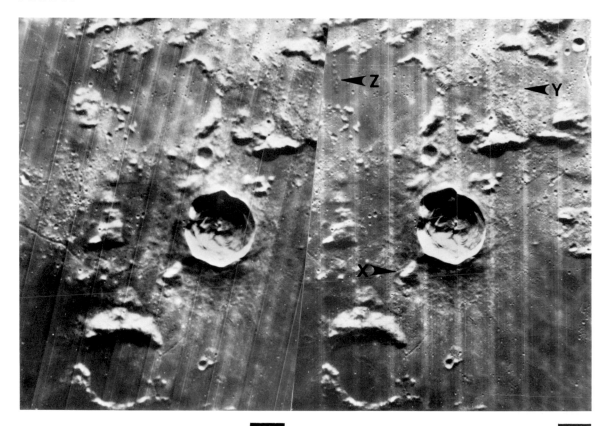

a. IV-109-H1   12 km

b. IV-108-H3   12 km

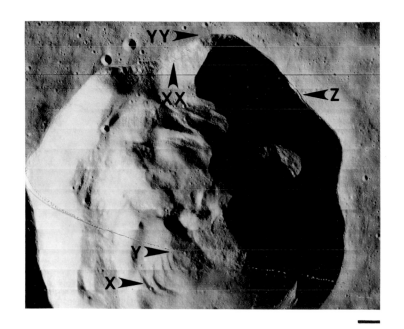

c. III-107-M   1.4 km

d. III-107-H2   0.18 km

Slumped Walls

Bürg (Plate 66, a and b) is a 40-km-diameter crater in the mare-filled basin Lacus Mortis and has a wall similar in appearance to that of Petavius B (Plate 30a). Typically, the upper walls of Bürg have relatively narrow and continuous (not extensively cut or broken) terraces. The upper terrace commonly is surrounded by a sizable scarp, which indicates significant displacement of material from the rim. Numerous minor faults on the upper terrace (arrow X) yield a generally convex appearance in section. Portions of some craters (see Copernicus, Plate 218) have no large displacements along these minor faults in the upper terrace, which results in a large inner platform beneath the rim crest. Within other craters (Aristarchus, Plate 236), most of the terraces are approximately the same width from the rim to the floor, yielding a roughly concave profile of the wall.

The lower portion of the northeast wall of Bürg (arrow Y) appears to be relatively continuous down to the floor-wall contact. This is typical of several large craters that have a major plains-forming unit on their floors. The western wall of Bürg (arrow Z), however, exhibits numerous thin terraces from near the base of the major scarp to the floor. This portion of the wall apparently failed as a large spoon-shaped slump zone, in contrast to the concentric terraces on the facing wall. Although this slide may have postdated considerably the initial crater-forming event, it clearly predates floor inundation.

The lower wall sections of numerous craters (generally greater than 30 km in diameter) are crossed by furrows and flows that extend from near the upper terrace or rim to the floor. In a few large craters, such as Humboldt (see Plate 42c), the base of the wall has been so highly furrowed that the floor-wall boundary has the appearance of fingerlike projections.

Narrow wall terraces commonly look like concentric fractures filled with smooth marelike material (see Aristarchus, Plate 236) and could have been created by the following processes:

1. Multiple slump blocks that follow faults approaching the horizontal with depth, combined with the original outward slope of the rim to produce back slopes of slumped blocks with relatively high angles
2. Funneling or collapse of essentially debrislike wall material into underlying concentric fractures
3. A complex graben-horst-slump arrangement caused by multiple synthetic and antithetic faults

The lower wall-floor region may be hummocky and poorly defined (Plate 66, c and d) or make an abrupt contact (Plate 66, a and b), depending on the extent and time of floor flooding as well as the particular process involved in wall modification.

In addition to Bürg, note:

1. The scarp (arrow XX) connected to Rima Bürg II (arrow YY)
2. Small craters (arrow ZZ) on the western rim and wall of Bürg (possibly related to the tangential wrinkle ridge)
3. The two large craters south of Bürg upon which a large extrusive (?) mass (arrow XY) apparently has been superposed

Plate 66, c and d, is a stereo pair of a highland crater, Newcomb, in the same size range as Bürg (approximately 40 km in diameter). The estimated rim-to-floor elevation difference is about 2.8 km (LAC 43), but the 40-km diameter testifies to the exaggerated view. The enormous scallop to the south (arrow X) probably was originated by collapse and may have been triggered by renewed faulting along the NE-trending grabenlike rille in the lower right (arrow Y). The rille is about 10 km in width and can be traced across the interior slump blocks (arrow Z) and the floor (arrow XX). It clearly predates the emplacement of the plains-forming floor units, but displacement continued after the formation of the large scallop.

Newcomb and Bürg are distinctly different in their over-all appearance. Perhaps the most obvious difference between these craters is the absence of a well-preserved or similar ejecta blanket around Newcomb. This includes not only the radial striations around Bürg but also the bulge around it created by both ejecta and uplift. Newcomb, however, has not been extensively smoothed, as is clear from the abrupt rim-wall transition and the wealth of topographic detail on its walls and floor. If it is assumed that Newcomb originally resembled Bürg, then slumping and other modifications have been extensive. The poorly defined rim facies around Newcomb may reflect an intrinsic difference between highland and mare impact craters that reflects regional stratigraphy. Thus, the paradox of a crater with a seemingly old rim zone yet recent floor and rim zones might be explained. Moreover, this difference could be enhanced by slumping that consumed the inner rim regions. Alternatively, Newcomb indicates volcanic modification of the floor that triggered extensive slumping considerably after crater formation.

PLATE 66

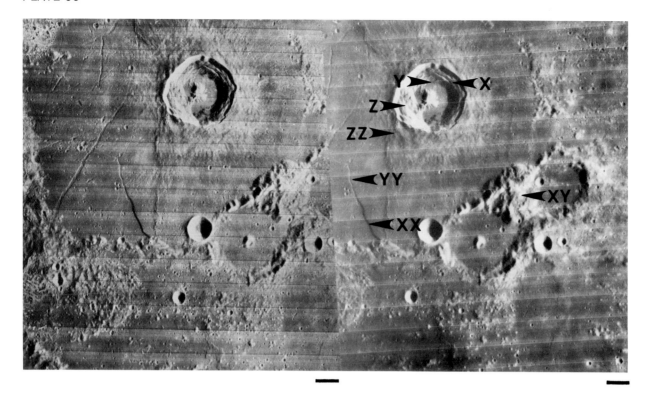

a. IV-091-H2   13 km   b. IV-086-H2   13 km

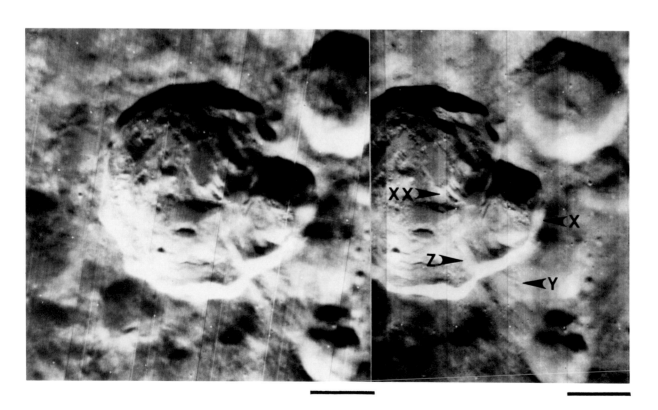

c. IV-067-H1   13 km   d. IV-066-H3   12 km

Slumped and Dissected Walls

Tycho (Plate 67) illustrates complex wall formations of a recent crater in the 20 km–100 km size range (see Plate 31, *a* and *b*, for stereo). For comparison and additional information, see the discussions (Appendix A) of Copernicus (Plate 218) and Aristarchus (Plate 236). Tycho is in the lunar highlands, whereas Copernicus is in mare units south of Mare Imbrium and Aristarchus is on an insular platform. Copernicus is about 95 km in diameter (3.8 km elevation difference from rim to floor); the diameter of Tycho is almost 90 km (4.5 km difference in elevation), and that of Aristarchus is 40 km (3.6 km difference). Any differences in their wall development, therefore, could be attributed to their regional placement and/or size. The former includes complexity in local stratigrapy, and the latter implies control by the intensity of the event and the subsurface zone of influence. The wall of Tycho generally can be characterized by the following features:

1. High outer scarps adjacent to the rim (arrow X)
2. Numerous terraces and minor scarps (arrow Y)
3. Small smooth-surfaced units, "pools" (arrow Z), and flow features (arrow XX)
4. Generally hummocky but dissected lower wall (arrow YY)

Arrow ZZ identifies the selected region shown in Plate 68.

PLATE 67

V-125-M 7.2 km

Slumped and Dissected Walls

Plate 68 is a high-resolution section of the northern wall of Tycho (Plate 67, arrow ZZ) and shows the wall formation from the rim (top) and outer scarp to the floor (bottom right). Two flow features (arrows X and Y) exhibit successive flow fronts, transverse ridges, and fractures (parallel to the direction of flow and probably later features). A breach in the scarp to the northwest (arrow Z) may be a channel through which molten rim material flowed to form this front. Close examination of the rim reveals considerable pooling of material and patterns indicative of flowlike movement. In some instances, these regions on the rim were sources for numerous smooth-surface units observable on the wall (see Aristarchus, Plates 240–241a, and the discussion on crater rims accompanying Plates 101 and 105).

The numerous cascades of originally molten material are extremely impressive. The lower left portion of the photograph (arrow XX) provides excellent examples of pools on a terrace interconnected by flow features. Flow levees (arrow YY) are well displayed, and smooth-appearing flows with well-defined termini typically fill local depressions.

The volcanolike features displayed on the walls and floor do not require that Tycho was volcanic in origin. The heat generated by an impact would have been more than sufficient to induce melting, although it is not known how much of this material would have been ejected. Once-molten material in depressions on the wall could be explained readily either by pooling from surface flows, some of which may be derived from fall-back material, or by extrusions following faults related to slumping.

PLATE 68

V-126-H2   0.91 km

Subdued Slumped Walls

Delambre (Plate 69) is a 52-km-diameter crater in the lunar highlands to the west of Mare Tranquilitatis. The stereo pair (Plate 69, *a* and *b*) exaggerates the 3.5-km elevation difference from the floor to the rim (LAC 78) and reveals the subdued terraces lining the wall. Plate 69*c* is an Orbiter III view of the northern wall region (Plate 69, *a* and *b*, arrow X) and shows that the subdued nature is intrinsic and not a problem of resolution. Plate 69*d* presents a small portion of the wall (see Plate 69*c*, arrow X) and should be compared to the highly pitted floors shown in Plate 43, *c* and *d*, as well as the examples to follow (Plates 70, *a* and *b*, 72–73). Note that many of the subdued craters are dimplelike, a result of the degradation of impact craters or of the drainage of the particulate overlay into subsurface cavities.

The tree-bark pattern and generally subdued character support the interpretation that Delambre is an ancient crater relative to the age of such craters as Tycho (Plate 67) and Aristarchus (Plate 236) but is more recent than the formation of the Imbrium basin. It should be noted, however, that degradation of the walls may have occurred through an extensive fall-back of ejecta during the formation of Delambre. This process is inferred from the crater Cichus (Plate 70*d*), which exhibits a well-preserved ejecta blanket but subdued slump blocks.

PLATE 69

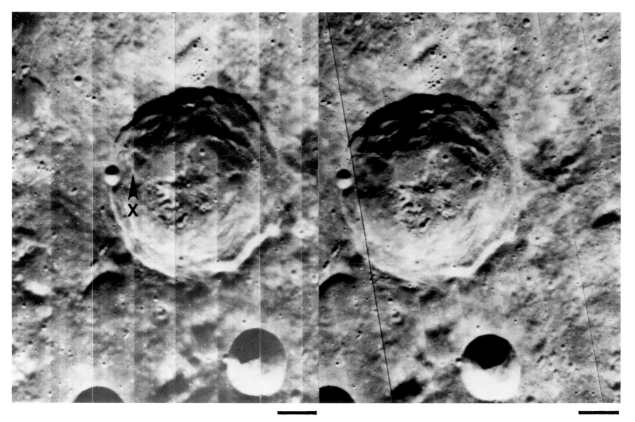

a. IV-090-H1   12 km

b. IV-089-H3   12 km

c. III-079-M   1.6 km *(top)*   1.4 km *(bottom)*

d. III-079-H2   0.20 km

Subdued Slumped Walls

The floor of Abulfeda (Plate 70a) is about 3.2 km beneath the rim and covers most of this 62-km-diameter crater. The wall is more subdued than that of Delambre and exhibits meager terracing, which may have been removed or buried during floor flooding. Plate 70b is a high-resolution view of the southern wall and includes the contiguous crater chain (Plate 70a, arrow X). At this resolution, the terrain is typical of the lunar highlands, with gently rolling topography and patterned ground that can be attributed to meteoroid bombardment over an extended period, combined with slow mass movement, such as creep.

It is possible that Abulfeda was formed during early lunar history, when degradational processes were much more active. For instance, the particle flux, both large and small, could have been much greater. Hartmann (1970) suggested that the rate might exceed two hundred times that for the average postmare rate. This increased flux would have an effect on the pulverization rate by smaller projectiles and on the extent of blanketing by the larger ones. In addition, Abulfeda probably predates the formation of the Imbrium basin, as indicated by the surrounding striated and stippled terrain associated with Imbrium ejecta. The Imbrium catastrophe included innumerable secondary impacts, ejecta flow, and impact-induced Moonquakes. These effects must have far outweighed the effects of long-term degradation by smaller impacts.

Plate 70, c and d, provides a stereo pair of the crater Cichus, about 65 km southwest of Mare Nubium. The ejecta blanket of Cichus clearly overlies the mare plains of Palus Epidemiarum to the west (arrow X), and the small-scale lineations associated with the ejecta illustrate the relatively good state of preservation. In contrast, the wall slumps do not exhibit the crisp appearance shown previously in Tycho (Plate 67). The 10-km-diameter crater on the western rim of Cichus may have contributed to this degraded state, but the floor and floor-wall contact do not appear to have been affected. Consequently, the subdued wall slumps may reflect either properties of the local stratigraphy, which includes the ejecta blankets of the Humorum and Nubium basins, or degradation associated with the formation of Cichus, such as fall-back ejecta.

PLATE 70

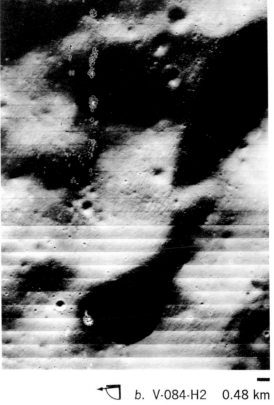

↖⬦ *a.* V-084-M   3.7 km   ↖⬦ *b.* V-084-H2   0.48 km

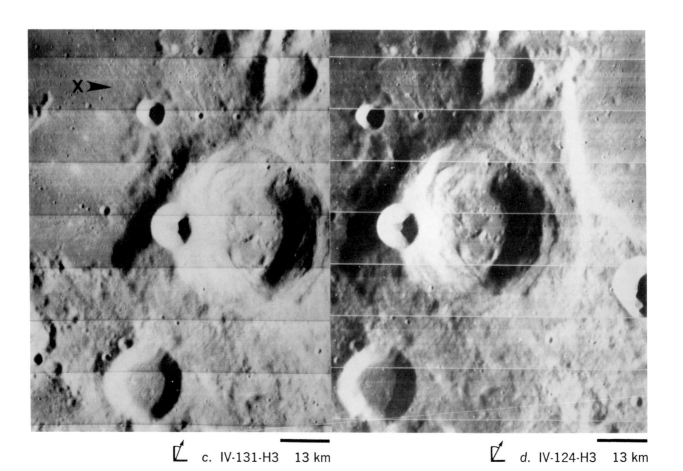

⬐ *c.* IV-131-H3   13 km   ⬐ *d.* IV-124-H3   13 km

Subdued Slumped Walls

The three immediately preceding descriptions of the craters Delambre, Abulfeda, and Cichus illustrate the degradation of walls. In Delambre and Cichus, floor filling by subsequent units had a minor effect, whereas in Abulfeda a major portion of the original wall was removed or covered. The crater Geber, shown in stereo (Plate 71, a and b), is approximately 45 km in diameter and illustrates an extreme example of wall degradation similar to that of Abulfeda, but without extensive rejuvenation or blanketing of the floor. It is in the Southern Highlands about 120 km south of Abulfeda.

The size of the crater indicates that slumping originally must have been extensive, but now only subdued vestiges of the slump blocks remain. If the crater does not become altered catastrophically, evolution of the wall could be visualized as proceeding one step further, in which it would become essentially continuous. The crater Parrot C (Plate 44, arrow Z) might be considered at such a stage. Although Geber is approaching a similar state, it, as well as Parrot C, still retains a raised rim. This indicates that any backwasting of the wall has not enlarged it sufficiently to include all the rim and that the smoothing of any wall slumps into a relatively continuous slope has not been sufficient to destroy the raised rim. The present stability of the wall is indicated by the preservation of several 0.5 km–1.0 km craters on it, as well as by several bright-haloed craters on its slopes.

Geber is situated on a stippled unit that dominates the region, including Abulfeda. The stippling represents the outer ejecta zone from the Imbrium basin, and this event may have been responsible for some of the extensive smoothing. Although this stippling is not well preserved on the walls, there are numerous subdued wall grooves, which are probably related. The crater Parrot C, however, probably was formed after the Imbrium event. The distinction between craters degraded catastrophically and those degraded by a continuous influx of meteoroids is not always obvious, even at such a large scale.

Plate 71, c and d, shows an unusual but not rare relationship between adjacent craters. From the stereo view, it can be established that a slide or flow encroached onto the floor of the northern crater from the south (arrow X). The numerous craters and crater chains are the effects of secondary ejecta from Tycho, which is to the west. The proximity of the Tycho event suggests that seismic waves initiated movement, but note that this movement must have been complete prior to the secondary impacts. A similar example can be found near Tsiolkovsky (Plate 11a) on the lunar farside.

More likely, the slide was directly related to the event that formed the crater to the south. The relative freshness of the displaced mass with respect to the assumed original ejecta blanket is rationalized by either more efficient processes acting on the ejecta blanket or the continued slow movement of the slide.

PLATE 71

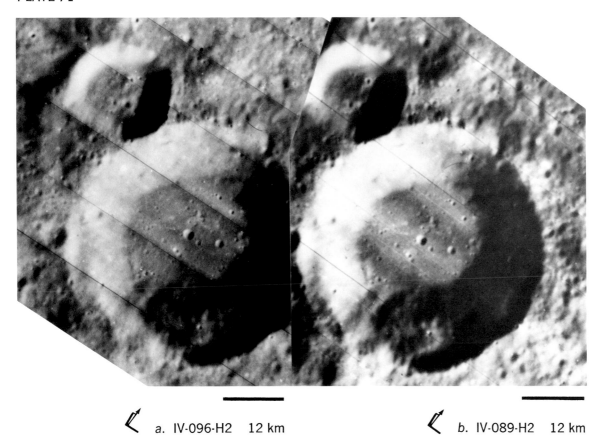

a. IV-096-H2   12 km

b. IV-089-H2   12 km

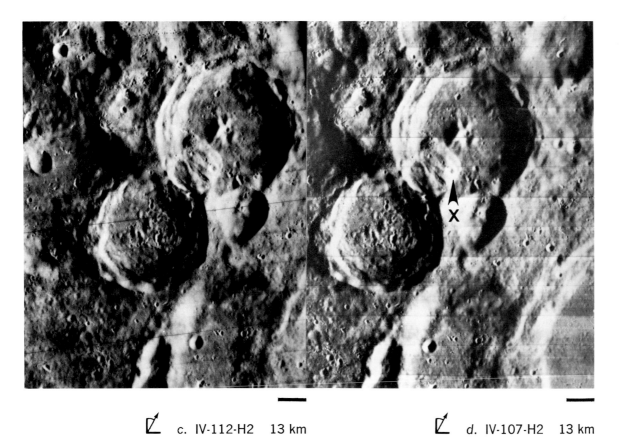

c. IV-112-H2   13 km

d. IV-107-H2   13 km

Subdued Slumped Walls (Highland Craters)

Alphonsus (Plate 72) provides an example of subdued crater walls of a large (110 km in diameter) crater in the Southern Highlands. Plate 44 shows an overview of the region, and, in the accompanying discussion, it is noted that Alphonsus may have been volcanically rejuvenated along weaknesses created by the Imbrium event. Relatively recent endogenetic activity also is indicated by the dark-haloed elongate craters adjacent to fractures on the floor (Plate 72a, arrows X and Y; also see Plates 186–187).

Portions of the wall of Alphonsus were modified by furrows trending radially from the Imbrium event (arrow Z). In addition, terraces on the wall were capped by relatively high albedo and cratered plains-forming units (arrows XX and YY) that postdate Imbrium sculpturing. These units appear to be emplaced in local depressions that may be remnants of slumps.

Plate 72b shows a high-resolution view of the wall southeast of the area marked by arrow XX in Plate 72a. The fully illuminated scarps are washed-out, but, in other areas (for example, arrow X), the wall surface is highly textured, probably the result of slow mass movement. Close inspection of the area reveals an absence of large blocks on the wall, in the ejecta of craters, and at the floor-wall contact (arrow Y). This suggests a surface that has been extensively pulverized or, more important, blanketed by deposits from the Imbrium event. Note the grid pattern on the wall of the subdued crater (arrow Z).

PLATE 72

a. V-119-M   3.3 km (*left*)   4.2 km (*right*)

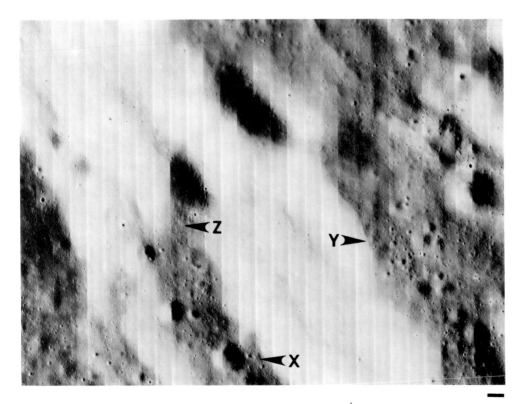

b. V-116-H2   0.49 km

Subdued Walls: Ringed Plains

Plate 73, a and b, shows the ringed plains of Fra Mauro (left), Bonpland (bottom right) and Parry (top). They form a triad (see Plate 121c for an overview) and have common wall-rim boundaries. In addition, Parry shares its eastern boundary (not shown), with an elongate ringed plain, Parry M, that is part of another complex, including the crater Guericke. The Imbrium sculpture clearly has modified these "craters," an indication of an origin prior to the formation of the Imbrium basin. These are generally considered to be highly degraded and inundated craters that once resembled Copernicus. The apparent clustering with common wall-rim boundaries, however, is possible but weak evidence for an endogenetic origin.

The wall-rim borders between the craters shown in Plate 73, a and b, have platformlike profiles at several locations (arrows X and Y), and their surfaces are cratered to a degree similar to that of the crater-fill material. This profile and appearance are not unusual for ringed plains and suggest mass wasting of a previous surface that was capped, in part, by a plains-forming unit. The wall-rim profile of the most recent crater, Parry, is typically more crested.

Grabenlike rilles cross the wall-rim borders, and arrow Z identifies a crater on the common wall of Bonpland and Parry that is near the intersection of two of these rilles. A similar occurrence is found on the floor of Bonpland (arrow XX). These small craters are possible endogenetic forms. More convincing volcanic features include the low-albedo irregular crater (arrow YY) and a volcanic ridge with a chain of summit craters (arrow ZZ). Closer views are provided in Plate 131.

Plate 73c reveals the highly textured wall-rim between Fra Mauro and Parry where the grabenlike rille crosses it (Plate 73a, arrow XY). The tree-bark pattern (arrow X) and grid pattern (arrow Y) are apparent. In general, the surface appears similar to the walls of Abulfeda (Plate 70b) and Alphonsus (Plate 72). The relatively low crater density is indicative of a mobile particulate surface.

Consequently, this complex of ringed plains is similar in appearance to other large subdued craters but also exhibits evidence for volcanic modifications, such as the dark-haloed crater form (arrow YY), the crater at the intersection of the grabenlike rilles (arrow Z), and the volcanic ridge (arrow ZZ). Plate 111 shows a similar complex that has had more extensive volcanic modification.

PLATE 73

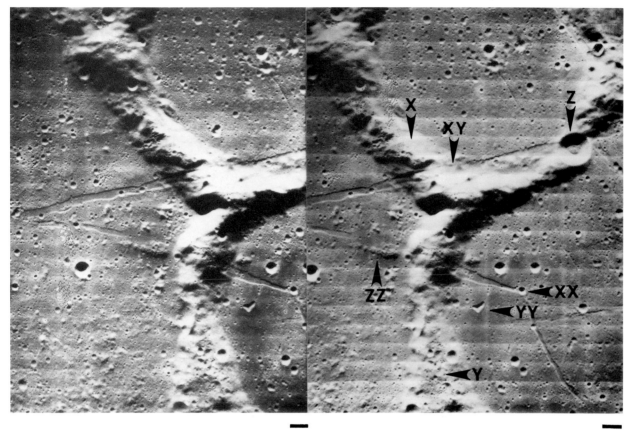

a. V-140-M   3.3 km

b. V-138-M   3.3 km

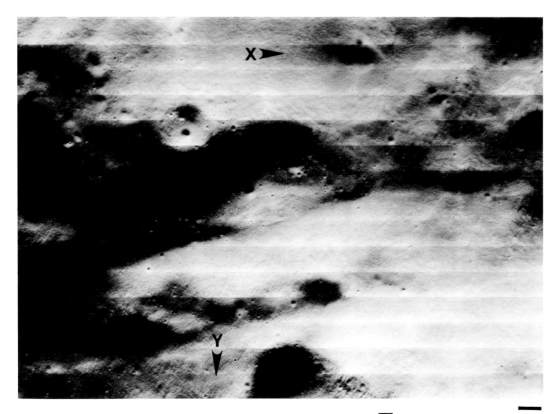

c. V-138-H1   0.43 km

Modified Walls

The examples presented in Plates 74 and 75 provide insight into the variety of events that could lead to the obliteration of lunar craters. Plate 74a shows an area east of Littrow on the eastern "rim" of Mare Serenitatis. The region exhibits extensive stippling of relief, with a similar effect on the walls of craters. The process responsible is perhaps related to the formation or development of Mare Serenitatis or to the outer ejecta from Imbrium. The stippled appearance is enhanced by low-albedo material that has migrated toward topographically low areas. This low-albedo material is probably derived from volcanic eruptions on the periphery of Mare Serenitatis (see Plate 138).

The backside crater Gagarin (230 km in diameter) is shown in Plate 74b and illustrates the combined effects of continuous and catastrophic degradation of an enormous crater. The catastrophic events were the formation of Tsiolkovsky, less than 600 km to the southwest, and Keeler, less than 500 km to the northeast. Ejecta from these craters have masked or destroyed much of the wall detail. Note the more recent development of a mare-like floor within the smaller interior crater (arrow X).

The effect of even larger events is shown in Plate 74, c and d. The crater Shickard (Plate 74c) is southeast of the Orientale basin and has been partially sculptured by this event. This sculpturing is a combination of long continuous furrows and ridges (arrow X) and is the result of an extensive base-surgetype blanket, gouging by streams of low-angle secondaries, and horsts and grabens, which are radial to Orientale. Later processes include emplacement of two contrasting plains-forming units (arrows Y and Z) over part of the floor. The lighter plains-forming unit (arrow Z) is also found on the wall and outside the crater, and arrow XX identifies a sinuous rille on this surface that extends from the rim area to the floor. Note that the lineations (arrow YY) on the northeastern wall are perpendicular to the Orientale sculpture and perhaps related to the very old Humorum basin to the northeast. The walls appear to be terraced in this region, but unit slump blocks comparable to those in Copernicus are poorly defined.

Flammarion (Plate 74d, arrow X) displays extensive sculpturing from the Imbrium event. Its wall is broken into ridges that align with Imbrium and gives an en échelon plan to the crater. The floor of Flammarion was later filled with a plains-forming unit that buried any interior lineations. Note the hummocky platform (arrow Y) with lobate boundaries connected to the northern part of Flammarion (see Plate 65, a and b, for a stereo view). This unit may have been formed extrusively or may be a remnant of a massive flow unit related to the formation of the Imbrium basin.

PLATE 74

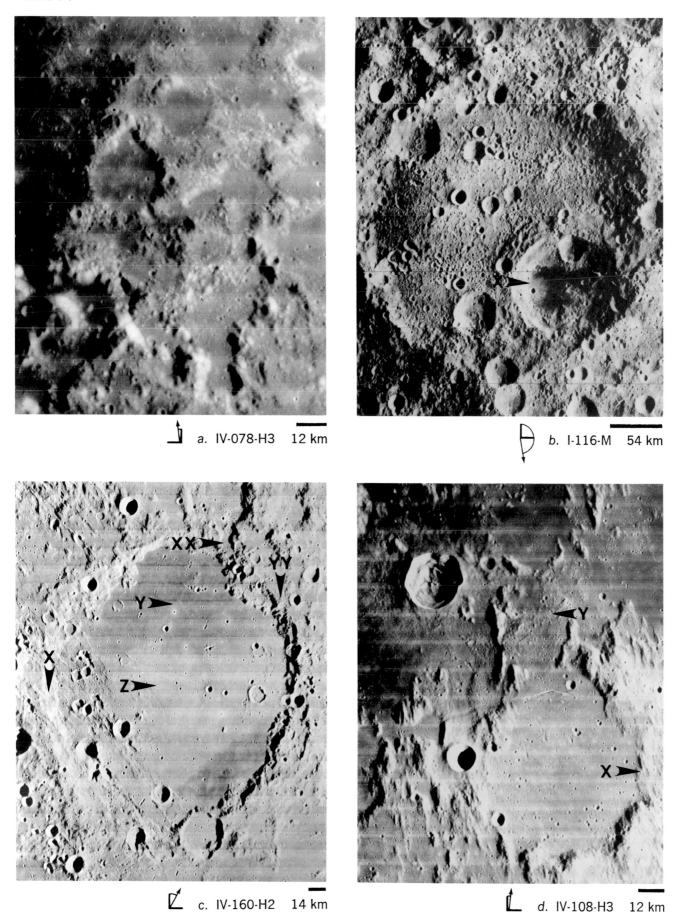

a. IV-078-H3   12 km

b. I-116-M   54 km

c. IV-160-H2   14 km

d. IV-108-H3   12 km

Dissected Walls

As discussed in connection with the walls of Tycho (Plates 67–68), the walls of large craters may be extensively crossed by furrows extending from the rim to the floor. Mare Ingenii (Plate 75a) and Tsiolkovsky (Plate 75b) illustrate extreme cases of this phenomenon.

Mare Ingenii is a mare-filled backside crater about 350 km in diameter. Within it are two large adjacent rings, and to the northeast (lower left) is a larger elongate crater, Van de Graaff (the oblique view shown here is inverted). Furrows (arrow X) characterize all the wall regions of Ingenii, portions of the interior rings, and portions of Van de Graaff. They also extend outward on the rim of Ingenii. In contrast, the Copernicus-like crater on the western rim (arrow Y) displays well-defined slump blocks without extensive dissection.

The furrows must considerably postdate the formation of Ingenii, for a preserved ejecta blanket is absent. For craters such as Tycho, the furrows are apparently the result of landslides and impact melt cascading to the floor from sources along the upper wall or rim. This appears to be exhibited on a grand scale in Ingenii and numerous surrounding craters. Ingenii is within a large region that is interpreted as an extremely ancient basin that rivals Imbrium in size. Massive slides and extrusions perhaps were triggered during the epoch of mare flooding, which was incomplete but nevertheless extensive in this region (see Plate 135). It is also possible that landslides were induced by the reinforcement of crustal shock waves produced during the Imbrium basin-forming event, which was antipodal to the region shown in Plate 75a.

Plate 75b is an enlarged view of the southern wall of Tsiolkovsky. The northern parts of the wall display extensive slumping (see Plate 40b). The southern sections shown in Plate 75b also show some slumping but are extensively crossed by furrows. Lavalike units can be identified along the rim (arrow X; also see Plate 39b) and, as in the case of Tycho, might be sources for material that crosses the walls. One furrow in particular is actually a sinuous channel extending from the rim (arrow Y).

PLATE 75

a. II-075-M   (oblique)

b. III-121-M   49 km

DISCUSSION AND SUMMARY

Crater walls are describable in general by their profiles and the detailed nature of their surfaces. Wall profiles are typically discontinuous as a result of terraces, scarps, and outcropping strata. Crater wall surfaces may be either rubbled or patterned, which includes the tree-bark and grid textures.

Small craters (less than 1 km) rarely have wall scarps greater than approximately 30° in slope and commonly are rubbled. Concentric craters, which are less than 0.2 km in diameter, display multiple scarps and are covered by blocky rubble. Gradually sloping walls are concave, convex, or straight in profile. The last two profiles result in a funnel or dimple appearance (see Plate 20).

The surfaces of subdued craters typically have wrinklelike textures termed the tree-bark pattern that approximately reflect the topography. In addition, the surface may exhibit a grid pattern that describes parallel and intersecting sets of closely spaced lineations. These sets predominantly trend NE and NW and intersect on the crater walls tangent to the direction of solar illumination and on the wall facing solar illumination. There also appears to be an imaginary focus in the opposite wall shadow. In general, the lineations are not straight but are arcuate; therefore, they may depend on slope, which further implies a dependence on the local elevation of the Sun. As a rule, the grid pattern does not occur on surfaces with rubble or with very coarse tree-bark textures. This may indicate that such a pattern develops only on relatively stable slopes or that there is a relationship between particle size and detectability. Craters with patterned walls typically have upper walls (and even their rims) studded with blocks, whereas the lower wall merges continuously with the floor. In several craters, rubbled high-albedo patches are identified along the upper wall. These patches are commonly depressed relative to the surrounding smoother wall and may be bounded on all sides by relatively well defined lobate borders.

Rubbled crater walls are not diagnostic of crater origin; they indicate either the physical nature of the subsurface in which they occur or the amount of fall-back debris during crater formation. Subsurface layering is believed to be expressed by wall terraces in concentric craters. Relatively well defined scarps within subdued craters also may be an expression of subsurface structure or indicate slope failure of debris slopes.

Subdued craters with gradually sloping walls are interpreted as degraded impact craters, blanketed volcanic vents, or collapse features. Ross (1968) and Soderblom (1970) constructed models for the degradation of impact craters that appear applicable for many small subdued craters less than 100 m in diameter. They assumed the primary mechanism to be repetitive impacts by meteoroids much smaller than the crater being degraded. Some of their conclusions appear inconsistent with the existence of subdued craters on well-preserved units, perhaps because these initial and idealized models neglected to take into account the effects of surface creep.

Although the small-scale grid pattern may be a lighting phenomenon, the tree-bark pattern is probably an indication of slow mass movement of an unconsolidated surface regolith. It is possible that the coarseness of the tree-bark pattern is directly proportional to the average creep velocity and particle size. This slow migration of material from the upper wall exposes outcrops and rubble beneath the rim crest. It is not clear, however, how blocks are exposed on the rim area beyond the rim crest of some craters (discussion of this point can be found accompanying Plate 78). In addition, a slow but continuous buildup of regolith by repeated impacts and the slow transfer of this debris are not consistent with the existence of bordered patches of rubble on several crater walls, such as those on the floor of Hyginus (Plate 161b). Consequently, it is suggested that these craters are collapse features and that the rubbled exposures might be produced by release of trapped gas along peripheral fractures. The well-defined borders around several of these zones require either that their formation was very recent or that the surface is extremely well preserved. Either conclusion strengthens the interpretation that some subdued craters are modified, if not formed, by collapse or sagging rather than by progressive degradation of impact craters.

Medium-size craters (1 km–20 km in diameter) typically have single scarps. Their wall surfaces are divisible into segments defined by relative differences in albedo, surface textures, and slope angles. In numerous highland craters, the upper fourth (or less) of the wall is low in relative albedo, whereas its lower two-thirds has a higher albedo. The transition zone in several craters is marked by outcropping strata, but in numerous other examples it is poorly defined. This transition commonly corresponds to irregular patches of low-albedo material that extend around the upper wall at approximately the same elevation (see Secchi X, Plate 57). The lower wall typically displays streaks of material with contrasting reflectivities, and the base of the wall may have a diffuse low-albedo zone that ends abruptly at the floor. A preliminary survey indicates that there is a reversal of this sequence of wall reflectivities in craters on the maria. For example, the lower three-fourths of the wall of Dawes (Plate 62) is slightly lower in albedo than the zone above the outcrops, an arrangement that is particularly apparent in regions having a generally low albedo (see Plate 166). These subdivisions in albedo contrasts are not restricted to subdued craters but also occur in craters having well-preserved ejecta blankets.

Detailed examination of walls of medium-size craters reveals textures or patterns that also divide the wall into zones. Pristine crater walls commonly display debris cones, boulder trails, boulder trains, flow features, and other phenomena indicative of rapid mass transfer. The crater Dawes exhibits debris cones beneath breaches in outcropping strata, whereas the lower wall slopes appear screelike and have several flow termini, presumably from landslides.

Craters with subdued ejecta blankets typically have on their upper walls patches of smooth material surrounded by blocky slidelike zones. The middle wall in several such craters has a complex rough surface that merges with the finely textured lower wall. Isolated blocks commonly are found at the base of the wall and floor.

Numerous craters in this size range also exhibit changes in the slope profile of their walls. The uppermost segments of crater walls merge with or abruptly meet the rim crest. The apparent abruptness of this transition depends, in part, on photographic resolution.

Craters with subdued or no ejecta blankets can be found with a relatively abrupt transition from the wall to the rim crest (Gambart C, Plate 59; Ritter D, Plate 60a; Plate 85b). Others have a convex element over the same distance (Secchi X, Plate 57; Hipparchus N, Plate 58; Plate 86a). Below the convex segment, the wall commonly has a straight slope to the floor. Several craters, however, appear to have two straight-wall segments with different slopes between the convex element and the base.

There appears to be a correlation between the three parameters (albedo, surface texture, and slope angle) that divide crater walls, but further studies are needed because views of the same crater under different solar illuminations are required. In addition, albedo depends on both block size and lithology, and the separation of these factors is not straightforward if the resolution is inadequate.

There are several possible sources for the wall debris:

1. Fall-back from the initial cratering event
2. Pulverization of outcrops by meteoroid bombardment and other forms of weathering, such as the effects of particle emplacement by solar wind
3. Debris from underlying incompetent units exposed by the crater
4. Regional blanketing by ash or ejecta
5. Retreat or reduction of the upper wall slopes that transfers unconsolidated surface material, such as ejecta, or material from underlying incompetent units

Wall divisions in pristine craters apparently reflect different types of mass transfer. Material from the upper wall is transferred to the straight-wall segment by rock slides that are revealed by debris cones, braided textures, boulder trails, and boulder trains. The straight-wall segment probably represents transfer by shearing of the scree slope and unitlike rock slides derived from farther upslope. The latter movement is indicated by well-defined flowlike units on the wall and the existence of high- and low-albedo streaks. Shearing of the scree slope is indicated by the general absence of large debris cones and the smooth appearance of the wall. The contact between the floor and wall may be abrupt or gradual. The former contact indicates either that the floor is more recent or that the wall is continuously advancing over the floor. Continuous advancement suggests that the slope of the wall is reduced rather than backwasted, and the absence of talus in this type of floor-wall contact supports the idea of such a history. The gradual transition typically is accompanied by contrasting albedos and identifiable talus. Backwasting in some craters apparently occurs above outcropping strata where material is removed and deposited on the lower slopes.

The rapidity of mass transfer on the wall is indicated by the preservation of mass-wasting features and the absence of small craters. Rapid mass transfer can continually rejuvenate the appearance of the crater, despite the apparently partial or complete destruction of its ejecta blanket. This assumes that there is knowledge of the original appearance of the ejecta blanket, and such an assumption is not always warranted. Wall rejuvenation develops at the expense of the rim and outcropping strata. Wall debris also is derived from ejecta of nearby impacts and from ash deposits. The latter two mechanisms may be, in part, responsible for the similarity in albedo between the lower screelike wall element and the terrain outside the crater.

The foregoing discussion leads to the consideration of more subdued-appearing craters with diameters in the 5km–20 km size range. The smooth patches on some upper walls may be small intact plates derived from the rim. This is indicated by the jagged borders of such units along the upper wall of Mösting. Although the diameter of Mösting is greater than 20 km and exhibits slumping, the large upper scarp is considered similar to the single scarp of smaller craters. More typically, the convex element of the upper wall displays smooth unconsolidated material that resembles the rim. This probably represents a relatively immobile regolith on the small local gradient. The lower part of the convex element exhibits rapid transfer of material, but this merges with a textured wall element that is either concave or straight in profile. The texturing most likely results from creep of sand-size (or larger) wall material. The stability of this surface is indicated by preserved impact craters and isolated blocks. Consequently, it is possible that wall material is fragmented

in place by micrometeoroid bombardment, although smaller size particles from the upper wall also contribute. The blocks at the base of the wall indicate the relatively slow advancement of the wall over the floor: blocks are transferred to this site more rapidly than they are buried.

If the craters examined represent degraded impact craters, then it appears that with time the wall is reduced in slope. This reduction is, in part, the result of the net downslope transfer of ejecta from small meteoroid impacts. The existence of subdued craters with relatively large severed craters on their rims indicates, however, that the wall remains screelike for a significant fraction of its history. The reduction in slope is perhaps triggered by Moonquakes, the probability of which should increase with time because of the greater likelihood of a large nearby impact. Titley (1966) suggests that this is an important agent for surface modification. He calculates that an impact crater of 1 km in diameter produces seismic energy equivalent to an Earthquake of magnitude 4.1 to 5.5 (Gutenberg-Richter scale) and concludes that this could result in failure of debris-composed slopes. Results from the Apollo seismic experiments indicate that seismic-energy absorption is extremely low, owing to the absence of volatiles (Latham et al., 1971). As a result, the effects of an impact-produced seismic event of a given magnitude on the Moon will be greater and more widespread than those from an equivalent event on the Earth. Thus, Titley's evaluation is probably underestimated.

There are subdued craters with narrow shelves on their lower walls. If the shelves are outcrops of pre-existing strata, then their persistence is difficult to rationalize. It was suggested (see discussion accompanying Plate 28c) that they might represent terraces from previous floor levels, but even this explanation requires relatively recent events or relatively stable walls.

The foregoing discussion does not apply to all crater walls in the 1 km–20 km size range. It is possible that apparent differences in age may be expressions of differences in crater origin. Photogrammetric studies from Apollo photographs should aid in this distinction as well as in understanding the evolution of sloped surfaces.

Large craters (20 km–100 km) typically have walls with multiple scarps that clearly result from slumping. In general, slumping occurs in craters greater than 15 km–20 km in diameter and typically produces scalloped crater plans. Six different wall profiles are identified. These are best revealed in oblique views from medium-resolution Orbiter IV photographs and are defined as follows:

1. Moderate-size upper wall scarp below which are many slump blocks; over-all straight or concave profile of crater wall: see Aristarchus (Plate 236)
2. Moderate-to-large upper scarp with a highly faulted wide terrace that retains a relatively flat surface; crater-within-crater appearance: see Copernicus (Plate 94) and Petavius (Plate 215a)
3. Moderate-size upper scarp with a large terrace broken by numerous minor faults because of synthetic and antithetic faulting or the formation of terracettes; over-all convex profile of slumps: see Bürg (Plate 66, a and b)
4. Large upper wall scarp (relative to total wall height) with slump blocks or scalloped remnants on the lower wall or floor; considerable floor debris and toe rubble: see Dawes (Plate 62), Römer (Plate 45, c and d), Mösting (Plate 65)
5. Large upper scarp with hummocky floor that typically cannot be identified as slump blocks: see Gassendi (Plate 36) and Callipus (Plate 28, d and e)
6. Small-to-moderate upper-wall scarp with smoothed terraces; over-all concave profile: see Geber (Plate 71, a and b)

The concave, terraced, and convex profiles (first three in the foregoing list) have wall slumps that commonly extend around large arcs of the upper wall. Slump blocks with large terrace widths and small lateral extent are also found in these craters but are more characteristic in the fourth profile listed.

Major slumping in craters having wall profiles with the first three descriptions apparently occurred prior to the last stages of emplacement of floor units but after ejecta had scoured their rims. This is revealed in the first case by floor materials that overlap the wall and in the second case by rim remnants on the upper terrace of the slump blocks.

Slump terraces in many craters greater than 20 km in diameter are capped by smooth-surfaced units, "pools." These units represent once-molten flows from sources on the rim, fall-back ejecta, and/or minor eruptions along slump faults. Greeley and Gault (1971) compared the size-frequency distribution of craters on such pools in Copernicus with the distribution of collapse craters on terrestrial basalt flows. They concluded that the pools are probably basalt lava flows containing a significant number of collapse craters.

The walls of many of the large craters exhibit numerous furrows that may extend from the upper rim to the floor but more typically cross only the hummocky lower wall. These furrows commonly contain flow material that contributes to the floor units (see Tycho, Plate 67). In several large craters, they are a more predominant wall form than slumping.

The furrows might represent channels cut by molten ejecta. This sequence is consistent with the interpretation of Dence (1968) concerning complex terrestrial impact craters. The furrows in many craters, however, are restricted to the lower wall zones, and this suggests that they result from landslides or lavalike flows from sources near the upper terrace or terraces and perhaps along the slump faults.

The cause for the different over-all concave, terraced, and convex profiles of highly slumped and recent craters is not understood. The terraced wall profile may indicate well-ordered competent strata included in the slump blocks, such as those strata produced by successive layers of mare basalts. It is possible that the convex and terraced profiles are the result of later wall modifications accompanying emplacement of floor units. The convex profile might be the result of several processes:

1. Numerous minor slumps developed on an originally terraced wall (Type 2 profile) due to later processes, such as floor subsidence

2. Later and more rapid landslides from the upper scarp that rest on an originally terraced wall (Type 2 profile)

173

3. Slumping that reflects different competence or layering of the material comprising the slump block

Mösting (Type 4 in the foregoing list) displays large scallops of its rim that result in a crater plan distinctly more irregular than those of very recent craters, such as Petavius B (Plate 30a). Mösting is not unique in this respect and may indicate later enlargement of the crater by mass wasting. Although such craters typically have slump material on their floors, they also have more recent plains-forming units that are characterized by numerous subdued craters and a higher albedo than the maria. This suggests that slumping occurred soon after crater formation and that the difference in the size of slump blocks in Petavius B and Mösting reflects regional differences in crustal properties. It is also plausible, however, that later crater enlargement was triggered during emplacement of molten floor units accompanying floor rejuvenation. It is suggested that this may have been responsible for slumping in such craters as Dawes (Plates 62–63) and Newcomb (Plate 66, c and d), whose slump materials rest on their floors. It might also apply to craters having a wall profile falling under Type 5 in the foregoing list. In addition, such craters as Kunowsky (Plate 92a) have crater plans similar to that of Mösting, but their floors have been completely inundated by marelike units. These craters apparently had further floor eruptions in which the wall slumps were completely engulfed.

Pike (1971) suggests from preliminary studies of photographs returned from Apollo 10 that the rate of slumping determines whether slumps are retained on the floor (rapid) or wall (slow). This is consistent with my discussion of craters similar to Mösting, Dawes, and Callipus. Pike's suggestion might apply also to larger craters that exhibit large upper wall scarps but retain considerable slump material on their walls. In contrast, walls without such scarps may reflect relatively slow mass movement. However, floor units in several large craters having this type of wall overlap the lower wall rubble; therefore, the floor must have remained active over the same time period as wall movement, unless this movement was restricted to the upper wall region. Sharpe (1938) observed in terrestrial landslides that

slump blocks remain intact although the slide occurs, in effect, instantaneously. This observation suggests that extensive wall slumps may have formed rapidly yet were retained on the crater wall. These two views of slumping are not necessarily contradictory but probably reflect two different modes of landslides, both of which can occur in the same crater.

Type 6 of the slumped-wall profile can be interpreted as degraded products of the first four types. The crater Delambre (Plate 69) has very subdued slumps and appears to be the result of considerable smoothing. The crater Geber (Plate 71, a and b), then, might represent a very late stage of degradation. There are numerous craters in the Southern Highlands that exhibit gradually sloping walls with only minor indications that they once were extensively slumped. Many of these were formed after, or perhaps during, the Imbrium basin–forming event. In addition, their shapes in plan remain remarkably circular, without large scallops produced by slumping or by overlapping craters. Consequently, degradation must have been performed by relatively small impacts and/or extensive blanketing. There are three other hypotheses:

1. These craters never had slumps. If this is correct, then the minimum size for slumping must have been much greater in the past.
2. These craters are not extensively degraded but represent calderas that were formed soon after the Imbrium event.
3. These craters represent large-scale secondary impacts produced during the formation of the Imbrium basin and modified by basin-induced seismic waves.

The last hypothesis in this list is based on the observation that secondary craters 1 km–2 km in diameter are typically subdued in appearance (see Plates 96 and 115). The Imbrium event must have ejected much larger yet relatively low velocity projectiles (below the escape velocity, 2.4 km/sec), and considerable variety in crater appearance could result, particularly if these secondaries impacted prior to the termination of enormous Moonquakes and deposition of ejecta. It is possible that the high crater density in the lunar highlands is largely an expression of such secondary impacts from Imbrium and other large impact basins rather than an expression of a high premare cratering rate from primary projectiles.

In conclusion, the walls of craters are impressive sites of mass wasting. The degree of wall preservation is not always a reliable indicator of crater age, for renewed mass movement can renew wall appearance. Other modifications active in larger craters include emplacement of plains-forming units and catastrophic burial by ejecta from the formation of large basins or nearby craters. The importance and significance of the various processes require further semiquantitative studies.

4. Crater Rims

INTRODUCTION

The rim of a crater can be described by its morphologic features either in plan or by profile. The in-plan description is most frequently used in the following discussions. The rim is broadly defined as that area beyond the wall. The inner rim boundary typically corresponds to a break in surface appearance due to the increased slope of the upper wall, but this criterion is not applicable to extremely subdued examples where the wall-rim transition is poorly marked. Utilization of crater profiles to describe the rim quantitatively must await detailed topographic analysis from stereo pairs. Reference, however, will be made to such profiles wherever they are revealed by local differences in sun angles and exaggerated stereo views.

Smoothing processes acting on the rim zones require assumptions for the initial state or genesis, which cannot be anticipated to be the same for all craters in all size ranges. The character of the ejecta and rim for impact craters depends on the particular projectile and target parameters and is expressed differently for craters of different sizes. Craters occur that exhibit well-defined raised rims but little or no evidence for ejecta blankets. Such differences may be assigned to continuous degradational processes alone, but, because the Moon has undergone considerable volcanic modification, such an assignment is dangerously simplistic. The danger resides not only in disregarding catastrophic changes but also in neglecting to take into account the possibility that the initial structure was volcanic.

The following preliminary discussion permits inspection of possible endogenetic craters asociated with definitely endogenetic forms, such as mare ridges, domes, platforms, peaks, and rilles (see their separate sections). In many respects, these craters resemble what might appear to be modified impact structures. To appreciate this problem, consider the comparative terrestrial examples—the Canyon Diablo meteorite crater and the Elegante crater in the Pinacate volcanic field, Sonora, Mexico (see Ulrich, 1966). Caution in both quantitative and qualitative interpretations cannot be overstated.

The inner rim-wall region is characterized by a variety of profiles. The narrow raised-lip profile describes a relatively well defined segment of the innermost rim char-

acterized by a slope greater than that of the region farther from the rim crest and occurs in circular impact-like craters (Censorinus [Plate 81a] and Mösting C), some of which do not display extensive ejecta blankets. In general, such a profile is restricted to craters less than 15 km–20 km in diameter. Craters larger than 20 km in diameter typically do not exhibit a pronounced raised lip appearance; rather, they have a generally continuous slope up to the rim crest (Plat 92a). This absence of the lip is attributed to slumping, which removes the innermost rim region. In several craters, however, an inner liplike profile appears to have been produced by slumping (Plate 93).

Symmetric profiles can be identified that are ridgelike (Plate 91), high-relief bulbous (Plate 180d), or low-relief bulbous (Plate 76c). The first is generally restricted to crater sizes with diameters less than 50 km; the second, to less than 10 km; and the last to less than 1 km. Structures similar to the Flamsteed Ring (Plate 154) also could be described as roughly symmetric in wall-rim profile and are illustrated by the Eddington-Russell-Struve complex (Plate 111). Several large craters having relatively broad rims give the appearance that the craters rest on large domal complexes. This profile is accentuated under low solar illumination or by encompassing mare material and is illustrated by Copernicus (Plate 94), Seleucus (Plate 110a), and Bürg (Plate 66, a and b).

The following photographs and discussion exclude the very low rimmed or rimless dimple craters, examined in Plate 20. Further related discussions can be found in the chapters "Craters: Shape in Plan," "Floors," "Walls," and "Albedo Contrasts."

Pristine Forms

Craters less than 1 km in diameter typically display ejecta composed of numerous blocks, which are largest near the crater. Small-scale radial lineations can be identified around some craters, but the rim sequence is generally characterized by the following distinctions:

1. Bright and/or dark ray material
2. Size distribution of blocks in ejecta
3. Symmetry or asymmetry in the spatial distribution of ejecta
4. Crispness of detail in ejecta blanket

Plate 76 offers a variety of examples of small (less than 0.3 km) craters. The crater shown in Plate 76a is south of Maestlin R. Its very irregular plan, small domed floor, asymmetric distribution of ejecta, and bulbous rim profile are unusual but not unique to this area or within the Flamsteed Ring.

The diameter of the crater is about 230 m, and local inhomogeneities in subsurface structure may have been responsible for its appearance. The energy distribution from impacting projectiles with small masses and low velocities is influenced by such irregularities. Experiments by Gault, et al. (1968) demonstrated the complex phenomena that accompany an impact in interbedded competent and incompetent layers. In particular, they noted that the tangential flow of particulate material will be restricted by competent layers and may produce upwarping rather than an overturned rim. Reverse or overthrust faults were found to dominate near the surface, whereas underthrusting was typical in the lower levels. Such events may have been responsible for the peculiar rims shown in Plate 76a. The local occurrence of similar craters reduces the probability that such craters were the result of fluke impacts, one inside the other. The existence of small craters in the rim region of the crater shown in Plate 76a suggests that this crater may be in a degraded state, despite the preservation of blocks. I feel, however, that these small craters, excluding those on the inner rim, reflect the transparency of the ejecta blanket.

An alternative is that such craters were formed by endogenetic processes producing low-velocity ejecta. More specifically, they may be analogous to blister craters that form on lava flows. Craters with rims similar to that seen in Plate 76a commonly occur in regions that have evidence for recent or well-preserved features associated with the emplacement of mare units (see Plates 151–152 and 196). It is possible that small low-relief domes (see Plate 125e) are related incipient forms. The crater shown in Plate 76a, however, requires preservation of meter-sized blocks since local mare emplacement. This implies that these mare units were emplaced without extensive flow patterns, which is possible if a non-impact-produced regolith was formed. An endogenetic origin is more plausible if the crater formed at a time considerably postdating mare emplacement. More likely this crater represents a low-velocity secondary impact into an interbedded surface.

In contrast, the 0.085-km-diameter crater shown in Plate 76b is approximately 5 km north of the crater shown in Plate 76a and displays much more symmetry both in rim profile and plan. Consequently, it is suspected that the impact occurred in a more homogeneous target, even though the inner terrace indicates a competent subsurface layer. The crater types in this region generally range from that shown in Plate 76b to that in 76a, with increasing crater diameter.

Plate 76c includes a crater (146 m in diameter) that exhibits a bulbous rim profile and a relatively symmetric plan—relative, that is, to the crater seen in Plate 76a. Although within the Flamsteed Ring, it is very similar to craters intermediate in size between the two examples just discussed that are found near Maestlin R. Note the generally meager display of ejecta (see Plate 78, a and b), and, as discussed above, the presence of small craters up to the rim reflects either postformation impacts and/or the relative thinness of the ejecta blanket.

Plate 76d permits examination of flat-floored craters; the largest example shown is approximately 150 m in diameter. The flat floor emphasizes the bulbous profile of the rim. Again, large ejected blocks are few, perhaps as a result of an incompetent subsurface or subsequent erosional processes.

PLATE 76

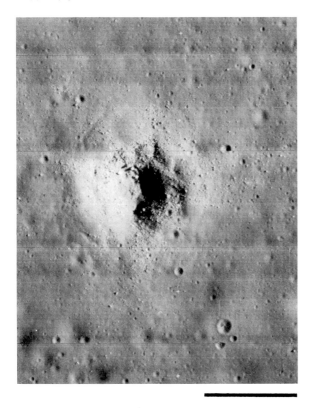

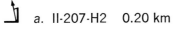

a. II-207-H2   0.20 km

b. II-207-H3   0.20 km

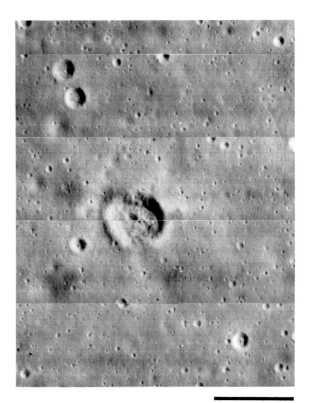

c. III-193-H2   0.23 km

d. II-091-H3   0.22 km

Pristine Forms

Plate 77 shows craters of slightly larger dimensions than those shown in Plate 76. The first example, in Plate 77a, is approximately 0.43 km in diameter and is adjacent to a mare ridge (lower left corner) within the Flamsteed Ring. Numerous blocks comprise a major component of the ejecta within a crater diameter from the rim. The size distribution of blocks is characteristic of crater ejecta in this region and suggests underlying competent layers. Several craters on the floor of Copernicus (Plate 227a) have a similar density and show a comparable distribution of sizable blocks. Such craters could be products of secondary projectiles, which have low impact velocities.

The example seen in Plate 77b is also within the Flamsteed Ring (see Plate 154) but is about 50 km to the west of the crater shown in Plate 77a. The crater is 0.55 km in diameter and displays a wide range of block sizes. Several larger blocks (arrows X, Y, and Z) are associated with small craters. Although the crater plan is not so well defined as that shown in Plate 77a, its approximately concentric plan implies subsurface layering. This may correspond to the base of the regional regolith, estimated to extend only 3.3 m (Oberbeck and Quaide, 1968). Such a figure is only an average (median) and could be quite different for local areas. However, another competent layer, besides the base of the regolith, could be responsible for the bench.

Note the low-albedo fan extending to the southwest (arrow XX). Farther from the crater, it curves toward the west. This could be the result of a zone of avoidance (owing to the clump of boulders [arrow YY] on the crater rim?) or a deposit of darker material.

The crater shown in Plate 77c is approximately 0.42 km in diameter, and, although it displays multiple terraces, the overall plan is much less irregular than that of the crater shown in Plate 77b. In addition, the ejected blocks appear to be smaller on the average. The fact that both dark and light rays extend radially from the crater may be due to differences in particle size or composition. Farther from the crater, one can note radial lineations (arrow X), which result from secondaries ejected at low angles. This crater is north of the crescent-shaped ridge or ghost crater in Oceanus Procellarum to the southeast of Flamsteed.

The rim region of the 0.61-km-diameter crater seen in Plate 77d is more complex than the three preceding examples. From the rim, there is a blocky zone (arrow X), a hummocky or dune-like zone (arrow Y), and a zone characterized by isolated worm-like lineations (arrow Z). Such ordering is typically displayed by larger craters, and this example represents a transition. The origin of this ejecta sequence is discussed with Plates 80–83.

PLATE 77

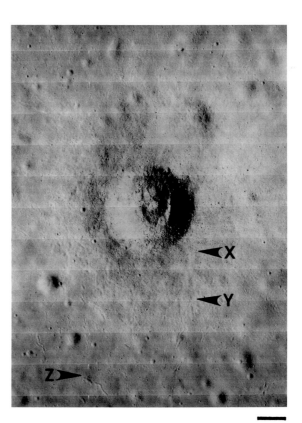

a. III-200-H1   0.23 km

b. III-189-H2   0.23 km

c. II-177-H2   0.24 km

d. II-171-H2   0.23 km

Subdued Forms

In contrast to the foregoing examples, the craters shown in Plate 78 are very subdued but exhibit interesting anomalies. Plate 78a includes a small crater pair about 75 km to the west of the large crater Reinhold. The northern crater displays numerous large blocks around the rim, whereas the southern member displays only a few. Both craters, however, appear to have similar rim profiles.

If the subdued southern crater was once surrounded by a similar blocky blanket, then this pair illustrates the rapid destruction of this type of blanket relative to the rim profile. Their proximity is consistent with this initial assumption, but the destruction of 20-m blocks with relatively minor alteration of the profile seems anomalous.

Alternatively, the craters formed differently, which resulted in different types of ejecta blankets. Important in this respect is the influence of, or relation to, the ejecta blankets of Copernicus and Reinhold. As shown in Plates 97 and 100, groups of secondaries typically display a nonblocky and subdued ejecta blanket. The southern component of this pair could be an isolated member of a group of secondary impacts which occur nearby (not shown). The subdued crater also may represent blanketing by ejecta from either Copernicus or Reinhold prior to the formation of the blocky crater.

The crater shown in Plate 78b is 1.5 km west of the craters shown in Plate 78a. It exhibits an irregular plan, relatively well defined rim-wall transition, and asymmetric distribution of blocky ejecta, found either in clusters (arrow X) or isolated (arrow Y).

The preservation of a few blocks on the outer slopes of an otherwise subdued rim could result from transfer of the finer-size particles of the regolith. This transfer may be due to sand-blasting by ejecta from later impacts. Preservation of certain blocks also could result from the different erosional properties of different rock types, such as crystalline rock versus breccia.

Meteoroid bombardment (Shoemaker et al., 1970) and emplacement of solar wind particles (Lunar Sample Preliminary Examination Team, Apollo 14, 1971) appear to contribute to the destruction of blocks exposed at the surface. Only the blocks most durable to these processes might survive. There is evidence that thermal fluctuations are not important weathering agents on Earth (Blackwelder, 1933; Griggs, 1936) and Moon (Ryan, 1962), but Cloud et al. (1970) find possible evidence for this process in close examination of Apollo 11 samples.

Approximately 5 km to the north of the craters shown in Plate 78a is the isolated rimless but pristine crater seen in Plate 78c. If this crater is compared to craters seen in Plate 76, then the absence of a rim and ejecta blanket indicates that it is morphologically subdued; however, the wall-rim transition is abrupt and completely unlike the subdued crater seen in the lower right in Plate 78c. The rimless crater is interpreted as an internally formed pit and is evidence that relatively recent endogenetic activity was a local formative process. This origin is also suggested by the presence of nearby larger structures, such as low-relief domes, which are believed to be volcanic in origin. An endogenetic origin could account for the irregular plan of the crater shown in Plate 78b, as well as for its limited but preserved ejecta blanket.

Plate 78d shows a crater with a broad rim, revealed by the concentric tree-bark pattern. The boundary of the rim appears lobate and can be traced (in the original photograph) into the subdued crater to the north (arrow X). This crater is near a wrinkle ridge (shown in Plate 141) in Sinus Medii and is probably genetically related.

Rimless pits and craters with broad subdued rims commonly occur on the once-molten floors of large craters. For example, note the pits within Copernicus (Plates 226a, 228, a and b, and 229a).

PLATE 78

a. II-161-H2   0.18 km

b. II-161-H2   0.18 km

c. II-161-H2   0.18 km

d. II-106-H2   0.17 km

Subdued Forms

Plate 79a shows a very subdued crater (arrow X) on a broad dome or rim, which exhibits both tree-bark and grid patterns. The presence of numerous blocks on the rim seems inconsistent with the over-all degraded appearance of the rim area. The region shown is about 1.5 km to the southwest of part of the Flamsteed Ring (see left center, Plate 154, b and c) and several nearby craters resemble the crater seen in Plate 76a.

The crater shown in Plate 79a is on an extension of a wrinkle ridge (a positive-relief form illustrated in Plates 136–150). Such ridges are, in part, volcanically constructed and should be plausible sites for other volcanic forms (see Plate 78d). If this crater formed volcanically as a maarlike eruption with collapse, then the absence of a raised lip but the presence of well-preserved blocks could be rationalized. Other craters in the area (not included in Plate 79) also show evidence for internal origin. Several are similarly subdued but have relatively well defined wall-rim transitions. Alternatively, the crater shown in Plate 79a

represents a relatively recent impact into a thick layer of unconsolidated material overlain by a consolidated flow unit.

Plate 79b illustrates the association of very subdued craters (arrows X, Y, Z, and XX) with a wrinkle ridge to the south of Arago in Mare Tranquilitatis. The association is revealed not only by placement but also by the continuation of surface texturing onto the crater rims. The doughnutlike rim of the crater marked by arrow Y is characteristic of craters on the highland plains-forming units (see Plates 51 and 200) and several zones along the mare-highland contact. Larger examples are shown in Plate 86.

As considered in the discussion accompanying Plate 78, the craters shown in Plate 79b could be highly eroded impact craters, but I prefer an internal origin associated with the construction of the ridge. Cameron and Coyle (1971) have reached similar conclusions in their study of the size distribution of blocks around selected craters.

PLATE 79

b. II-047-H3   0.21 km

a. III-182-H3   0.25 km

Pristine Forms

The ejecta sequence of numerous larger craters becomes more complex but displays an ordering of characteristic morphologic features outlined below.

1. Zone I (extending outward from the rim-wall boundary for a distance of approximately 0.2–0.5 of the crater radius):
   a. Boulders, boulder fields, and boulder streams strewn over apparently unconsolidated debris
   b. Minor annular fracturing or lineations
   c. Radial troughs, with possible debris flows on the wall
2. Zone II (extending beyond the end of Zone I for an additional crater radius):
   a. Radial and subradial lineations
   b. Ill-defined corrugate and lavalike units
   c. Few identifiable leveed or smooth-surfaced flow units.
   d. Relatively subdued but rolling topography
   e. Concentric ripplelike (transverse dunelike) and crescentic (barchanlike) forms
   f. Boulders and boulder associations
3. Zone III (extending beyond approximately two crater radii from the crater rim):
   a. Herringbone craters with horns extending away from the main crater (also in clusters and chains)
   b. Rillelike features that possess rims and may exhibit unusual V-shaped or W-shaped plans
   c. Threadlike radial-trending (in general) lineations that "smear" terrain
   d. Bright rays

Mösting C (about 3.5 km in diameter) clearly illustrates the zones here described and is shown in the oblique views of Plate 80, a and b. The inner rim zone, Zone I (arrow X), displays numerous blocks and block associations, most of which appear to be partially buried or destroyed (Plate 80c, arrow X). The block clusters, in several cases, may have been large projectiles that broke apart at impact. The absence of trails suggests that the blocks along the inner rim zone were ejected with low velocities and came to rest before burial by a "shower" of finer debris. Blocks can also be identified at greater distances.

Plate 80c reveals the subdued ridgelike features (arrow Y) that are characteristic of Zone II and encircle the block-strewn raised rim of Zone I. These merge with a complex terrain that exhibits numerous radial lineations and approximately concentric ripplelike ridges (arrow Z). At greater distances from the crater (note Plate 80, a and b), the ripple ridges commonly show adjacent negative relief.

The features in Zone II are perhaps the result of the base surge, the term applied to surface density flows from underground nuclear explosions (Young, 1965; Roberts, 1968). The ripple ridges and overlapping crescentlike forms might be depositional features from this quickly moving cloud of hot gas and particles (Moore, 1967; Masursky, 1968). The Zone I blanket appears to overlie these features. At greater distances from Mösting C, Zone II forms clearly have eroded the pre-existing surface and appear to be separated from the bulk ejecta deposit. Note the scarring of the crater-facing slopes of the hills seen in Plate 80, a and b, and the crossing of the crater at left (arrow Y). Clearly, more than simple deposition has occurred. In the original photographs the lineations can be seen to continue in and out of craters, which is consistent with the interpretation of a density flow.

PLATE 80

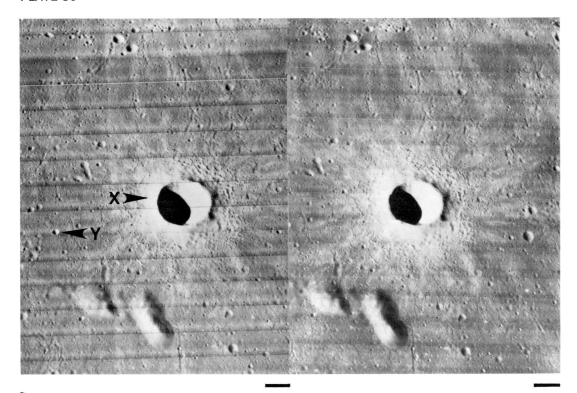

a. III-114-M 2.5 km (top) 1.3 km (bottom)   b. III-113-M 2.5 km (top) 1.3 km (bottom)

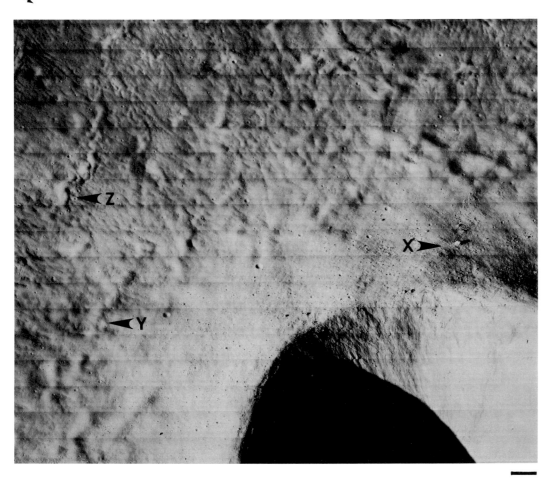

c. III-113-H2   0.23 km

Pristine Forms

Censorinus is the 4.5-km crater shown near the center of Plate 81a (arrow X, oblique view) west of a larger conelike crater, Censorinus A. Although relatively small, Censorinus is easily identified telescopically by its extensive bright-ray system. It illustrates the effect of pre-existing topography on ejecta (Plate 82) and the nature of the bright rays (Plate 83).

The area shown in Plate 81b (an enlargement from Plate 82, arrow Z) is west-northwest of Censorinus on the rim of Censorinus A. Four blocks, and possibly six (see arrow X), can be identified with arcuate trails and associated head craters (arrow Y). If the boulders were ejected from Censorinus with a high trajectory, then they must have struck an ENE-sloping surface after the bulk of the ejecta blanket had been deposited. Bouncing out of their impressions, they encountered a change in slope (facing northwest) produced by the rim of Censorinus A. However, the regional shadowing is not consistent with an ENE-facing slope of sufficient angle to encourage rolling boulders, some of which barely display a "head" crater. Perhaps the projectiles were heaved from a nearby crater other than Censorinus, but no crater can be found that is close enough to produce the low velocities necessary for their survival upon impact.

A third hypothesis is that the projectiles were ejected from Censorinus at low trajectories with relatively low velocities and impacted on a NW-facing slope. Upon impact, the boulders rolled and skipped uphill under the momentum preserved by the low angle of incidence. The blocks clearly postdate the striated Zone II ejecta blanket. This time sequence is also suggested by the blocks and finer debris in Zone I (see Plate 82). Consequently, the inner ejecta probably represent low-velocity material, and the blocks seen in Plate 81b are thought to be outlying members of this inner blanket of material. Unless the formation of Censorinus was multiphased, such a depositional sequence illustrates the rapidity with which the outer ejecta blanket was emplaced. This sequence also raises a question concerning the analogy between an ejecta cloud enveloping a lunar impact and the base surge from a nuclear explosion. If these blocks represent ejecta from Censorinus, then they impacted after the passage of the proposed base surge. Such a sequence suggests that the blocks were suspended until this time, a situation that does not seem realistic because of the inferred low-to-moderate angles of impact (less than 50° from the horizontal).

PLATE 81

b. V-063-H2   0.50 km

a. V-063-M   4.7 km (top)   2.9 km (bottom)

Pristine Forms

The effect of a nearby crater, Censorinus A (Plate 82, top) produces interesting modifications on the various ejecta types of Censorinus (Plate 82, bottom). Transverse ridges (arrow X) appear to be more common (or perhaps better delineated) on the facing rim of Censorinus A, and radial lineations appear to veer away from its northwestern rim (arrow Y). The absence of radial lineations immediately to the east of Censorinus probably is a lighting effect, as both lighting and lineations are directed east-west. The eastern wall of Censorinus A was the target of some of the ejecta, but shadows prevent a more complete description. The western wall of Censorinus A is overexposed, but there are narrow threadlike patterns of low-albedo material that extend toward the floor. These patterns probably represent mass movement after the formation of Censorinus.

As noted for Mösting C (Plate 80), numerous large blocks (up to 70 m across) are scattered over the inner rim, indicating a late stage in ejecta deposition. Arrow Z locates the region shown in Plate 81b where several impacting secondaries left trails. Radial lineations in Zone II commonly appear to be focused within small (1 km in diameter) subdued craters (arrow XX). In one example (arrow YY), a flowlike feature appears to have accumulated in a fanlike form on the Censorinus-facing wall. This feature illustrates the bulk nature of portions of the Zone II blanket, in contrast to the ejecta blanket of craters smaller than 1 km in diameter. The focusing of radial lineations, the veering away from the rim of Censorinus A, and the flowlike feature indicate topographic control of the ejecta and, therefore, a ground phenomenon.

Although these observations seem consistent with the base-surge analogy, they are also consistent with a model of crater formation based on laboratory impact studies made by Gault (personal communication, 1973). High-speed motion pictures reveal angles of ejection near 45°, and the ejecta are contained within the sheetlike surface of an inverted cone. As the crater enlarges, the conelike sheet of ejecta enlarges, but the angle of ejection remains approximately the same. An ejecta plume and the base surge are not developed. At later stages in crater formation, the excavated debris is ejected at rapidly decreasing velocities and therefore closer to the rim.

By analogy with these small-scale cratering experiments, the bulk ejecta near the rim (Zone II) of a lunar crater impact the surface, and the resulting debris from these impacts and the additional incoming secondary material rapidly travel downrange as a massive ejecta cloud. Thus, the density flow is envisioned as a result of ballistic ejecta and their accompanying ejecta. This process is unlike the formation of the base surge for nuclear and explosive volcanic eruptions as suggested by Moore (1967). In this analogy, the base surge develops from an outward expansion of gases and ejecta rushing over the upturned crater lip. Both models result in flowlike ejecta clouds, but the base surge seems to require a longer development time, which may not be consistent with the blocks shown in Plate 81b. It is entirely possible, however, that small-scale cratering cannot be extrapolated to craters that are four orders of magnitude larger in diameter.

PLATE 82

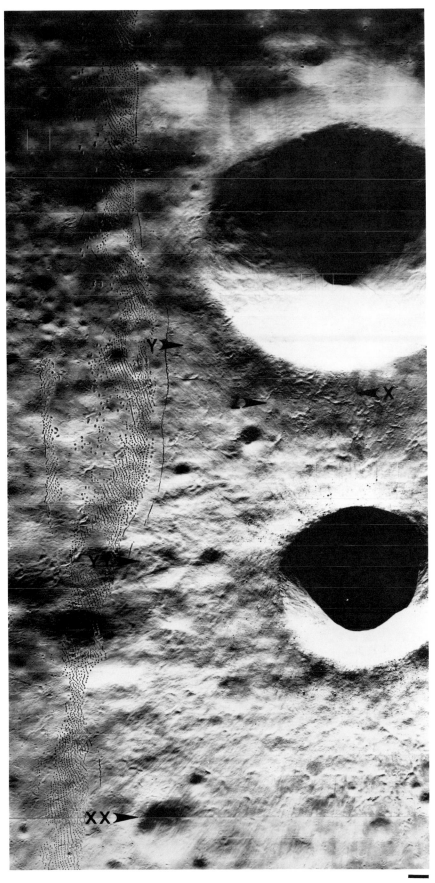

V-063-H1,H2   0.53 km (*top*)   0.44 km (*bottom*)

Pristine Forms

Plate 83 shows the area west of Censorinus and the Zones II and III type of ejecta. The association of the "smearing" effect and lineations of Zone III with the bright-ray material is quite apparent (arrow X). In some areas, the "smearing" becomes so extensive that the photograph appears to be out of focus, but close inspection reveals many small, sharply defined craters.

Two general types of ejecta features are recognized in Plate 83. The first is the high-albedo streamerlike ray pattern (arrow Y) that commonly has an ill-defined leading edge uprange. Highly subdued craters and parallel lineations occur within such rays. The second type includes high-albedo patches that exhibit elongate or irregular plans in contrast to the long raylike patterns. Several of these patches display a well-defined leading border characterized by highly subdued craters (arrow Z) or groovelike features (arrows XX, YY) that typically are arranged concentric to the crater. Such a leading edge, however, does not always occur (arrow ZZ). The surface within these patches is striated or smooth and may contain highly subdued craters downrange from the leading boundary. The subdued morphology associated with these patches and rays is not the result of long-term surface processes, owing to the recent formation of Censorinus. Rather, the subdued appearance is thought to be the result of relatively diffuse secondary ejecta, interactions of the secondary impact with an encompassing "cloud" of secondary debris, and/or interaction of ejecta from adjacent secondary impacts. These events should produce turbulent "tertiary" ejecta that blanket the newly formed secondary crater complex.

The smeared patches and rays indicate zones of extensive alteration of the original surface. The mechanisms responsible for such alteration are impacts by turbulent or unconsolidated debris from Censorinus and downrange tertiary ejecta from secondary impacts. The patches exhibiting well-defined uprange borders support the latter mechanism, whereas patches without such borders are consistent with the former mechanism.

Oberbeck (1971) suggested that the bright rays of large craters, such as Copernicus (Plate 94), are products of well-defined secondary craters excavating subregolith material. This is most likely one contributing process. I do not believe, however, that it is generally responsible for the high-albedo rays and patches seen in Plate 83, because the boundary of such regions commonly corresponds to the boundaries of the striated or smeared terrain and not to clusters of well-preserved bright-haloed craters. If the subregolith material is responsible for the local high albedo, then I believe it was excavated by both the impacts represented by subdued craters and the scouring action of the regolith by secondary and tertiary ejecta clouds.

There are at least two other contributions to the albedo of secondary impact regions. First, this albedo may correspond to blocky material ejected from the primary crater, in this case Censorinus. This is obviously demonstrated by the high-albedo inner rim and is inferred from dark-ray systems of large craters (Plate 89). Finally, it is convincingly shown by Apollo 16 photographs of blocky ray material in the lunar highlands (Muehlberger et al., 1972).

A second plausible contribution to the albedo is the alteration of the physical structure of the surface. Specifically, the scouring of the regolith, whether or not it is penetrated, may increase the albedo through a change in surface roughness. This is suggested by the bright halo that surrounds the lunar landing sites. Hinners and El-Baz (1972) speculate that this increase in albedo relates to a decrease in the surface porosity produced by compaction from the dynamic pressure of the descent-engine exhaust. This speculation is based on photometric models by Hapke (1963). However, such compaction should occur directly beneath the descent module. As the module descends, the area surrounding the direct exhaust will experience considerable horizontal flow, as illustrated in films of the lunar landings. The surface will be altered, but compaction does not seem to be the appropriate modification for the broad area surrounding the module, which is shown in the orbital photography. Rather than an increase in surface compaction by dynamic pressure, the decrease in porosity may be performed by a smoothing of the fragile microsurface, owing to windlike erosion. This may be analogous to an impact by a tenuous cloud of secondary ejecta. Such an inferred erosive cloud is consistent with theoretical calculations by Rehfuss (1972). His work suggests that impacts greater than 20 km/sec produce considerable vaporization, which results in a rapidly expanding explosive cloud extending to eleven crater radii.

PLATE 83

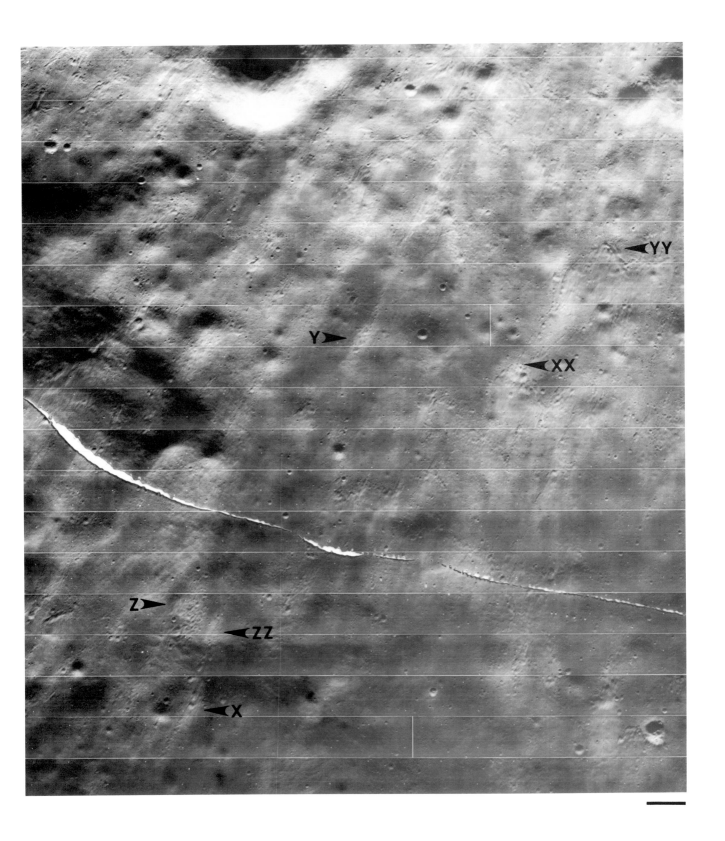

V-063-H3   0.45 km

Subdued Forms

Secchi A (Plate 84, *a* and *b*, arrow X) is 5 km in diameter and is surrounded by a complex ejecta blanket that might be the result of intermixing with the blanket from the crater to the north. Alternatively, it reflects regional processes associated with the subdued linear depressions that cross these craters (arrow Y).

The higher-resolution photograph (Plate 84c) shows that the surface is relatively smooth,. except along the rim of Secchi A. Numerous blocks are clearly preserved, but many appear to rest on mounds of particulate debris, which were presumably created by the destruction of the blocks (arrow X). Smaller blocks also occur at greater distances from the rim crest and along the inner and outer rims of several dimple craters (arrows Y and Z; identified in original photograph).

If the ejecta characteristics of Secchi A, Censorinus, and Mösting C were once similar (perhaps an unjustified assumption), then the outer rim zone has been smoothed more conspicuously than the inner zone. The ropy Zone II deposits were largely destroyed, whereas the blocks within Zone I remain. This indicates more rapid smoothing of the hummocks of unconsolidated debris than the *in situ* destruction of blocks.

Such differential smoothing also may be inferred by the numerous subdued craters that presumably formed after Secchi A. If the smoothing process was solely due to the redistribution of material by small meteoroid impacts, then it is intuitively disturbing that relatively small blocks could remain. This possible paradox may be the result of a high density of endogenetic craters, impacts into a highly unconsolidated target, or an efficient redistribution process other than small meteoroid ejecta (such as creep). Note in the lower-resolution photographs (Plate 84, *a* and *b*) that the subdued dimple craters appear sharp rimmed and recent in contrast to the view of Plate 84c.

These observations might be evidence for an erroneous assumption that the ejecta blanket of Secchi A was once similar to that of Mösting C or Censorinus. The reason some craters display an extensive Zone II deposit, whereas others do not, is not just the effect of differences in degradation states. It is also dependent on the initial velocity of the projectile (primaries versus secondaries), physical structure of the target, and crater origin in general.

PLATE 84

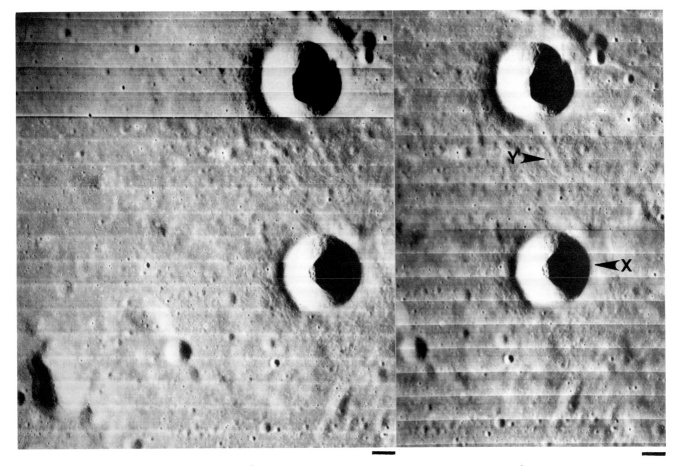

a. II-022-M   1.6 km

b. II-024-M   1.6 km

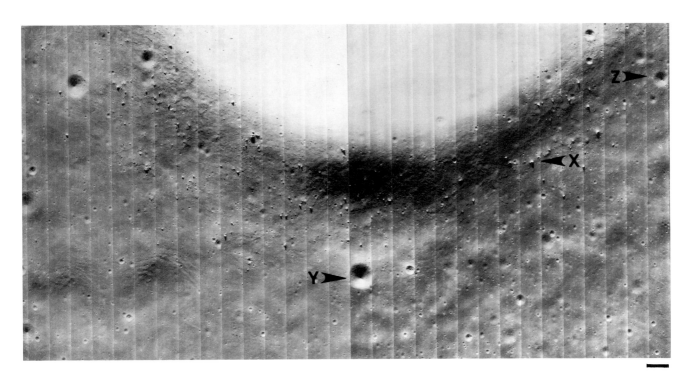

c. II-024-H3,H2   0.21 km

Miscellaneous Forms

Plate 85 illustrates variations in the appearance of the ejecta blanket and rim profiles presented in the previous five plates. The high-albedo crater on the inner Imbrium ejecta sequence, shown in Plate 85a, is west of the Sulpicius Gallus rilles (bottom left of Plate 166, a and b). In contrast to the ejecta of Censorinus (Plate 81a) and Mösting C (Plate 80), the high-albedo outer ejecta are not organized into filamentary rays, and the crater appears subdued, i.e., without sharp breakaway scarps along the wall-rim transition. Both characteristics are also displayed by other nearby craters. Such craters appear to be pristine with respect to the preservation of their ejecta but subdued with respect to their profiles. This may not be a contradiction, however, but an indication of a highly unconsolidated target, which results in a subdued profile. Such a target also might be responsible for the absence of well-defined ray patterns through either the type of ejecta or relatively rapid destruction by mass movement, which subdues any sharp contacts between ray material and adjacent surfaces.

As discussed with Plate 78, a and b, small-scale craters may have extremely subdued over-all appearances yet display numerous block along the outer rim. It was suggested that such an anomaly could result from incompetent surface layers, differences in erosional properties, or nonimpact origin. Plate 85b presents a larger example on the Flamsteed Ring (see upper center of Plate 156, a and b, and Plate 157). The very subdued topography and funnellike profile seem inconsistent with the large blocks on the rim (arrows X and Y) and its high-albedo halo (shown in other photographs of the same region). If the crater was the result of an impact, then it illustrates the very rapid degradation by creep (in this instance) over meteoroid bombardment. This interpretation is consonant with the preservation of the bright rim deposits over the adjacent mare. It is also plausible that the crater was endogenetically formed.

Alpes A, the crater shown in Plate 85c, is on the northern edge of Mare Imbrium within the mountain range Montes Alpes. In contrast to the rim area of the two preceding examples, its rim area displays a concentration of small craters. These might be a secondary field from another crater, but it is fortuitous that they clustered so nicely around Alpes A. If these are secondaries related to Alpes A, then they were ejected at high angles. Alternatively, Alpes A might be volcanic in origin, and the nearby craters were the result of low-velocity volcanic bombs. The crater shows an asymmetric and irregular low-albedo halo. This might be used as evidence for internal origin, which would be consistent with other nearby endogenetically produced forms (see Plates 179–180).

Plate 85d shows an area on the western rim of Alphonsus (Plate 186, b and c) and is representative of the numerous small subdued highland craters that display poorly defined rim sequences or none at all. The rim profile is difficult to discern, and typically there is only a small gently raised rim. Such craters appear as pits whose rims cross a variety of topography. Just as typically, the rim may be bulbous in profile. Both rims could be envisioned as degraded impact craters. The first form could result from enlargement of the crater by slope decline of a screelike wall. The second could represent the degraded state of the type of crater shown in Plate 85a.

The crater Kies A, Plate 85e, is approximately 15 km in diameter and displays an unusual distribution of ejecta. The hummocky, striated, and highly reflective Zone II region is generally absent to the west, approximately corresponding to the region subtended by the hornlike ridges (arrow X). As shown in Plates 96 and 116b, ridges commonly occur with probable simultaneous impacts, a phenomenon described and reproduced in the laboratory by Oberbeck and Morrison (1974). For secondary craters surrounding large impact craters, this arrangement is termed the "herringbone" pattern, but typically the horns subtend the secondary crater. The studies by Oberbeck and Morrison demonstrated, however, that the pattern shown in Plate 85e also can be produced by superposed and nearly simultaneous impacts. In this case, either a small impact occurred simultaneously but slightly to the west of the main impact, or two slightly nonsimultaneous impacts of equal size occurred, the first impact being to the east. The crater Cauchy in Mare Tranquilitatis (IV-073-H2) shows a very similar ejecta distribution and perhaps reflects a similar origin.

PLATE 85

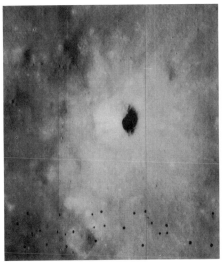

a. V-090-M   3.9 km

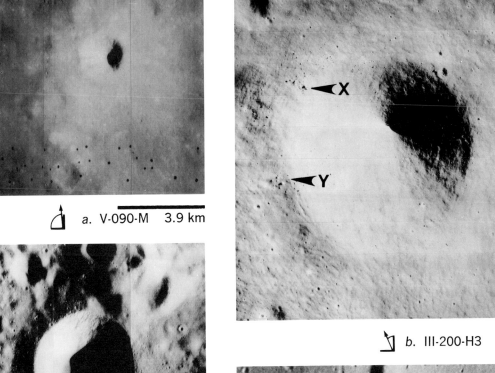

b. III-200-H3   0.23 km

c. V-132-M   8.1 km

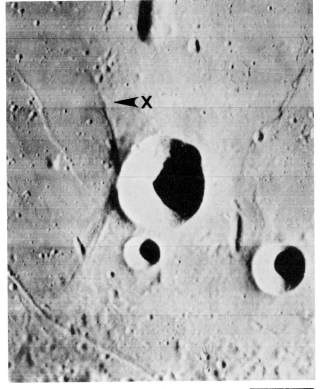

e. IV-125-H1   12 km

d. V-116-M   4.1 km

Subdued Forms

The crater rims shown in Plate 86 are different from those in the preceding examples in that they have broad bulbous profiles and highly textured surfaces.

The craters shown in Plate 86, a and b, are on the floor of Gassendi. The first crater can be found also to the far right of Plate 36, b and c (arrow X), adjacent to the moat of mare material. It is about 1.5 km in diameter and appears to be very subdued. The surface texturing is a combination of the tree-bark and grid patterns, which generally are extensive on the floor of Gassendi. The transition from the rim to the wall is not well marked. No blocks are identifiable, on either the rim or the wall.

A similar description applies to the crater seen in Plate 86b (arrow X; identified in Plate 36, b and c, arrow Y) except that here the rim is less pronounced. In addition, there are two narrow ropelike extensions to the south (arrows Y and Z). This feature is east-northeast of the first example, adjacent to the wall of Gassendi.

The floor of Gassendi and similar craters with fractured floors abounds with small craters open to similar descriptions. These could be viewed as degradational products of impacts, but they also may represent relatively pristine volcanic structures. If the ropelike features from the crater shown in Plate 86b are extrusive units, then their preservation requires a similar preservation for the crater. A close-up of a comparable feature is shown in Plate 251a on the floor of Vitello. It is possible, however, that these ropelike forms are compressional ridges related to relatively recent differential movement. This interpretation is inferred from their common occurrence on the floors of larger floor rilles (Plate 38c, arrow Y) and their apparent disregard of local topography.

The feature shown in Plate 86a also might be interpreted as a tephra cone, which is composed of pyroclastics and perhaps particularly vulnerable to lunartype "weathering" (also see Plate 130d). As is discussed in the chapter on crater floors, there is other evidence to suggest relatively recent volcanic resurgence on the floors of floor-fractured craters, and, thus, the interpreta-

tion of these examples as volcanic forms is reasonable. Consequently, the interpretation of similar craters outside such domains as volcanic forms seems plausible but becomes less convincing.

The area shown in the stereo pair in Plate 86, c and d, is in eastern Mare Tranquilitatis to the south of Maskelyne F. The region exhibits irregular platforms, small domes (500 m in diameter) surrounded by moats, ghost rings, irregular depressions, and ridges (see Plate 124, a and b). In addition, there are indications of regional blanketing (possibly volcanic), revealed by flat-floored craters (Plate 15, b and c). The textured crater rim displays a bulbous profile similar to those of the preceding examples. In contrast, however, the innermost rim exhibits a narrow depression, or "moat," concentric to the wall-rim transition (arrow X), and the wall itself is screelike. Such a profile with a concentric moat near the crater rim crest is not unique to this region, and similar features are found in the Littrow Rille region (Plate 140a) and within the Flamsteed Ring along wrinkle ridges (III-177-MR). Their locations near recent endogenetic features suggest either an endogenetic origin for these craters or later modifications associated with this activity. For example, the moats may be analogous to those surrounding domes (Plate 128), where they appear to be produced by previous lava levels. The subdued irregular depression (arrow Y) may be a collapse feature related to this local inundation by lava flows. Under this interpretation, the funnel-shaped crater (arrow Z) overlapping the rim requires considerable erosion of the smaller crater, provided it is impact produced, yet preservation of the rim moat. Another possibility is that the rim moat was produced by a previous lava level within the crater, in analogy with terrestrial volcanic cones. The occurrence of such moats on the rims of nearby craters of various sizes, however, is consistent with interpretations involving regional stratigraphy. If a competent bedrock is heavily blanketed by an incompetent unit, such as ash, then the moatlike ring might develop through mass wasting of this overlay down the outer crater rim.

PLATE 86

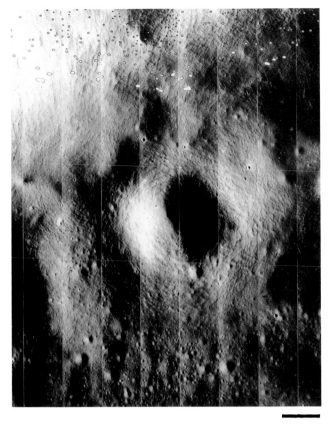

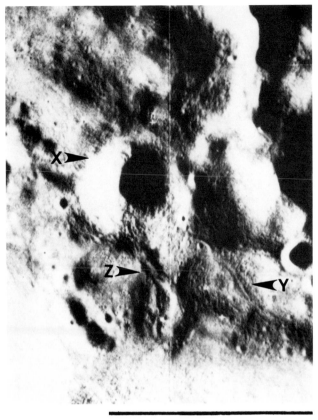

a. V-177-H2   0.55 km

b. V-177-M   4.1 km

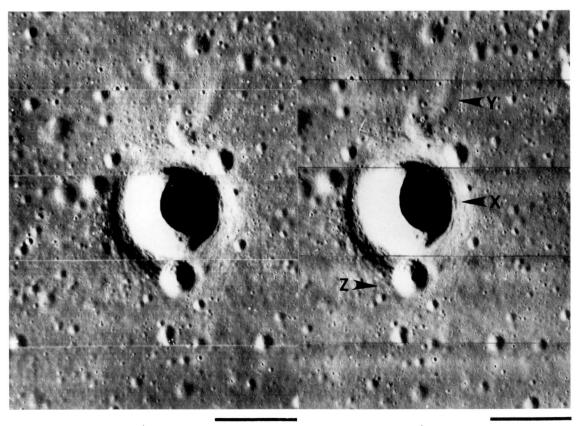

c. III-009-M   1.8 km

d. III-013-M   1.8 km

Subdued Forms

Plate 87 shows two craterforms northeast of the crater Vitello (see Plate 249). The crater seen in Plate 87*a* (arrow X) is subdued and displays a low rim. It is in a complex terrain with numerous well-preserved irregular scarps that are interpreted as flow termini, boundaries of subsided lakes, or collapse structures. The last interpretation opens a possibility of formation considerably postdating mare emplacement. But the breadth of several of these flat-floored depressions suggests roof support insufficient to prevent collapse relatively soon after the formation of a shallow subsurface cavity. In addition, the terrace within the crater (arrow Y) resembles a high-level lava ring, analogous to terraces in terrestrial calderas. I prefer the interpretation that this complex terrain was produced by a combination of flow termini and collapse following lava withdrawal. Both mechanisms indicate that these features have been preserved since

the time of mare emplacement. Consequently, the subdued crater (arrow X) must be considered a relatively well preserved volcanic form and not an impact crater subjected to extensive meteoroid bombardment. Otherwise, the much smaller topographic transitions associated with the terrain, on which this crater occurs, should have been destroyed.

The crater shown in Plate 87*b* (arrow X) is 20 km southwest of the area shown in Plate 87*a* and is adjacent to a rectilinear rille (see Plate 178*b*). It has an elongate plan and a crested ridgelike rim profile. Several larger craters display a similar rim profile (Plates 91—92). I suggest that this crater also was formed volcanically through eruptions of tephra with subsequent collapse. Its placement adjacent to a rille that is probably structurally formed strengthens such an interpretation (see Plate 119).

PLATE 87

a. V-168-M   5.6 km

b. V-168-M   5.7 km

Leveed Channels

Plate 88*a* shows a crater (arrow X) 13 km in diameter on the rim of the large (75 km in diameter) crater O'Day, which is on the rim of Mare Ingenii (Plate 75*a*). The smaller crater (Plate 88*a*, arrow X) is enlarged in Plate 88*b*. Its rim displays an ejecta blanket (arrow X) and at least four sets of double ridges, interpreted as constructional flow levees (arrows Y, Z, XX, and YY). Such well-developed features are common on the rims of much larger craters, such as Tycho (see discussion accompanying Plate 104), but are not so typical for craters smaller than 20 km in diameter (but see Plate 22*b* and Plate 118, *a* and *b*).

The flows appear to be directed by the local surface slope and are derived from the crater interior. In particular, note the contiguous crater chain connected with the channels (arrow ZZ). If these flows were emplaced at the time of an explosive crater-forming event, then they would have been directed radially from the rim crest. Their directional dependence on surface slopes suggests later formation. There are at least three plausible sources for these rim flows. First, this crater represents a volcano, and the channels were constructed from molten material that spilled over the crater rim. The location of the crater on the rim of a larger crater prompts comparison to parasitic eruptions typical of terrestrial calderas. Second, the crater shown in Plate 88*b* was produced by an impact that tapped a pre-existing magma chamber. Third and most likely, it is an impact crater, and the low crater rim permitted impact melt to escape after crater formation. The last suggestion requires that a significant portion of the impact-generated melt was not ejected during the early stages of crater formation. Further examples and discussion of these possibilities accompany Plate 105.

PLATE 88

a. II-075-H3   (oblique)

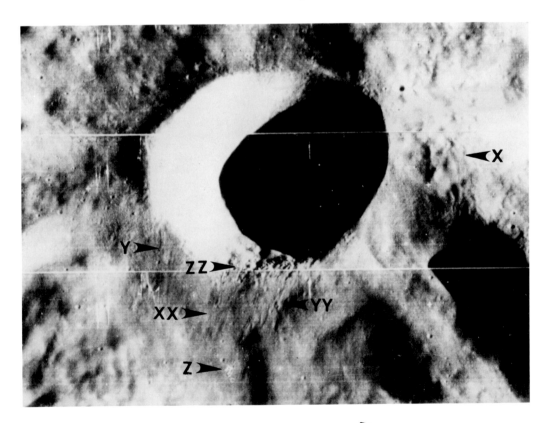

b. II-075-H3   (oblique)

Pristine Forms

Dionysius (Plate 89; 18 km in diameter and 2.5 km in depth) is considered in the chapter "Crater Floors" (Plate 26), but here the attention is focused on the rim, which typifies "fresh" craters having a diameter of approximately 20 km. The stereo pair (Plate 89, *a* and *b*) clearly shows the raised rim, which rises between 0.3 km and 0.4 km above adjacent terrain (Apollo Intermediate Chart [AIC], 60D). Small concentric scarps immediately adjacent to the rim crest and an annular corrugate pattern farther downslope on the rim characterize the Zone I region (arrow X). In addition, this inner rim zone exhibits numerous blocks and radial grooves (arrow Y). Zone II is generally similar to Mösting C (Plate 80), with complex radial patterns and wormlike ridges approximately concentric to the parent crater. Dark mottled regions occur in this zone (arrow Z) and extend into dark filamentary rays. The existence of a dark-ray system suggests that the albedo is derived, at least in this example, from the primary crater rather than from numerous secondary impacts that excavate the subregolith, as proposed by Oberbeck (1971) for the rays of Copernicus. This dark-ray system is expressed more clearly under full solar illumination.

Plate 89c is a high-resolution view of the northeast rim (rotated clockwise 90° from Plate 89, *a* and *b*). Most of the blocks appear to be partly buried by debris, suggesting active erosional processes. Alternatively, the emplacement of the boulders was followed by deposition of finer ejecta (perhaps a sorting process due to released or generated gases). Despite the rather well preserved appearance given by the stereo view, at high resolution the ejecta blanket appears to be subdued and generously spattered with small subdued craters, which are probably secondary craters from Dionysius or drainage pits of the unconsolidated ejecta blanket into subsurface cavities.

PLATE 89

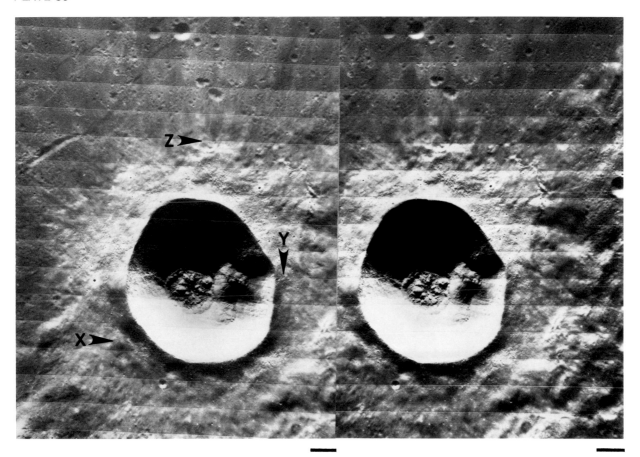

a. V-083-M   3.2 km   b. V-081-M   3.2 km

c. V-082-H3   0.41 km

Subdued Forms

Just as Secchi A can be contrasted to Mösting C and Censorinus, so can Gambart C (Plate 90) be contrasted, at a slightly larger scale, with Dionysius (Plate 89). Plate 90a provides an index for the high-resolution view, and Plate 59a should be inspected for more detail at medium resolution.

Gambart C is in a hummocky surrounding that is partly its own ejecta blanket and partly that of secondaries from Copernicus. The Copernicus ejecta, however, are not always separable, for nonradial curvilinear lineations commonly comprise the outer rim component for medium-size craters. Less than 20 km to the northwest of Gambart C is an arc of secondary craters from Copernicus. As illustrated in Plate 100, such craters typically have extensive lineations directed away from the parent crater, and Gambart C rests in this zone. Several characteristic Copernicus secondaries are to the north of Gambart C (Plate 90a, arrows X and Y). The regional importance of Copernicus is also inferred by telescopic views of this area under full illumination, which reveals its bright-rayed deposits. Under the same lighting conditions, Gambart C is a bright-walled crater without a bright halo. A low-albedo region extends to the southeast, but it is difficult to assign this any importance since similar zones are encountered elsewhere within the Copernicus ejecta sequence.

In contrast to the rim of Dionysius, the inner rim has a tree-bark pattern (Plate 90c, arrow X). In addition, the south and northwest rims are crossed by myriads of narrow NW-trending grooves (5m–10 m widths) that are identified in Plate 90a (arrow Z), 90b (arrow X), and 90c (arrow Y). A minor NE trend is also present (arrow Z). The lineations northeast of Gambart C (Plate 90c, arrow XX) might be assigned to either the NW-trending grooves or a system concentric with Gambart C. Extensive blanketing by Copernicus perhaps covered or erased the inner-rim sequence of Gambart C and altered its outer-rim features. The NW lineations align with Copernicus ejecta, but the NE trend does not.

The origin of the NW and NE systems could be related to the lunar grid. But in this example the NW system is clearly not a lighting phenomenon, because the grooves can be mapped continuously onto the wall of Gambart C (Plate 90b, arrows Y and Z) and have related elongate craters (arrow XX). Note also Plate 59b, which shows the wall of Gambart C in detail and reveals complex patterns produced by mass movement of fragmental debris.

If these lineations result from Copernicus ejecta, then they are extremely well preserved in contrast to other ejecta impacts. This is particularly apparent on the wall of Gambart C, where slopes are much greater and should be subject to rapid degradation.

In addition, a blanket from Copernicus sufficient to mask effectively the small-scale ejecta products of Gambart C would seem sufficient to have covered all blocks. Clusters of blocks, however, exist along the rim of Gambart C (Plate 90c, arrow YY) as well as along the rims of smaller subdued craters (not shown). They do not appear to be related to Copernicus secondaries and are not necessarily at abrupt breaks in slope (also see Plate 59b). Perhaps these blocks were exposed through a sorting process from the migration of particulate material downslope or from compaction. Such processes must have been relatively rapid if one attributes a large fraction of the particulate material to be from the Copernicus blanket, but this is inconsistent with the preservation of the NW-trending grooves.

The possibility that Gambart C was formed after Copernicus can be ruled out because there is little evidence for blanketing of the Copernicus ejecta. Consequently, problems in interpretation must focus on the apparent inconsistencies that have been pointed out. As a pre-Copernicus crater, the primary effect of the Copernicus event on Gambart C was seismically induced phenomena that include creep (tree-bark pattern), landslides (inner crater wall), compaction (subdued hummocks), and drainage of particulate material into fractures (grid pattern) and subsurface cavities (subdued craters). Ejecta from nearby Copernicus secondaries apparently had little effect on Gambart C, although there is evidence for subdued lineations (100 m wide) to the north and south of it.

An alternative interpretation is that Gambart C is a secondary product from Copernicus or is the result of a primary impact associated with Copernicus. This suggestion implies that many of the surrounding smaller subdued craters are also secondaries that were blanketed during the formation of Gambart C and the passage of the downrange surge of material from the secondaries to the northwest. The existence of only a few large blocks on the rim of Gambart C remains an enigma unless the subsurface is composed of relatively unconsolidated material.

PLATE 90

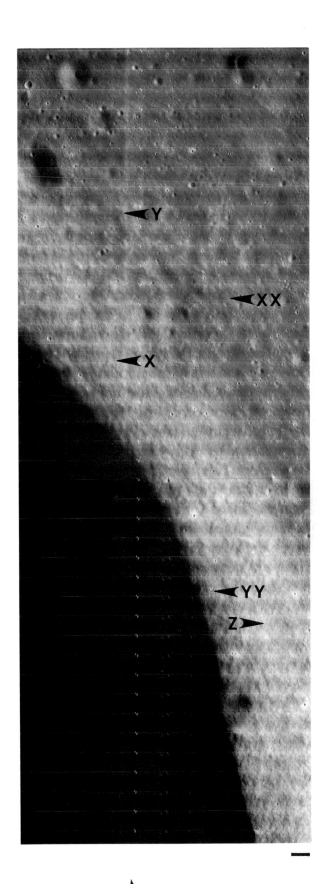

a. II-112-M   1.3 km

b. II-112-M   1.4 km

c. II-112-H3,H2   0.18 km

Ringed Plains

Gambart (Plate 91, *a* and *b*) illustrates a typical form found on the lunar maria and on similar but higher-albedo plains-forming units in the highlands. The crater is characterized by a marelike floor and a sharp-crested rim, which appears symmetric in profile relative to other craters in its size class. The ringlike appearance of the rim is accentuated by the interior marelike floor. The interior wall of the ring typically has a much higher albedo than the outer face (telescopic views). The over-all plan of such rings is commonly scalloped or polygonal and may be breached.

The stereoscopic overview provided by Plate 91, *a* and *b*, reveals that Gambart is in a region that displays crater-topped cones (arrows X, Y), rilles (arrow Z), and small tephra-ringlike features (arrow XX), all of which are interpreted as volcanic in origin. The large number of small flat-floored craters also might be evidence for regional blanketing (see the closer view in Plate 91, *c* and *d*). This blanketing is perhaps due to Copernicus, 300 km to the northwest (see Plate 94). Clusters of secondaries from Copernicus are found within Gambart (Plate 91*c*) and across the hilly terrain to the north (Plate 91*d*). Blanketing also may be due to local extrusions of low-albedo ash, which appears to be characteristic of this region (see Plate 189).

At higher resolution with stereo (Plate 91, *c* and *d*), the marelike floor is recognized as slightly lower in elevation than the exterior mare level. Note the two positive-relief features on the floor. The concentric pattern around the larger feature (arrow X) suggests surges of extrusive activity associated with its formation or with a varying floor level of Gambart.

The rim rises to a maximum 1.1 km above the central floor and about 0.6 km above the adjacent hummocky terrain (AIC 58C). The crater on the northern wall (arrow Y) apparently formed on a cone, and ejecta or extrusives extend onto the interior and exterior surfaces.

The exterior face of the ring exhibits an apron and an interior moat that are most easily recognizable on the northeast base (arrow Z). A possibly related ropelike apron collars the base of the western exterior wall (arrow XX) and has partly buried a small crater (arrow YY). This apron may be a product of mass wasting.

The evolution from a raised-lip profile to a relatively symmetric ridgelike profile has been aided by the inundation of the crater floor. The new floor level permits a more rapid decrease in wall slope because of the reduced surface area over which wall talus can be distributed. Eventually the slope of the wall matches the outward-dipping slope of the rim, producing a relatively symmetric profile (see Plate 153*a*). Crested rims of some inundated craters appear to have resulted from collapse of the rim

region along grabens concentric to the crater. This rim rejuvenation process is illustrated by the crater Doppelmayer (Plate 191*b*, arrow YY) and is discussed at the end of this chapter.

The scalloped plan of Gambart, as defined by the rim crest, is reflected in the contact between the floor and inner wall as well as the contact between the outer mare surface and the outer rim. One explanation might be that the rim represents a tephra ring, which would be large by terrestrial standards. Another explanation is that the scalloped plan was produced by premare slumping that crossed a variety of topography, as noted in Chapter 1 (Discussion and Summary). Equivalent wall slopes cutting irregular terrain will produce a scalloped plan. Although this scalloped plan defined by the rim crest may be reflected by the outer rim-mare contact, it will not be reflected by the inner wall-floor contact unless the wall slope along a concave scallop is greater than the adjacent wall slopes or unless premare slumping preferentially occurred where the rim crest crossed the greatest elevations. The latter occurrence is a plausible coincidence because of the greater likelihood that this rim region would experience overburden, which is conducive to slumping.

Extrusive forms typically surround, or are actually related to, Gambart-like rings. The features already noted within and around Gambart are not uncommon; however, these cannot be used as proof of the volcanic nature of Gambart-like rings themselves. The role of such ring structures could readily be envisioned as rejuvenated volcanic centers, regardless of the initial crater-forming event (see Plate 92*a*).

The interpretation of Gambart and Gambart-like craters having relatively symmetric rim profiles as endogenetically produced is supported by the example shown in Plate 199*c* (arrow YY). The raised rim makes a relatively abrupt contact with the surrounding terrain without evidence for other associated ejecta. It was formed after the Orientale event and is adjacent to a thick lobate flow unit believed to be associated with Orientale ejecta, but its ejecta blanket has not visibly affected either well-preserved surface. Therefore, this crater is interpreted as either a volcanic feature without the large component of radially directed ejecta typical of an impact event or a large Orientale secondary crater. It resembles Gambart except for the absence of floor filling and comparable scallops. Note that a basal apron similar to that around Gambart surrounds the crater seen in Plate 199*c*.

Multiple symmetric rings similar to the single ring of Gambart also occur. A small example is shown in Plate 91*b* to the southeast of Gambart (arrow YY). A more impressive and enigmatic pair is shown in Plate 2*c*.

PLATE 91

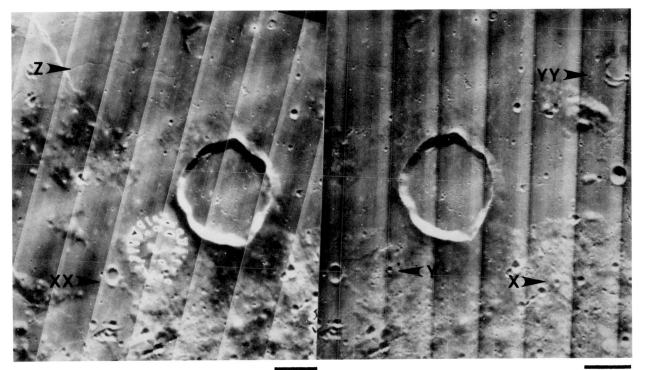

a. IV-121-H1   12 km   b. IV-120-H3   12 km

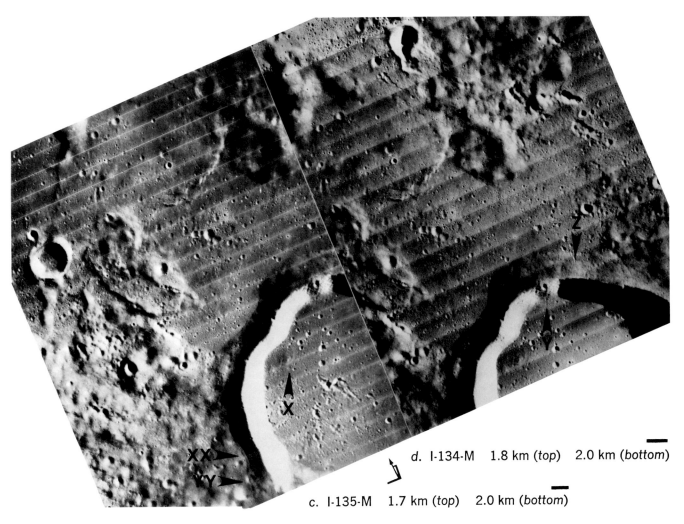

d. I-134-M   1.8 km *(top)*   2.0 km *(bottom)*

c. I-135-M   1.7 km *(top)*   2.0 km *(bottom)*

Ringed Plains and Assorted Forms

Kunowsky, about 18 km in diameter (Plate 92a), is in southeastern Oceanus Procellarum, approximately 300 km southwest of Copernicus. In contrast to Gambart, it displays a hummocky ejecta blanket that has been partly destroyed by the encroachment of mare material. This example supports the interpretation that some Gambart-type craters are volcanically modified impact craters, which might initially resemble Dionysius (Plate 89). The internal origin of such forms, however, cannot be excluded, for an ejecta sequence, similar to the blanket around Kunowsky, could be the product of ash eruptions. Note the annular-ridge arrangement of the central peak complex, which represents either a resurgent central volcano or collapse of the central peak region, not necessarily related to volcanic processes (see the discussion with Plate 47).

Plate 92b shows two other Gambart-like craters on the floor of Sacrobosco, a very subdued crater 100 km in diameter in the Southern Highlands. The east crater is about 12 km in diameter and the west one, about 17 km. Both appear to rest on platforms bounded by a scarp and surrounded by the typical high-albedo plains-forming unit within Sacrobosco. This islandlike appearance is most pronounced near the area marked by arrow X.

The two craters might be impact in origin, with the platform appearance created by processes accompanying the emplacement of the plains-forming material. This is also illustrated in Plate 92c, but in the example shown here the outward-facing scarp to the north (arrow Y) is not in contact with the plains-forming unit. Consequently, it is proposed that the craters represent summit calderas on volcanically constructed platforms.

Plate 92c shows the crater Ramsden in Palus Epidemiarum, which is adjacent to the southeastern border of Mare Humorum (see Plate 191a). This crater, 19 km by 26 km, is similar to Gambart in that both have been inundated by mare material, which produced a somewhat symmetric appearing ring with a screelike inner wall. The eastern outward-facing scarplike border

of Ramsden (arrow X) is similiar to those of the craters shown in Plate 92b. The interior mare floor is about 1 km below the surrounding mare (LAC 111), and both surfaces display numerous shallow craters. The scoured appearance to the south and east of Ramsden is due to an overlapping ejecta ray from Hainzel A.

The isolation of Ramsden from the nearby shores might indicate:

1. It was formed on a local topographic high prior to mare flooding.
2. It was formed on a mare surface emplaced earlier than the present surface.
3. The process of encroachment by mare material involved both removal and burial of topographic features.
4. Ramsden is a caldera on a volcanically constructed mound.

The interpretation of an endogenetic origin is favored because Ramsden occurs along three grabens. A neighboring nested-ring form, also interpreted as a volcanic feature, interrupts a nearby graben less than 100 km to the northeast (see Plate 3, c and d). Plate 169 shows analogous relations between similar-appearing craters and grabens.

Plate 92, d and e, includes the irregular ring Burnham, southeast of Albategnius in the Southern Highlands. The feature is much more subdued than the preceding examples shown in this plate and is breached at two locations (arrows X and Y). A low-relief groove crossing the crater floor coincides with the breach at arrow Y. The interior is hummocky rather than marelike, and the exterior is covered by the high-albedo plains-forming unit. Burnham is interpreted as either a multiphased volcanic collapse structure that predates the emplacement of the surrounding plains or a low-velocity Imbrium secondary crater containing remnants of the secondary projectile.

210

PLATE 92

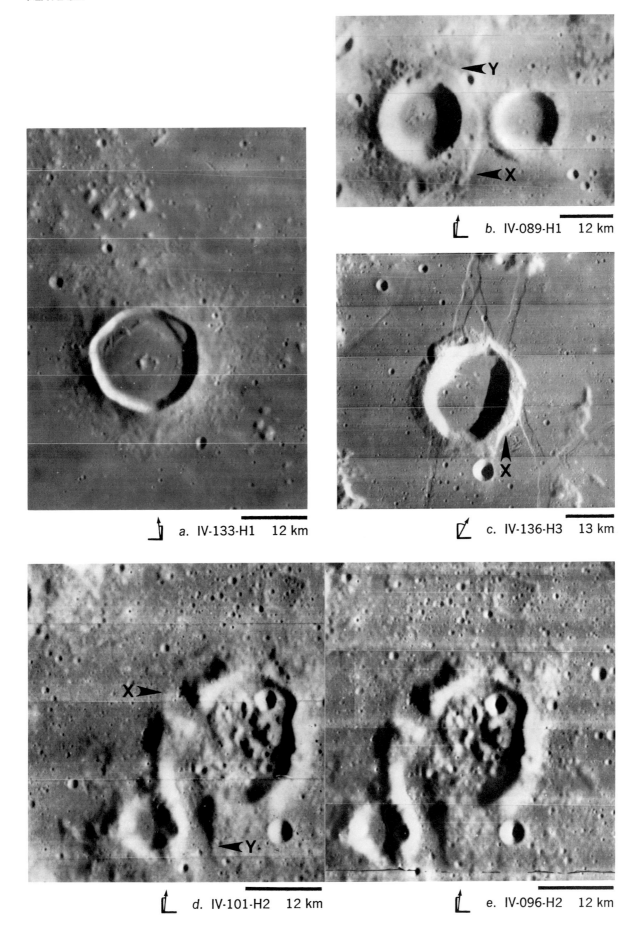

a. IV-133-H1   12 km

b. IV-089-H1   12 km

c. IV-136-H3   13 km

d. IV-101-H2   12 km

e. IV-096-H2   12 km

Pristine Forms

The rim of Theophilus, 95 km in diameter (Plate 93), rises about 6.8 km above its floor and is clearly shown in this oblique view. To the right (east) the rim appears hummocky and slopes gently away from the crater, but, toward the foreground and farther to the left, high-albedo plains-forming units fill depressions adjacent to the rim (arrow X). In some regions (arrow Y), the outer slope of the rim is steeper than the inner wall. Such a profile and moatlike region adjacent to the rim crest could be the result of massive slump blocks, following faults with relatively low dips. In addition, it must be assumed that the slope of the rim prior to slumping was sufficient to make the crest of the block the highest elevation of the present rim. In other regions, synthetic and antithetic faulting may be responsible for graben-like depressions concentric to the rim (arrow Z).

The smooth-surfaced units, "pools," along the outer rim are typical of recent craters larger than about 25 km. As illustrated in Plates 102–104, flow material on the rims of such craters displays various degrees of viscosities. Flows commonly appear to originate on the rim and extend down the crater wall onto the floor. The material around Theophilus displays numerous low-rimmed and rimless craters, some of which could be of internal origin. The smooth-surfaced rim units are thought to be once-molten country rock fluidized by the impact and either tapped by faults, which are revealed by the concentric depressions, or ejected from the crater during the last stages of crater formation.

The view provided by Plate 93 is a good starting point for the photographs to follow in Plates 94–104. Despite the three-dimensional effect provided by stereo pairs, the apparent relief is commonly exaggerated, and the effect erroneously enhances the notion that large lunar craters represent enormously deep pits surrounded by vertical walls. Such a notion results from the necessity of examining features under low sun angles. As examples, consider the stereo pairs of Theophilus (Plate 46, *b* and *c*) and Tycho (Plate 31, *a* and *b*).

PLATE 93

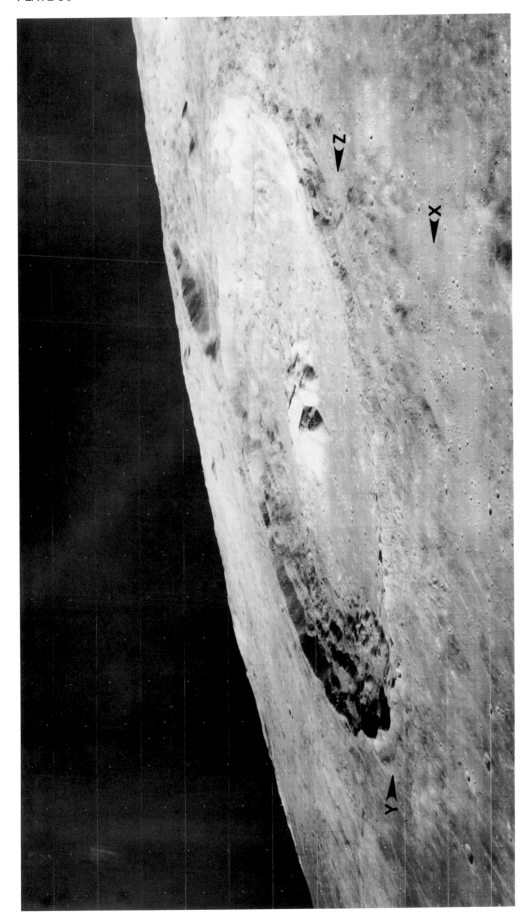

III-078-M (oblique)

213

Pristine Forms

Copernicus (Plate 94) illustrates the three general zones of rim and ejecta blanket for craters larger than 25 km and reveals the increased complexity from the sequence described for Mösting C (Plate 80). These zones are outlined below:

1. Zone I (extending outward from the rim crest a distance of 0.25–0.5 the crater radius, but the extent may depend on postformation scalloping by gravitative transfer):
   a. Extensive annular fracturing and corrugated patterns
   b. Isolated boulders, boulder fields, and boulder strings
   c. Generally rugged and high-albedo topography
   d. Pools of smooth plains-forming units in some local depressions
   e. Flow units typically bordered by levees and directed radially from the crater with a wide range of implied flow viscosities
   f. Radial channels that commonly are traced onto the crater wall as flow units or downslope on the crater rim
2. Zone II (extending beyond the end of Zone I for an additional crater radius):
   a. Radial and subradial lineations
   b. Limited flow units with subdued but corrugated surfaces
   c. Pools of smooth plains-forming units in local depressions
   d. Relatively subdued hummocky forms with few blocky exposures
   e. Barchanlike relief, i.e., crescentic relief, with horns directed away from the crater
3. Zone III (extending beyond approximately two crater radii from the crater rim):
   a. Herringbonelike craters with tips extending away from the main crater
   b. Herringbonelike crater chains that occur radial, subradial (tangential), and concentric to the main crater

   c. Herringbonelike crater clusters (may be as far as four to five crater diameters from the main crater)
   d. Threadlike lineations trending approximately radially from the crater and subduing or "smearing" the surface (typically accompanied by a higher albedo that defines the "rays")
   e. Clusters of subdued craters also associated with the ray material and lineation
   f. Patches of plains-forming material that show no extensive scarring but are identified by small flat-floored craters and relatively few craters

The boundary between the various zones is not well defined, but in general the boundary between Zone I and Zone II is characterized by a change in both albedo (Zone II being darker) and relief forms, possibly accompanied by plains material. The irregular boundary between Zone II and Zone III is more easily identified on a mare surface, where the interconnecting features of Zone II can be separated from the isolated forms of Zone III. Zone II commonly corresponds to a low-albedo halo identifiable under full illumination.

The sequence applicable to Copernicus can be extended to Tycho (Plate 101) and Aristarchus (Plate 236) but does not apply in all details to enormous structures, such as Orientale (Plate 201), or smaller craters, such as Mösting C (Plate 80). Furthermore, subdivisions of the zones may be visually identified only in part around more subdued craters, such as Eratosthenes (Plate 106). There can obviously be considerable variety that reflects the pre-existing composition, structure, and topography of the region as well as crater origin and later modifications.

The arrows in Plate 94 refer to features referenced in the discussions with Plates 96 and 100.

PLATE 94

IV-121-H2,H1   12 km

Pristine Forms

Regional features of Zone I and II for Copernicus are shown in Plate 95. Parallel sets of fractures of Zone I are evident along the northwestern rim (arrow X), but ridges and furrows that trend radially from the crater are more prominent along the northern and northeastern rims (arrow Y). The relative absence of parallel fractures in this latter area is due in part to unfavorable lighting conditions for east-west lineations. Fractures on the northern rim (arrow Z), however, are not concentric to the crater but are parallel to the northwest wall or scarp. A similar but poorly expressed pattern aligns with the irregular northeastern scarp. These nonconcentric sets of lineations indicate a crater-forming event affecting a prestressed medium. At several locations, they cross the radially trending furrows and ridges, and this implies that the release in stress was not complete until after the formation of the radially trending lineations.

Zone II (top of Plate 95) shows relatively subdued hummocky forms with pools of smooth-surfaced units in closed depressions (arrows XX and YY). The transition between Zones I and II occurs across the middle of the plate and merges without a well-defined contact. The cluster of craters seen in the upper left and right of Plate 95 may be secondaries from the crater Aristarchus, rather than ejecta associated with Copernicus.

PLATE 95

V-157-M   3.2 km (left)   3.5 km (right)

Pristine Forms

Plate 96 illustrates the herringbone pattern within the outer (Zone III) ejecta blanket of Copernicus. Although Zone III is characterized at lower resolutions (Plate 94) by an absence of bulk ejecta, the higher-resolution stereo pairs in Plate 96 reveal extensive lineations and blanketing of the intercrater areas. The herringbone pattern observed in Plate 96, a and b, accompanies the long furrow (arrow X) and complex crater groups (arrow Y). As shown in Plate 94 (arrow X), the furrow is not radial to Copernicus and dissolves into contiguous crater chains and aligned herringbone craters uprange and downrange from the region shown in Plate 96, a and b. The horns of the herringbone patterns associated with these features are directed away from Copernicus.

Note the wrinkle ridge (Plate 96, a and b, arrow Z) that crosses a secondary crater, presumably associated with Copernicus. In contrast, the secondary crater (arrow XX) clearly overlaps a southeast extension of the ridge. These relations indicate that the ridge was formed over a time span predating and postdating the Copernicus event.

Plate 96, c and d (see Plate 94, arrow Y), includes an area south of that shown in Plate 96, a and b, and illustrates the numerous and complex craterlike high-rimmed depressions (arrow X) that also have overlapping "wings." In this stereo view, the floors of the crater appear to rest below the level of the surrounding country and hence are not just pseudocraters created by built-up rims. As noted in the preceding stereo pair, lineations on the relatively uncratered surface (arrow Y) suggest that the pre-existing mare surface is covered by an extensive blanket of ejecta. This is not readily apparent in Plate 94 (arrow Z) except that its albedo is higher than that of the unscathed surfaces. From the discussions of the smaller crater Censorinus (Plate 83), these lineations may represent the effects of a cloud of secondary debris, low-angle ejecta from uprange secondary impacts, or remnants of a base-surge cloud.

There are at least five possible mechanisms responsible for the formation of herringbone craters. Plates 94 and 96 show that the pattern resembles "frozen" shock waves. O'Keefe et al. (1969) also were impressed by this comparison and suggested that the pattern was formed during the passage of the base surge, which was attributed to an explosive expansion of trapped gas into a vacuum. Guest and Murray (1971) hypothesized that the interaction between the base surge and ejecta from contemporaneous secondary impacts could produce bow dunes analogous to the herringbone pattern. I can imagine bow-duning without the passage of the base surge, in which an interaction occurs between a relatively tenuous cloud of ballistic secondary debris or tertiary ejecta from uprange secondary impacts and the ejecta from secondary impacts produced by solid projectiles imbedded in this cloud. A similar but independent interpretation by Oberbeck and Morrison (1974) evolved through laboratory studies of multiple impacts, which resulted in ridges, or "septa," separating the component craters. These ridges were convincingly demonstrated to be the result of the interaction between the enlarging conelike ejecta sheets from each crater. A V-shaped plan of the ridges resulted when either the relative time of formation or the relative size of these craters was varied. The former situation commonly occurs in oblique impacts, a characteristic of secondary impacts. Further discussion of these plausible mechanisms follows additional examples of Zone III features (see the discussion with Plate 100).

Plate 96, a and b, also includes arrow YY, which locates the region shown in Plate 97 at high resolution; similarly, arrow Z in Plate 96, c and d, locates the region shown in Plate 97b.

PLATE 96

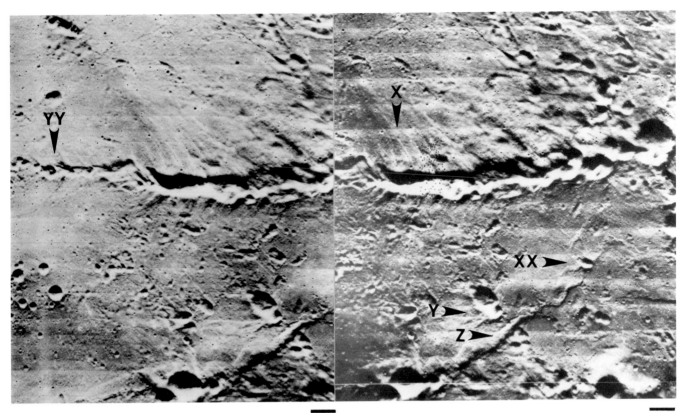

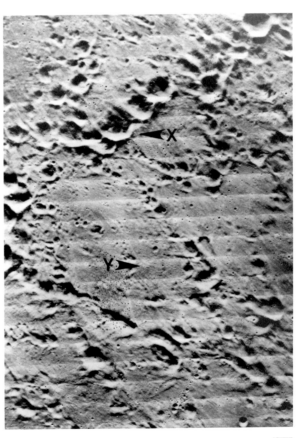

a. V-145-M   3.9 km (top) 3.4 km (bottom)

b. V-144-M   3.8 km (top) 3.3 km (bottom)

c. V-143-M   3.7 km (top) 3.3 km (bottom)

d. V-142-M   3.7 km (top) 3.2 km (bottom)

Pristine Forms

The features examined in Plate 96 appear relatively subdued, as shown at high resolution in Plate 97. Plate 97a is in part shown on the left edge of Plate 96, a and b (arrow YY); the coverage of Plate 97b is on the left edge of Plate 96, c and d (arrow Z).

These close-ups reveal the subdued appearance of the profile and topography of the secondary complex. Such an absence of surface detail is in contrast to isolated craters of comparable age and size as well as to features on the floor and inner rim region of Copernicus. If the subdued appearance is the result of degradation by long-term smoothing processes, then the extrusive activity on the floor of Copernicus apparently has extended over a long period. A more likely explanation is that secondary craters were subdued at the outset. If the herringbone pattern is produced by the interaction of debris, then the turbulence associated with this interaction may be followed by a general debris fall-out, rather than a well-ordered sequence of deposition inferred from small primary impact craters. This complex mixing of ejecta could explain the general absence of visible blocks, which characterize primary craters of comparable size, and the exaggerated rim profiles typical of some secondary craters.

In addition, the secondary complex displays numerous small craters, most of which are subdued with respect to their ejecta blankets and form. The high crater density can be attributed to a long-term exposure to random primary impacts, numerous late-arriving secondaries from Copernicus, or collapse features. The subdued appearance from long-term processes perhaps reflects a relatively unconsolidated target and continuous degradation by small meteoroids. This does not seem consistent, however, with the relatively low crater density within Copernicus unless, as previously suggested, extrusive activity continued over a long period on the crater floor. It also does not seem consistent when comparison is made to other level surfaces that are presumably older than Copernicus and are apparently composed of unconsolidated material (see Plate 166c). Therefore, these small subdued craters are believed to represent secondary impacts and/or collapse features. Late-arriving secondary ejecta may have impacted prior to complete deposition of the surrounding cloud of debris associated with the larger secondaries, thus producing small subdued craters.

There are also several bright-haloed craters visible in Plate 97a. Their high albedo might be the result of the composition or physical properties of the Copernicus ejecta blanket. Larger (4 km) dark-haloed craters, such as Copernicus H (see Plate 5b), may have punctured the ejecta blanket and revealed the underlying darker mare material. The minimum size of such dark-haloed craters perhaps indicates the thickness of the blanket.

PLATE 97

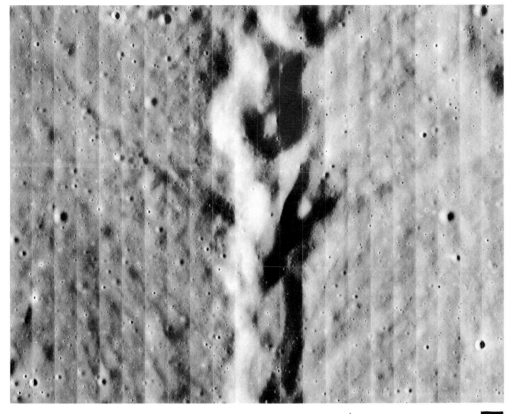

a. V-145-H3   0.47 km

b. V-143-H2   0.45 km

Pristine Forms

Aristoteles (more than 90 km in diameter) is shown in Plate 98*a* (lower far right) and exhibits an extensive ejecta blanket that illustrates contiguous crater chains having the herringbone pattern. The chain (arrow X) extends across the outer Zone III region almost continuously for more than 200 km to the west. Along its length, three major offshoots (arrows Y, Z, and XX) diverge at approximately a 30° angle from the main chain (directed away from Aristoteles). The main chain (arrow X) is tangential to the north rim of the crater. Perhaps the formation was a result of filamentary ejecta, which have been reproduced in the laboratory (Gault et al., 1968).

Note that the herringbone pattern occurs at the most remote extension of the chain, and that isolated craters also display such a pattern. It does not appear obvious how a base surge could be responsible for this pattern so distant from Aristoteles, especially when the surrounding surfaces show no evidence of an onrush of material. Consequently, the pattern in this example probably originates from properties of the ejected mass separated from the bulk ejecta.

Similar features are associated with Krafft and Cardanus, Plate 98, *b* and *c* (top and bottom, respectively). The stereo pair reveals an apparent vertical displacement of the areas to either side of the furrow (arrow X). Earth-based photographs under different solar illumination confirm this impression. As is clearly shown, the structure crosses the rim, wall, and floor of Krafft, which makes this cratering event an unlikely parent. The feature is not radial to Cardanus (nor is the smaller chain to the west) and extends onto the wall of this crater (arrow Y). Although the furrow might be attributed to another crater, the only recent and plausible candidates are too distant for this type of feature, at least in comparison with similar furrows associated with other major craters, such as Copernicus (Plate 94, arrow X). Therefore, a structural origin or structural modification related to Cardanus is tentatively favored.

PLATE 98

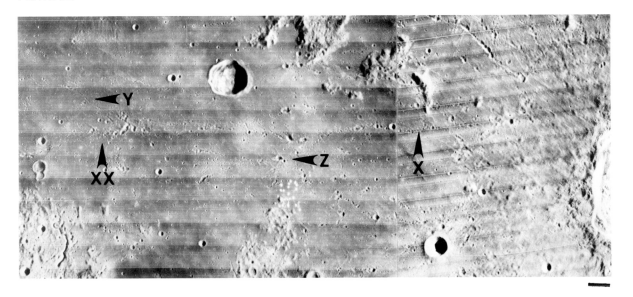

a. IV-115-H3; IV-110-H3   13 km

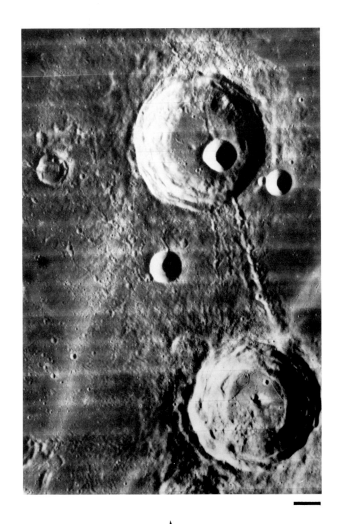

b. IV-174-H2   12 km

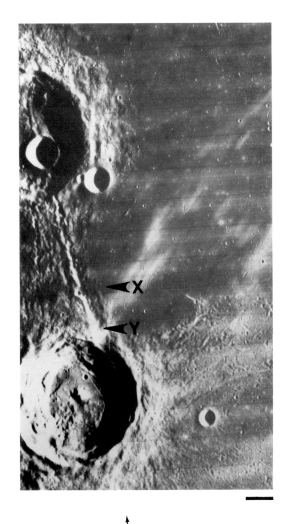

c. IV-169-H2   12 km

223

Pristine Forms

West and southwest of the crater Aristillus (Plate 99a), are numerous ENE-trending ejectalike lineaments (arrow X) that are clearly not radial to Aristillus. Another nonradial NNW-trending set exists to the southeast of Aristillus (not shown). The ENE-trending lineaments may represent ejecta from the craters Theaetetus, Callipus, Cassini, Archimedes, Eudoxes, Aristoteles, and Copernicus. The first two named are small craters and either do not have an ejecta blanket (Callipus) or exhibit a very limited one (Theaetetus). The ejecta blankets of Cassini and Archimedes are buried by mare material. Portions of the ejecta blanket from Archimedes remain, but the sets shown in Plate 99a are not radial to it and occur on the mare units. The last three examples named are distant craters, and, as sources for the lineaments, they require alignment nonradial to the source.

As can be verified by re-examining the overview of Copernicus (Plate 94), the alignment of a crater chain typically is not radial but tangential and possibly annular to the parent crater, even at great distances. The herringbone pattern, however, commonly gives an indication of the appropriate parent crater. Therefore, the objection to the association of the ENE-trending lineaments in Plate 99a with Eudoxes, Copernicus, or Aristoteles might not be convincing. There is an absence, however, of similar secondary impacts near the region shown in Plate 99a as well as any comparable groups of secondaries occurring at such distances from the primary. Hence, it is suspected that the pattern may have been related to the formation of Aristillus and expresses pre-existing stresses strongly controlling the ejecta.

Plate 99, b and c, shows in stereo a crater chain that has a poorly expressed herringbone pattern and is arranged in a spiral-shaped plan around the crater Piazzi C. The contiguous chain appears to have a well-defined rim north-northwest of the crater, but the chain and its rim become less pronounced along the northern arc, in part the result of unfavorable solar illumination. Shoemaker (1962) described loops of secondaries around Copernicus and attributed the pattern to the influence of pre-existing structural weaknesses as the projectiles were torn from the lunar crust. This may be applicable to this example as well. It is also possible that the feature is a large lava channel, collapsed tube, or fracture with minor eruptions. The crater Dawes (Plate 62a) exhibits a similar but less well defined spiral structure.

PLATE 99

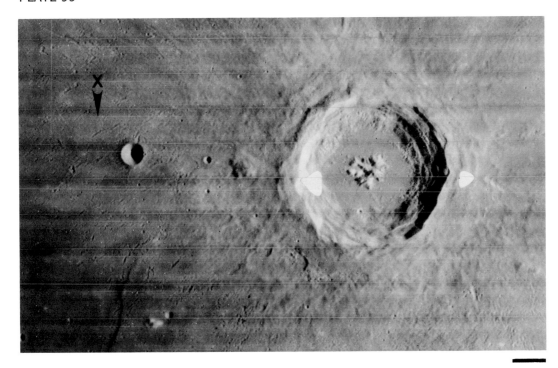

a. IV-110-H1   13 km

b. IV-167-H3,H2   13 km

c. IV-160-H3   13 km

Pristine Forms

Plate 100 illustrates features in the outer rim sequence (Zone III) of large craters. Plate 100a covers a portion of the marelike floor of Prinz (Plate 255) and shows the "smearing" of pre-existing surfaces by ejecta from the crater Aristarchus, which is 100 km to the southwest. The boundary of this complex nearest the parent crater commonly, but not necessarily, is accompanied by a craterlike or irregular depression (Plate 100a, arrow X). Farther to the northeast (away from Aristarchus), the surface is characterized by the following features:

1. Higher albedo than unaffected surfaces
2. Narrow ridges and furrows (30m–70 m in width) generally trending away from Aristarchus
3. Subdued craterform depressions that typically have very few blocks, if any, on their rims
4. Deep gouges that do not necessarily trend radial to Aristarchus and that commonly exhibit the herringbone pattern
5. Stippled terrain that commonly is predominate on that part of the complex farthest from Aristarchus

Farther from the primary crater, the ejecta complex typically merges with the mare surface without a well-defined transition.

Smooth-surfaced patches also occur in some Zone III regions (Plate 100a, arrow Y; also see Plate 258). Their surfaces typically are without lineations radial to the primary crater, and craters within these patches are subdued and flat floored. Commonly, the material within the flat-floored craters is piled against the crater wall facing the primary (Plate 100a, arrow Z). The albedo of such patches typically resembles that of the mare. It is suggested that these regions have been blanketed by a fluidized portion of the ejecta (also see Plate 189).

The ejecta complex seen in Plate 100b is approximately 600 km to the southeast of Aristarchus near the crater Tobias Mayer and can be identified in Plate 134, a and b. The features in this complex are similar to those shown in Plate 100a.

The ejecta shown in Plate 100c are 7 km south of the area shown in Plate 100b. It also appears to be related to Aristarchus, but its borders are poorly defined. Although narrow NW-trending lineations are important surface forms, the complex appears highly stippled with small subdued craters 0.05 km–0.3 km in diameter. Similar stippling is associated with Aristarchus near the crater Prinz (Plate 100a, arrow Y; also see Plate 257) and with Horrocks (Plate 43d).

Plate 100d shows a simpler pattern than the previous three examples. It is near the ejecta complexes shown in Plate 100, b and c, but is probably related to the crater Kepler, 350 km to the southwest. This complex is composed of several parallel furrows. The widest furrow (arrow X) has a narrow rim and a craterform depression nearest Kepler (arrow Y).

The examples shown in Plate 100 reveal that ejecta complexes in Zone III are found at considerable distances from their primaries but that they retain many of the morphologic features in Zone II of the ejecta blanket. The separation from the bulk ejecta deposits of Zone II seen in Plate 100, b, c, and d, suggests that the base surge cannot be responsible for these features. In

addition, the existence of isolated herringbone patterns at considerable distances from their primaries (Plates 94, 98a) does not seem consistent with the hypothesis of base-surge-produced shock waves as suggested by O'Keefe et al. (1969). At such distances this mechanism requires passage of a base surge that produces the herringbone pattern around secondary craters only and not around other topographic features. Base-surge duning (Guest and Murray, 1971) also seems inapplicable at great distances from the primary crater because this process not only implies passage of the base surge without modification of the surrounding surfaces but also requires that this surge cross the region at the same time as the impact by secondaries.

The existence of surface scouring from secondary impacts without large or well-defined craters (see Plate 15a, arrow X, and Plate 83) suggests that some isolated herringbone craters are the result of the interaction between secondary debris clouds and ejecta from secondary impacts. At impact, the horizontal velocity component of the relatively tenuous cloud spreads the material in-line with the primary crater. The smearing behind the leading edge is attributed to the skidding secondary complex and downrange ejecta from its initial impact. If the horizontal component of the impacting complex is greater than the velocity of the ejecta from the secondary craters, then the herringbone pattern might be produced.

As noted in the discussion with Plate 96, Oberbeck and Morrison (1974) suggested a similar mechanism involving nearly simultaneous impacts. This mechanism is most likely responsible for the formation of the septa that accompany many doublet and multiplet craters and perhaps the herringbone pattern. As the spacing between adjacent impacts decreases, a single but slightly irregular crater was formed in their laboratory experiments. In addition, the intervening septa remained as ridges extending from the rim, and they subtended an angle dependent on the relative time (or size) of formation as well as the angle of adjacent impacts. As the spacing between impacts increases, the interacting ejecta sheets continued to produce intervening ridgelike forms. Secondary impacts with relatively low angle trajectories, in turn, produce considerable low-angle downrange ejecta. In this case, the hypothesis of interacting secondary or tertiary clouds with secondary impact ejecta described above merges with the mechanism suggested by Oberbeck and Morrison (1974). Within the inner Zone III of the rim sequence (Plate 94, arrow XX), these two mechanisms as well as base-surge duning are the most probable causes for the herringbone pattern that collars the uprange secondary crater rim, rather than closely spaced simultaneous impacts. This conclusion is based on both the evidence for blanketed and modified surfaces immediately uprange (Plate 96) and the highly developed appearance of the herringbone pattern, which leads the secondary complex. Farther from the primary crater, where a nearby but separate complex of uprange secondary impacts is absent, the herringbone pattern is more likely the result of either simultaneous impacts, as Oberbeck and Morrison suggest, or the interaction of the secondary impacts and an enveloping cloud of secondary debris.

PLATE 100

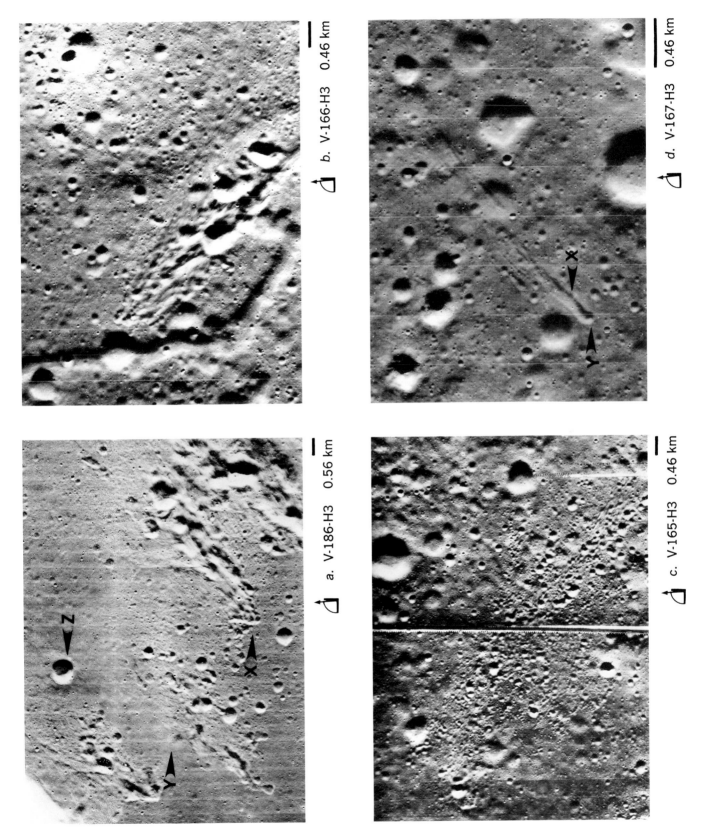

b. V-166-H3   0.46 km

d. V-167-H3   0.46 km

a. V-186-H3   0.56 km

c. V-165-H3   0.46 km

Pristine Forms

Tycho (Plate 101) is considered in Plates 101–104 with the attention directed to its rim zones (see Plate 31, *a* and *b*, for stereo). Whereas Copernicus (Plate 94) formed in the mare plains, Tycho (Plate 101*a*) occurs in the Southern Highlands. The herringbone pattern around secondary craters is shown to be characteristic in this region as well as near Copernicus. Under full illumination, Tycho exhibits an asymmetric bright-ray pattern resembling two whiskers and several dark rays. In addition, a dark halo surrounds the rim in an annulus approximately corresponding to the Zone II region (Plate 101*a*, arrow X). Arrow Y identifies a depressed region that collars the inner rim and that contains numerous smooth-surfaced units (see Plate 103, *a* and *b*).

Plate 101*b*, a general view, provides an index for subsequent coverage. Plate 101*c* reveals, along the northern rim (Plate 101*b*, arrow X), a prominent annular corrugated pattern (arrow X) that is probably a fracture system created by the explosive Tycho event. The precise mechanism of formation—e.g., the initial shock wave or later release of stress—has not been determined. Fractures can be traced into the hummocky terrain to the west (arrow Y). In many areas, however, they have been locally covered or destroyed by later processes, such as massive flow units from rim eruptions or molten ejecta (see Plate 104). The fractures also disappear at the borders of smooth-surfaced pools, which are shown in detail in Plate 102.

Other characteristic features of Zone I shown in Plate 101*c* are the radial lineations (arrow Z), numerous blocks (arrow XX), and generally rugged topography. Arrows YY, ZZ, and XY refer to regions shown in Plate 102, *b*, *c*, and *d*, respectively.

228

PLATE 101

a. IV-119-H2   13 km

b. V-125-M   7.2 km

c. V-126-H1; V-127-H1   0.91 km

Pristine Forms

Plate 102 displays small smooth-surfaced pools of material on the northern rim of Tycho. These units typically occur in local depressions and commonly lack termini expressed as large topographic relief (see Plate 103 for impressive exceptions). Their surfaces have relatively low albedos and low crater density. Craters on these units are typically either rimless or rubbled, with irregular plans.

Plate 102a shows polygonal fractures (arrow X) that commonly are found on the pools. In addition, the individual cells appear to be domal. These were probably produced by the cooling of once-molten material, which may have been lavalike or ashlike. The sunlit borders of a pool (arrow Y) may represent termini or are additional indications of shrinkage. Fracture patterns on other pools suggest differential movement through settling or lateral transport (Plate 102b, arrow X).

There is not an obvious source for the unit shown in Plate 102a. Arrow Z (Plate 102a) identifies a terrace that represents either a later unit or perhaps a previous level. Arrow XX locates a separate unit of similar appearance that apparently is not connected surficially to other pools. The problem of the source for such units is reconsidered after discussion of Plate 102d.

Plate 102b includes a region closer to the rim (Plate 101c, arrow YY) and shows another smooth-surfaced pool that surrounds numerous large blocks. This unit also exhibits fracturing (arrow X) but not the polygonal pattern illustrated in Plate 102a. Note the linear depressions (arrow Y) that extend down the outer slopes of the rim away from Tycho. They merge to a single linear feature that is interpreted as a channel for material from the pool seen in Plate 102b to another smooth-surfaced unit to the north (beyond the edge of Plate 102b).

The pool shown in Plate 102c (Plate 101c, arrow ZZ) displays lobate fronts (arrows X and Y) that indicate a source for the unit from the west. Arrows Z and XX identify possible channels from an elevated pool to the west (outside the area shown in Plate 102c).

Adjacent to the northern rim are several smooth-surfaced units that do not fill local depressions but have positive relief on the pre-existing surface. Plate 102d shows such a feature (arrow X; Plate 101c, arrow XY). It was perhaps derived from an irregular depression (arrow Y) that is linked to the unit by a channellike form (arrow Z). Note the numerous small units that fill local topographic lows as well.

The following observations summarize and supplement the examples shown in Plate 102:

1. The pools near the rim crest typically are linked to units on the walls through leveed channels.
2. The sources for some pools appear to be irregular depres-

sions that may be linked to the pools by leveed channels. Some of these channels are not continuous and typically do not cut into the pre-existing surface.
3. The pools commonly occur not more than 40 km from the rim crest and generally are restricted to the Zone I region and Zone I–Zone II transition of the ejecta sequence.
4. Transitional areas occur that appear to be links between blocky and smooth-surfaced units.
5. Crater density generally is low on the pools despite the apparent increase in density on the surrounding terrain farther from the rim crest.
6. Craters on the pools generally fall into two classes: subdued rimless pits and rubbled craters. The subdued pits were probably formed endogenetically, for these surfaces are well preserved. The rubbled craters are generally larger and commonly have irregular or concentric plans. These are either craters impacted in competent units or explosive volcanic forms.

These observations indicate that the smooth-surfaced pools were emplaced after the bulk ejecta were deposited. The cascades down the wall indicate that they also postdate major slumping, which in turn predates the last stages of floor flooding. The low viscosity of the pool material is illustrated by:

1. The general absence of large flow fronts
2. The transferral to topographic lows, leaving relatively empty leveed channels that commonly are discontinuous
3. The widespread filling of local depressions

Shoemaker and Morris (1968) distinguished between those pools, or "playas," with and without the branching grooves, some of which have polygonal patterns. They interpreted both units, however, as deposits from low-viscosity gas-charged ejecta analogous to nuées ardentes. It is not clear whether they consider these to be the same type of units as those illustrated in Plate 102, b, c, and d. Although nuées ardentes have the low viscosity necessary for my observations listed here, it is doubtful that they would act as single units, portions of which cascade to adjacent topographic lows through leveed channels. Units like the one seen in Plate 102a may have a different origin from those shown in Plate 102, b, c, and d, but the distinction in appearance between them is obscure in many cases, except for the absence or presence of fractures. As noted already, several examples along the inner rim have fracture patterns that appear to be the result of differential movement rather than cooling, and this interpretation is more consistent with once-molten flows.

PLATE 102

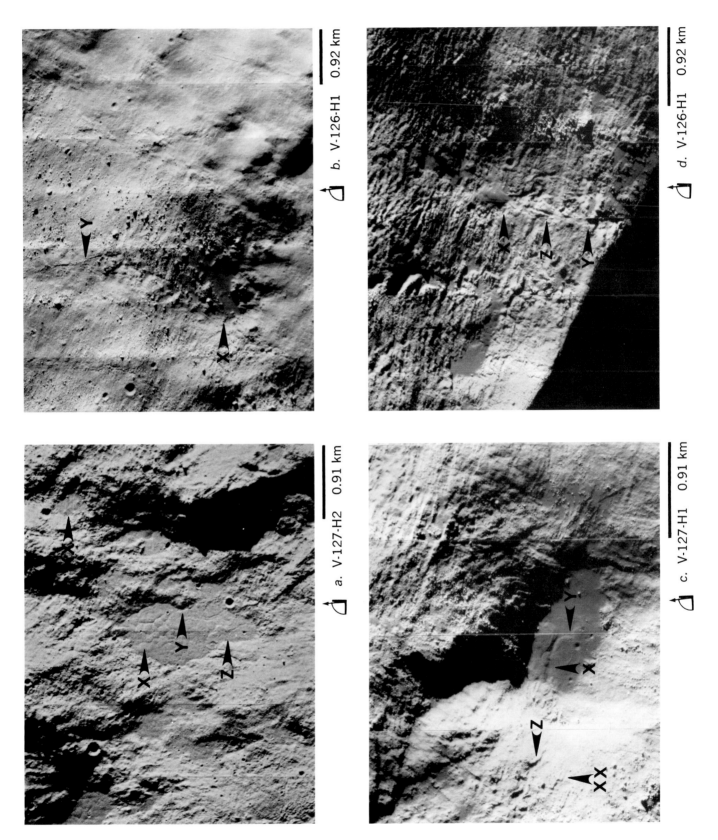

a. V-127-H2   0.91 km

b. V-126-H1   0.92 km

c. V-127-H1   0.91 km

d. V-126-H1   0.92 km

Pristine Forms

The eastern rim of Tycho (Plate 103, *a* and *b*; see Plate 101*b*, arrow Y) displays an extensive system of low-albedo plains-forming units larger than those illustrated in Plate 102. Arrow X identifies a poollike unit displaying well-defined borders that stand in relief. These units are in topographic lows, which commonly are closed depressions and have scarplike boundaries (arrow Y). The pools clearly postdate the hummocky and corrugated terrain along the rim.

Shoemaker et al. (1968) suggest that similar-appearing units (see discussion with Plate 102) represent deposits from gas-charged molten ejecta analogous to a *nuée ardente*. Their study was restricted to the area in the immediate vicinity of the Surveyor VII landing craft (Plate 103, *c* and *d*, arrow X) and may not apply to these units, although the distinction between their proposed deposits and these units is not always obvious from Orbiter photography.

Several pools indicate low-viscosity lavalike flows that may have been derived from molten ballistic ejecta or melt tapped by the concentric fractures peripheral to Tycho. Studies by Carr (1964) indicate that a large impact probably would not induce extensive volcanism alone but could alter a pre-existing thermal gradient that might result in later eruptions. Specific eruptive centers are not always obvious, and it is suggested that several pools are the result of overspills from other levels. Plate 103, *a* and *b* (arrow Z), locates a fracture parallel to the rim of Tycho that is one plausible source for these units. In addition, a cone-like form with a summit pit is on-axis with this proposed fissure (Plate 103*b*, arrow XX). The closed depressions with inward-facing scarps (arrow Y) are evidence that they may also be eruptive centers. Supporting evidence for once-molten flows includes cascadelike regions (near arrow XX), fracture patterns, and ridges that suggest slow movement beneath a crusted surface (not shown in these photographs). Strom and Fielder (1970) reached similar conclusions from their independent study.

Plate 103, *c* and *d*, is a stereo view of the northern rim area of Tycho and reveals the assortment of ejecta units. Note that mismatches of framelets require shifting of the stereo viewer, and relative changes in relief are most reliable along individual framelets and not across them. Shoemaker et al. (1968) described eight distinguishable units in a selected region near the Surveyor VII landing site (arrow X). Two of these units are the smooth-surfaced pools already discussed. The remaining six are listed here in order of the proposed emplacement:

1. Patterned debris (arrow Y)
2. Smooth flow (arrow Z)
3. Divergent flow (arrow XX)
4. Ridged lobate flow (arrow YY)
5. Steep-fronted flow (arrow ZZ)
6. Patterned flow (arrow X)

In this sequence the smooth-surfaced pools were emplaced concomitantly with the smooth-flow unit. These units are better identified in Plate 104, which includes the same region at higher resolution.

Shoemaker et al. (1968) proposed that the entire sequence was emplaced on the order of 10^3 seconds. The absence of secondary craters on the smooth units they suggested to be the result of the semimobile nature of these units during cooling. It is suggested here, however, that at least some of these pools are the result of later extrusions or ballistic impact melt and should be placed at the end of the foregoing time sequence.

Plate 103, *c* and *d*, covers an area larger than that considered by Shoemaker et al. (1968) and shows additional examples that fit their classification. For instance, arrow XY marks another steep-fronted flow. In addition, note the textured-flow unit (arrow XZ) that fills the topographically low areas in the chaotic terrain between the rim and the steep-fronted flow identified by arrow ZZ. This chaotic region is within a synclinal or moatlike area that encircles much of the rim of Tycho (see Plate 101*a*, arrow Y). The pools shown in Plate 103, *a* and *b*, are included in this area but lack the chaotic appearance. The outer border of the moatlike ring shown in Plate 103, *c* and *d* (arrow XZ), is approximately coincident with the apparent origin of the leveed channels (steep-fronted flows). This region is considered in more detail with Plate 104.

232

PLATE 103

a. V-128-M   6.9 km   b. V-125-M   7.2 km

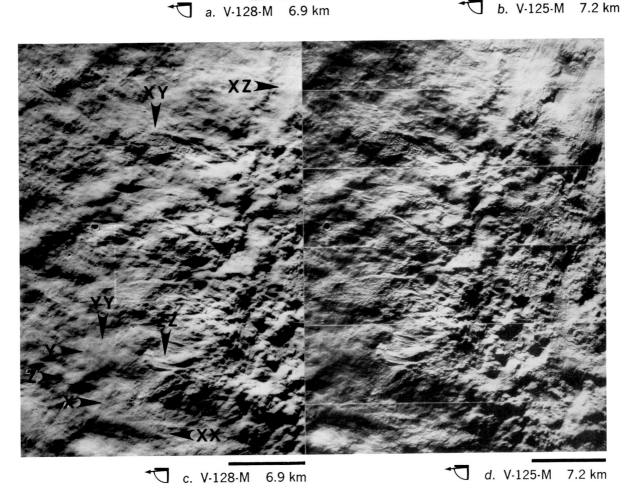

c. V-128-M   6.9 km   d. V-125-M   7.2 km

Pristine Forms

Plate 104 includes a higher-resolution portion of the area shown in stereo in Plate 103, c and d. The units identified by Shoemaker et al. (1968) are more easily recognized:

1. Patterned (corrugated) debris (arrow X)
2. Lunar "playa" material (arrow Y)
3. Smooth-patch material (arrow Z)
4. Smooth flow (arrow XX)
5. Divergent flow (arrow YY)
6. Ridged lobate flow (arrow ZZ)
7. Steep-fronted flow (arrow XY)
8. Patterned flow (arrow XZ) with numerous fissures

For additional descriptions and contacts for these units, reference should be made to the original paper.

The chaotic terrain is shown in the lower one-half portion of Plate 104. The interstitial units include the smooth-surfaced ("playa") material (arrow YX) and highly textured flow units similar to the floor of Tycho (arrow YZ). The chaotic appearance is the result of numerous irregular depressions and craterforms (2 km in diameter) that commonly have breached walls facing away from the crater rim (arrow ZX and area north of arrow YX). This area may have been battered by late-arriving molten ejecta. Such an event requires that the impact-generated melt was not completely ejected during the earliest stage of crater formation, when the pressures and temperatures were the greatest (see Plate 245 and the discussion with Plate 105). Alternatively, the chaotic zone indicates a complex series of volcanolike eruptions along the concentric syncline and fractures that were formed during and after deposition of the bulk ejecta. The latter deposits are believed to be illustrated by the dunelike topography in Zone II (top of Plate 104).

The concentric moatlike depression includes this chaotic zone. It is suggested that this zone is the result of large peripheral slumping, whose surface of rupture extends beneath the crater floor and perhaps was a path for molten material. This stage might be roughly analogous to peripheral eruptions along ring fractures in the Glencoe type of caldera on Earth (Williams, 1941; Williams and McBirney, 1968). However, it is not clear what might be a comparable driving force for rim extrusions around an impact-produced crater. The steep-fronted flow, ridged lobate flow, and smooth flow are interpreted as originating in the chaotic zone. Moreover, the highly textured flow units remaining in this zone either are the result of late-stage extrusions or are localized remnants of ballistic impact melt.

PLATE 104

V-127-H2; V-128-H2   0.91 km

Plains Units

Plate 105a shows two nested craters on the lunar farside. The larger crater is about 40 km in diameter, the smaller interior one about 25 km. The interior crater is more recent and is off-set in such a way that the rim crest at the left coincides with the outer crater rim. Patterns (arrow X) from the formation of the inner cratering event are clearly nonradial. If these patterns are the result of secondary impacts, then they may have been controlled by the walls of the exterior crater. The area shown at lower right (arrow Y), however, has the least obstruction yet is relatively devoid of the patterns. Consequently, either the ejecta distribution was asymmetric, or these features are structural in origin. Plate 105b is an enlarged segment of the wall and rim area of the interior crater. It shows a ropy lavalike unit that was derived, at least in part, from the leveed channel or channels (arrow X), which extend from the rim crest.

The crater Ölbers A (Plate 105c) also exhibits an extensive plains-forming unit that overlies its ejecta (arrow X); however, related well defined channels on its rim are absent. The bordering of this unit by irregular inward-facing scarps (arrow Y) suggests partial containment within an irregular depression. Under full solar illumination, this unit is considerably darker than the bright-ray system. It is also darker and has a smoother surface (at this resolution) than the plains unit on the crater floor.

Leveed channels on crater rims are common around recent impact craters (see Plates 104 and 245). In addition, several large craters exhibit extensive plains-forming units beyond the rim crest (Plate 9a; also see King crater, AS16-1578). If these and similar units represent impact-generated melts, then a significant amount of undispersed melt is retained within or around large impact craters relative to smaller craters. This has been suggested by Dence (1971) through geologic comparisons with large terrestrial impact craters. The unscathed appearance of, and the superposition by, such units clearly require emplacement of these melt units after deposition of the rest of the ejecta. Thus, several sequences for the formation of the lavalike rim units can be pictured:

1. The units were "sloshed" out of the crater owing to peculiar geometry of the impact crater.
2. The original crater floor level was much higher, and the smooth-surfaced rim units represent spillover followed by floor subsidence.
3. These rim units are derived from rim or wall fractures extending beneath the crater floor.

The first suggestion is consistent with the observation that craters having extensive plains units on their rims typically exhibit an adjacent rim region that is depressed relative to the opposite rim. For example, the placement of the crater seen in Plate 105a within a larger crater has resulted in a high interior wall opposite the rim region exhibiting the plains unit. The northwestern rim of Ölbers A (Plate 105c) crosses a depressed area relative to the northeastern rim, which crosses the elevated rim of a larger crater. Similar configurations are true of Rutherford (Plate 9a), King (AS16-1578), and the smaller crater shown in Plate 88b. Plate 245 shows that smooth-surfaced units appear to extend from or beneath bulk ejecta deposits, which may indicate a separation of the more fluid fraction of the ejecta. Large-scale analogs appear to exist around Orientale (Plate 199). It is also significant that the outer ejecta blankets of several of these craters exhibit strong asymmetries or peculiar patterns. The pattern associate with the crater seen in Plate 105a has been noted, and the swirllike deposits called Reiner γ are interpreted as possible ejecta products from Ölbers A (see Plate 183).

The suggestion that the rim units were from spillover of a much-higher floor level may be applicable to some craters, but this situation typically requires considerable subsidence, which seems possible for an impact crater only if considerable devolatization of the floor unit occurred or if the floor melt was connected with a magma chamber beneath the crater.

The hypothesis of tapping magma or impact-generated melt from the crater floor is believed to be a possible alternative to the first suggestion. The containment of portions of such rim units within closed or partially closed depressions suggests a mechanism involving internal processes. Note the extensive slumping of the northwestern wall in Ölbers A (Plate 105c) that may have been triggered by removal of underlying melt during this proposed peripheral eruption. In addition, the volcanically modified crater Marius (Plate 110b) indicates that eruptions occurred in the rim region. Although it is plausible that such rim vents were established prior to volcanic resurgence and that they were sites for eruptions immediately following crater formation, the mechanism responsible for elevating an impact-generated floor melt along fractures to the rim and forcing extrusions is not understood.

It is interesting and significant that these rim units typically differ in albedo and appearance from the floor units. Such differences can be rationalized in all the suggestions listed here. As either splash-out or spillover impact melt, the rim units represent higher temperature and thus more fluid melts than the residual floor melts. As tapped material, the rim units reflect the chemical evolution of the magma (internally derived or impact generated).

PLATE 105

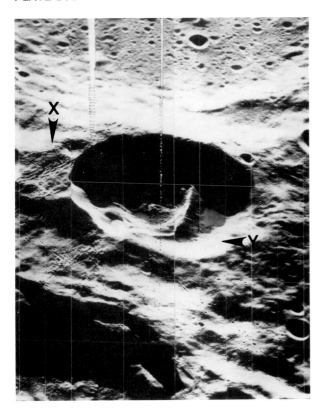

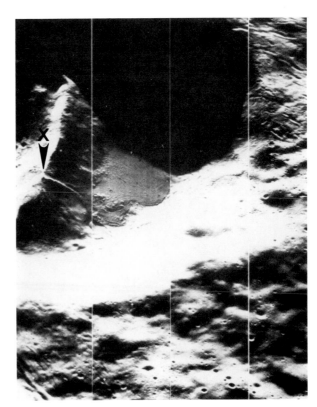

a. V-103-H3,H2 (oblique)

b. V-103-H3,H2 (oblique)

c. IV-174-H2   11 km

Subdued Forms

Plate 106 shows the ejecta blanket of Eratosthenes, 60 km in diameter, which will be compared to that of Tycho (Plate 101a). Both craters display a similar outer-rim zone, but the blanket from Tycho is spread over the highland regions in contrast to the mare surfaces surrounding Eratosthenes. The radial lineations around these craters appear to be relatively unaffected by obstructions. The Apennine Mountains apparently have not blocked a significant portion of the Zone III ejecta from Eratosthenes (Plate 106a, arrow X), and herringbone craters can be identified on the farside of several peaks (arrow Y). This observation is not consistent with the analogy between the bulk ejecta and density flows (at this distance) and implies deposition from ballistic debris. Blockage of the Zone II ejecta cannot be determined with certainty, owing to the coincidence of the peaks and the typical boundary of this ejecta zone.

A prominent bright-ray system extends from Tycho but is very obscure around Eratosthenes. Bright-ray systems are believed to darken slowly with apparent crater age (Shoemaker, 1962; Hapke and Van Horn, 1963). Several mechanisms may alter the lunar-surface albedo, and these are listed and discussed in the chapter "Albedo Contrasts" (Plate 185). In addition, photographs having full solar illumination of Tycho and Eratosthenes reveal dark haloes that approximately correspond to the Zone II ejecta blanket. Consequently, this albedo contrast appears to be preserved, whereas the bright-ray system has been destroyed. It is also possible that not all of the craters originally possessed high-albedo rays. The crater Picard (Plate 185d) displays a broad dark halo. The crater Aristillus (Plate 99a) exhibits large deposits of low-albedo material, and Dionysius (Plate 26a) has low-albedo rays. These are not unique examples and they comprise evidence that bright-ray systems may not have been produced or that these rays were later blanketed by low-albedo ashlike deposits.

In contrast to the "fresh" appearance of Tycho and Copernicus, Eratosthenes (Plate 106, b and c) displays at higher resolution comparatively little detail on the rim except for numerous craters, some of which may be the result of ejecta from Copernicus. The subdued appearance of many of the small craters could indicate a uniform erosional process over a long period or could be due to regional processes associated with regional events or composition. It also must be reiterated that these small subdued craters might be of internal origin, such as collapse into underlying cavities. The high-resolution photograph (Plate 106c), however, shows that the annular fracture pattern of Zone I has been either destroyed or covered. Hence, it is suspected that erosional processes were responsible for both this phenomenon and the subdued nature of the small craters. The Zone I patterns also could have been removed by mass transfer, such as slumping. Consequently, the subdued appearance may have been augmented as slumping expanded the crater rim toward the Zone I–Zone II boundary. The rim of Eratosthenes could easily be imagined to have once been similar to that of Copernicus or Aristarchus but has apparently been subjected to a variety of degradational processes.

PLATE 106

a. IV-114-H2   12 km

b. V-136-M   3.9 km (*left*) 3.3 km (*right*)

c. V-136-H2   0.46 km

Ray Patterns

Not all craters display symetrically distributed ray patterns like those found around Mösting C (Plate 80) and Censorinus (Plate 81a). Petavius B (35 km in diameter) is shown in Plate 107a and displays a missing sector in the higher-albedo ejecta to its north (arrow X). This zone is more clearly revealed under full solar illumination. Whereas the sector subtends only 115° in the case of Petavius B, the zone of Proclus (to the southwest of Mare Crisium) subtends about 140°. Tycho (Plate 101a) is also asymmetric with respect to its outer ejecta, the eastern array being much more extensive than the western. Messier A and its twin "whiskers" (see Plate 7b) are extreme examples.

The boundary of the sector is typically abrupt, and Plate 107b permits closer examination of this zone near Petavius B (Plate 107a, arrow Y). The radial pattern, which typifies Zone II, blends with the darker mare material (arrow X). The lineations also disappear near the base of the hill (arrow Y) but reappear on the other side (arrow Z) without obvious scarring of the hill itself. This could indicate active mass movement on the hill. Several other areas around Petavius B also exhibit low albedos, including those to the east and west (Plate 107a, arrows Z and XX, respectively). In the latter region, ray patterns (as revealed through Earth-based photographs) continue beyond this interruption of bright-ray material.

The question arises whether the sector around Petavius B is the result of asymmetry of the ejecta (Plate 85e) or local blanketing with low-albedo material (see Plate 185e). The question is not directly answerable in this example. In support for ejectal asymmetry are the laboratory experiments producing oblique and multiple impacts, large-size missile impacts, and the observation that the ray-to-nonray transition occurs abruptly over a wide variety of topographies. However, the large number of small flat-floored craters that characterize the region shown near arrow X in Plate 107a suggests blanketing. It is possible that this missing sector of the higher-albedo ejecta represents deposition of a low-albedo component that is analogous to the smooth patches associated with Aristarchus (Plate 258).

It should be noted that the asymmetry, at least around Petavius B, is primarily in the outer portion of the ejecta blanket, Zone II and Zone III. The sector lacking obvious ejecta debris is enhanced by the typically longer and brighter rays marking its boundary. Around Tycho, Zone II and part of Zone III are preserved, with the asymmetry resulting from the two long rays to the west and the more extensive Zone II and Zone III pattern to the east. Based on laboratory cratering studies of oblique impacts by Gault (personal communication, 1968), the direction of impact for Tycho might have been from the southwest; for Petavius B, the north; for Proclus, the southwest; and for Messier A (almost a grazing impact), the east. Alternatively, these asymmetries could reflect closely spaced simultaneous impacts resulting in the deflection of the ejecta (see Plate 85e) as reproduced in laboratory cratering by Oberbeck and Morrison (1974).

PLATE 107

a. V-037-M   3.9 km

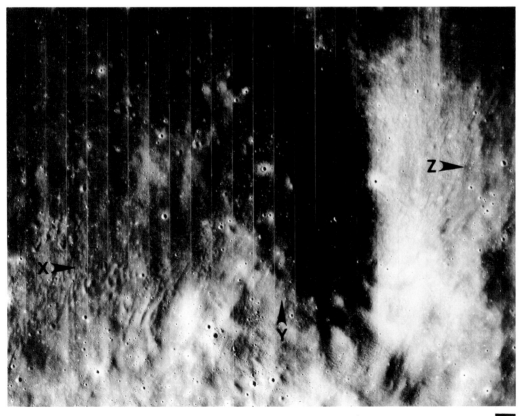

b. V-037-H1   0.51 km

Ejecta Flows

Plate 108 shows the large farside crater Tsiolkovsky, which has an asymmetric ejecta blanket and a massive striated flowlike unit (Plate 108a, arrow X). As is illustrated by other large craters, ejecta patterns are not consistently radial. This is particularly apparent for ejecta from Tsiolkovsky, where the NE-trending wall can be extrapolated across the surrounding terrain as a boundary between extensive deposition and little deposition of ejecta (arrows Y and Z). Northwest of this boundary, the surface appears to be relatively devoid of bulk ejecta (with the exception of the flowlike unit), whereas to the southeast it is heavily blanketed. The hummocky terrain seen in the lower right of Plate 108a is derived from the crater Langemak (arrow XX). Guest and Murray (1969) also noted the small quantity of ejecta to the northeast, which is clearly revealed by crater counts, and suggested this to be the result of a low-angle trajectory of the presumed projectile. Crater counting, however, does not reveal the apparent absence of the striated and barchanlike ejecta sequence to the northwest.

Another interesting feature of Tsiolkovsky is that, in contrast to the ejecta blankets of the other large and recent impactlike structures, its ejecta blanket does not display a high albedo; however, the inner wall does (see Plate 190b). The crater-forming process may have excavated lower-albedo material at depth that blanketed the higher-albedo component, which composes the ray system of smaller craters.

Guest and Murray (1969) have reviewed the geomorphology of the Tsiolkovsky region and, in particular, the characteristics of the striated unit on the northwest rim, which is shown in detail in Plate 108, b and c. They calculated the flowlike unit to be more than 100 m thick at its terminus (Plate 108b, arrow X) and thicker toward the crater rim. Furrows that are directed away from the crater (Plate 108c, arrow X) characterize this unit beyond 30 km from the rim, but transverse ridges (Plate 108c, arrow Y) are more typical nearer the rim. Guest and Murray commented that these grooves appear to be radial to Tsiolkovsky. I, however, feel that lineations farther to the west, see Plate 108a, arrow YY; Plate 108 b, arrows Y and Z) indicate a relatively parallel flow direction from a source to the southwest, which is displaced considerably from the center of Tsiolkovsky.

Guest and Murray also pointed out the peculiar curvature of the transverse ridges along the margins of the flow (Plate 108c, arrow Z). Ridges on a lava flow typically curve with their convex side downslope. Guest and Murray suggested that such a pattern was the result of slumping along the rim and that the flatter part of the flow unit "may have been emplaced by a less viscous flow developed at the foot of the slump and possibly as a density current of the base surge type" (p. 133). Note that the ejecta blanket (Plate 108a, arrow ZZ; Plate 108b, arrow XX) must have been deposited after such an event.

PLATE 108

a. III-121-M   48 km

b. III-121-M   48 km

c. III-121-H1   6.1 km

Ringed Plains

One of the objections to the interpretation that some lunar "craters" are the result of ring fissures is that there are very few examples in which one ring pattern continues within another. This might be understood if it is supposed that the original circular plan resulted from an ancient impact and that rejuvenated activity followed deep fractures originally expressed by the inner crater wall or scarps. Subsequent overlap by impacts of sufficient size would be expected to destroy the original fault zone of the previous event; therefore, the absence of intersecting extrusions along such zones might be rationalized. A possible exception is shown in Plate 109a, in which the wall and rim of a large crater appear to continue through a smaller crater (arrow X). In this case, the brecciated zone of the later and smaller impact did not extend deep enough to disrupt the rejuvenated fault zone defined by the plan of the larger crater.

Plate 109, b–d, includes more complex interconnections of ringed features in Mare Smythii. The individual forms resemble Gassendi (Plate 36) or Posidonius (Plate 2b), which were partly inundated by mare material. Plate 109, c and d, provides stereo and higher-resolution views of a complex found on the right side of the overview (Plate 109b, arrow X). Note that the overview of Mare Smythii is illuminated from the west, the illumination opposite in Plate 109, c and d.

The largest ring boundary shown in Plate 109, c and d, has a scarp on its eastern side (arrow X). Within it are two smaller ringlike forms: a polygonal crateral form adjacent to the eastern scarp (arrow Y) and a mare-filled ring adjacent to the western edge (arrow Z). The western ring member (arrow Z) may overlap the boundary of the largest ring, and two small rings are superposed on this overlap (arrow XX). Within the mare-filled region are annularly aligned hills and ridges as well as a central subdued peak. To the south is an arcuate ridge (arrow YY) that overlaps the eastern scarp and possibly the mare-filled structure to the north, as well. The same ridge appears to form part of the eastern boundary of another smaller discontinuous ring.

If each ridge represents an old impact-crater rim, then a plausible sequence of overlapping impact events is not obvious. Consequently, these ring structures are interpreted as either calderas or resurgent impact craters. The complex ring arrangement results from multiple periods of caldera formation or structural rejuvenation of faults related to ancient craters.

Note the wrinkle ridges (adjacent to arrow YY) that cross the mare-filled ring and possibly are connected to the scarp on the mare region to the south. Because these ridges cross a variety of terrain and because they appear to be related to a fault, the ridges are interpreted as structurally produced features (see Plates 136–150 for further examples of wrinkle ridges). Also note the asymmetry in the bright-ray pattern of the small but prominent crater on the southern edge of the mare-filled ring. To the north, the pattern is radial, but, to the south, it is chaotic. This is probably due to impacting on the edge of the ridge, resulting in low-angle trajectories to the north but high angles to the south (see Plates 183–184).

244

PLATE 109

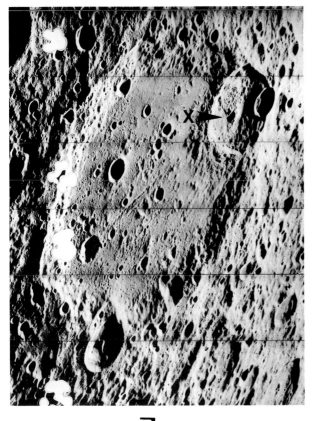

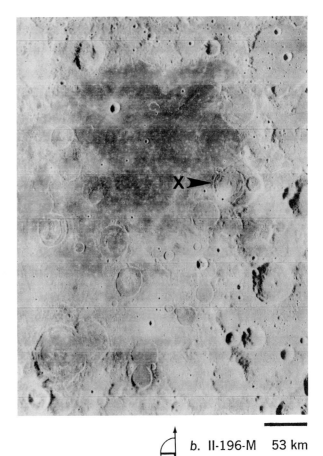

a. V-079-M (oblique)

b. II-196-M   53 km

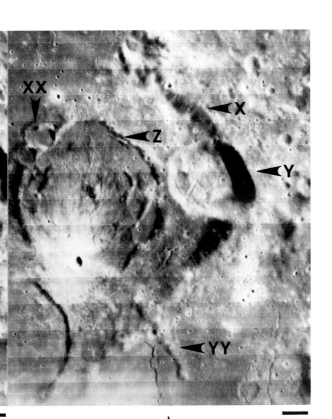

c. I-017-M   6.8 km

d. I-019-M   6.8 km

Ringed Plains (and partly inundated craters)

Plate 110 illustrates further variety in crater rim appearances that result from constructional processes or later modifications. Seleucus (Plate 110a) is in western Oceanus Procellarum about 75 km east of the crater Eddington (Plate 111). It rests on a mound of hummocky and stippled material that has been surrounded by mare material. This mound exhibits traces of radial lineations (arrow X), which may represent remnants of ejecta patterns analogous to the inner Zone I rim region of pristine craters. Consequently, the moundlike appearance may be simply these inner ejecta deposits and has been enhanced by the surrounding mare units. Such rim profiles are typical around large pristine craters but are not noticeable unless examined under low solar illumination. Note that the contact between the mare and mound is commonly abrupt, with small outward-facing scarps (arrow Y). Also note the rim-wall crater chain (arrow Z) that, it is suggested, was endogenetically produced.

The crater Marius (Plate 110b) is also surrounded by mare material. Islandlike remnants of its Zone III ejecta are recognized to the northeast (arrow X); therefore, Marius is interpreted as a premare Copernicus-like crater. In contrast to Seleucus, however, Marius exhibits marelike units on the inner rim area (arrow Y). These units typically are topographically above the surrounding mare plains; consequently, they are thought to be either remnants of rim pools or mare material extruded along rim fractures. The latter interpretation is supported by the irregular rimless depression (arrow Z) along the southeastern rim that connects with the mare units.

Although the mounded rim appearance of Seleucus and Marius is certainly enhanced by the Zone I and Zone II ejecta, some of this relief is thought to represent uplift of the inner rim region by stratigraphically controlled lateral flow of subsurface material during crater formation. This process was suggested for smaller craters (Plate 76a) by analogy with impacts produced in the laboratory. Such uplift would be favored for craters formed on the maria where a well-ordered stratigraphy with interbedded competent and incompetent units is likely. A mounded inner rim appearance will be enhanced by slumping that extends beyond the sharp wall-rim break.

Plate 110c shows the crater Lavoisier A, which is in western Oceanus Procellarum. It illustrates the outward-facing scarps (arrows X and Y) that commonly develop in association with the emplacement of mare and other plains-forming units (see also Plate 92). These scarps may have been formed by mass wasting, such as slumping, that was triggered by undercutting due to movement of the extremely low viscosity mare units.

The irregular craterform Wolf (Plate 110d) does not exhibit the rim appearance of Marius or Seleucus. It is surrounded, in part, by a flat-topped rim (arrows X and Y) that typically exhibits considerable relief relative to the platformlike outer-rim area (arrow Z). The southeastern rim appears to be breached by an ill-defined NW-trending depression (arrow XX) that is reflected by a smaller rille on the mare. The cloverlike plan is probably the result of multiphased formation along such structural weaknesses. Wolf is interpreted as a caldera that was formed prior to the last stages of mare emplacement.

PLATE 110

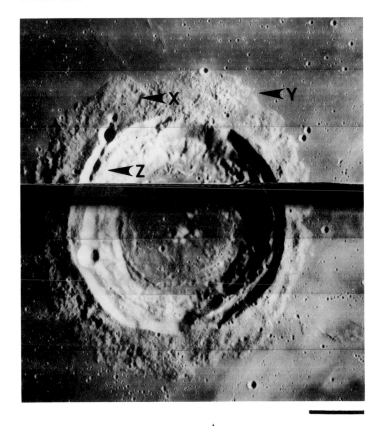

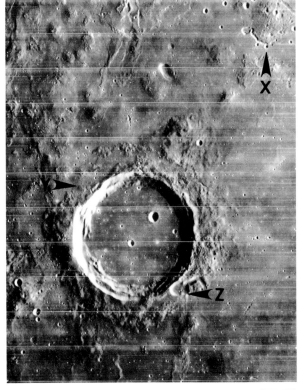

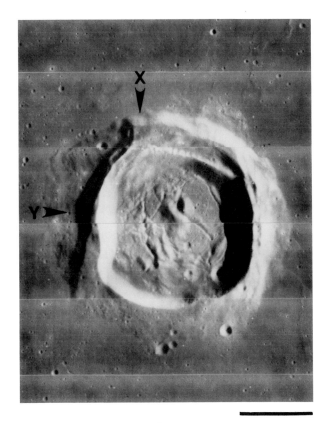

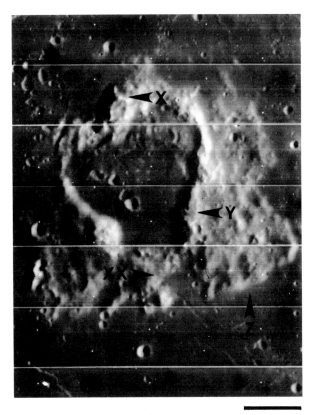

a. IV-169-H3 11 km

b. IV-150-H2 11 km

c. IV-183-H2 12 km

d. IV-120-H1 12 km

Ringed Plains

Ringed plains generally are interpreted as partly inundated craters. Plate 111 shows the craters (starting at the top and proceeding clockwise) Russell, Eddington, and Struve. The rims of these old craters have been heavily modified by ejecta and faulting associated with the Orientale event to the southwest. In particular, note the furrow (arrow X) and the large ridge (arrow Y), both of which are radial to the Orientale basin. The large crested ridge forms the boundary between Struve and Eddington and does not resemble an ejecta product. It is interpreted as fault-produced relief. The remaining portions of the rims are platforms bounded by relatively well defined scarps. Within topographic lows on these platformlike rims are highly cratered high-albedo plains-forming units (arrow Z), which resemble the unit on the floor of Struve (lower left). Arrow XX identifies a large secondary crater from Orientale. Aside from the obvious modifications by the Orientale event, the rims may have been sites of extrusive activity. In particular, the sinuous rille (arrow YY) is interpreted as a possible lava channel (see Plates 171–177), and the low-albedo scalloped crater (arrow ZZ) may represent a volcanic or volcanically modified form.

If this crater complex had been degraded or inundated more extensively, it would have been classified as a ring structure—which does not have a genetic connotation—similar to the Flamsteed Ring (Plate 154). As an aside, note the irregular and polygonal platforms adjacent to Russell (arrow XY). Such features are illustrated and described in the chapter "Positive-Relief Features" (Plates 122–124).

PLATE 111

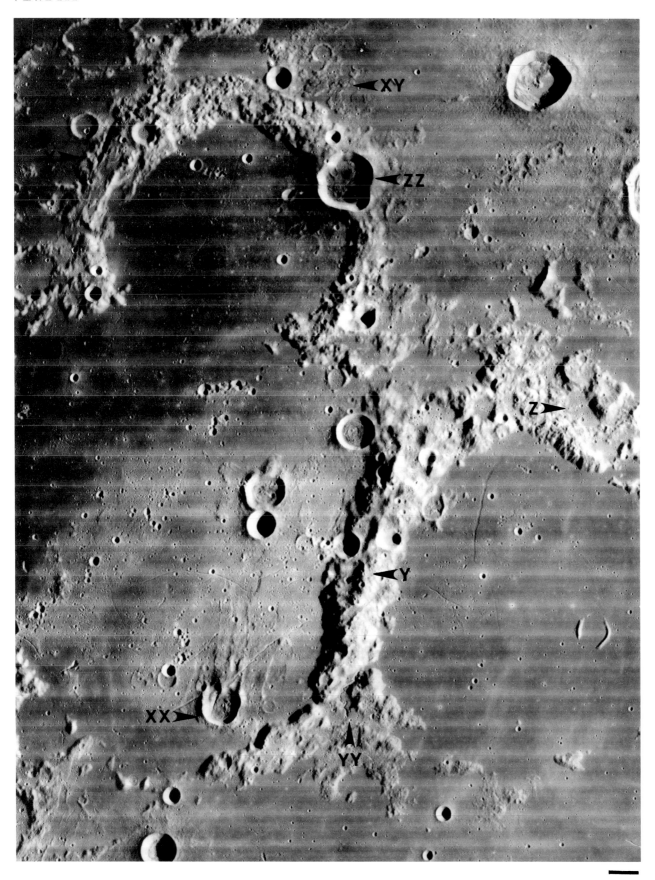

IV-174-H3   12 km

Lunar Basins

The Apennine Mountains (Plate 112; also see Plate 192a) comprise the eastern scarp bordering Mare Imbrium and can be viewed as the rim of a gigantic basin. Such basins typically exhibit multiringed boundaries; therefore, their outer scarps may not represent a "rim" with the same connotation associated with smaller (less than 200 km in diameter) craters. It is plausible that the initially formed basin was much smaller but that it rapidly enlarged through large-scale slumping along a plastic zone, which is perhaps the lunar mantle (see Mackin, 1969). Consequently, ejecta may be mantle material erupted along concentric zones of weakness as well as debris heaved by the initial impact. In addition, the outer scarp may represent the crown that borders interior slumps, rather than being an overturned lip.

Plate 112 shows the region around Mt. Bradley (arrow X), which is typically 2.4 km above the mare plains to the northwest. Both transverse (arrows Y and Z) and radial corrugations (arrows XX and YY) characterize the area beyond (southeast) the scarp. These may be surface features of massive base-surge flows, SE-directed slumping, or annular faults related to the major scarp. Farther from the major scarp, massive units have been deposited that resemble the bulk ejecta Zone II region surrounding smaller craters. Secondary craters from such enormous events must have formed in Zone III, but their identification is difficult, owing to confusion with degraded primary craters. I believe, however, that the high crater density recorded in the highland regions is largely the result of secondary impacts (see Plate 121a) from such basins. In some examples (Plate 111, arrow XX), these secondaries resemble those surrounding smaller craters, such as Copernicus, and are more easily identified.

For additional illustrations and discussions of features associated with large basins, refer to Plate 133 (Imbrium) and Plates 198–199, 209, and 213 (Orientale).

PLATE 112

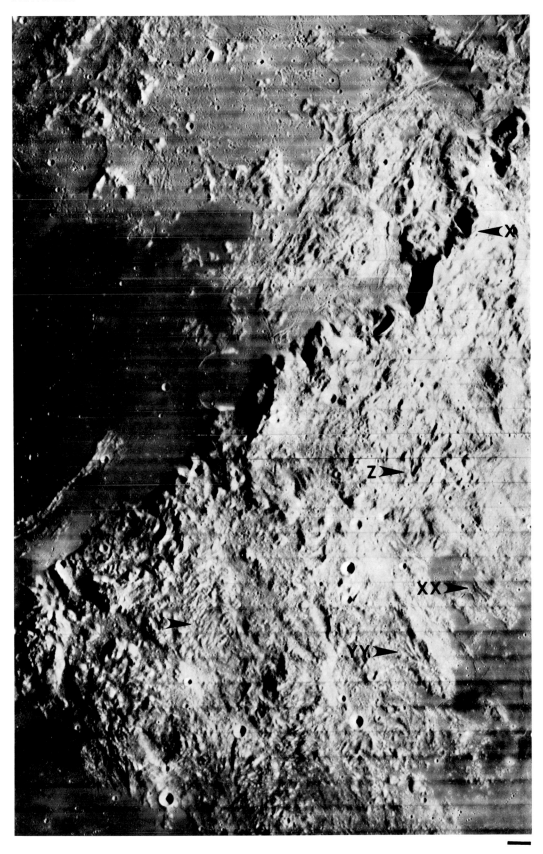

IV-109-H3,H2   12 km

Lunar Basins

The southern and northern borders of Mare Crisium are shown in Plate 113, *a* and *b*, respectively. They illustrate the platform-like rim of a large (320 km by 450 km) mare basin.

Plate 113*a* reveals the highly stippled appearance of the southern rim area, with intervening relatively high albedo and highly cratered plains-forming units. There is a marked absence of extensive lineations radiating from Crisium except along relatively restricted zones near the southeastern (upper left, Plate 113*a*) and northwestern borders (lower right, Plate 113*b*). Such zones, however, also align with a regional trend that is tangential to the eastern mare border.

Plate 113*b* shows the separate platforms that compose the northern rim. The stippled terrain (Plate 113*b*, arrow X) is similar to that along the opposite rim (Plate 113*a*, arrow X), and appears to overlap the adjacent inundated crater.

The Crisium basin is probably an ancient form analogous to Mare Imbrium (see Plates 112 and 192*a*), and the general absence of radial lineations—characteristic of Imbrium—might be an indication of its age. It could also be, however, an indication of extensive volcanism that overlaps these ejecta, and such activity is perhaps represented by constructional platforms. The stippled units are similar to the Gruithuisen domes (Plate 130, *a* and *b*) and could be viscous extrusive units. They also resemble portions of the interior of Mare Orientale (Plate 205), but preservation of an analogous unit in Crisium is inconsistent with the absence of radial lineations, if these have been removed by secular erosional processes.

Smaller craters similar to Crisium include Pitatus (Plate 42, *a* and *b*) and Regiomontanus (Plate 47, *a* and *b*). They exhibit the islandlike platforms that comprise the rim. These platforms might be remnants of rims, rather than volcanic constructions, that have been isolated by dropped terrain and the emplacement of high-albedo plains-forming units. Such a complex structural history is possibly the result of isostatic adjustment of the crust along fractures created by the basin- or crater-forming events.

PLATE 113

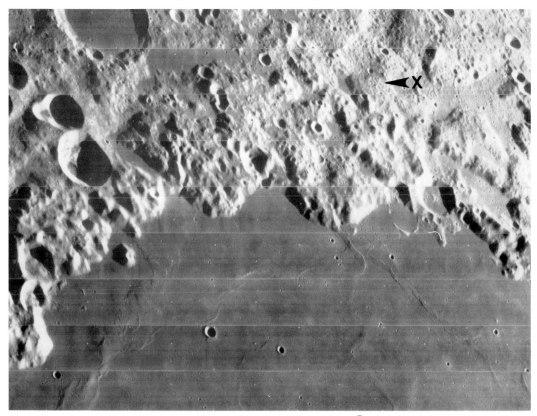

a. IV-191-H3   (oblique)

b. IV-191-H3   (oblique)

Lunar Basins

The Altai scarp (Plate 114) is the outer ring of a concentric set of scarps surrounding Mare Nectaris. Plate 114, a and b, shows approximately the same region of the Altai scarp with nearly vertical and oblique views, respectively. The scarp represents typical elevation differences of 0.3 km–0.6 km. Plate 114c is a high-resolution photograph of the region marked by arrow X in Plate 114, a and b. The rugged appearance given by terrestrially based telescopes, as well as in the photograph shown in Plate 114a, is primarily the result of poor resolution. The scarp is highly textured and has a relatively low crater density. Its base makes a poorly defined contact with the downthrown regions (Plate 114c, arrow X). Plains-forming units characterize both sides of the scarp (Plate 114, a and b, arrow Y).

The Altai scarp might not be an eroded remnant of an Apennine-like scarp (Plate 112). Although the ejecta blanket of Nectaris basin has been largely removed, several radial features remain (Plate 114a, arrow Z). The Rheita Valley is a prominent radial structure (not shown) thought to be related to Nectaris basin; it is similar to the Alpine Valley (Plate 179a), associated with the Imbrium basin, and to Bouvard Valley (Plate 209), with the Orientale basin. If this interpretation is correct, then it seems inconsistent that large mountainous peaks, which once encircled Nectaris basin, could be degraded to the extent shown in Plate 114 but that relatively minor relief, such as Rheita Valley, could remain preserved. In addition, the Altai scarp appears to be relatively continuous, despite the inferred degradational history. Hence, it is suggested that this scarp was rejuvenated or formed in partial adjustment to the emplacement of the relatively high density mare basalts, resulting in subsidence of the inner basin region. Western Mare Humorum (Plate 191) shows evidence for a similar structural history. The extensive blanketing by Imbrium ejecta must conceal, in part, the older lineations from Nectaris in the region shown in Plate 114, whereas the Rheita Valley was probably less affected, owing to its greater distance from the Imbrium event. The persistence of the Rheita Valley also may reflect continued structural development.

PLATE 114

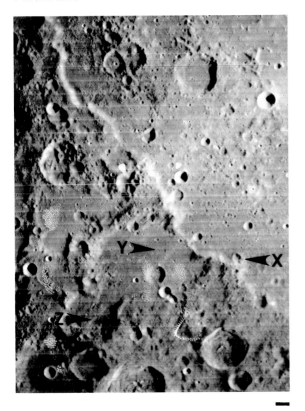

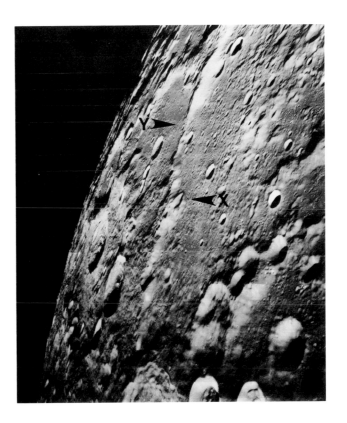

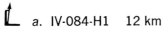 *a.* IV-084-H1   12 km

b. V-054-M   (oblique)

c. V-054-H2   (oblique)

DISCUSSION AND SUMMARY

Crater rims are described by their profiles and ejecta. Differences among craters reflect differences in origin, in impact parameters, or in subsurface structure. Considerable variety in rim appearance is observed throughout the selected size ranges.

Small (less than 1 km) pristine craters can be broadly classified in two types, depending on their ejecta blanket. The first type exhibits extensive small-size ejecta that commonly extend as a continuous deposit over a crater diameter from the rim crest. Craters less than 100 m in diameter typically lack identifiable structuring of an ejecta sequence, but this is in part the result of inadequate resolution. Larger craters (100 m–1 km) of this rim type display numerous radial patterns that are produced by grooves, block strings, and high- or low-albedo rays. In addition, they exhibit larger blocks (greater than 7 m) near their rim and secondary craters. Concentric groove patterns also occur beyond the continuous ejecta blanket, but these are more typically found surrounding similar-appearing craters larger than 1 km.

The outer rim for the second type of pristine craters is characterized by large blocks, some of which are larger than 20 m for a crater 0.5 km in diameter, but by a general absence of an extensive continuous blanket of ejecta. Blocks are generally restricted to within a crater radius of the rim crest and are asymmetrically distributed in azimuth. The rim profiles of both crater types are raised, but, for most, if not all, there is no sharp transition to the wall with an exposed rock face.

Craters without visible ejecta blankets are termed "subdued," but there are notable subdued-appearing craters that have large isolated blocks scattered across the rim. The rim area commonly displays an extensive tree-bark texture. The rim profile may be crested or rounded, and, in some examples, the crater is essentially rimless. In addition, craters with subdued rounded rims occur that have relatively sharp wall-rim contacts.

Several craters exhibit unusual rim appearances. Plate 16c shows a small crater with a platformlike rim area, a feature that also occurs around several other craters in the region. Near this crater are others that have radiating ridges and blocks along the inner rim but

show little evidence, other than the existence of raised rims, for bulk ejecta deposits (Plate 130, a and b). Very complex bulbous rim areas and scattered blocky ejecta are found in the region of Maestlin R (Plate 76a).

The variety in the rim appearance of small craters is attributable to several mechanisms:

1. Variations in the visible ejecta blanket due to:
   a. Subsurface competency and subsurface structure
   b. Projectile velocities or properties (secondary versus primary impacts)
   c. Crude sorting of ejecta debris resulting from the interaction of ejecta with either a surrounding debris cloud or ejecta from a nearby simultaneous impact
2. Subsequent surface processes:
   a. Secular erosional processes, such as meteoroid impact and creep
   b. Catastrophic processes, such as blanketing by nearby large impacts or volcanic events
3. Differences in origin

Fundamental laboratory work by Gault and Heitowit (1963) and Gault (personal communication, 1973) has revealed the mechanics of crater formation at small scales (less than 1 m) and provides insight into the possible impact mechanics of larger craters (less than 1 km) and their resulting rim appearances. During the earliest stage of an impact by a high-velocity projectile (6.5 km/sec), material believed to be fused matter is ejected at relatively low angles from the horizontal (less than 30°) with velocities three times greater than the projectile velocity. Most of the subsequent ejecta, however, leave the crater at considerably lower velocities and at higher angles. In profile, ejection angles are concentrated at approximately 45°. In three dimensions, these ejecta are found within a sheet that forms an inverted cone. As the crater enlarges, the apex of the cone extends to a greater depth while the ejection angle remains at approximately 45°. Ejected fragments increase in size and the ejection velocities rapidly decrease with time. In the later stages of crater formation, large chunks are ripped from the crater rim and are deposited a short distance from the final crater rim. A nuclear explosion buried at depth, on the other hand, expands as a spherical shock wave and ejects material within a wide range of angles. Whereas the end result of an explosion and impact may resemble each other, the mechanics leading to their formation are notably different.

The studies by Gault and Heitowit (1963) for impacts into solid basalt suggest that ejecta near the rim of a crater produced by a high-velocity projectile should resemble those around a low-velocity projectile—provided the targets are equivalent. This result is due to essentially similar last stages of the two cratering events. Therefore, unless a projectile has a very low velocity, a secondary and primary impact might be indistinguishable from each other. If this extrapolation to craters 1 km in diameter is valid, then only the nature of the target, i.e., subsurface material, will determine the crater appearance.

The key problem is the validity in extrapolating these laboratory-produced impacts to much larger craters, between 200 m and 1 km in diameter. If such an extrapolation is appropriate, then craters with extensive blankets of fine debris appear to be impacts into a surface that is relatively incompetent. This does not necessarily imply a thick regolith. Rather, it implies a target that may be interbedded with competent layers, but that, on the whole, acts as an incompetent unit. Where the subsurface contains a larger proportion of greater thickness of incompetent layers, the ejecta debris becomes blockier, and the crater plan may be affected. Additional variety results where the stratigraphy contains two contrasting layers—for example, a deep layer of incompetent material capped by a competent flow unit.

Öpik (1969) theoretically derived equations relating crater diameter and projectile penetration to parameters describing the projectile (velocity, density, angle of impact, dimensions, compressive strength) and the target (density, lateral crushing strength). These equations indicate that, for a crater of a given diameter, a small solid projectile will have a significantly greater depth of penetration than that of a nonsolid low-density projectile—for example, a cloud of debris. In contrast, as the velocity of a solid projectile increases, the depth of penetration remains approximately the same, whereas the diameter increases. Gault (1973), however, has

raised serious questions about Öpik's theory, based on empirical studies of high-velocity (0.8–7.3 km/sec) impacts of small (3×10^{-10}g–4 g) spheres of different densities and masses into basalt and granite targets. As Gault cautions, such questions may not apply for larger impact craters, e.g., 1 km in diameter. It seems intuitively reasonable that an impacting cloud of low bulk density (a secondary complex?) will not penetrate as deep as a solid projectile. If this is the case, then some of the peculiar crater rim deposits may reflect ultimately properties of the projectile.

Most well-established and well-preserved secondary impacts occurring in crater groups exhibit both ill-defined and relatively nonblocky (at available resolutions) ejecta blankets (see Plates 97, 100, 115). I believe this seemingly anomalous appearance of established "recent" craters indicates a low-velocity impact by either a solid secondary projectile embedded in a cloud of debris or a cluster of solid secondary projectiles. Such impacts probably will result in a complex interacting tertiary ejecta cloud. This situation opens the possibility for sorting of debris, or at least an alteration of the depositional sequence from that accompanying an impact in a noninteracting vacuum, which should be characteristic of a single primary impact. This alteration will produce not only the intervening septa, as demonstrated by Oberbeck and Morrison (1974) for multiple impacts, but also the destruction or burial of the complex patterns that commonly accompany isolated lunar craters.

Paradoxes result when a single mechanism is applied to all subdued craters. In several examples, the preservation of blocks on the rims of subdued-appearing craters is inconsistent with secular erosional processes. The proposal that some of these craters have been endogenetically formed is confirmed by their association with other endogenetic features, such as wrinkle ridges or flow features. Plate 78c illustrates the existence of a recently formed pit on an otherwise typical mare surface. In addition, the variety of rim appearances of small craters on the floor of large well-preserved craters indicates the possible variety in crater rims at the outset in other regions, including the strong possibility that some subdued craters in the maria are pristine forms.

Recently formed medium-size craters (1 km–25 km) with extensive ejecta blankets typically exhibit numerous blocks overlying a relatively smooth (0.10 km wavelength irregularities) surface. Zone II displays a complex array of radial furrows and ridges (less than 50 m wide); transverse ripplelike ridges (100 m wide); and smaller-scale corrugations. The transverse ripple ridges are commonly accompanied by parallel grooves trending away from the crater rim. These ridges are not strictly transverse in that they are not necessarily concentric with the crater rim. The outer Zone III region has elongate rimmed depressions that typically have major axes concentric with the primary crater. Around some craters, these depressions display unusual patterns, such as **V**- or **W**-shaped plans that are not concentric to the crater. This zone also is characterized by high- or low-albedo rays, topographically expressed as striated islands similar in appearance to portions of Zone II.

These observations apply to craters on both highland and mare terrains. Therefore, they probably reflect phenomena that are functions of impact parameters rather than surface and subsurface structure. Zone III exhibits isolated ejecta impacts that include patches of lineations not accompanied by craters. These are likely the result of secondary impacts by unconsolidated masses. Zone II represents deposition of bulk ejecta that indicates large horizontal velocities and a near-surface phenomenon that is revealed by topographic control of lineations. The transparency of this deposit to small pre-existing craters reveals that it is relatively thin. Zone I displays late arrivals of large blocks. If such blocks had high trajectories that returned after deposition of the ejecta blanket, then it seems remarkable that relatively few have associated craters. These blocks indicate low-velocity ejecta that were heaved during the last stages of crater formation from either deep within the crater or ripped-off portions of the crater rim, in analogy with laboratory-produced hypervelocity impacts. Regardless of these details, the blocks generally postdate the rapid deposition of Zone II bulk ejecta. The numerous small craters on the Zone II blanket also may be representatives of late-arriving secondaries. Around Mösting C (Plate 80), the small-crater density is notably lower in Zone I relative to the density in Zone II. Some

of the highly subdued craters probably are secondary impacts formed and smoothed soon after deposition of the bulk ejecta. The idea of extensive mobility of this deposit, however, is not realistic where narrow furrows and ridges have been preserved. Several subdued and rimless craters with funnellike profiles were probably formed by collapse into underlying cavities.

Characteristics of crater rims without these zonal subdivisions are summarized in the following outline:

1. Narrow, crested rim (raised lip):
    a. Blocky ejecta with subdued hummocky outer ejecta sequence (Secchi A, Plate 84)
    b. Apparently well preserved blocks found in widely separated clusters on a textured inner rim; subdued hummocky outer ejecta sequence (Gambart C, Plate 90)
    c. No apparent ejecta sequence (Plate 199c, arrow YY)
2. Broad, crested rim:
    a. Subdued ejecta blanket with numerous small craters (Alpes A, Plate 85c)
    b. No apparent ejecta sequence (Ritter D, Plate 60a; Bode E, Plate 23)
3. Ridgelike rims (Plates 87b, 91, 132e)
4. Broad, bulbous rim:
    a. Low-relief rim: tree-bark pattern on rim; commonly with narrow concentric depression adjacent to wall-rim contact (Plate 86, b, c, d, and e)
    b. High-relief rim: tree-bark pattern; over-all conelike profile (Plate 86a)
5. Broad, platformlike rim (Plate 92b)
6. Low rim or rimless:
    a. With ejecta blanket
        (1) High-albedo diffuse ejecta blanket and scattered blocks (Plate 85b)
        (2) Low-albedo diffuse ejecta blanket (Plate 188b)
    b. Without ejecta blanket (Plate 87a)

Inundation of the floors and surrounding terrains adds further variety to the rim profile. For example, Plate 92c shows outward-facing scarps along the mare-rim contact.

The different rim types at these scales reflect differences in origin as well as in surface processes. Some craters with narrow crested rims but without extensive ejecta sequences may be formed by secondary impacts. If they represent degraded primary impact craters originally similar to Mösting C, then the outer hummocky ejecta sequence has been degraded extensively without removal or destruction of blocks near the rim. The existence of such enigmatic craters should not detract, however, from the large number of premare craters that seem to be products of long-term degradation. On the other hand, the existence of enormous impact basins suggests—through analogy with smaller (100 km in diameter) impact craters—that secondary craters as large as 25 km in diameter may be formed. These secondaries probably represent a major portion of the crater density in premare highland surfaces (Plate 121a).

The remaining rim types outlined above are suggested to be products of endogenetically formed or modified craters. The greater breadth of these crater rims reflects volcanic eruptions that are less violent than a meteoid impact. Whereas most of the ejecta from an impact event are thought to be ejected at approximately 45° from the horizontal, ejecta from a volcanic eruption are vertically directed by the conduit. This fact, coupled with lower ejecta velocities (100 m/sec), gives reasonable fits from the classical ballistic formula to observed lunar volcanic-appearing craters and cones. This calculation predicts, however, that cone-like forms will not have flanks at the angle of repose because ejecta will be more widely dispersed under the lower lunar gravity. Wright et al. (1963) concluded that this dispersion would be so great that lunar volcanic cones could not be formed. McGetchin and Head (1973) re-examined this problem and constructed theoretical rim profiles based on angles and velocities of ejecta from terrestrial cones. Their results also indicate broad-rimmed vents. Therefore, the volcanic origin for some of the craters categorized in the outline seems to be a reasonable hypothesis. It should be noted, however, that there are lunar volcanic cones that do not have the extremely broad low-relief rims as predicted by these calculations (see Plate 130, e and f) but have instead rims resembling terrestrial cones. Such features must indicate less violent and extended eruptions. They also in-

dicate that the existence of a raised rim does not exclude the possibility that it is volcanically formed.

The rim area of pristine craters greater than 25 km in diameter is similar to the area around smaller pristine craters but is typically more complex. The inner zone commonly exhibits corrugated patterns, some of which are fractures, concentric to the rim, with overlying flows and pools that indicate a range of viscosities. Flows and pools around some craters may represent extrusions along concentric faults or molten ejecta. There are typically three broad types of flows: thin lobate units (Plate 245), thick lobate flows that are associated with large leveed channels (Plates 104 and 245), and thick ill-defined flows resembling rubble (Plate 245). These units generally postdate deposition of the Zone II blanket. Some thin lobate units commonly have no apparent source and appear to be genetically related to the thick rubbled flows. These thin units may have been molten ejecta entrapped in bulk ejecta but released near the termini of the thick flows. Other thin lobate units and some thicker lobate flows can be traced back to breached depressions, and the flow-direction is clearly controlled topographically rather than directed by ejecta impact (Plate 247a). These are possibly extrusions along the rim that followed deep-seated fractures. Alternatively, they represent once-molten ejecta originally entrapped in the bulk ejecta deposits, and their release produced the breached depressions.

Pools of material within Zone I have smooth surfaces with relatively low crater density and commonly display lobate borders. Some pools are derived from cascades of molten material (Plate 103, a and b) that can be traced to pooled depressions. They are interpreted as extrusions associated with annular fractures or remnants of ballistic impact melt that sought the lowest local elevation. The difficulty with interpreting such pools as extrusions is the unknown driving force necessary for rim eruptions. However, pools of ballistic impact melt require that the melt remained relatively undispersed after impact. Other pools are in closed depressions and have polygonal fractures that presumably are related to cooling (Plate 102a). These might represent extrusives or deposits of the *nuée ardente* type of ejecta. Around Tycho, the most extensive array of such flows and pools is either within or adjacent to an annular moat.

Although these phenomena are rarely found around small craters, they do exist. In particular, Plate 105b shows a large leveed channel extending down the outer rim of an interior nested crater. Material that flowed through this channel forms a large lava unit. Plate 88b includes a smaller crater with several small leveed channels originating near the rim crest.

The hummocky and lineated Zone II region around smaller craters is replaced by dunelike forms and craters with herringbone plans for large craters. The transition between Zone I and Zone II annuli is marked by similar but more subdued and hummocky forms. The Zone II region represents deposition of the bulk ejecta. The mechanism responsible for the dunelike forms is believed to be the result of the interaction between secondary impacts and one or a combination of the following: a base-surge event producing base-surge duning (Guest and Murray, 1971), later-arriving secondary debris, and adjacent nearly simultaneous impacts (Oberbeck and Morrison, 1974). The absence of herringbone secondary craters around smaller craters probably indicates small-sized secondary ejecta that did not form craters large enough to compete with the bulk ejecta or accompanying debris cloud.

Blocks and other low-velocity ejecta debris in Zone I appear to overlie the Zone II sequence; furthermore, there are typically a large number of small (less than 1 km) craters in this zone. If these debris represent ballistic ejecta, i.e., ejecta tossed out during crater formation and not from fallout from an ejecta plume, then the Zone II sequence must have been emplaced very rapidly, a situation that may not be consistent with the base-surge analogy described by Moore (1967). Oberbeck (personal communication, 1973) also has questioned the base-surge analogy on different grounds. He notes that the base surge in nuclear explosions develops when the explosion occurs at depth, and such conditions are not analogous to impact cratering. However, owing to the topographic control of linear patterns within Zone II and inner Zone III of lunar craters, it is clear that a ground phenomenon has occurred. But it is unclear whether these lineations and the Zone II ejecta blanket in general indicate one or a combination of a base surge as envisioned by Moore (1967), secondary and tertiary ballistic debris, or a turbulent gas cloud produced by volatization (Rehfuss, 1972).

Isolated secondary craters and ray deposits characterize Zone III. The herringbone pattern associated with secondary craters shows that the pattern occurs outside the influence of the Zone II bulk ejecta, thus eliminating the possibility of base-surge duning in this region. It is suggested that such secondaries were complexes of liquid, gas, and solids. The interaction between the rapidly expanding tertiary ejecta from the impact of embedded solid material and the horizontal velocity component of the surrounding matrix is one plausible mechanism for the formation of the herringbone pattern. A similar interaction has been proposed by Oberbeck and Morrison (1974) in which ejecta from multiple secondary impacts interact.

The rays around smaller craters are commonly associated with a scouring of the surface rather than with numerous individual secondary craters. Their high albedo perhaps indicates mineralogic composition but also might reflect changes in surface roughness, glass content, and mean particle size by this scouring. Oberbeck (1971) has suggested that rays emanating from Copernicus are produced by numerous small (typically 10 m–20 m in diameter) bright-haloed craters, which were interpreted as secondary impacts from Copernicus that have excavated high-albedo bedrock from beneath the regolith. Interesting problems arise from this explanation. If these small craters are indeed Copernicus secondaries, then they generally are in an extremely well preserved state. Such preservation requires that most other features associated with the formation of Copernicus be similarly preserved, but this condition reveals an extremely wide variety of pristine crater types, including highly subdued forms (Plates 227–229). In addition, preservation of these secondary craters, which typically have concentric plans, requires that they impacted after formation of the large subdued secondary complex on which they commonly occur and that the supposed highly reflective subregolith material occur in this region as well.

In view of these restrictions, I feel that additional mechanisms must be operative. For ray systems around larger craters, a combination of surface alteration and ejected material from the primary, as well as excavated subregolith material, is inferred. Surface alteration is believed to be performed by tenuous clouds of secondary debris, rather than discrete projectiles. Rays associated with larger subdued secondary crater fields are interpreted as excavated subregolith material due to the formation of these complexes and not the smaller blocky concentric craters. Much-later impacts may repeatedly exhume this highly reflective material and thus lessen the rate of darkening. Further discussion of crater rays and the parameters affecting surface reflectivity is found in the chapter "Albedo Contrasts."

Large craters having these rim subdivisions typically display interconnected crater chains that extend for enormous distances with arclike, spiral, or relatively straight plans. It is suggested that several such chains are structural phenomena, owing to their tangency with the parent crater and topographic offsets across their widths. The majority of examples, however, appear to represent multiple secondary impacts from large-scale filamentary loops of ejecta, as proposed by Shoemaker (1962).

Degradation of Copernicus-like rim sequences may be inferred from low photographic resolutions by the following parameters:

1. Zone III
   a. Disappearance or darkening of the bright-ray systems
   b. Removal or smoothing of outer isolated secondary impacts
2. Zone II
   a. Smoothing of hummocky terrain and dunelike forms
   b. Increased crater density
3. Zone I
   a. Disappearance of raised rim
   b. Degradation of concentric fractures
   c. Removal or degradation of pools and flow features

Each of the parameters listed relies on important assumptions that are not consistently valid. Bright-ray systems in Zone III do not always accompany large craters, and an apparently darkened ray system could be in a pristine state. Smoothing of outer secondary impacts might be a reliable parameter at higher resolutions, but such impacts are characteristically subdued at the time of formation. In Zone II, smoothing of dunelike forms requires the assumption that all craters of this size produce similar-appearing features. Furthermore, a high crater density in this zone may indicate a large number of late-arriving secondaries rather than a long exposure to primary impacts. The absence of a raised rim in Zone I may be the result of crater expansion through slumping (see Plates 12b and 133a). This also may remove the zone of concentric fractures. Removal of pools and flow features requires, of course, their initial existence, but their apparent degradation seems to be a reliable parameter. These are perhaps moot points, but they should be considered when an apparent age is assigned simply on the basis of rim appearance.

In general, the surface features of the bulk-ejecta deposits in Zone II appear to be most consistently destroyed relative to other crater features. The raised rims of some craters are found to persist even though the wall and ejecta blanket are highly subdued. If this is actually a result of noncatastrophic degradation, then it reflects the differential smoothing of mobile bulk-ejecta deposits relative to the competent bedrock of the rim crest. Clearly, quantitative studies of crater profiles and ejecta forms are needed.

The most reliable indicator of relative crater age is the classical approach of superposition. Burial of the ejecta sequence by mare units (see Plates 110, 185e, and 195–196) and the overlapping of ray systems are good illustrations.

The gradual destruction of a Copernicus-like crater can be readily confused with volcanically formed or rejuvenated craters. This is illustrated by ringed plains that display numerous indications of volcanic processes. Ringed plains in this size range have several general rim-wall appearances:

1. Relatively continuous or discontinuous rims with crested profiles (Plate 109, b, c, and d; Plate 28f)
2. Discontinuous rims that are composed of numerous

261

mamelon-type domes (Flamsteed Ring, Plate 154)
3. Flat-topped rims with inward- and outward-facing scarps (Pitatus, Plate 42, *a* and *b*)
4. Hummocky and highly complex massifs (Plate 111)

Plate 110*d* illustrates complex rim regions that possibly imply a volcanic origin for the craters themselves. For example, flat-topped rim profiles are interpreted as the combined result of an encircling ring graben and interior floor rejuvenation. Such volcanic resurgence either enhanced an encircling rim syncline developed during an initial impact event or is an expression of rapid isostatic adjustment in response to a shallow regional layer of molten or plasticlike material. An end result is an exhumed platformlike rim region, the inner raised portion having been removed previously during crater slumping or masked by surrounding topography.

Rim rejuvenation also can produce a crested rim profile, such as that for the northwest rim segment of the floor-fractured crater Doppelmayer (Plate 191*b*, arrow X). This portion of the rim is bordered by a moat exterior to the crater, and the local rim profile is in distinct contrast to other portions of the rim as well as to the generally subdued rim morphology. The moat is believed to represent a collapse region analogous to the encircling grabens responsible for the platform rim profile. Such a process and the end result indicate a plausible explanation for the relatively symmetric profile of ringed plains.

Thus, these rim types simply may represent faulted and inundated craters that initially resembled Copernicus. They have been extensively modified, however, by volcanic processes.

5. Crater Groups

INTRODUCTION

Craters commonly occur in groups that are defined and illustrated by two subgroups termed "crater chains" and "crater clusters." They are divided further into the following classes:

1. Aligned craters (open chain or cluster)
2. Adjacent craters (contiguous chain or cluster)
3. Connecting craters

In each subclass, craters with various degrees of separation are included, and a particular chain or cluster may display all three classes. In some examples, the degree of separation is reduced with age because of degradational processes. Uncertainty can arise at either end of the separation scale. Several aligned craters compose a possible open chain near Secchi X (Plate 57a), and, if it were not for the related grabenlike rille, the designated classification would be questionable. At the other extreme, Rima Plato II (Plate 173c) might be regarded as a contiguous chain.

At least two craters must obviously compose a crater group. The origin of such pairs could be merely a statistical coincidence of impacts, but genetic relationships should also be considered. The distinction between fortuitous placements and actual causal relationships is not a trivial matter. Terrestrial volcanoes typically possess multiple conduits where the major vent has been tapped by "parasitic" routes for magma escape. This has prompted several investigators over the history of selenology to propose that large craters with associated smaller craters are lunar volcanoes (see, for example, Plate 88). The interpretation is plausible, but in many cases the arguments become little more than philosophical because of a lack of definitive data.

A similar problem is encountered where gigantic crater associations, such as the Ptolemaeus, Alphonsus, Arzachel "chain" (Plate 44) or the Fra Mauro, Bonpland, Parry "cluster" (Plate 121c), are considered. Statistics of such groupings cannot offer proof of the existence or nonexistence of possible genetic relationships. Morphologic studies may yield plausible arguments for selected examples, but such proposals cannot be generalized. Thus, the solution must await inspection by the field geologist.

Groups of more than three craters also do not escape multiple interpretations regarding their exogenetic or endogenetic nature. Ejecta from large events, for example, may yield a long string of secondary impacts. In several examples to follow, ambiguities have been largely discounted owing to their particular appearance or relation to nearby structural features. Others—for instance, the composite chain shown in Plate 117d—are not so obvious.

Small-Crater Clusters and Chains

Plate 115a shows small crater chains and clusters with N (arrow X) and ENE (arrow Y) trends. The slight herringbone pattern associated with at least one crater suggests that they are isolated secondaries from Copernicus (N trend) and Aristoteles or Eudoxus (ENE trend). The origin of the intervening septa and herringbone ridges is discussed with Plates 96 and 100.

Plate 115b displays a close view of an open cluster with several connecting craters. Of particular interest, however, is the tree-bark pattern associated not only with the craters themselves but also with the area around the cluster. Blocks can be identified along the walls of several of the craters, and several blocks occur on their outer rims (arrow X). The area is more hummocky that the surrounding mare material and is similar to the terrain associated with ray material and secondary ejecta. In addition, the craters are dispersed in an ENE direction, and parallel lineations can be identified (arrows Y and Z). As discussed in the chapter "Crater Rims" (Plates 78 and 96, discussion, and summary), these features are characteristic of secondary crater complexes. This enigmatic association of subdued craterform and large angular blocks is thought to indicate complex deposition from a turbulent tertiary ejecta plume. Such a departure from the depositional sequence found around primary craters is created by interacting ejecta between adjacent simultaneous impacts as well as between these impacts and a debris cloud accompanying the secondary projectiles.

The cluster shown in Plate 115c contains connecting and contiguous craters. In addition, the cluster boundaries are relatively well defined and the crater density does not diminish markedly toward the borders. Axes of connecting craters appear to be aligned preferentially with a NW trend (arrow X). This cluster is another product of secondary impacts but apparently is not derived from a nearby source to the northwest or southeast. A secondary cluster from Kepler is less than 8 km to the south, and, if this is also from Kepler, then the northwest trend does not have significance for its origin since Kepler is to the northeast. The bright vertical streaks in the photograph are blemishes.

Plate 115d presents an extreme case wherein the crater density is so great that it becomes difficult to distinguish between small craters and rolling topography.

PLATE 115

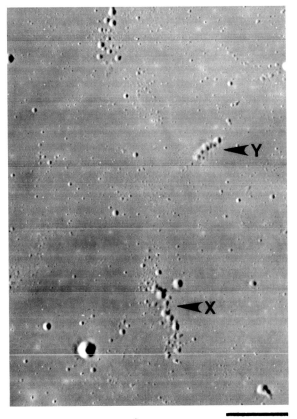

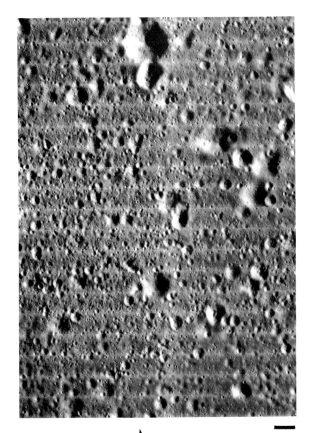

⬐ *a.* IV-122-H2   12 km

⬐ *b.* II-032-H2   0.21 km

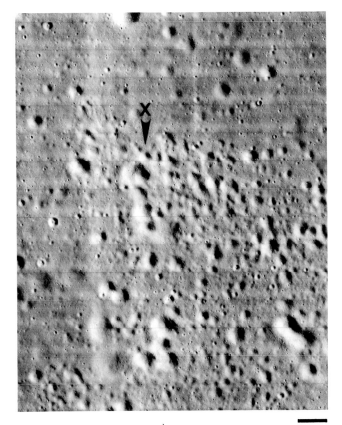

⬐ *c.* III-185-H2   0.23 km

⬐ *d.* II-180-H2   0.19 km

Medium-Size Crater Chains

The three types of crater chains are illustrated in Plate 116 on a scale larger than that of the crater groups considered in Plate 115. The first example (Plate 116a) shows an open chain of rimless craters. The similarity in appearance suggests a common, probably internal, origin. They are south of Hortensius, which has nearby several low-relief domes with summit pits (see Plate 125d).

The distinction between genetically unrelated and related craters in chains is illustrated by the crater pair shown in Plate 116, b and c. As the number of members increases, the probability that they are genetically related also increases. In considering only two craters, comparable genesis must rely heavily on morphologic similarities between them or other pairs.

Plate 116b shows a crater pair whose rims are not quite contiguous. The intervening ridge or septum (arrow X) is not unique; similar ones occur between the contiguous craters to the west (arrow Y). This configuration is probably diagnostic of simultaneous events, either impact or explosive. The eastern member of the pair is larger and appears to be more recent; at least its ejecta blanket is better developed. Under full illumination this impression is confirmed by the high-albedo halo around the eastern crater and a dark halo around the western. If the pair is the result of simultaneous impacts, then the cause for differences of ejecta appearance and photometry is not clear. This particular arrangement of craters and ridges is possibly due to chance, but that is not the case for the adjacent triplet and other similar-appearing chains.

The pair of craters shown in Plate 116c is more likely the result of two independent impact events. Ejecta from the eastern member clearly blanket the floor of its companion. Their similar sizes and lighting conditions result in a peculiar rim in which the two craters appear to have interlocking rims.

The last example in this sequence (Plate 116d) shows the remaining subclass of crater chains consisting of connecting craters. A genetic relationship among members is much more probable than in the previous two examples, but the activating mechanism is not so obvious. There is no prominent ejecta blanket, except possibly a few lineations. Consequently, the chain is interpreted as a result of nearly simultaneous impacts by secondary ejecta, an event that typically produces ill-defined ejecta facies.

PLATE 116

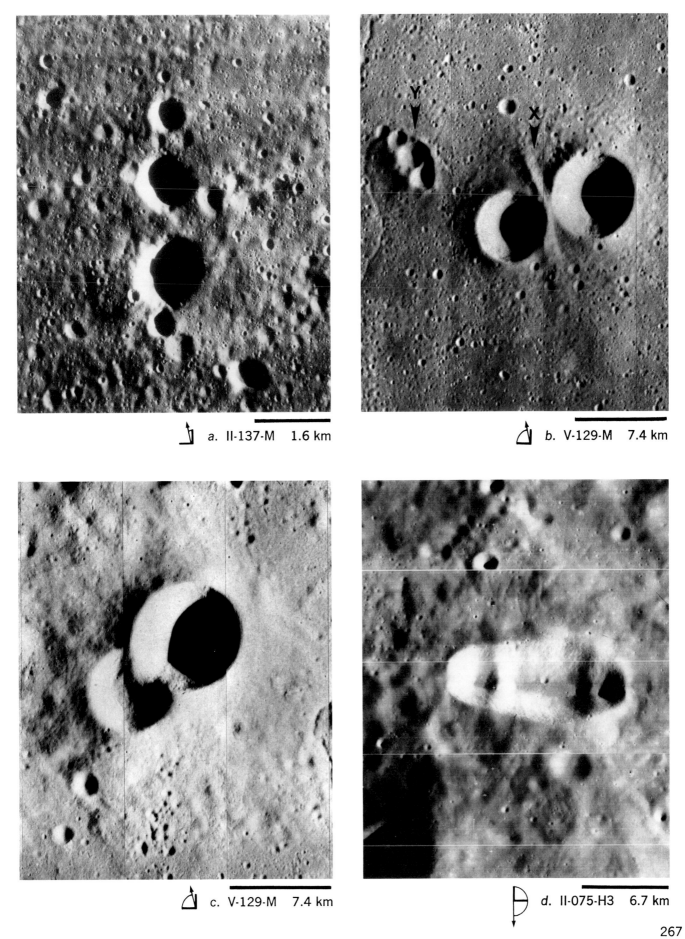

a. II-137-M 1.6 km

b. V-129-M 7.4 km

c. V-129-M 7.4 km

d. II-075-H3 6.7 km

Medium-Size Crater Chains

The stereo pair shown in Plate 117, *a* and *b*, exhibits all three subclasses of crater chains. The chain near the bottom of this pair is unconnected at its northwestern extension (arrow X) but becomes contiguous and connecting farther to the southeast (arrow Y). Since the chain members are without rims and have "pinched" ends, the origin of such an array is most probably collapse. Note the linear depressions within each "beaked" crater. Also note the funnellike centers of several flat-floored craters nearby (arrow Z). This chain is part of an incipent rectilinear rille in eastern Mare Fecunditatis (see Plate 178c) that appears to begin in a breached maarlike crater having a low-albedo halo.

The elongate raised-rim depression (arrow XX) seen in Plate 117, *a* and *b*, is similar to the chain shown in Plate 116d, except that in this example the floor is relatively continuous. At the other extreme, the elongate depressions along the grabenlike rille at right are designated an open chain because of their association with the rille.

Plate 117c shows the transformation in plan from rimless beaked craters into a rille. This particular example is in the Marius Hills (Plate 265, *a* and *b*, arrow ZZ) and also must be considered as the result of collapse into an underlying cavity, which was created either volcanically or structurally.

Plate 117d illustrates another composite chain having both connecting (arrow X) and open (arrow Y) associations of craters. In contrast to the three previous examples, this chain is at a larger scale, is in the lunar highlands, and exhibits craters with raised rims. If several of the component craters occurred separately from such a chain, it would be difficult to distinguish them from other isolated craters. A similar problem is illustrated near Secchi X on a mare (Plate 57a). Consequently, the interpretation that such craters have internal origins illustrates the difficulty in assigning endogenetic or exogenetic origins for isolated craters. The chain shown in Plate 117d could be the result of faulting and associated volcanic activity, or it could be degraded remnants of a string of secondary impacts. The crater seen in the upper left is Abulfeda, shown at higher resolution in Plate 70b.

268

PLATE 117

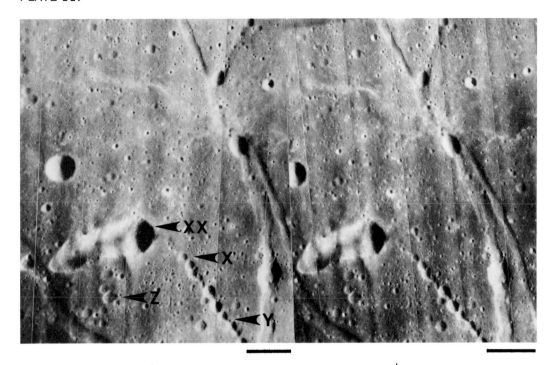

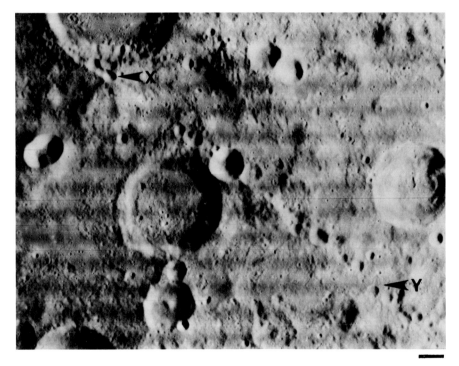

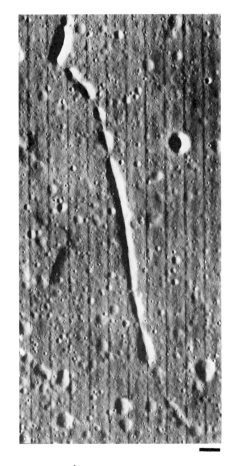

a. V-048-M   4.6 km

b. V-044-M   3.4 km

c. V-211-H3   4.8 km

d. IV-089-H2   12 km

Medium-Size Crater Chains

The complex association shown in Plate 118, a and b, occurs along the eastern border of Mare Serenitatis adjacent to the Littrow Rilles (see Plate 138). The appearance is similar to that of chains associated with large craters, such as Copernicus (Plate 96, c and d). The chain seen in Plate 118, a and b, however, appears to be more complex, and it is isolated from any reasonable parent crater. The stereo pair reveals lobate flow fronts (arrows X and Y) extending from either side of the chain. The crater seen at the western end can be examined in more detail in Plate 23c. Evidence for extensive and recent endogenetic activity occurs in the adjacent mare, and the chain shown here may have been associated with these events. If the high-albedo ejecta blanket represents pyroclastics from a volcanic eruption, then it implies an interesting evolution in the composition of the erupted material from previous activity, represented by the surrounding low-albedo shelf unit. If the chain is the result of simultaneous impacts, then the unit flows (arrows X and Y) extending from the shallow crater suggest an unusual control of the ejecta. Owing to the general similarity to other secondary craters, an impact origin is favored, but the choice between an endogenetic and exogenetic origin should be aided greatly by the detailed photography from Apollo 17.

The Davy Rille is included in Plate 118c and illustrates the transverse ridges that separate members of some crater chains (see also Plate 116b). The trend of the Davy Rille does not appear to follow any major regional stress system. It extends from near the crater Davy to the west (not included in Plate 118c) onto the highlands to the east, where it terminates in a large crater.

If the Davy system is volcanic, then it should be contrasted to other volcanolike features, such as the low-albedo subdued craters in Alphonsus (Plate 187). The dark-haloed craters in Alphonsus and similar large craters typically occur along fractures that are restricted entirely to the crater floor and probably represent extrusives derived from shallow depths related to floor rejuvenation. The Davy chain crosses floor boundaries; therefore, it may indicate more deeply seated stresses. In addition, it does not display a low-albedo veneer. Consequently, erupted material from this chain might be different in composition as well. If the Davy chain is accepted as volcanic in origin, then the existence of ridges or septa between the craters in this chain shows that such intervening septa are not diagnostic of simultaneous impacts. In support of this interpretation is the similarity in size of approximately twenty-one of the twenty-five identifiable crater members. Such similarity seems to require an unusual cluster of projectiles, in both size and configuration prior to impact. In addition, note the doublet craters that parallel the main crater chain (arrow X). A similar association is illustrated in Plate 260 (arrow XZ), in which the chain is thought to represent an incipient rille.

Plate 118d reveals a very narrow but long chain (arrow X, Y, and Z) that crosses a variety of forms and relief in the northern highlands near the crater Mouchez. The craters are typically interconnected and produce a fracturelike appearance. In contrast to the foregoing example, there are few, if any, transverse septa. This chain is interpreted as a result of fracturing rather than multiple impacts.

PLATE 118

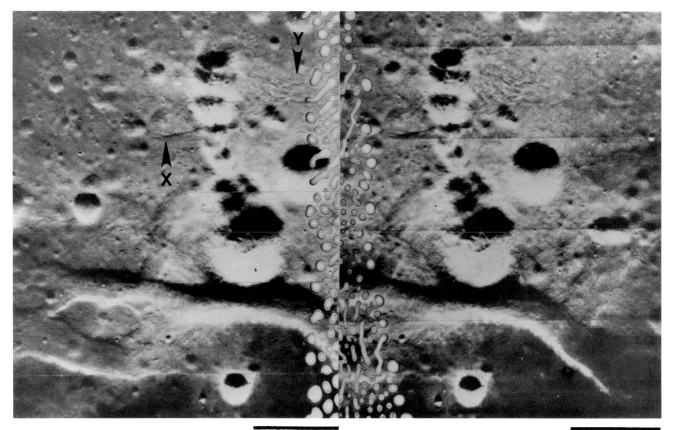

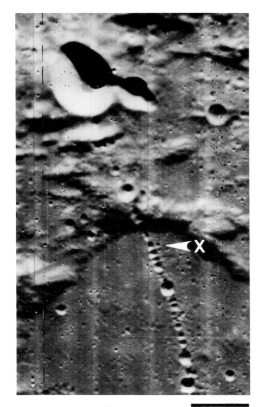

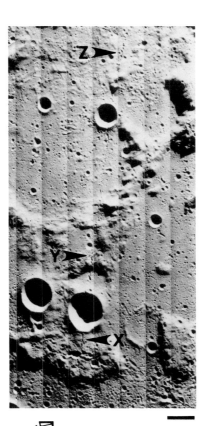

a. V-069-M   4.0 km

b. V-067-M   3.9 km

c. IV-108-H2   12 km

d. IV-164-H3   14 km

Medium-Size Crater Chains

The rimless depression (arrow X) shown in stereo (Plate 119, a and b) and at high resolution (Plate 119c) is adjacent to Rima Bode II. Of interest are the connecting craters (arrow Y) that extend to the northwest from the main depression. The spatial relation of an irregular depression adjacent to a rille is not unique; commonly such structures are adjacent to, rather than overlapping, the rille. The crater chain occurs along a shallow NW-trending rille, an extension of which is noted by arrow Z. Numerous outcrops can be identified along the wall in Plate 119c, and blocks have accumulated at the base of the V-shaped valley. The region immediately surrounding the depression has a slightly lower albedo, which implies some ejecta, but the absence of an extensive blanket indicates collapse without a violently explosive eruption.

Several other interesting features shown in Plate 119, a and b, include the following:

1. Generally low albedo of the highland relief, with high-albedo specks seen near the top of the plate (also see Plate 188a)
2. Flat-floored crater Bode E (arrow XX) in the upper right (see Plate 23, a and b)
3. The irregular depression and rille to the north of Rima Bode II (arrow YY)
4. The irregular depression (arrow ZZ) and wrinkle-ridge system on the mare (see the discussion accompanying Plate 181)
5. General NE-trending lineations radial to the Imbrium basin and expressed by wrinkle ridges, one of which extends into the highlands as a scarp (arrow XY), and grabenlike rilles (east of Bode E)

PLATE 119

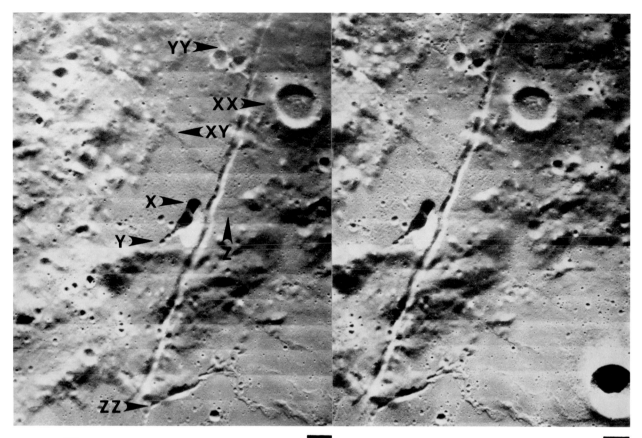

a. V-123-M 3.8 km (top) 3.2 km (bottom)   b. V-121-M 3.8 km (top) 3.1 km (bottom)

c. V-122-H2   0.45 km

Medium-Size Crater Chains

Plate 120 illustrates groupings of ringlike craters. The stereo pair includes contiguous and interconnecting forms (arrows X and Y) similar to those shown in Plates 91 and 92, d and e. The NE-trending chain (arrow X) seen at right in Plate 120, a and b, reflects a regional trend expressed by other chains and rilles in the highlands. Volcanic activity is indicated by the multiple domes on the floor of the northernmost member of the chain at the right. This chain does not resemble those commonly ascribed to secondary impacts. It is proposed that the chain represents interconnecting calderas, with indications of late-stage construction of volcanic domes.

The interconnecting doublet and triplet craters seen to the left (arrow Y) are filled with dark marelike material, which was emplaced on the eastern edge of Deslandres, a large ancient crater within which these chains are found. They also have pronounced rims that are exaggerated by the surrounding high-albedo Deslandres floor material. Conelike features (arrow Z) overlap their rims at several locations. Note the elongate summit pit on the volcanolike ridge that extends along the axis of a doublet chain (arrow XX). The absence of an intervening wall between component craters suggests either foundering of these features into a molten floor or caldera collapse for the entire complex. Very similar structures are found at other highland sites and on the maria.

Plate 120c reveals a cluster of Gambart-like craters (arrow X) on the floor of Walter, which is adjacent to the southeast edge of Deslandres. The clustering of these craters appears anomalous relative to the exterior of Walter, which obviously must be at least as old as its floor. In addition, the clustering occurs along an acentral peak and platforms. Thus, caldera formation of these examples is also a viable working hypothesis. Note the double-rimmed structure shown at left (arrow Y; also see Plate 4d).

In contrast to the previous chains, Plate 120d shows four typical highland craters adjacent to eastern Mare Fecunditatis that compose a contiguous chain. Each member has a low-relief rim, and their floors are partly flooded by cratered low-albedo marelike material. The low-relief rims and relatively steep walls imply collapse origin or extensive enlargement by slumping. I prefer an endogenetic origin of the complex.

PLATE 120

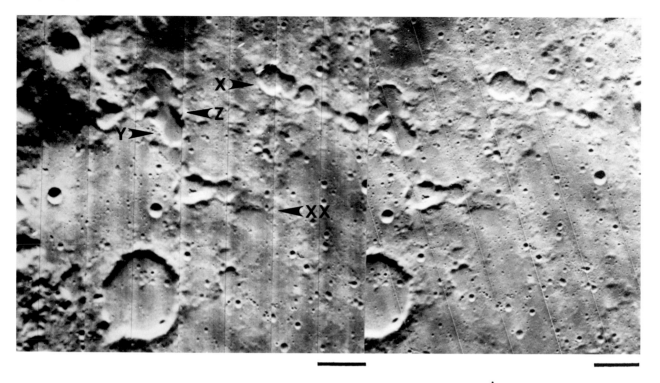

a. IV-113-H1   12 km

b. IV-112-H3   13 km

c. IV-107-H3   14 km

d. I-029-M   7.2 km

Large-Crater Clusters and Chains

For craters at much larger sizes than the foregoing medium-size examples, the significance of crater chains becomes much more dubious. Highland areas are more likely to display such groupings. If such regions represent an early epoch of much higher influxes of meteoroids, then statistical overlaps and genetically related chains may become indistinguishable from each other. Arguments on grounds of similarity of morphologic features usually become frustrating discussions of personal bias. Plate 121a illustrates the problem with a segment of the Southern Highlands (south of Clavius).

In analogy with secondary cratering known to have occurred around craters the size of Copernicus, basin-forming events probably produced numerous secondary projectiles of much larger sizes. It is plausible that some of the 10-km–25-km craters shown in Plate 121a were produced by such a catastrophe. More generally, the high crater density in the lunar highlands may be in part the result of a short period of secondary cratering, rather than a long period of primary impacts.

Plate 121b shows the crater Thebit and its two smaller companions, Thebit A and L (proceeding northwestward). The oldest member is Thebit, temporally followed by Thebit L and Thebit A. With additional degradation, all members might eventually appear to be contemporaneous. Even at their present states of preservation, proof of origin by either chance overlaps or volcanic relationships seems impossible without field samples.

The last example (Plate 121c) includes the ringed plains Fra Mauro, Parry, and Bonpland (starting at the top and proceeding clockwise). They are shown at higher resolution in Plate 73 and are used in that context to illustrate walls of craters that have been inundated by plains-forming units and possibly volcanically modified along their remaining rims. Hence, they illustrate the cluster arrangement of very large ringlike structures. From the appearance of relative degradation, Fra Mauro appears to be the oldest, followed by Bonpland and then Parry.

PLATE 121

a. IV-094-H3   16 km

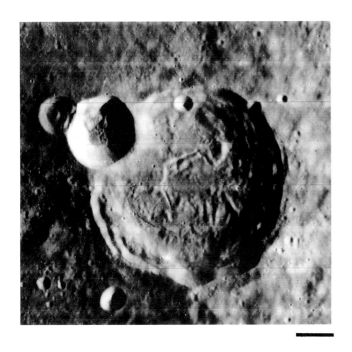

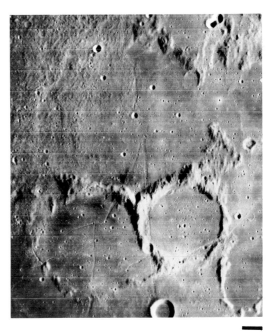

b. IV-108-H1   12 km

c. IV-120-H3   12 km

277

DISCUSSION AND SUMMARY

Crater groups can be important indicators for crater origin and modification. However, the statistical or morphological significance of the spatial association of craters must be established; the morphological, not the statistical, is the purpose of this study. As shown in this chapter, craters may occur in chains or clusters at various separations. As the separation between craters increases, the significance of their being members of a group becomes questionable unless morphological criteria can be established. Two criteria are used implicitly in the identification of such groups:

1. Their close spatial or morphological relationship with other forms of known origin, for example, craters aligned along a grabenlike rille
2. A particular appearance that is unique to member craters, which may have a wide range of diameters

The second criterion is not always reliable for determining crater origin. For example, floor rejuvenation commonly occurs on a regional scale and may affect craters having various origins. The appearance of the rim zone is perhaps more significant with respect to common crater-forming processes.

Crater groups are recognized throughout the size range for craters: from those less than 100 m in diameter to the enormous lunar basins. At small scales, numerous crater chains apparently were formed endogenetically (see Plates 116 and 56 [top]). If formation occurred at the time of mare emplacement, then crater chains can be used as indicators of surface processes. If they were formed more recently, then they provide evidence for continued production of endogenetic features.

At larger scales, crater chains interpreted to be endogenetically produced provide clues on the origin of other craters that have similar appearances but are not members of a chain. Craters along the Hyginus Rille (Plate 159) and near Secchi X (Plate 57) illustrate the significance of this approach.

Thus, a detailed study of crater chains should provide important clues for not only crater origin but also surface processes.

6. Positive-Relief Features

INTRODUCTION

Local irregularities on the lunar surface are very generally described as either positive-relief or negative-relief structures. In the broadest sense, many craters also should be included in this grouping, but the discussions that follow exclude such features except where an understanding of their genetic history is pertinent to the topic under discussion. Positive-relief forms are interpreted as either constructed or relict relief. Constructed forms include those formed by

1. Igneous processes (intrusions, extrusions)
2. Structural events (faulting, compression)
3. Impact ejecta

Relict-relief features are pre-existing structures that have been surrounded by plains-forming units or whose apparent relief is enhanced because of removal of surrounding terrain by mass wasting or fluidlike withdrawal. The distinction between relict and constructed forms is not always obvious.

Positive-relief features easiest to identify are on the maria or marelike regions. Similar relief on the highlands may be confused wth superpositions of crater forms. Many features, however, appear to be characteristic of the maria, and may represent volcanic constructions formed during the last stages of the mare-flooding epoch.

Two broad subclasses further refine the descriptions. The first, labeled "isolated forms," includes:

1. Platforms
    a. Polygonal (horstlike)
    b. Irregular (volcanolike)
2. Domes
    a. Low relief
    b. Mamelonlike
    c. High relief
3. Cones

The second includes extended and interconnected, or composite, forms:

1. Irregular ridges
2. Septa
3. Extended insular massifs
4. Wrinkle ridges
5. Ring structures

The wide variety within the subclass indicates different constructive processes, stages in construction (either chemical or mechanical), or degradational processes.

Flow features and central crater peaks are obvious omissions and can be examined in Plates 192–193, 198–199, and Plates 44–49, respectively. In addition, one should refer to the separate section on the Marius Hills (Plates 263–268) and the low-albedo features shown in Plates 188–189.

Polygonal Platforms

Platforms with polygonal plans commonly are found on mare borders, such as those along the western edge of Mare Fecunditatis (Plate 122). The peak on the western border of the stereo pair (arrow X) rises about 580 m above the mare plain immediately to the east. The surface of the platform complex is similar to the surrounding maria in albedo, but there are fewer craters 0.1 km–0.2 km in diameter and more of 1 km in diameter. The anomalous size distribution could indicate:

1. A recently emplaced veneer that erased the small craters but only subdued the larger ones
2. An endogenetic component in the crater count
3. The structure and composition of the platform, which may be conducive to a more rapid destruction of small craters— e.g., a relatively unconsolidated surface subject to seismic compaction

Surrounding the platforms is a basal apron, identified by a change in slope and texture and typically convex in the vertical section (Plate 122, a and b, arrow Y). Plate 122c permits a more detailed examination of the eastern end of the platform complex. The same textural characteristic can be traced into the NE-trending furrows that cross the northern part of the plateau (Plate 122c, arrow X). Above the apron, a coarser surface texture predominates and can be identified on the platform walls (Plate 122c, arrow Y), within the subdued craters, and around significant changes in relief where the lighting conditions are favorable. This textural feature may be present over most of the surface, but unfavorable lighting conditions prevent identification. The basal apron is also shown in Plate 60, b and c, and is examined more closely with domelike forms (Plates 126–128) and the Flamsteed Ring (Plate 154).

Of interest are the shallow flat-floored craters present on both the mare and platform surfaces. Several of these craters contain central dimplelike depressions, which are exaggerated by the low sun angle, and exhibit relatively well defined boundaries between the crater floor and wall (Plate 122, a and b, arrow Z). Such features may be the result of local blanketing by impact or volcanic ejecta with later compaction or gravity stoping of this material, yielding the unusual profile. Alternatively, they are the result of maarlike eruptions. Similar examples are discussed in the chapter "Crater Floors" (see Plate 20).

The NW-facing walls of the platform align with the regional trend of three grabenlike rille systems: Rimae Goclenius, Gutenberg, and Cauchy. The NNE-facing walls parallel another unnamed rille system. These alignments indicate that the polygonal platforms are remnants of features that were crossed by regional structural weaknesses. If, however, shallow craters with dimpled floors indicate regional eruptions, then the structural development of the platform apparently was accompanied by volcanic activity.

PLATE 122

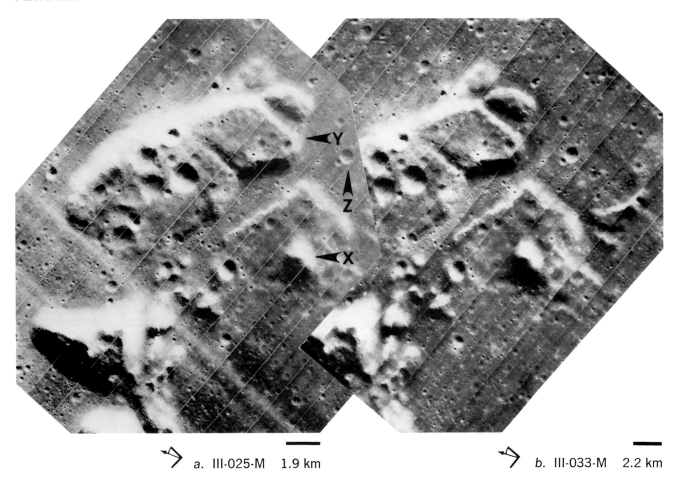

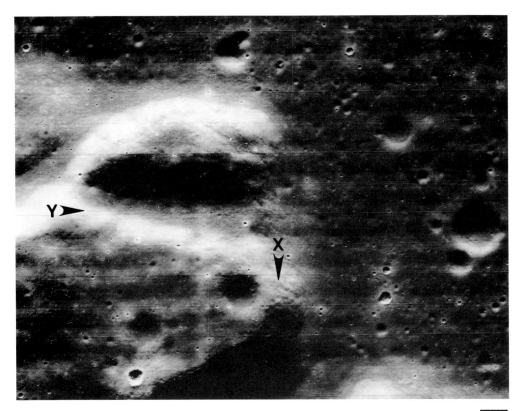

a. III-025-M   1.9 km

b. III-033-M   2.2 km

c. V-048-H3   0.42 km

Irregular Platforms

The irregular platform shown in Plate 123a is near the end of the western extension of the Carpathians. The entire feature rests on a low-relief bulge, which is outlined by possible flow termini similar to the examples shown in Plate 125c. The surrounding region (outside Plate 123a) displays results of endogenetic activity that include sinuous rilles, symmetrical domes, and irregular depressions. In addition, the region has numerous crater groups and rays that probably were derived from Copernicus; hence, crater statistics on and off the platform do not provide meaningful interpretive data. The area surrounding the platform has a comparatively low albedo, with numerous flat-floored and dimple craters (arrow X) similar to those shown in Plate 122, a and b. The appearance and the abundance of other nearby engodenetic features support the hypothesis that this platform formed volcanically rather than structurally. It may represent a buildup of extrusions or remnant relief associated with a complex volcanic history.

Arrow Y identifies a small-scale flat-topped form that is isolated from the main platform complex. A larger yet similar feature is noted by arrow Z; however, its summit is bordered by a discontinuous low-relief rim. Similar structures are recognized in Mare Fecunditatis (Orbiter I-33-M) and Mare Smythii (Orbiter I-17-M). I interpret many of these small isolated platforms as volcanic constructions formed during mare emplacement. The flat-topped profile is created by filling an old vent to the brim with either lava or late-stage floor resurgence, whereas the low-rimmed feature (arrow Z) indicates incomplete inundation. The larger irregular platform shown in Plate 123a may have had a similar origin, in which an old crater rim has been overrun by late-stage extrusions. The complex history represented by such masses is inferred from the identification of a highly modified and breached ring structure (5 km in diameter) interpreted as an ancient caldera (arrow XX). Arrow YY locates another probable vent. Other platforms interpreted as volcanic constructions are shown in Plates 46d, 148d, and 263.

The final expression of topographic relief is enhanced by the drainage of surrounding lava or subsidence of the surrounding terrain through subsurface lava removal. Such a process is inferred from remnant platforms within terrestrial calderas as observed by Jaggar (1947). On the Moon, it also is suggested by the terraces and aprons bordering the mare plains and surrounding other premare relief (see Plates 128, 249). In addition, note the platform within the breached crater shown in Plate 130a and b (arrow YZ). The surrounding moat is thought to have been produced by the removal of lava.

Plate 123, b and c, is a stereo view of flat-topped relief 25 km south of the crater Censorinus (Plate 81a); it is also shown in Plate 124a. The platform complex is in a highland region between Mare Fecunditatis and Mare Nectaris that displays an assortment of volcanolike domes and cones. It is higher in relief than the example shown in Plate 123a, and the difference in elevation along the southern scarp (arrow X) appears to be greater than that along the northern scarp (arrow Y). The southern scarp is highly scalloped, whereas the northern border is more irregular, owing to furrows and irregular pits.

This platform is interpreted as a mass-wasted remnant of either a complex volcanic plain or molten ejecta associated with the formation of the Nectaris basin. The relative relief was, in part, produced by subsidence within and around ancient craters. This subsidence was probably triggered by volcanism accompanying the emplacement of the mare plains, which surround this peninsular region.

PLATE 123

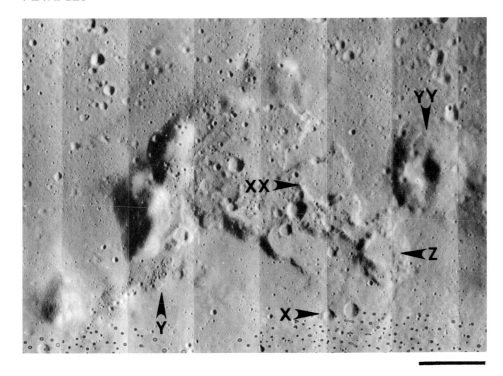

a. V-166-M   3.4 km

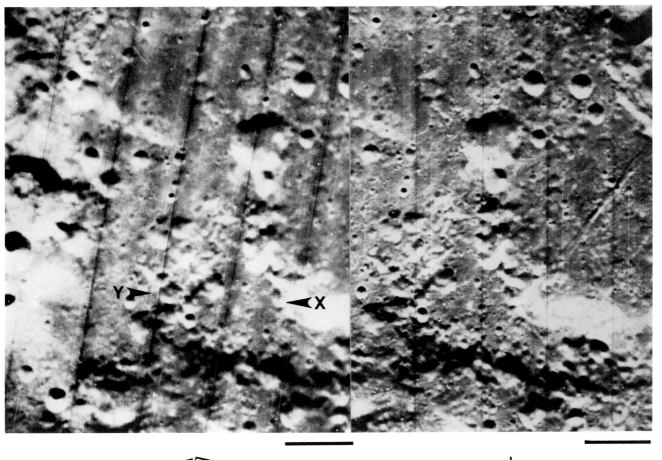

b. IV-073-H1   12 km   c. IV-072-H3   12 km

Irregular Platforms

Plate 124 illustrates additional examples of irregular plat-forms. Arrow X (Plate 124a) identifies the northern edge of the platform complex shown in Plate 123, b and c. Arrow Y locates the platforms that are shown at higher resolutions in Plate 124, b, c, and d. They are within a marelike region in east-ern Mare Tranquilitatis near Maskelyne F (arrow Z; also see Plate 153a).

Plate 124b reveals a complex terrain characterized by funnel-like craters with broad rims (arrow X; also see Plate 86, c and d), elongate depressions (arrow Y; also see Plate 8a), numerous positive-relief features, and a mottled pattern due to albedo contrasts and differences in small-scale surface textures (see Plate 15, b and c). Many of the positive-relief forms have flat-topped profiles and are bordered by relatively abrupt scarps (ar-rows Z, XX, YY).

The stereo view (Plate 124, c and d) shows several platforms having irregular plans. The feature identified by arrow X is rec-ognized in Plate 124a (arrow Y) as part of a breached crater-form that appears to be related to sculpturing from the Imbrium or Serenitatis basins. The platforms are surrounded by broad basal aprons (arrow Y), and, aside from their irregular plans, they are similar in appearance to the platforms shown in Plate 122. Note the insular forms separated from the platforms (ar-rows Z and XX), the numerous small flat-floored craters (arrow YY), the general absence of pristine craters larger than 0.2 km in diameter, and the patches (arrow ZZ) that lack a crater count comparable to other mare surfaces (compare to crater density shown in Plate 192, b and c).

The flat-topped profile suggests the existence of a previous plains-forming unit. The irregular plans and related insular forms indicate a complex history, with the removal of the sur-rounding terrain. These features are not consistent with a simple degradational history. Hence, the platforms are interpreted as relict-relief features. Their local relief probably was established by processes associated with the emplacement of the surround-ing plains-forming units. These processes include subsidence or compaction of the plains-forming units and lateral mass wasting of the platforms.

PLATE 124

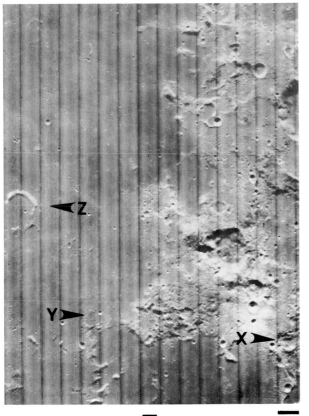

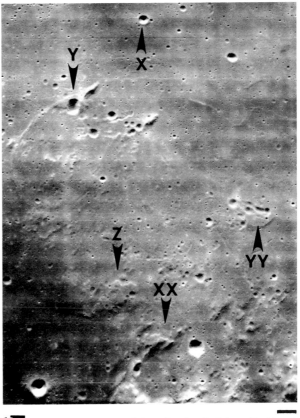

a. IV-073-H1   12 km

b. V-061-M   4.3 km (*top*)   3.0 km (*bottom*)

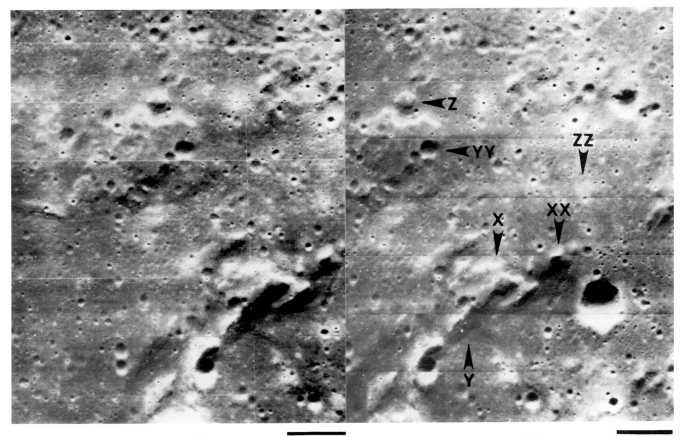

c. V-062-M   3.0 km

d. V-059-M   3.0 km

Irregular Platforms and Low-Relief Mare Domes

Rümker (Plate 125, a and b) is a large platform (70 km across) in northwest Oceanus Procellarum that is surrounded by mare material. It exhibits numerous low-relief domes (arrows X, Y, and Z), and at this resolution is similar to the Aristarchus Plateau (Plate 171a). Although the boundaries appear well defined, it is possible that they would lose much of their identity at higher resolution. Several wrinkle ridges converge toward the feature and are traced onto the platform as sinuous relief. Lineations in the northwest corner (arrow XX) are aligned with those from Sinus Iridum, which is thought to have formed after the Imbrium event but before mare flooding.

Rümker is interpreted as premare relief that was subjected to Iridum ejecta, with partial erasure through later volcanism. The domes may be due to laccolithic intrusions or are constructions of low-viscosity flows analogous to small-scale shield volcanoes. Note that a low-relief ridge forms a large ring (arrow YY) that contains most of the Rümker Plateau. This suggests that the entire volcanic complex may be related to an ancient cauldron, the remnants of which are delineated by this ring (see Guest, 1971).

Isolated domes analogous to those on Rümker are illustrated in Plate 125, c and d. The crater-topped dome in Plate 125c is approximately 120 km to the south of the irregular platform shown in Plate 123a. The resolution is not sufficient to reveal more about the lobate pattern extending along the base of this dome (arrow X), but it may be a flow feature or a terrace from a previous mare level. Also note the elongate volcanic cones (arrow Y; see also Plate 130f) and the rimless depression (arrow Z). These features indicate multiple vents in the region.

Plate 125d shows a cluster of at least five low-relief domes southwest of the Carpathian Mountains. These domes, along with another seen beyond the left edge of the photograph, are arranged along a regional NE trend that also is reflected by the crater-topped ridge (arrow X) and the rille (arrow Y). Each of these domes has a marelike surface and summit pit.

Mare domes may be surface expressions of laccolithic intrusions. Alternatively, they are small-scale lava shields—i.e., lava domes—that were constructed by repeated extrusions of low-viscosity flows. By analogy, the summit pits are calderas that formed after removal of the supporting magma. Elongate or rillelike summit depressions on several mare domes indicate either elongate vents or extensional fractures.

Telescopic observations of the lunar surface under very low solar elevations reveal larger mare domes, or mounds (about 20 km in diameter), that apparently have no summit pits. Apollo photography (for example, AS15-91-12372) has confirmed the absence of such a pit on a dome in western Mare Serenitatis. Low-relief domes commonly are associated with crested-relief structures (see Plate 134, a and b).

Plate 125e shows a very small domelike form (arrow X) within the Flamsteed Ring (Plate 154). A similar feature is recognized about 3 km to the south-southeast (also see Plate 15, b and c, arrow ZZ). A moatlike ring surrounds the dome and may be analogous to the forms on the floor of Copernicus (Plate 227). These small-scale domes may represent blisterlike swellings, called tumuli, which occur on terrestrial basalt flows. As a feature related to mare emplacement, the small dome seen in Plate 125e and similar domes found elsewhere indicate the degree of preservation of small-scale features on the original mare surface. As a much later feature, it suggests that endogenetic processes may have modified the mare surface over a long time period and may represent a significant formative mechanism, which is generally ignored (also see Plate 194).

PLATE 125

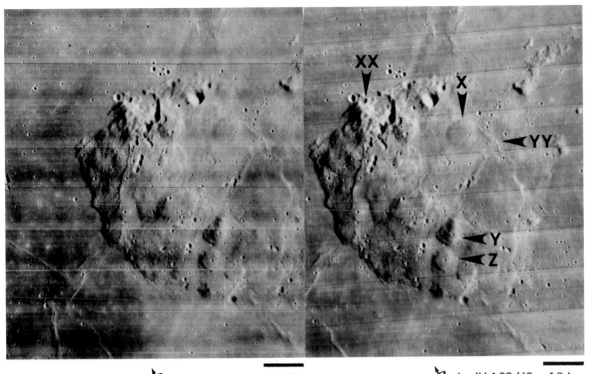

a. IV-170-H2   12 km

b. IV-163-H2   12 km

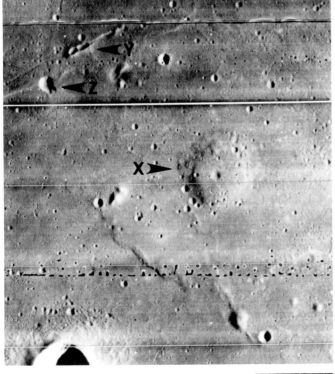

c. IV-133-H2   11 km

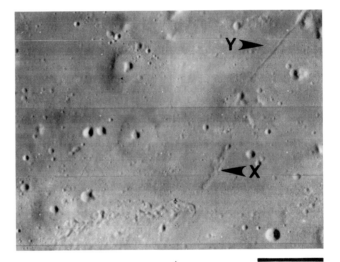

d. IV-133-H1   11 km

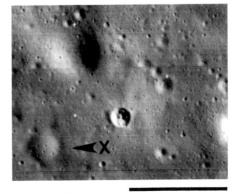

e. III-189-H1   0.23 km

Mamelonlike Domes

In contrast to the low-relief mare domes (Plate 125, c and d), Plate 126 illustrates higher-relief mamelonlike domes without summit pits. The oblique view in Plate 126a shows the puffy appearance of several such domes about 300 km south of Copernicus. They rest on or nearby the mare surface and commonly occur in clusters, although isolated examples are not rare. Their plans are not always symmetrical, and several varieties of domal features result from the coalescence of two or more domes.

Plate 126, b and c, shows greater detail of a dome about 30 km to the south of the region shown in Plate 126a. This class of domes is characterized by a basal apron, identified by a change in slope and surface texture (arrow X). Similar aprons occur within mare-filled craters (see Plate 40b and Plate 60, b and c). Above the finely textured apron, the surface exhibits the coarser tree-bark pattern (arrow Y) with considerably fewer craters that the surrounding mare surfaces. Plate 127 shows a high-resolution view of this dome.

Mamelonlike domes might represent relict or constructed relief. As relict forms, the basal aprons are interpreted as either the result of mass wasting or a phenomenon that accompanied mare emplacement. As constructed relief, these domes are volcanic in origin and perhaps historically postmare. Evidence for these alternatives is presented in Plates 127–129.

PLATE 126

└→ *a.* I-137-M (oblique)

↘ *b.* III-124-M 1.5 km   ↘ *c.* III-128-M 1.5 km

Mamelonlike Domes

Plate 127 is a high-resolution view of the dome shown in Plate 126, *d* and *e*. The inset shows an oblique view of the same dome and provides a feeling for the relief. For reference, arrow X in the inset identifies the region located by arrow X in the main plate. The texture of the basal apron is finer than that on the upper slopes of the dome, and the change in slope at the top of the apron is well marked. In contrast to the more typical tree-bark texture, the southern slopes of the dome display a cross-hatch or grid pattern across both the apron and upper surfaces (arrow X). There are numerous blocks on the dome (arrow Y), and these may be residual accumulations from the movement of a debris layer.

Even at this higher resolution, the origin of the dome or its basal apron remains disputable. The general boundary between mare and dome can be recognized by a transition in texture or relief, rather than by the abrupt termini of mare material overlapping the dome or the dome overlapping the mare material.

PLATE 127

III-120-M   (oblique)

III-125-H3,H2   0.19 km

Mamelonlike Domes

Plate 128 provides additional clues to the formation of mame-lonlike domes and the basal aprons. The stereo pair of Plate 128, a and b, shows a dome less than 50 km to the south of the region shown in Plate 126a. The surrounding area is character-ized by a highly cratered mare surface that is mottled with nu-merous relatively uncratered patches. A dominant NE trend is reflected by small-scale ridges. Approximately 30 km to the east of the area shown in the photograph is a hummocky highland-like surface believed to represent ejecta from the Imbrium event. In general, the domes display greater relief than the in-dividual hummocks of that terrain.

The apron in this example is relatively well marked at the mare and dome contacts. The mare contact along the north-western side of the dome is indicated by a ridge, whereas the apron-dome contact is marked by a narrow moat. The narrow apron-dome moat cuts several small craters (arrows X and Y) and appears to have formed after the craters. In addition, the subdued crater (arrow Z) at the base of the northwestern slope of the dome provides insight into the formation of the apron. If that crater was formed by an impact temporally after the apron, then its very subdued appearance is in marked contrast to that of the adjacent contacts between apron and mare or apron and dome. Consequently, the crater might represent a pre-existing form, which was smoothed by the apron-forming process, or it was internally formed.

The features seen in Plate 128, a and b, suggest that mare material overlapped the dome. Further support for this interpre-tation is shown in Plate 128, c and d, which includes a small area adjacent to Maskelyne D in southeastern Mare Tranquilita-tis. The dome (arrow X) has an apron that is reflected on one side of a nearby dome (arrow Y) as well as on the marelike sur-face (arrow Z). These aprons are interpreted as "bathtub rings," produced by subsidence of the marelike units.

Plate 128, e and f, shows two features west of the crater Lansberg. These small domes (arrow X in both photographs) are surrounded by relatively large aprons. The contact between the dome and apron appears as a moat (arrow Y in both photo-graphs). This is an unusual profile for a mass-wasting phenome-non. It is suggested that it is analogous to lips around domes on the floor of Tycho (Plate 33), which are thought to be produced by the encroachment of molten material.

PLATE 128

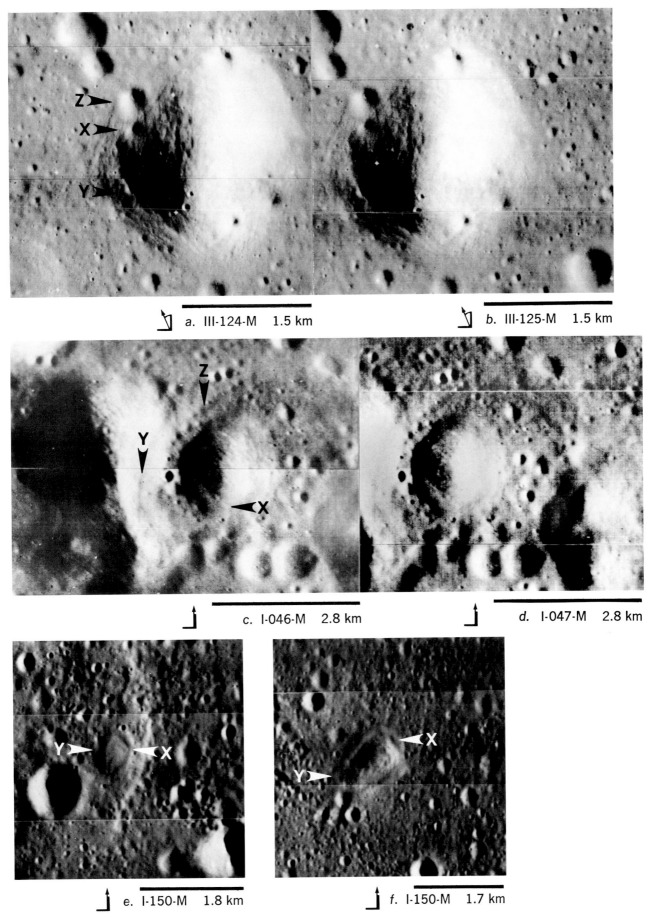

a. III-124-M   1.5 km

b. III-125-M   1.5 km

c. I-046-M   2.8 km

d. I-047-M   2.8 km

e. I-150-M   1.8 km

f. I-150-M   1.7 km

Mamelonlike Domes

Plate 129, *a* and *b*, shows a not uncommon relationship between a wrinkle ridge and a dome (also see Plates 46*a*, 60*b*, and 125*c*). It is difficult to determine with certainty whether the features were constructed simultaneously, whether the dome was built over the wrinkle ridge, or whether the wrinkle ridge was formed after the dome. Portions of the wrinkle ridge overlap the base of the dome but then disappear (or cannot be detected at this resolution). Some wrinkle ridges cross positive-relief features (see Plate 145, *a* and *b*), but such a relationship does not seem to be present here. The impression is that the dome is a later feature.

It is possible that the dome seen in Plate 129, *a* and *b*, and some other mamelonlike domes are volcanic in origin. Plausible terrestrial analogs are cumulo domes that lack a visible central orifice. They are believed to be formed by extrusions of viscous lava and develop, in part, by internal growth, i.e., by lateral expansion of the body as viscous lava is extruded from within. This is contrasted to external growth, in which the dome is constructed by less viscous lavas successively flowing over previous flows. Under this suggestion, the aprons of some mamelonlike lunar domes may be coulees. Comparisons to terrestrial cumulo domes suggest that additional weathering would have been

necessary, and this could have been accomplished by later meteoroid bombardment or rapid fragmentation.

If the volcanic origin is acceptable for the dome in Plate 129, *a* and *b*, then this hypothesis must be equally as reasonable for portions of certain "ghost" craters (Plate 154) as well as for central peaks (Plate 46*a*) that exhibit similar spatial relationships to wrinkle ridges.

The fact that structures other than domes display basal aprons is evidence for either an erosional process or a phenomenon associated with the relief-mare contact. It is also a fact, however, that not all relief features exhibit such an apron. Hence, such features may have no particulate overlay or may have resulted from different processes reacting on material of various composition and competency.

Plate 129*c* contains part of a domelike feature (shown in Plate 84*a*, lower left) in eastern Mare Tranquilitatis. This example is included here to show a surface pattern different from that illustrated in Plate 127. In particular, note the lobate pattern (arrow X), which is distinctly different from the tree-bark pattern previously encountered around domes. Clearly, a meaningful classification of domal structures requires more detailed photography of these features.

PLATE 129

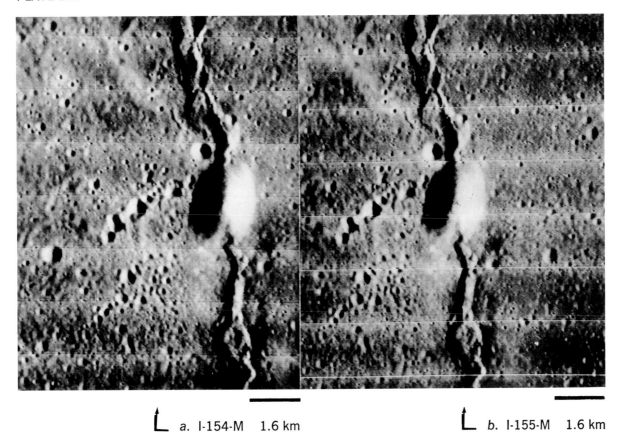

a. I-154-M 1.6 km

b. I-155-M 1.6 km

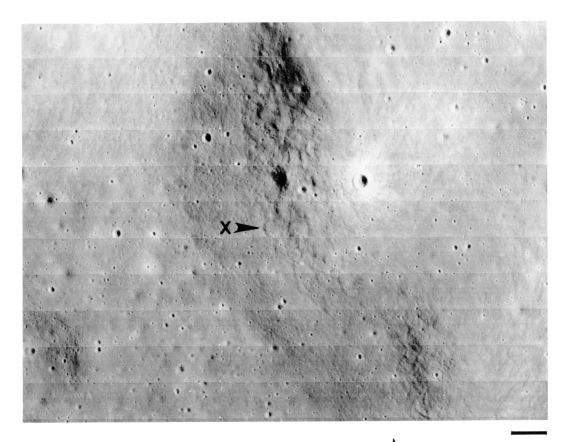

c. II-021-H1 0.21 km

High-Relief Textured Domes and Cones

Gruithuisen γ and δ are two domelike forms illustrated in the stereo pair (Plate 130, a and b); the former is identified by arrow X, the latter by arrow Y. In contrast to the preceding examples of domes, they have larger over-all dimensions and large-scale surface textures. The low-relief mare domes (Plate 125, c and d) are typically 8 km in diameter and 0.2 km in local relief, and the mamelonlike domes (Plates 126–129) have diameters less than 5 km and relief of 0.5 km. The dome type shown here, however, is commonly greater than 15 km in diameter and 1 km in height.

The Gruithuisen domes rise about 1.7 km above the mare surface and are conspicuous features in oblique views of the region. Their surface texture consists of numerous discontinuous ridges and furrows extending downslope and is much coarser than the texture displayed by the preceding examples. The difference in the texture between the pitted domes and the smooth highlands permits identification of the contacts of the two units.

The domes are interpreted as volcanically constructed forms that are superposed on the surrounding highlands, with a flow unit (arrow Z) extending onto (or buried under) the mare surface to the southeast from Gruithuisen γ. Gruithuisen δ has a broad basal terrace (arrow XX), which is interpreted as a flow or a high-level mark of mare flooding. The flowlike features indicate that these domes developed, at least in part, by external growth, but their convex profile leads to the analogy of internal growth as well. It is suggested that they are large cumulo domes (a similar conclusion was reached by Smith [1970] from comparisons to Mono Craters).

The crater on Gruithuisen γ (arrow YY) might be related genetically to the dome, but its appearance is similar to numerous other craters in the area. There are several small craters on these domes that have broad rims and may represent endogenetic forms. Note the relatively subdued crater on the eastern slope of Gruithuisen γ (arrow ZZ) that has high-albedo ejecta. A separate high-albedo deposit is recognized at the base of Gruithuisen δ (arrow XY) and may be related to a crater in shadow or to an event associated with more recent volcanic activity.

Other domes similar to those shown in Plate 130, a and b, are shown in Plate 190d (arrow X), in which the dome displays a high albedo, and Plate 66, a and b (arrow XY), in which a dome complex overlaps two craters.

On the mare southwest of Gruithuisen γ is an E-trending linear structure that displays a texture similar to that of the domes (arrow XZ). The structure is aligned with a regional trend and probably is the result of extrusions along this subsurface weakness (see Plates 131–132 for similar-appearing features).

The complexity and volcanic nature of the region can be illustrated further by nearby features shown in Plate 16c; Plate 181, a and b; and Plate 25. Note the curvilinear rille and depression adjacent to the peninsular form at left center (arrow YX). Also note the narrow moat within the breached crater (arrow YZ). This suggests removal of lava, leaving an interior platform. Recall that the broad terrace at the base of Gruithuisen δ also was interpreted as a result of different levels of the mare plains.

Plate 130, c–f, shows conspicuous isolated positive-relief features, including crater-topped conelike forms.

The stereo pair (Plate 130, c and d) reveals a symmetric cone called Mairan T, west of Mairan in northwestern Oceanus Procellarum. The structure is dissimilar to features in the surrounding region because of its cluster of summit craters and isolation. The cone shows a high albedo and numerous small craters around its base.

In contrast, Plate 130, e and f, shows low-albedo conelike forms that are analogous to terrestrial tephra cones. The example shown in Plate 130e (arrow X) is isolated from any similar structure, but nearby endogenetic features include wrinkle ridges and shallow dark-floored craters (if these can be interpreted as endogenetic). The double feature (arrow X) in Plate 130f is also analogous to terrestrial cones and is isolated from nearby volcanic forms, except possibly a N-trending wrinkle ridge and two breached ringed structures (arrows Y and Z).

The relatively solitary placement of these two examples illustrates the problem in interpreting lunar features as volcanic forms on the basis of their proximity to other endogenetically produced forms. Especially severe is the interpretation of craters that might represent calderas or maars but are removed from a volcanic setting.

PLATE 130

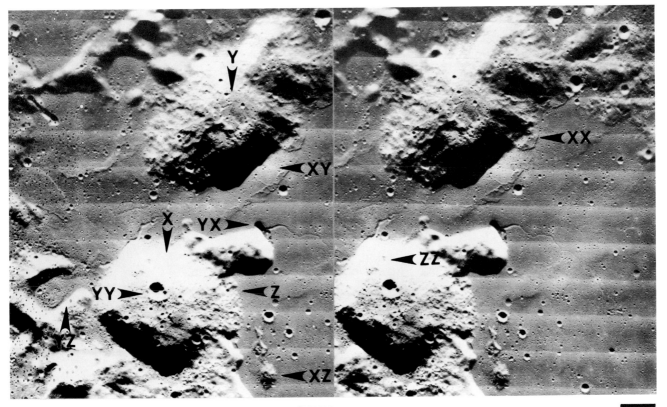

a. V-185-M   5.4 km

b. V-182-M   5.2 km

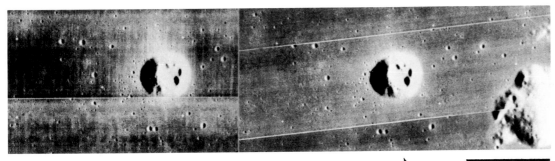

c. IV-163-H2   12 km

d. IV-158-H2   12 km

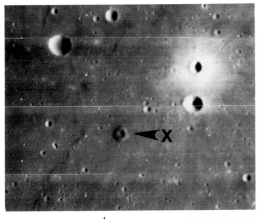

e. IV-113-H2   12 km

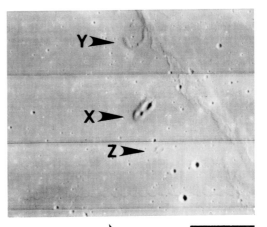

f. IV-163-H1   12 km

Irregular Ridges

The irregular-plan ridge shown in Plate 131 illustrates interconnected or composite positive-relief features. This example (arrow X) is on the southern edge of the ringed plain Fra Mauro (also see Plate 73). Plate 131, *a* and *b* (stereo pair), shows its alignment with an adjacent grabenlike rille (Rima Parry V) that crosses the highly pitted floors of Fra Mauro and Bonpland and their common wall-rim. The rille disappears in an area marked by low-albedo material, which surrounds the ridge.

The high-resolution view (Plate 131c) shows more clearly the interconnecting craters near the summit of the southern end of the ridge (arrow X). Blocky outcrops, some of which exhibit relatively low reflectivities, can be identified (under high magnification) along the upper walls and rims of craters, as well as along local crests on the ridge. The absence of boulder trails and blocky accumulations along the base of the ridge indicates that there has not been any recent dislodgment. This is in contrast to numerous trails identified in high-resolution views of a scalloped segment of the Fra Mauro wall about 10 km to the west (Plate 131, *a* and *b*, arrow Z). The ridge surface has a wrinkled texture that generally reflects local contours. This merges with the surrounding surface without a well-defined contact.

The Fra Mauro ridge resembles the platforms and hills of the Marius Hills region (Plate 264b). It is interpreted as a volcanic construction along a fissure related to the graben system. The absence of well-defined termini and the presence of a low-albedo deposit suggest that ash was an important late extrusive unit. This may have been responsible for the subdued contacts and the shallowness of the nearby craters. Shallow craters, however, are also characteristic of the region outside the low-albedo zone. If their fill was the result of ash as well, then either the albedo has been altered or it was initially higher. Besides the ridge-forming and ash units, extrusions from a bocca along the base of the ridge may have been responsible for the inundation of the adjacent graben.

The construction of the ridge may be related temporally to other endogenetic features in the region, which include isolated volcanolike hills, dark-haloed craters, and irregular depressions. The numerous shallow craters mentioned also may have internal origins, perhaps related to the ridge. This is indicated for the small crater along the graben to the south of the ridge (Plate 131, *a* and *b*, arrow Y; also see Plate 73, *a* and *b*), where the graben has been inundated by material that apparently issued from the crater.

PLATE 131

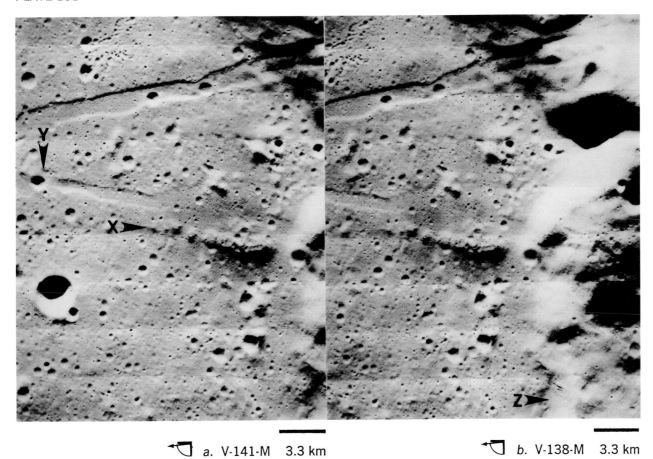

a. V-141-M   3.3 km

b. V-138-M   3.3 km

c. V-139-H2   0.43 km

Irregular Ridges

Plate 132a shows a discontinuous ridge in western Mare Fecunditatis near Secchi X; this area is included also in the overview of Plate 57a (arrow Y). In that overview, the ridge is clearly aligned with a regional NW trend, which is revealed by crater chains, grabenlike rilles, and other discontinuous ridges. It is also surrounded by a diffusely distributed deposit, darker than the local mare.

Plate 132, b and c, permits closer inspection of two segments of the ridge. The first (Plate 132b) reveals the relatively smooth transition from mare to ridge. Below the peak is a subdued crater (0.2 km in diameter) that displays reflective patches along its upper wall (arrow X). This feature is similar to the features exhibited on the floor of Hyginus (Plate 161b) and in the area to the south of Arago (Plate 194d).

The second close-up (Plate 132c) isolates an enigmatic feature of dark and light material near the summit of the western segment of the ridge (arrow X). This appears to be a summit crater, with a breached eastern wall and asymmetrically deposited low-albedo ejecta.

The suggestion of volcanic origin for this ridge is based primarily on the surrounding low-albedo deposit. A similar association is illustrated in the preceding example in Fra Mauro (Plate 131), and these two selections are not unique. The appearance of the ridge alone is not distinctive enough to warrant a conclusive statement of genesis, but it is suggested that it is a degraded version of volcanic forms similar to the ridge shown in Plate 131.

In contrast, Plate 132d shows a ridge composed of slightly offset coalescing cones with well-developed summit pits. This feature is 50 km to the northeast of Mösting and is identifiable in the upper left corner of Plate 65a (arrow Z). Telescopic views under full solar illumination show that the ridge has a low albedo and is surrounded by a smaller diffuse deposit, such as that around the example shown in Plate 132c. The NE trend defined by the ridge is approximately parallel to Rima Flammarion, about 80 km to the south.

At Orbiter IV resolution (see Plate 65a), its component members appear to be very similar to the isolated cones shown in Plate 130, e and f. At the higher resolution displayed by Plate 132d, the general form is lost in the increased detail. Such loss is a general rule and may, in part, be the result of subsequent smoothing at small scales.

The summit pits, low albedo, and alignment prompt the conclusion that this represents a series of large spatter cones, constructed along a NE-trending fissure. From a purely qualitative approach, it is not clear whether the structure predates or postdates Mösting. The eruption appears to be associated with volcanic activity around the periphery of Sinus Medii that resulted in low-albedo margins. The absence of visible flow termini indicates a long degradational history, thin lavalike units, or pyroclastic eruptions. The departure from the classic conelike profile also could indicate degradation, but the numerous summit pits imply a complex eruptive history.

The example shown in Plate 132d is not unique; a notable field of similar structures can be found in Oceanus Procellarum around the crater Kunowsky (Plate 92a). Plate 132e shows one of these ridges, and it should be noted that it is not composed of low-albedo tephra. Although a small elongate cone (arrow X) clearly postdates the crater Hortensius D, the ridge appears to be partly buried (arrow Y) by rim material. Had Hortensius D (arrow Z) been formed by an impact event that postdated this portion of the ridge, then its ejecta would have been much more extensive and destructive by analogy with the ejecta of other craters of similar size. Consequently, a tephra ring is a more plausible explanation of its formation. Note the isolated conelike peak that is approximately on-axis with the ridge (arrow XX).

PLATE 132

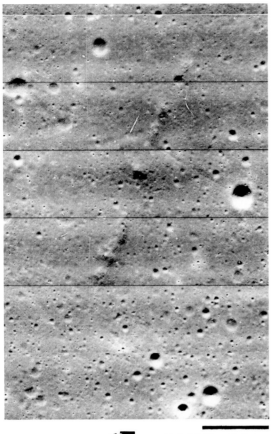

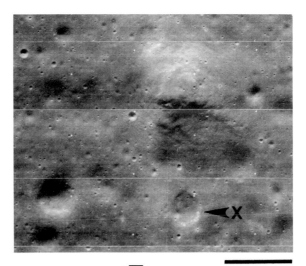

a. V-047-M   3.1 km

b. V-047-H2   0.42 km

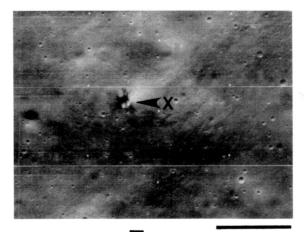

c. V-047-H2   0.42 km

d. III-106-M   1.3 km

e. IV-133-H1   12 km

Septa

Another continuous positive-relief feature is illustrated by the Imbrium basin sculpture, shown in Plate 133, *a* and *b*. It is composed of linear ridges, or septa, arranged radial to a major basin, such as Imbrium or Orientale. Septa are broadly defined as straight ridges with sharp crests. They are distinct from the volcanolike ridges described with Plates 131–132, which have irregular borders and profiles. Plate 133*a* is an oblique view of septa from the Imbrium basin and includes the crater Ukert (arrow X), with the west-facing wall of Murchison in the foreground. Ukert clearly postdates the septa but predates the emplacement of the nearby plains-forming units (arrow Y). The severed crater on its northern rim (arrow Z) indicates that Ukert became enlarged by later slumping, but outcrops associated with the septa are not visible on the walls.

Plate 133*b* shows a region to the northeast and beyond the horizon of the area shown in Plate 133*a* and demonstrates that the lineated appearance is not simply a product of perspective. The bright-rayed crater near the top is Menelaus, which clearly postdates the lineations. Septa shown in Plate 133*b* occur as insular features on the mare material (arrow X), as small-scale (1-km width, 0.2-km amplitudes) corrugations on the uplands (arrow Y), and as larger-scale (5-km widths, 1-km heights) discontinuous hogbacklike ridges on both surfaces (arrow Z). They typically form offset walls of craters and the resulting plan becomes en *échelon* or polygonal (see Plate 74*d*).

If Orientale is reasonably assumed to be an exposed analog to the Imbrium basin, then many septa visible in Plate 133*a* appear to be relict forms of massive ejecta flows (see Plate 201). This near-surface phenomenon is implied by the slight detour of lineations due to obstructions and the pile-up of material within craters on the basin-facing wall. Subsequent encroachment of the furrows by plains-forming units, such as the mare basalts and higher albedo units, accentuates the hogback appearance.

Not all septa, however, can be interpreted as relict flow ridges. Notable exceptions are the larger-scale ridges (Plate 133*b*, arrows Z and XX). They appear to be rims of elongate basins that commonly are filled with high-albedo plains-forming units or very low albedo mare material. The interior scarps typically exhibit a higher albedo than that of the surrounding terrain and appear analogous to talus slopes within craters. The parallel septa composing the opposite rims of these structures may be slightly scalloped or arcuate and thus are interpreted as highly elongate or interconnecting craters. Hartmann (1963, 1964) reached similar conclusions from telescopic observations. After partial inundation by mare material, septa that are flow ridges and septa that are parallel rims of elongate craters become difficult to distinguish.

Further comparison with the Orientale basin provides possible analogs to these elongate structures. Plate 209 shows a large furrow, Bouvard, that lacks the extensive mare fill but displays well-defined rims. Such "gashes" of the lunar crust are discussed with Plate 179 and are interpreted as grabens formed by extensional fracturing radial to the large basins. The con-

struction of rims implies that graben formation was accompanied by venting along its length or by the construction of levees from the guided ejecta flow. Several examples along southern Mare Vaporum and those shown in Plate 133*b* are explained more convincingly by coalescing craters, which may have been formed by Imbrium ejecta or internal mechanisms. The very low albedo mare material and surrounding deposits suggest that if these large furrows were not endogenetically formed, they at least localized later distinctive volcanic units. This localization, however, suggests rejuvenated vents and more generally a volcano-tectonic origin for such depressions radial to Imbrium.

Smaller-scale graben and horst formation appears to be responsible for part of the "Imbrium sculpture," and close spacing could result in septa. A structural origin is indicated by the dropped appearance of some regions, none of which are shown in Plate 133. This origin is also inferred from the localization and typical isolation of the highland plains-forming units within the supposed grabens. Such interstitial areas cannot be explained as "spillovers" from larger plains but are anticipated logically along weaknesses reflected by grabens. However, such units may have resulted from local accumulations of dust, a *nuée ardente* type flow, or fallout following the Imbrium event, rather than from lavalike units (see Plates 199–200).

A final possibility is the volcanic construction of septa along fractures radial to the major basin. This interpretation is not plausible for the continuous hogbacklike septa, because of the simultaneity required for construction. It also may be inconsistent with the absence of exposed dikes on the wall of Ukert. It is plausible, however, for relatively small scale alignments of coalescing domes (see Orbiter IV-95-H3), but such forms are not septalike. In addition, a volcanic origin may be applicable to larger-scale broken septa whose components have summit pits. Poor resolution precludes an estimation of their relative contribution to septalike features.

In summary, four interpretations for the origin of septa are the following:

1. Remnants of scouring produced by low-angle projectiles
2. Ridges composed of bulk ejecta, either ballistic or base surgetype
3. Series of horsts and grabens
4. Extrusions along fractures

These hypotheses are all offered with respect to a major basin-forming event, and they all appear applicable, depending on location and relationships to other forms. In general, the small-scale corrugations appear to be ridges related to the massive base-surgetype flows. The large-scale septa appear to be remnant rims of elongate craters or levees constructed along the borders of elongate depressions (impact or structurally formed), which acted as channels for the surface flow. The consideration given to septa here (independent of the detailed discussion of basin formation) is warranted by the large number of examples on the lunar Earthside related to the Imbrium basin.

PLATE 133

a. III-085-M   (oblique)

b. IV-090-H2   12 km

Sharp-Crested Massifs

An example of a composite positive-relief form is the sharp-crested features that appear as multipeaked islands on mare surfaces. The description "sharp-crested" is a relative term to distinguish such features from round domes and flat-topped platforms; however, their profiles are relatively gentle, with only moderate slopes (20°–30°) in contrast to clifflike scarps. Furthermore, they are mostly elongate in plan and may be aligned with other similar features to form a pattern, such as a broken arc or ridge. Although more readily identifiable on the maria, such forms are found also in the highlands.

Plate 134, *a* and *b*, shows in stereo an example of this form southwest of the Carpathian Mountains in Oceanus Procellarum. The ridgelike plan is composed of several sharp-crested peaks, and the trend of the entire structure is approximately radial to Mare Imbrium. A low-relief mound with an elongate summit pit extends to the west of the massifs. Such a relationship is not unique. In several instances, the crested relief is at the center of a low-relief dome or platform. The region exhibits numerous indications of endogenetic activity, such as the rille, flow units, and volcanic platforms (one of which is shown in Plate 123a). Many craters also appear to have been endogenetically produced or to have been extensively blanketed. The three prominent sets of secondary craters superposed on the area are derived from Aristarchus, Kepler, and Copernicus. Some ejecta from Copernicus appear to be partially obscured by later volcanism.

The sharp-crested massifs are probably relict structures from the Imbrium basin-forming event. Their association with low-relief platforms indicates genetic relationships of the massifs to deep-seated fractures that yielded extrusive units. That is, some of these sharp-crested peaks were brought into relief not just as ejecta but as a combination of structural events and volcanism.

Plate 134c shows another example on the southern border of Mare Imbrium to the north of the Carpathians. It also displays the sharp-crested profile with elongate plan. The partly buried crater along the southeastern wall supplies clues on relative surface processes or origin of this type of relief feature. Although massif material has buried part of the crater, there is no apparent "pooling" on the floor. The crater has retained a raised rim but has no sculptured ejecta blanket.

The partial inundation of the crater could be the result of an advancing flow from a volcanically constructed ridge; however, "pooling" on the crater floor should be expected unless the extrusion was extremely viscous. More likely, the crested form is a remnant of a premare massif, which originally had a much steeper southeastern face. After emplacement of mare material and the subsequent formation of the adjacent crater, the scree-like slopes of the ridge advanced over the northwestern third of the crater. Consequently, the degradational rate for the massif appears much more rapid than that for the crater, whose rim remains well preserved. Under this hypothesis, the small crater density on the ridge relative to that on the adjacent mare is a further indicator of rapid movement. Such differential smoothing suggests a ridge primarily composed of unconsolidated material. It is possible that formation of the adjacent impact crater triggered slope failure of the massif. If part of the crater overlapped the base of the massif, the massif slope would have been temporarily oversteepened. At the end of crater excavation, this slope could have immediately buried part of the crater.

The stereo pair in Plate 134, *d* and *e*, shows the feature Montes Recti, which extends approximately 75 km along the northern border of Mare Imbrium. It is part of an arc of similar features concentric to Imbrium and outlined by wrinkle ridges, some of which are shown to the south. These insular forms represent remnants of the inner Imbrium scarp similar to the exposed features within Mare Orientale (Plates 201, 204, and 208). Extensive and complex dropping or destruction of such relief would seem to have been necessary to inundate the Imbrium basin and to leave the relatively few insular structures of relief comparable to Montes Recti. For further discussion and examples of this point, reference should be made to: Plates 49b (Schrödinger and Antoniadi); Plates 179–180 (Montes Alpes); Plates 112–114 (examples of major scarps); and Plates 201, 204, and 208 (Mare Orientale).

The general trend of an *en échelon* system of wrinkle ridges approximately parallels Montes Recti. A branch from this system is linked to the insular massifs. The association of wrinkle ridges with these relief forms is a common phenomenon, but its significance is a topic delayed until closer examination of wrinkle ridges is made (see Discussion and Summary at the end of this chapter).

PLATE 134

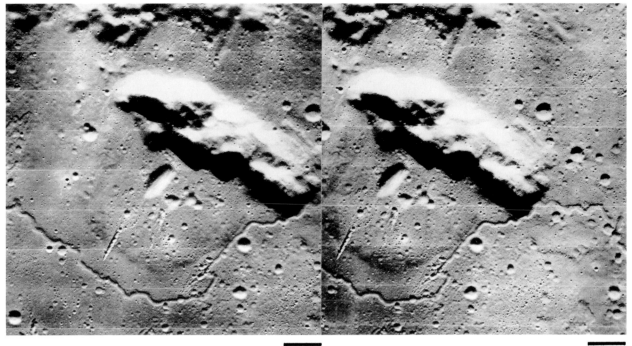

a. V-167-M   3.5 km   b. V-164-M   3.5 km

c. IV-133-H3   12 km

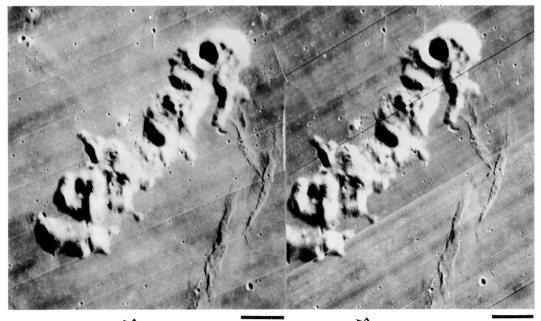

d. IV-139-H3   12 km   e. IV-134-H3,H2   12 km

Sharp-Crested Massifs

Plate 135a shows sharp-crested massifs (arrows X and Y) on the lunar farside northwest of the large basin Apollo. Plate 135, b and c, provides a stereo view of these massifs (arrows X and Y) and reveals their highly furrowed slopes. Without the stereo view and surrounding mare units, the massifs are not easily recognizable.

Such peaks normally are associated with large multiringed basins. The massifs are not believed to be associated with the multiringed Apollo basin, because they do not appear to be concentric with it. If these massifs are independent of a basin, then they represent important but previously unrecognized evidence for mountain-forming processes on the Moon. It is more likely, however, that the massifs are remnants of an enormous basin that rivals Mare Imbrium in size. Such a proposal is consistent with the regionally extensive marelike units, which include units in Mare Ingenii (Plate 75a), Leibnitz, and Apollo (see discussion on pages 461–462). It is suggested further that the chaotic sculpturing, typical of regions south of these massifs (Plate 135a, arrow Z; also see Plate 75a), is related to volcanic activity and seismically induced landslides localized within this basin. This region is approximately antipodal to the Imbrium basin, and seismic waves generated by its formation may have reinforced to produce catastrophic surface disruption and landslides.

PLATE 135

a. I-030-M (oblique)

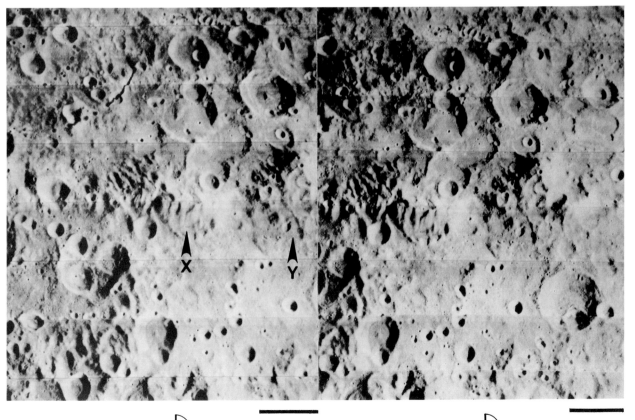

b. I-028-M   46 km

c. I-030-M   46 km

Wrinkle Ridges: Broad Marelike Surface

"Wrinkle ridge" is a descriptive label for the continuous positive-relief features that are distributed over much of the maria. The ridges exhibit considerable variety in their specific morphologic features, but, in general, the term "wrinkle ridge" is assigned to a sinuous form of relatively low relief (sometimes 300 m) and of considerable linear extent.

The examples seen in Plate 136a (oblique) show a typical ridge formation on Oceanus Procellarum about 150 km southwest of the crater Kepler (the crescent-shaped ridge in the foreground is Maestlin R). This view illustrates the lateral extent and low relief. The slight brightening of the surface toward the terminator is known as the "Heiligenschein"* that results when the phase angle (the angle between the observer and illumination) is less than 5°.

The stereo pair (Plate 136, b and c) is a nearly vertical view of another wrinkle-ridge system about 300 km southeast of the area shown in the preceding oblique view. It reveals several characteristics of such features:

1. Wrinkle ridges typically occur in systems following the same general trend.
2. They typically have narrow sinuous ridges on top of a broader rise.

* This term literally translates as "holy light." Such a phenomenon, and the word derivation, is particularly apparent if one looks at the bright halo surrounding the shadow cast by one's head on well-cut grass. It is produced by a strong backscattering of light near zero phase angle and indicates the physical structure of the surface material. In the case of the Moon, the Heiligenschein confirms the particulate nature of its surface.

3. The surface of the broad component appears very similar to that of the surrounding mare.

From these and other general observations, a variety of origins can be inferred:

1. The ridges are the result of compression as the mare sinks because of subsurface removal of magma (Baldwin, 1963).
2. They are surface expressions of intrusions with limited extrusions along zones of weakness created by compression, as in the preceding suggestion, or other causes (Quaide, 1965; Fielder, 1965; Strom, 1971).
3. The ridges are analogs to terrestrial pressure ridges and squeeze-ups of lava as crusted extrusions flowed over an obstruction (Morris and Wilhelms, 1967).
4. They represent the last stage of mare flooding and were the fissures through which the bulk of the mare basalts extruded (Fielder and Fielder, 1968).
5. At least some ridges might be uncollapsed lava tubes (also suggested by Oberbeck et al., 1969).
6. They reflect buried structures brought into relief by differential compaction of overlying units (Quaide, 1965; Mackin, 1969).

Perhaps each suggestion is applicable for different regions, and the plausibility of all with respect to specific examples is considered in the following text. Colton et al. (1972) review the overall morphology of ridges and possible theories for their formation as viewed from Apollo 16.

Arrows X and Y in Plate 136, b and c, identify the features shown in Plate 137.

PLATE 136

a. III-161-M   (oblique)

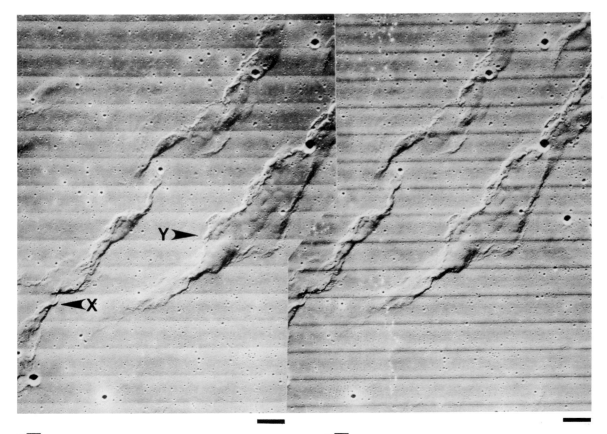

b. V-171-M 3.2 km (*top*) 3.6 km (*bottom*)   *c.* V-169-M 3.2 km (*top*) 3.6 km (*bottom*)

Wrinkle Ridges: Broad Marelike Surface

The stereo pair (Plate 137, *a* and *b*) is a high-resolution view of the Oceanus Procellarum wrinkle-ridge system shown in the lower left corner of Plate 136, *b* and *c* (arrow X). The photographs show the similarity of the crater density on the mare surface and on portions of the wrinkle-ridge surface. This correspondence implies that they have comparable ages and surface materials. Possibly the broad ridge surface is a raised portion of the mare surface. The difference in elevation may have been effected through subsurface intrusions or compressional stresses that preserved the original marelike surface during its elevation. In several instances, the mare surface slopes to a single scarp, and the broad component of the ridge appears as a tilted fault block.

Other portions of the ridges display a tree-bark or wrinkled texture and narrow ropy ridges. The wrinkled texture is pronounced along the borders of the ridges and probably would be recognized elsewhere if the lighting conditions were more favorable. The ropy ridges shown in both high-resolution stereo sets can be traced across the ridges in the overview of Plate 136, *b* and *c*. The ridge shown in Plate 137, *c* and *d*, forms a border for the northeastern edge and crosses the broad ridge to serve as a similar demarcation on the southwestern edge. Such a crossover is typical of the ropy ridge and generates the twisted overall appearance of the wrinkle ridge. The ropy relief is interpreted as the result of relatively recent buckling of the surface due to renewed faulting and compression.

PLATE 137

a. V-175-H2,H1   0.49 km   b. V-172-H2   0.43 km

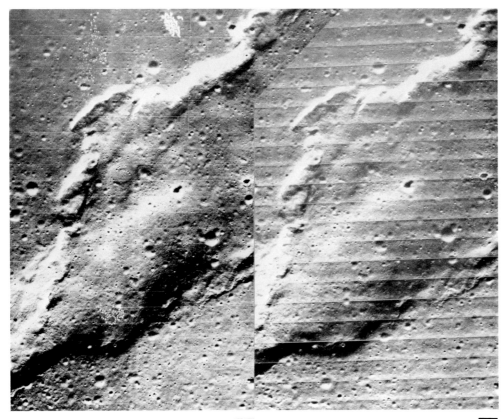

c. V-169-H1   0.43 km   d. V-173-H1   0.49 km

Wrinkle Ridges: Nonmarelike Surface

A complex of wrinkle ridges, rilles, irregular craters, and albedo contrasts bordering eastern Mare Serenitatis is shown in Plate 138. These features are in a regionally low albedo setting and characterize the periphery of Mare Serenitatis. Shallow graben-like rilles roughly parallel this eastern border and cross both the mare surface and a transitional shelf between the mare and highlands (upper portions, Plate 138, b and c). Many of the nearby small craters are shallow and flat floored (Plate 138a, arrow X) and are interpreted as products of regional blanketing (see Plate 15, b and c). In addition, several craters 3 km in diameter (arrow Y) display dimpled floors and narrow annular moats on the bulbous rim profile (similar to that shown in Plate 86, c and d).

Portions of the wrinkle ridges (Plate 138a, arrow Z) are similar to the Oceanus Procellarum system, but the ropy relief is much more evident and "crisp." In addition, the broad component does not display a crater abundance comparable to that of the mare surface, a divergence that suggests a less durable, different, or recent surface. Plate 138, b and c, shows in stereo the southern extension of the system included in Plate 138a. The stereo view strengthens the impression that these ridges are superposed over the mare rather than being merely ghost-like remnants of subsurface relief revealed by compaction. The branching and crisscrossing of the ropelike ridges are also evident.

Plate 138a includes the northern extension of the system that displays greater relief, with a nonmarelike surface (arrow XX; shown in more detail in Plate 139a). It appears to be an extension of the small sinuous ridges farther to the south, although the sinuous relief also appears adjacent to this larger ridge. Note the branching rille system that approximately parallels the plan of the wrinkle ridges; furthermore, note that one rille (Plate 138a, arrow YY) changes into a ridge, which has an appearance similar to that of adjacent wrinkle ridges.

The following two sets of plates provide more detailed views of the ridges. The northern end is examined in Plate 139, the southern in Plate 140.

PLATE 138

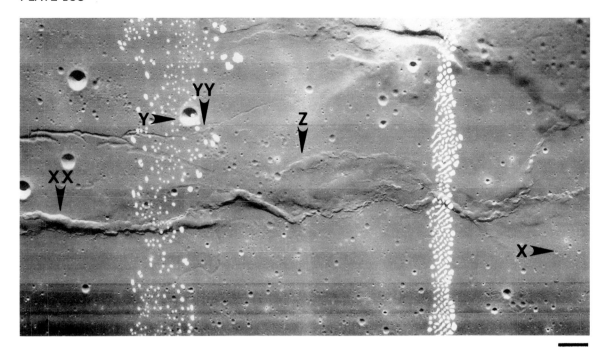

a. V-069-M   4.0 km

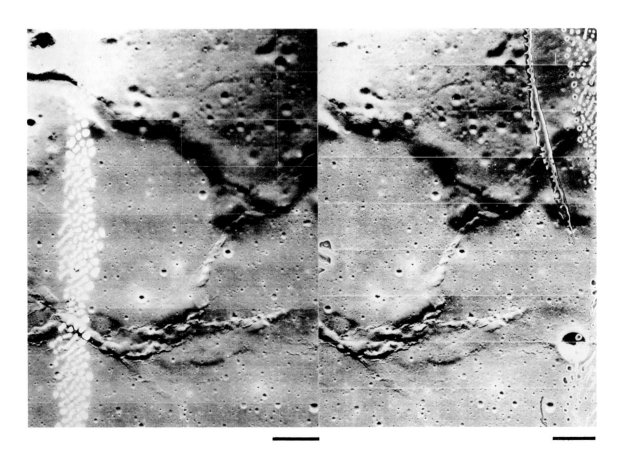

b. V-069-M   4.0 km   c. V-066-M   4.0 km

Wrinkle Ridges: Nonmarelike Surface

Plate 139 shows enlargements of the northern section of the main ridge shown in Plate 138a. The narrow but relatively high relief ridge shown in Plate 139a has a "sculptured" surface that is characterized by small-scale undulations and few craters. It is bordered on the east by a narrow basal apron (arrow X), which is likewise bordered by a narrow moat. The apron appears to be a local feature and is not present along the ridge south of this area. Narrow sinuous ridges (arrow Y) also border the main ridge.

Plate 139b shows a typical arrangement of wrinkle ridges: one ridge disappears and another develops adjacent to it (arrow X, top of the photograph). The eastern member extends south-ward as a relatively high relief ridge similar to the western member, part of which is shown in Plate 139a. These two ridges are separated by a narrow crevice, which is shown in Plate 138a to extend northward along the eastern border of the broad western ridge. The eastern member extends southward as a relatively high relief ridge similar to the western member. Two craters are clearly transected by the ridge (Plate 139b, arrows Y, Z), and their rims cross both the mare and ridge surfaces. This contact is accompanied by a narrow ridge on the floor of the crater at arrow Y and a scarp within the crater at arrow X.

The large ridge is shown to be very complex near the center of Plate 139b and is traced southward as an association of nar-row ropy ridges (arrow XX). These mark the western boundary of a much broader low-relief component.

Of interest for the interpretation of both ridges and rilles is the subdued rille shown crossing the ridge at the top of Plate 139b (arrow YY). It cuts the broader and higher ridge, but the narrow ropy ridge to the west is superposed on it. This relation-ship suggests that the large ridge is a compressional fold of the original mare surface and that the narrow ridge represents later extrusions or much more recent small-scale folding. The preser-vation of craters (Plate 139b, arrows Y, Z) on both ridge and mare surfaces, however, is more consistent with a structural origin for the narrow ridges as well. In particular, the trace of the scarp within the southern crater (arrow Z) suggests that in this region the inferred fault plane dips to the east. Because this scarp faces west, a reverse fault is indicated. In contrast, the trace of the contact across the crater at arrow Y suggests a normal fault with elevation of the western side of the crater. Ad-ditional examples are shown in Plate 140.

Numerous small (0.5-km diameter) low-rimmed craters with dimpled floors also are shown in detail in Plate 139. These may be the result of possible regional blanketing or may represent maars. But the inferred faulting and buckling in this region sug-gest that such craters may have been degraded rapidly by slope failure induced by Moonquakes.

PLATE 139

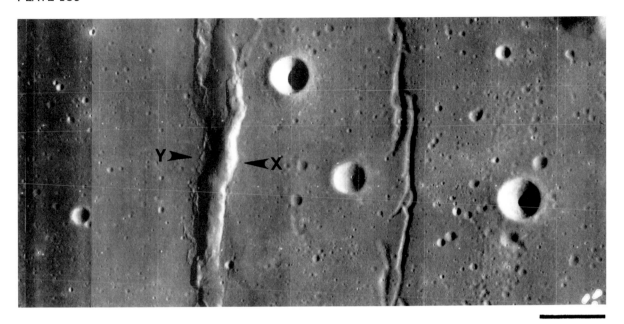

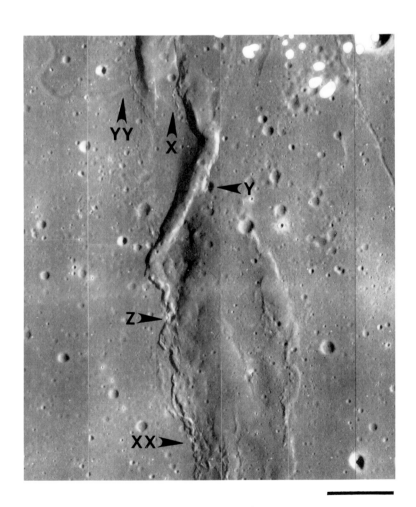

a. V-069-M   4.0 km

b. V-069-M   4.0 km

Wrinkle Ridges: Nonmarelike Surface

Plate 140a covers the southern extension of the Littrow ridge system and its numerous small sinuous ridges. As is exhibited along other parts of the wrinkle ridges, small ridges can be traced in and out of craters (arrows X and Y). Where a crater does interrupt the course of the ridge, remnants typically can be recognized on the crater wall. Further disregard of topography is illustrated by the extension of the eastern branch of the ridge onto the highlands (arrow Z). Thus the term "mare" ridge is occasionally a misnomer, and discussion of this point is made with Plates 144a–145c.

Plate 140b is a high-resolution view of the ridge where it crosses a small crater (arrow X; also see Plate 140a, arrow Y). This occurrence is consistent with the interpretation that ropy ridges are formed by compressional folds. Although the braidlike appearance may be the result of separate periods of viscous eruptions, the northern or southern ends of the ridges commonly merge with the surrounding surfaces, rather than making abrupt contacts comparable to the east or west ridge borders. Furthermore, the absence of preferred floor filling of adjacent craters overlapped by the ridges and the general disregard for topography are observations more consistent with compressional folds. Note the broad ridge (arrow Y) that overlaps a smaller ridge (arrow Z).

Plate 140c shows on a ridge the highly reflective patches (arrows X and Y) that commonly correspond to clusters of boulders. In several examples, boulders are not recognized and may indicate their smaller sizes. Such patches are thought to correspond to material exposed near or at the base of the regolith as a result of seismically induced mass movement.

Between the branching ridges is a generally smooth region, similar to much of the surrounding mare material. At high resolution (Plate 140d), a N-trending fracture system is found on this surface (arrow X), and such detail may be more widespread. The preservation of the fracture pattern indicates recent and perhaps even continuing activity in this region. These small-scale fractures probably have an origin similar to that of the larger-scale rilles paralleling the ridge farther to the north (Plate 139a). If the ridges are the result of compressional folds and low-angle thrust faulting, then the parallel system of fractures may indicate adjacent bowing of the mare.

Although not shown in these plates, there are several small features (less than 1 km in diameter) in this region that may have been produced volcanically. These include high-rimmed conelike features, small craters surrounded by moats, and circular platforms. Consequently, it is probable that at least the early period of regional deformation was accompanied by volcanic eruptions.

PLATE 140

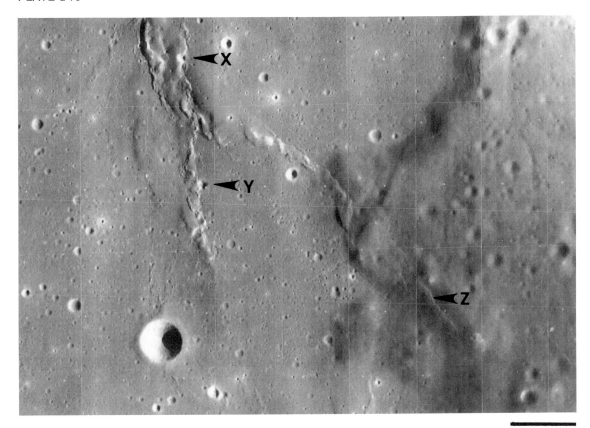

a. V-069-M 4.0 km

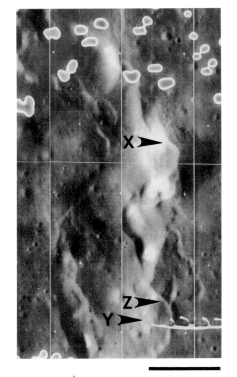

b. V-066-H3 0.50 km

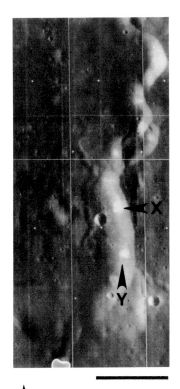

c. V-066-H3 0.50 km

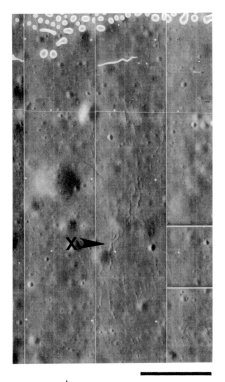

d. V-066-H3 0.50 km

Wrinkle Ridges: Surface Features

Plate 141 is a high-resolution photograph of the ropy component of a wrinkle ridge and adjoining areas in Sinus Medii. This and the following two sets of plates provide examples of surface texture and other small-scale features associated with wrinkle ridges. The twisted or braided-appearing ridge in Plate 141 has a characteristic stippled texture that is clearly identifiable along the surface of the slopes (arrow X) and that also covers a broad area to the north of the ridge (arrow Y). The texture does not resemble the tree-bark pattern, which typically has a wrinkled surface paralleling local contours. Part of this difference may be due to the direction of illumination, unfavorable for revealing small-scale elevation differences. In addition, numerous blocks are distributed across the sloped surfaces (visible in the original photograph).

Another textural feature is the prominent set of lineations trending NE (arrow Z), which may be ejecta scars from Mösting or Triesnecker. The radial alignments with these craters, however, are not exact; more likely, they are structural features.

The contact between the ridge and mare surface is typically distinct and in some areas appears as a moat (arrow XX). The northern contact near the center of the photograph is poorly delineated, except for abrupt changes in surface texture. This is partly the result of unfavorable lighting conditions.

Near the center of Plate 141 is a **D**-shaped crater (arrow YY) that either resulted from partial filling by the adjacent ridge or is a collapse feature associated with it. Another crater (arrow ZZ) that interrupts the ridge has several boulders strewn on its walls and rim, but the ridge-crater contact indicates either that the crater was formed before the ridge or that it is not an impact crater. The three aligned rimless craters on the ridge (arrow XY) are other plausible candidates for internal origin.

318

PLATE 141

II-105-H2; II-106-H2   0.17 km

Wrinkle Ridges: Surface Features

South of the crater Arago and west of the ghost crater Lamont (another wrinkle-ridge system) in Mare Tranquilitatis is a complex region crossed by an arcuate rille system and mare ridges. Plate 142, *a* and *b*, shows a section of the region included in the overview of Plate 194, *a* and *b*. The surface appears mottled because of varying densities of small (0.2 km) craters and varying surface textures. From the overview, the low density of 0.2-km craters generally is accompanied by a smoother surface. At higher resolution, however, the smooth surface is highly textured, with a wrinkled appearance (tree-bark pattern), reflecting topographic forms, and intersecting grid sets. Both surfaces exhibit a predominance of very subdued and shallow craters.

The tree-bark texture is typical of wrinkle ridges. In addition, the stereo view (Plate 142, *a* and *b*) reveals a set of furrows that does not reflect local contours but is aligned with, and perhaps genetically related to, the nearby arcuate rilles (arrows X and Y). The tree-bark texture is best developed on the ridge where the curvilinear furrows have spacings of about 20 m. These furrows are not found on the continuation of this ridge to the northeast. Similar furrows are recognized on ridges in the Marius Hills and in the Littrow region.

Plate 142c provides a high-resolution view of the ridge surface. It displays numerous small craters, of which the more subdued are elongate parallel to the lineations (arrow X). Another less prominent set of furrows trends perpendicular to the obvious northwest set (arrow Y).

The ridge apparently was formed concurrently with, or before, the large subdued rilles (see Plate 194, *a* and *b*) that cross it. The existence of these fine-scale lineations indicates either a well-preserved surface—despite the over-all subdued appearance—or recurring activity. Similar alternatives must be proposed for an unusual small-scale feature shown in Plate 194c, which includes an area about 16 km to the south.

The enhancement of these furrows along the ridges suggests processes either genetically associated with the formation of the ridge or localized and preserved on it. In the former case, a joint system may have formed during cooling of extrusions, a possible mechanism for ridge formation. Alternatively, the arcuate plan having the convex side directed northeastward might represent pressure ridges developed on a lava flow. However, such mechanisms require well-preserved surfaces on scales of less than 10 m, a requirement possibly in conflict with the subdued appearance of nearby craters. If these furrows are not genetically associated with ridge formation, then perhaps they indicate fractures on a surface covered by a relatively thin regolith. In particular, it is suggested that such fractures represent small-scale shears related to the much larger rille system to the west. Similar fracturing of the surrounding surfaces may not be revealed, because of their particulate overlay.

PLATE 142

a. II-044-M   1.6 km   b. II-048-M   1.6 km

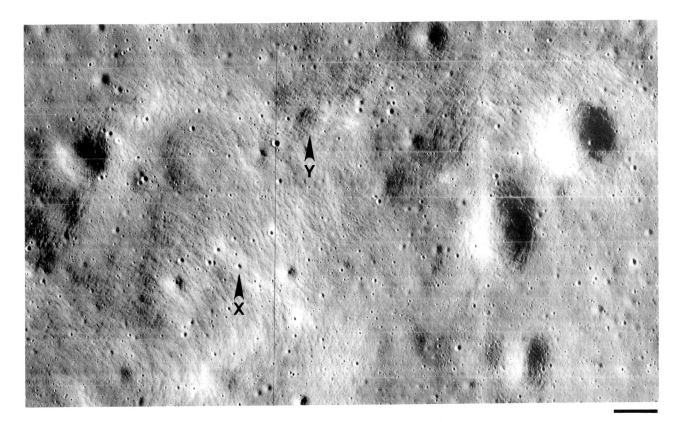

c. II-046-H3; II-047-H3   0.20 km

Wrinkle Ridges: Surface Features

The region shown in Plate 143a is to the southwest of Sabine E in western Mare Tranquilitatis (about 140 km southeast of the area shown in Plate 142 and 30 km east of the Apollo 11 landing site). The ridge does not show the twisted appearance and other detail noted in Plates 138–141. The western border of the ridge (arrow X) is expressed as a rounded scarp with a basal apron, whereas the eastern face (arrow Y) has a gentler slope. This impression is exaggerated, in part, by the eastern illumination.

At small scales, the sloped surfaces are highly textured with the tree-bark pattern. Under magnification, the original photograph reveals blocks in isolated clusters along sloped surfaces.

The ridge could be interpreted as a highly degraded remnant of the broad component shown in Plate 137 or the large ridge component of the Littrow system shown in Plate 139, a and b. The tree-bark texturing could represent, therefore, a creep phenomenon, as already suggested. The continuous western scarp does not appear to be constructed from a fissure eruption. The ridge plan (beyond the area shown in the photograph) displays en échelon offsets and a right-angle change in trend just to the south. Thus, it appears to be controlled structurally. This example may have been produced by either compressional folding soon after mare emplacement or subsidence of the mare over a buried structure. The subdued appearance is thought to reflect not only degradational processes but also the relative slowness of vertical displacement.

The very subdued 0.5-km crater adjacent to the ridge (arrow Z) suggests an interesting set of alternatives for its own origin and/or that of the ridge. If the crater was formed by impact prior to development of the ridge, then the plan of the ridge should not have been affected, as it clearly was. If the crater was formed after the ridge, then the crater and ridge have undergone extreme degradation, but the ridge has survived in better condition. Consequently, it is supposed that there has been differential erosion of the ridge and crater; also, either the crater was formed endogenetically in a subdued state or the entire region has been heavily blanketed. The correct model requires detailed studies of the various processes implied.

Of additional interest are the contrasting ejecta blankets of two craters 90 m in diameter, shown at right center of Plate 143a (arrows XX and YY). Both are on the ridge and have similar appearances. The southern member (arrow YY), however, has low-albedo ejecta, exhibited also by several other small craters not on the ridge (outside the photograph). This probably represents local differences in composition.

Plate 143b shows another wrinkle-ridge structure that exhibits one well-defined boundary and only traces of the other. As in the previous example, the "missing" border is under greater solar illumination and consequently less easily recognized. The definite expression of a boundary seen near the center of the photograph, however, suggests that its absence elsewhere may be the result of a gently sloping eastern border. Consequently, although the term "wrinkle ridge" might seem inappropriate, inspection of the feature under different solar illumination reveals that it is indeed a broad (2 km) ridge.

The ridge is on the relatively high albedo plains-forming unit (Cayley Formation) in the Southern Highlands between the craters Delambre and Agrippa (west of Mare Tranquilitatis). This unit postdates the Imbrium event and was emplaced in an irregular topographic low. Numerous subdued craters indent the surface, and there is also a significant density of convex-floored craters, most of which show no well-defined ejecta blankets. A comparable density was observed on the floor of Flammarion (Plate 18c).

The medium-resolution view provided by Plate 143b does not permit detailed comparison of surface textures. It illustrates, however, the disregard for relief by the well-defined western border, which can be traced across very subdued craters (arrows X, Y, and Z). This phenomenon also is exhibited by the ridges on eastern Mare Serenitatis (see Plate 140 a, arrow X).

Both surfaces (to the east and west of the west-facing border) are similar with respect to over-all appearance and crater abundance. If the ridge resulted from extrusive flows, then this similarity might suggest a similar age. The density of very old craters might imply further an ancient epoch of emplacement, but clearly the well-defined border obviates such a suggestion. If the border represents a reliable age indicator, then the presence of subdued craters on top of the flow hypothesized suggests either that the unit is "transparent" to the underlying topography or that such craters resulted from slow collapse. In the latter proposal, the trace of the supposed flow across craters implies that these craters were formed on both the pre-existing surface and the flow unit. If the ridge was the result of viscous extrusions, then the absence of pooling on the floors of the subdued craters may be rationalized; however, the apparent transparency to these pre-existing features becomes an enigma.

Perhaps the ridge is related to a nuée ardente type of flow that completely inundated pre-existing topography. Later settling and compaction may have exhumed these underlying features. The well-defined border shown in Plate 143b is inconsistent, however, with the interpretation that it represents the terminus of such a flow.

A structural origin for the ridge and western scarp provides the most reasonable explanation. The ridge may represent a thrust fault or a compressional fold from a westward-directed stress, as indicated by both the trace across the craters overlapped by the ridge and the pronounced west-facing scarp. Note, however, that the total displacement must not be large, for the intersected craters are not displaced or distorted geometrically.

The feature shown in Plate 143b is not unique. Plate 200 shows a possibly analogous feature that marks the contact between the highlands and the plains-forming unit. In addition, the floor of Hipparchus (Orbiter V-99-M) exhibits ropy ridges. The correct interpretation of these features may be fundamental for an understanding of subdued craters, the origin and evolution of the highland plains-forming units, and the relative rates of degradation of surface features.

322

PLATE 143

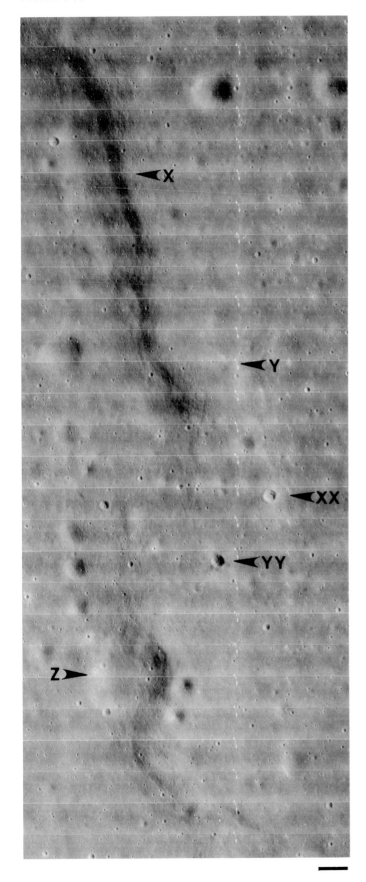

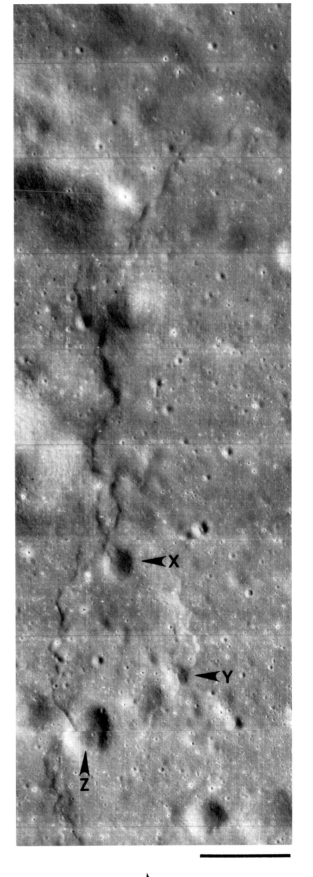

a. II-089-H3   0.21 km

b. II-094-M   1.6 km

Wrinkle Ridges: Association with Positive-Relief Features

Where a wrinkle ridge intersects another positive-relief feature, it may be cut off but reappear with a similar trend on the opposite side, as shown in Plate 129, a and b. At some intersections, however, the ridge continues either around or over the obstruction. In the former case, the wrinkle ridge appears to form a collar around the lower slopes of the intersecting relief, as shown in Plate 144, a and b (arrows X and Y). The obstructions in this situation are the insular features Montes Teneriffe E and W in northern Mare Imbrium. The resolution is not adequate to distinguish between a peaked ridge, broad platform, or ledge of the collar. Although the different views represent a 6.2° change in sun angle, the shadowing of the western slopes is partly a photometric effect and not a true shadow. Consequently, it is probably not significant that the feature is visible on the western slopes in both photographs. Note the low-relief circular dome extending from the west edge of the larger insular feature (arrow Z).

Plate 144, c and d, permits closer inspection of a similar collar around the border of the highlands southeast of Littrow B (see Plate 138a). The collar can be traced northward from the base of the scarp (arrow X) across the summit of the highland margins (arrow Y). The peaks immediately to the west of the collar (arrow Z) appear to be dropped. Note that the collar has a ridgelike profile, especially apparent along the extension onto the marelike surface (arrow XX).

If these ridges represent viscous extrusions, then it is not clear why such extrusions outline the lower slopes of the obstructions. The collar and ridge perhaps are related genetically but are not the same features. If the peaks shown in Plate 144, c and d (arrow Z), have been displaced vertically, then the collar may represent a renewed fault scarp, whereas the ropy ridge (arrow XX) may be extrusions along this weakness. This interpretation also appears viable for the collars and wrinkle ridges in Plate 144, a and b. More likely, the ridges are compressional folds and the collars represent material piled onto stable relief. This process might be viewed as analogous to ice ramparts, except that the cause of thrusting onto the shore is due to (a) a renewed source of lava beneath a cooling crust, (b) convective activity within a large pool of molten mare material, or (c) large-scale subsidence.

Regardless of the explanation, the preservation of these collars along the base of scarps indicates either recent formation or an absence of degradation. This is particularly significant when comparison is made with the insular positive-relief feature shown in Plate 134c, which has encroached into a crater at its base.

PLATE 144

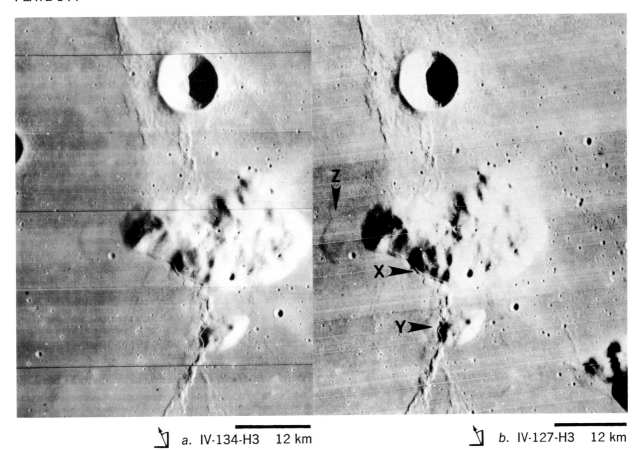

a. IV-134-H3   12 km

b. IV-127-H3   12 km

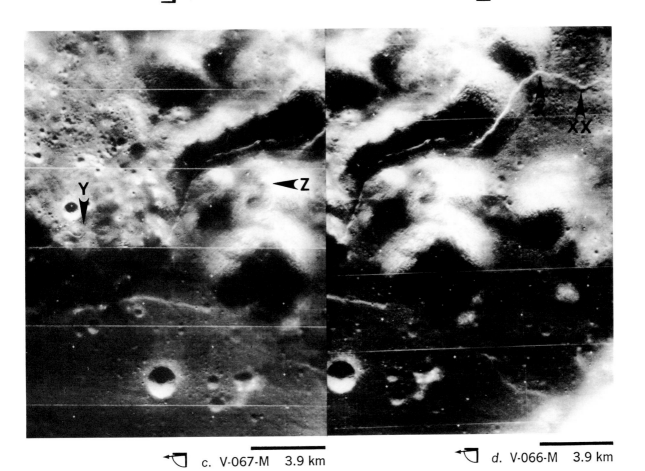

c. V-067-M   3.9 km

d. V-066-M   3.9 km

Wrinkle Ridges: Association with Positive-Relief Features

The collaring of positive-relief forms on the maria by wrinkle ridges is illustrated further in Plate 145a. This oblique view is in Sinus Medii, with the solar illumination from the left rather than the right. The ridge displays a well-defined scarp to the west (arrow X) but is breached to the west of the mamelonlike dome (arrow Y). The eastern border is the ropy ridge (arrow Z) that abuts against and collars the western slopes of the dome (arrow XX). The dome is a relatively isolated structure and does not appear to possess the characteristic apron. The base of its eastern slopes is bordered by a moat (arrow YY), revealed clearly in other views of the same feature.

The deflected plan of the ridge by the dome suggests obstructed movement from the right. This may have been produced by:

1. Termini brought into relief after settling due to migration of the underlying molten (and highly mobile) lava
2. The thrust of a cooled crust over a still-molten lava lake from a new surge of material
3. Elevation and subsidence related to Earth-produced tides, which may have been enormous during the epochs of mare emplacement

The first suggestion requires additional explanations to account for the absence of other well-defined lobate termini. There are other wrinkle ridges in the area, but they are not associated with this system and typically show strong structural influence. The second and third suggestions imply that the ropy ridges are rampartlike features or perhaps pressure ridges with squeeze-ups. The western scarp (arrow X) perhaps was formed after migration of the still-molten lava to another outlet. The survival of such features from the time of mare emplacement again creates a possible problem for currently popular degradation rates. This is also brought out by the moat at the eastern base of the dome (arrow YY), interpreted as a lava front.

Plate 145b initiates a new series of photographs that illustrates disregard for relief by some wrinkle-ridge systems. This first example is within the farside crater Aitken, also shown in Plate 24a. The ropy ridge is along the eastern (left) margin of the crater floor across the dark mare fill (arrow X). It is also traced, however, beyond the borders of the mare unit into the hillocky topography of the lower crater wall (arrows Y and Z). This ridge cannot be simply the result of compaction of material on underlying relief, nor can it be a pressure ridge on the crust of a mobile lava pool. More likely, it represents a fault that cuts a wide variety of topography.

Plate 145c shows a wrinkle-ridge system that is on the floors of mare-filled craters and appears to be related to a well-defined lineation crossing the lunar highlands. The lineation is a positive-relief feature that has a prominent east-facing scarp. It appears to be a fault, but along its length it typically assumes a ridgelike profile (arrow X). Thus the term "mare ridges" is not appropriate for all wrinkle ridges.

The apparent reflectivities of the ridges generally match the terrains that are crossed. This may indicate compositional changes of an extrusion that depend on the crust through which the magma travels. It also could indicate local surface processes that tend to mask such albedo differences. However, these compositional changes and surface processes are not required for the theory that the ridges are compressional folds. Across the marelike units within the crater, the surface responded to the compressive stresses as multiple folds rather than as a faultlike system in the highlands. It is likely that the mare units are either interbedded with incompetent regolith layers or underlain by ancient and debris-laden floor material. Such discontinuities will form a zone that will respond to lateral stresses by slipping, whereas the overlying mare units will buckle into complex folds. Similar stratigraphy in the highlands is generally absent, and lateral stresses will result in thrust faulting.

PLATE 145

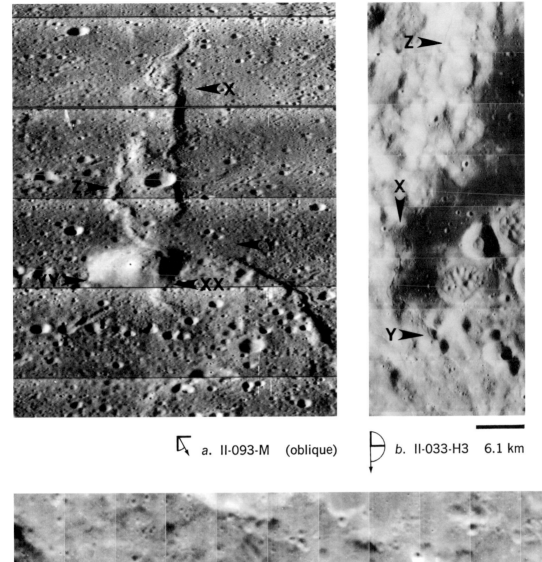

a. II-093-M (oblique)

b. II-033-H3 6.1 km

c. IV-010-H1 14 km

Wrinkle Ridges: Association with Positive-Relief Features

Plate 146 shows a phenomenon similar to that considered in Plate 145: the extension of a "mare ridge" across the highlands. The hummocky uplands are part of the Fra Mauro Formation, generally believed to be ejecta from the Imbrium event. The area shown in the foreground (bottom) of Plate 146a is about 40 km north from the northern rim of Fra Mauro.

The overview provided by Plate 146a shows that the ridge extends northward in a sinuous plan. Farther to the north (below the border of the photograph), the mare ridge appears to be related genetically to a rille that marks the contact between mare and uplands. At least two other minor lineations (not shown) parallel the ridge shown in Plate 146a.

Along the course of the feature, the profile changes. Plate 146b shows a segment that appears as a rille. At several loca-tions it is ridgelike, i.e., essentially analogous to the ropy mare ridge (Plate 146c, arrow X). Plate 146c shows scarplike segments (arrow Y), perhaps analogous to the broad component of mare ridges, and platformlike segments (arrow Z).

Most likely, this system of lineations represents a thrust fault that produced small displacements along part of its length. The rillelike segments could indicate drainage of a particulate overlay into a fracture associated with the faulting. The platformlike and ropy ridgelike segments may be compressional folds or extrusive features and are analogous to similar ridges on the maria. If small features undergo more rapid destruction in highland topography, then preservation of these features must indicate their recent formation.

PLATE 146

a. III-132-M (oblique)

c. III-132-M (oblique)

b. III-132-M (oblique)

Hogbacklike Ridges

Plate 147 illustrates a characteristic ridge type that, at low resolutions, commonly resembles a wrinkle ridge. At high resolutions, however, these hogbacklike ridges lose the twisted appearance and are generally continuous in plan. Plate 147, a and b, shows a stereo view of such a feature (arrow X) that forms part of the Flamsteed Ring. The elongate domes (arrow Y) are recognized in Plate 156, a and b (arrow X), and are aligned with the ridge. This ridge and similar ones have aprons (arrow Z) that appear to be analogous to the aprons surrounding the adjacent domes. High-resolution views of such ridges commonly show blocks along their crest.

The relatively continuous appearance of hogbacklike ridges in plan and profile seems inconsistent with the interpretation of their origin as extrusions. Some hogbacklike forms are suggested to be relict relief, but several examples appear too continuous for this origin to be acceptable. Others are thought to be products of instrusions along fractures. This view is supported by the observation that such ridges cross subdued craterforms (Plate 15, b and c, arrow YY) and are associated with grabenlike rilles (see below).

Plate 147, c and d, includes another hogbacklike ridge (arrow X) that is near the Littrow rilles (see Plate 138a). It extends from a NNW-trending grabenlike rille (arrow Y) and clearly is genetically related. This association is not unique (see Plate 165d, arrow Z; Plate 203, arrows YY and ZZ). Also note the narrow ridge along one border of a subdued rille in Plate 147, c and d (arrow Z). Such occurrences are consistent with the interpretation of Fielder (1965) that some wrinkle ridges are formed by intrusions along grabenlike rilles. Alternatively, the rille represents an earlier history of tension fracturing, followed by a reversal of the directed stresses to form a compressional ridge along the same crustal weakness.

PLATE 147

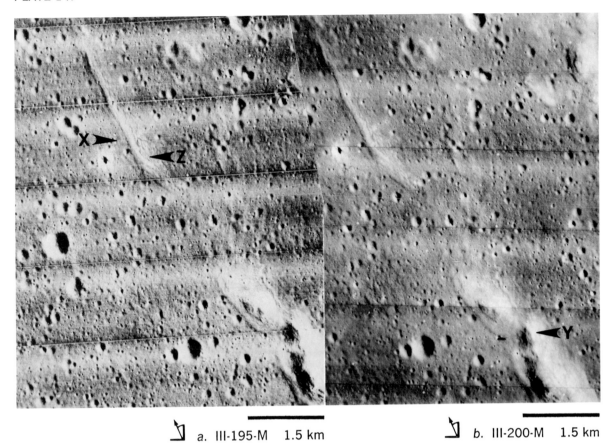

a. III-195-M   1.5 km   b. III-200-M   1.5 km

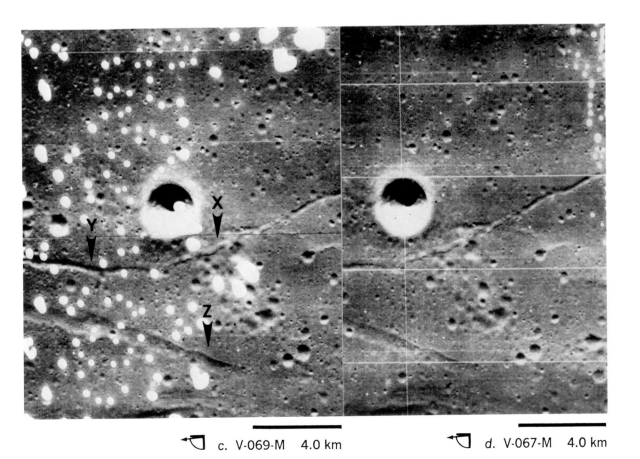

c. V-069-M   4.0 km   d. V-067-M   4.0 km

Wrinkle Ridges: Association
with Negative-Relief Features

Subdued craters are present along wrinkle ridges and may be related genetically to the ridge-forming process. Plate 148a shows an interesting spatial relation between crater (arrow X) and ridge (also see Plates 34a and 181, a and b). Note the collar along the southern mare-rim border of the large crater (arrow Y) that extends to the west from the intersection of the wrinkle ridge and crater (analogous collars were examined with Plate 144). Plate 181 shows more convincing examples of a genetic relation between craters, rilles, and ridges.

Plate 148b illustrates another interrelationship between wrinkle ridge and crater (arrow X). Obviously the term "crater" applies here only to the circular plan and shallow central depression, and this feature resembles a circular platform. Perhaps it represents a subsurface ringed fracture system produced by a premare impact crater. If the ridges are extrusions following major fracture zones, then the ringlike structure could have formed as these extrusions followed the local weakness defined by the buried crater. The absence of deformation of the circular structure is evidence that this ridge system was not formed by folding from a near horizontal stress—as suggested for other systems—unless the structure was formed after deformation.

The association between the wrinkle ridge and ghost crater shown in Plate 148c provides strong support for the interpretation that some wrinkle ridges are produced by near horizontal stresses. The wrinkle ridge appears to be obstructed by the western rim of the ring structure (arrow X). These features are in Mare Fecunditatis, east of the crater Taruntius.

The irregular platforms shown in Plate 148d are associated with NW-trending wrinkle ridges to the south of Flamsteed (see Plate 154a, arrow YY). Most of the ridges appear to be the ropelike component. The platforms shown in the upper left (arrow X) and within the ridge complex (arrow Y) are interpreted as possible flow units. The irregular ringlike form to the west of the ridge (arrow Z) is more complex and may be the remnants of a caldera. The interior of this structure resembles a settled lava lake because of the corrugated surface pattern.

Plate 148e reveals another interesting relationship in which at least one wrinkle ridge is clearly superposed on a sinuous rille (arrow X). This is not a superposition rule, for several examples (see Plates 173, a and b, 177b, and 182a) show sinuous rilles crossing wrinkle ridges. The feature is along the eastern peninsula called the Laplace Prominence, bordering Sinus Iridum. Close inspection reveals that little or no subsidence occurred on the ridge to the west (arrow X), and only questionable evidence exists for the next intersection (arrow Y) of rille and ridge to the east. The two rilles (arrow Z and XX) to the northeast appear to cross the ridge, but it should be noted that they probably belong to a separate grabenlike system.

The significance of Plate 148e can be directed to the origin of either the ridges or the rilles and depends on which feature was pre-existing. If the rille represents the pre-existing structure, then it may have formed according to almost any reasonable model. The ridge, however, must not have formed by gentle uplift or compaction, as the rille apparently has been destroyed at the junction with the ridge. On the other hand, if the ridge was pre-existing, the significance switches to the rille because the ridge could then be ascribed to a variety of mechanisms. Specifically, because of the obstructing ridge, the rille must not have formed as a surface flow. Thus, it must have formed through collapse along a fracture or lava tube (provided the tube and ridge were contemporaneous).

The existence of a rille superposed on a wrinkle ridge (Plate 177b) requires that more general statements must be considered cautiously.

Plate 148f (arrow X) shows a small breached crater (0.7 km in diameter) on a wrinkle ridge found on the mare plains adjacent to the crater Gassendi (Plate 36a). The rim of this small crater appears to terminate abruptly at the base of the ridge. It could represent a breached volcanic vent that was subsequently encroached by lava flows from the south. However, similar inundations in this region are absent. As a result, this crater is interpreted as a "hanging" crater that was elevated by faulting and folding related to the formation of the wrinkle ridge. The absence of a rim on the downthrown side is thought to reflect thrust faulting directed to the south. Moreover, the relatively small vertical displacement suggests that this thrusting occurred on a slip face with a very small dip.

PLATE 148

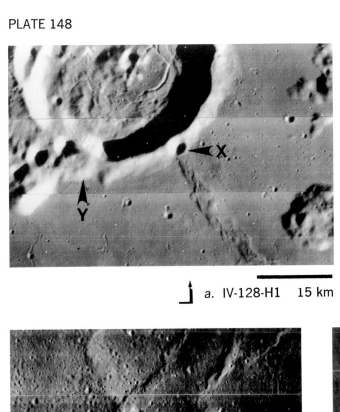

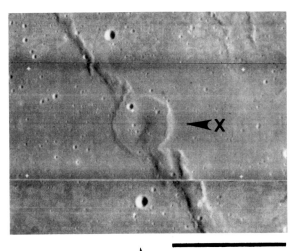

a. IV-128-H1   15 km

b. IV-162-H2   12 km

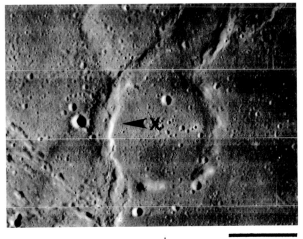

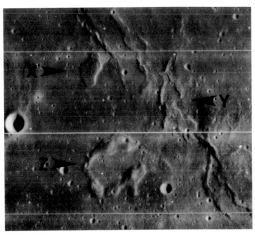

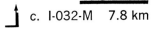

c. I-032-M   7.8 km

d. IV-143-H2   12 km

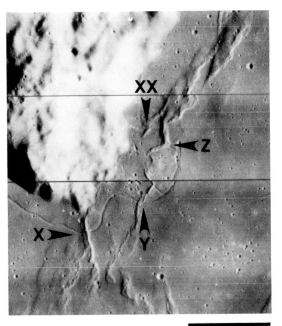

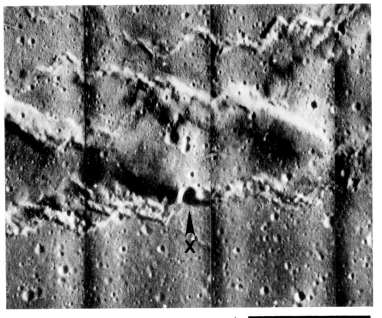

e. IV-134-H2   12 km

f. V-177-M   4.4 km

Wrinkle Ridges: Arrangements and Associations

Thus far, the discussion of wrinkle ridges has been restricted to describing individual ridges, including their plan. Plates 149–150 illustrate ridge arrangements, associations, and systems. These are defined according to increasing orders of size and complexity: a ridge arrangement describes the arrangement of separate or relatively closely spaced ridges; a ridge association is a well-defined combination (association) of ridge arrangements; a ridge system is a well-defined combination (system) of ridge associations. This scheme, outlined below, includes numbered plates showing specific examples. Included for completeness is the lowest order of description—the plan of individual ridges.

1. Ridge Plan
   a. Sinuous (Plate 149b)
   b. Angular (Plates 148e and 149a)
   c. Braided (Plates 139b, 141, and 149b)
   d. Branching (Plates 138, 140a, and 144a)
   e. Circular (Plate 148b)
2. Ridge Arrangements
   a. En échelon (Plates 136, 149b)
   b. Parallel (Plate 149b)
   c. Intersecting (Plate 150, c and d)
3. Ridge Associations
   a. Parallel (Plates 136a, 149b)
   b. Circular (Plate 150a)
   c. Polygonal (Plate 150c)
4. Ridge systems
   a. Concentric (Mare Humorum, Plate 191a; Mare Imbrium, Plate 192a)
   b. Linear (southwest Oceanus Procellarum system)

The angular arrangement of wrinkle ridges (Plate 149a) in Sinus Medii supports the contention that they correspond to fracture or fold zones and perhaps related extrusions. The two dominant linear trends—northwest and northeast—appear to be part of a wide-scale macrogrid pattern. The offset to the west (arrow X) could imply significant horizontal displacement along a strikeslip fault. The angular displacement of the ridges (arrows Y and Z) indicates a secondary stress pattern, but the alignment of these ridges is with the Imbrium sculpture (or lineation), which typically exhibits an en échelon arrangement (see Plate 133b). In addition, it appears that the secondary fracture zones become primary in the eastern part of the ridges shown in Plate 149a (arrow XX); consequently, major stresses along two separate directions were probably present.

Tjia (1970) studied the en échelon offsets of lunar wrinkle ridges and suggested that they result from strike-slip faulting. Specifically, he found that sinistral offsets typically trend between north and east. He interpreted the offset wrinkle ridges as either drag folds along primary and secondary shears due to east-west compression or, following Fielder (1965), dikes that fill tension gashes resulting from north-south compression. From studies of lunar lineaments, Strom (1964) finds different evidence for north-south compression.

The parallel ridge association in Plate 149b implies large-scale stresses if the ridges reflect a fracture system. It is part of a linear ridge system that trends approximately northwest at longitude 50°W. This system approximately corresponds to a positive gravity anomaly in Oceanus Procellarum (see the surface mass distribution calculated by Sjogren et al., 1971). The anomaly may be related to a thin lunar crust and fissurelike vents partly responsible for the inundation of Oceanus Procellarum (also see Plate 182a). The ridge patterns, however, suggest compressional folds; consequently, this region may indicate a subsidence, responding isostatically to the thick layers of relatively dense mare basalts built-up along once-active fissure vents.

334

PLATE 149

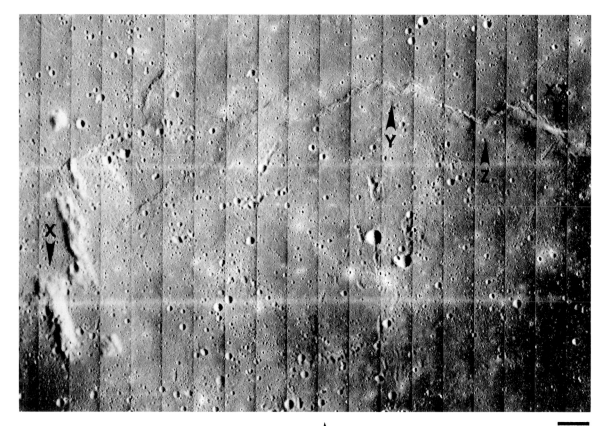

a. V-110-M   3.4 km (*left*)   2.9 km (*right*)

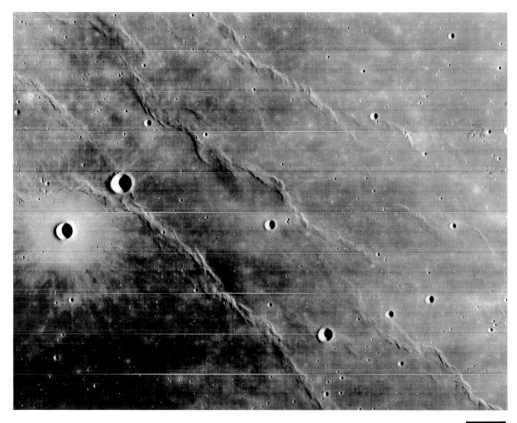

b. IV-163-H1   12 km

Wrinkle Ridges: Associations and Systems

Circular patterns formed by a system of wrinkle ridges are shown in Plate 150, *a* and *b*. The arrangement gives the impression of a buried crater and, hence, the term "ghost" crater (also see Plates 152–154). The first example is labeled Lambert R (Lambert being the crater to the north) and is approximately 60 km in diameter. Commonly, such structures are not completely closed, as illustrated by the gap along the eastern edge. Telescopic views of the southwestern segment of the ring reveal low-albedo deposits near the rimless pits (arrow X). The wrinkle ridge to the east continues southeastward and merges with a system of ridges (not shown) subconcentric with Mare Imbrium. The crater Lambert appears to be superposed on the ridges although terrestrial photographs and LAC 40 show that they can be traced through the ejecta blanket.

The circular plan shown in Plate 150b also can be examined in Plate 57a (upper right) with the discussion of Secchi X. The mare-ridge feature composes a major portion of the ring, but, to the west (arrow X), it merges with a ridge and dome complex (see the discussion accompanying Plate 129, *a* and *b*).

Plate 150c illustrates the effect of several intersecting wrinkle-ridge systems. The plan is roughly polygonal. Many of the patterns created by such ridges result from branching or merging of one or more systems, but in this region—eastern Mare Frigoris—overlapping is apparent. Superposition indicates discrete events at different times. Plate 150d shows a similar association in northwestern Oceanus Procellarum. The relief feature seen near the bottom of the photograph is Rümker (see Plate 125, *a* and *b*).

The ringed ridges shown in Plate 150, *a* and *b*, are interpreted as pressure ridges formed over buried crater rims during mare emplacement and accentuated during subsidence of the maria. The buried rim region corresponds to likely sites for late-stage eruptions, owing to both the locally thin mare crust and the probable vents associated with inundated craters (see Plates 35–39). I suspect that these vents resulted in a ring-dike network that in some cases stood out in relief following subsidence of the mare basalts during cooling and later differential compaction.

Guest and Fielder (1968) have suggested that some of these rings represent important volcanic structures from which large volumes of ignimbrite sheets and lavas were erupted. The ring system Lamont in Mare Tranquilitatis was cited as one possible example. In support of their theory, this structure corresponds to a positive gravity anomaly (Sjogren et al., 1971), a correlation suggesting a region of extensive mare fill. Another cited example is the interior ridge of the Flamsteed Ring (Plate 154). This ringed ridge, however, is thought to correspond to the old floor boundary—or faults associated with this boundary—in analogy with such craters as Gassendi (Plate 36). Because such rings typically occur in regions having relatively shallow total mare depths, it is reasonable to suspect that they represent buried impact craters. This suggestion does not imply, however, that the old impact craters were simply passive structures; on the contrary, they probably were sites of volcanic eruptions, perhaps as extensive as those envisioned by Guest and Fielder (1968).

The complex systems shown in Plate 150, *c* and *d*, may be related to extrusive vents, but the overlapping ridges strongly suggest thrusting of individual plates against each other. As suggested in the discussion with Plate 145a, such plates may have been formed during the cooling stages of the mare basalts and were driven by subsurface movement of still-molten lava or subsidence produced during cooling or once-enormous lunar tides.

PLATE 150

a. IV-126-H3   12 km

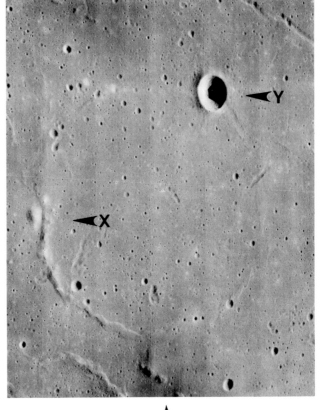

b. V-048-M   (oblique)

c. IV-091-H3   13 km

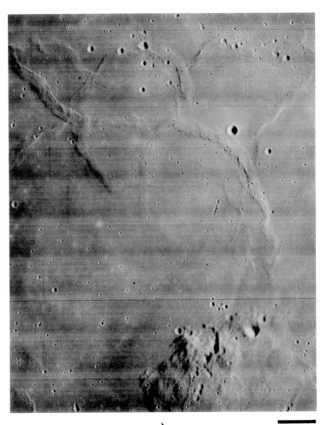

d. IV-170-H2   12 km

Ring Structures: Small-Scale Rings

Plates 151–158 illustrate another type of positive-relief feature, commonly termed "ghost" craters because they appear to be vestiges of partially or completely inundated craters. Such features are discussed in the chapter "Crater Rims" (Plates 91–92, 110–111), but their consideration here is perhaps more appropriate in order to preserve nongenetic implications. Further noncommittal nomenclature is provided simply by the term "ring" rather than "ghost" crater. The rings are present in a variety of forms and sizes and are expressed by wrinkle ridges (Plate 150a), crested relief, mamelonlike domes, and lineations. Again, the scheme is to consider the smallest examples at the outset.

The area shown in Plate 151, a and b, is southwest of Maestlin R in southern Oceanus Procellarum. The grid and concentric tree-bark patterns are evident in both examples. The annular pattern seen in Plate 151a is found more than a ring radius from the low-relief rim. In this first example, the ring also is marked by a narrow moat (arrow X). The interior and exterior surfaces exhibit a wide range of crater sizes and degrees of crispness of form. The high density of subdued depressions yields a rolling topography that is generally absent along the crest of the ring, which is dominated by the finer grid and annular textures. Numerous blocks line the outer rim of the ring and are particularly evident to the north (arrow Y).

The ring shown in Plate 151b is similar to that seen in Plate 151a, except that its interior is characterized by a marked change in texture. This textural change is due to an increased number of subdued craters and swales, along with NW- and NE-trending sets of narrow furrows (arrow X) that resemble those associated with crater rays. Such furrows, one of which is associated with Kepler, crisscross the region; therefore, the occurrence within this ring may be fortuitous. The dark-rimmed, flat-floored crater (arrow Y) within the ring also appears anomalous. Although flat-floored craters in this size range are not uncommon, the low reflectivity is not common and perhaps is an indication of local compositional difference.

The two structures shown in Plate 151 are not unique to this area; at least four similar rings are within a 25-km radius from the one shown in Plate 151b.

Three general times of formation are obviously possible for the ring features: prior to, during, or after the last emplacement of local mare material. The first possibility rests on the hypothe-

sis that the rings represent inundated impact craters. This is perhaps the most plausible suggestion because of the proximity of craters that apparently have been only partially encroached (see Plates 195–196). The small craters shown here are actually within the extrapolated boundaries of Maestlin R (Plate 195a), which is a large breached ring. Consequently, if these are premare craters, they (excluding Maestlin R) probably formed on the old Maestlin R floor that was later covered by one or a series of flows.

Three models of inundation can be envisioned. First, the mare material overran the entire crater perimeter and filled the crater to the brim. Second, a breach in the wall provided the primary path for entering material. Third, the crater itself acted as a conduit for the inundating material. The first suggestion allows for the absence of a well-defined mare-crater contact, but it does not account for the abundant blocks along the rim, unless they are the result of later processes. The second suggestion is plausible because of the breach of the eastern wall exposed in Plate 151a and perhaps of the western wall shown in Plate 151b. Later degradation then would have to be imposed to destroy the mare-crater contact, and, concomitantly, the survival of the narrow moat would have to be rationalized, perhaps as a result of continued settling. The third proposal is reasonable for larger craters but apparently did not happen in the nearby large craters shown in Plates 195–196. Thus, it is reasonable to surmise that such inundation did not occur for those shown in Plate 151.

The rings also could represent scars of impacts in a still-plastic mare unit. In this respect, they resemble laboratory models of craters formed in a medium scaled to simulate the lunar crust (Scott, 1967). The lifetime of such craters was interpreted, however, as extremely short, but perhaps the model is pertinent for a still-cooling mare surface.

The endogenetic origin of these rings also must be held tentatively viable. It is possible that these structures were related genetically to the last mare-flooding event.

The postmare impact origin for the rings is untenable under any hypothesis, owing to the selective degradation that would have had to be imposed. This leaves open endogenetic processes, such as small-scale intrusions that domed the surface and subsequently subsided.

PLATE 151

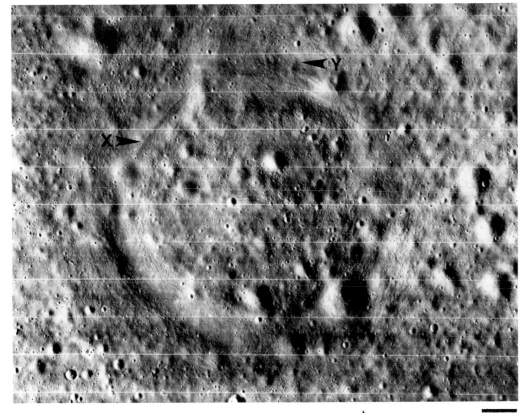

a. II-204-H3   0.20 km

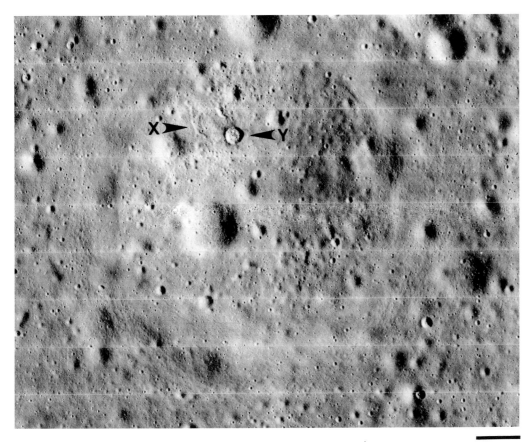

b. II-200-H2   0.20 km

Ring Structures: Small-Scale and Medium-Scale Rings

Plate 152, a and b, provides a stereo view of a small-scale feature similar to those shown in Plate 151. In this example, a narrow moatlike feature is evident within the ring above the base of the wall (arrow X). The northeastern part of the ring has been breached, and the moat can be traced to an exterior portion of the ring (arrow Y).

The interpretation that this is an inundated crater appears the most reasonable. The placement of the narrow moat above the ring base further suggests that the interior has undergone subsidence. If this moat has been preserved since the time of mare emplacement, then the nearby subdued craters must have formed endogenetically. Alternatively, the floor of this crater continues to subside.

The example shown in Plate 152c is a smaller ghostlike feature just ahead of the Imbrium flow front (Plate 192, b and c, arrow XX). A concentric texture (arrow X) is identified around the feature but not within it. Another ghostlike feature similar to that shown in Plate 152, a and b, occurs to the northwest (but not included). The same alternatives offered for the foregoing examples apply: it might indicate an intrusive or extrusive body; or it is a remnant of a crater that was invaded completely by mare material.

Plate 152d reveals a doughnutlike object 12 km in diameter in the northwestern section of Mare Nectaris. Again the question is whether the feature is pre- or postmare. The inner plain is not a smooth marelike surface, and the encircling ridge does not have the sharp-crested appearance of ringed plains, such as Gambart (Plate 91). Similar rings of comparable size do not occur nearby. Although the area displays shallow craters (including Daguerre) having domed or flooded floors, they occur on the mare margins, whereas the example shown in Plate 152d is well within the margin. In addition, wrinkle ridges and sheets of lavalike extrusions are distributed over this section of Mare Nectaris. Apparently, volcanic processes have been active and could have developed this structure (also see Plate 132e, arrow Z).

PLATE 152

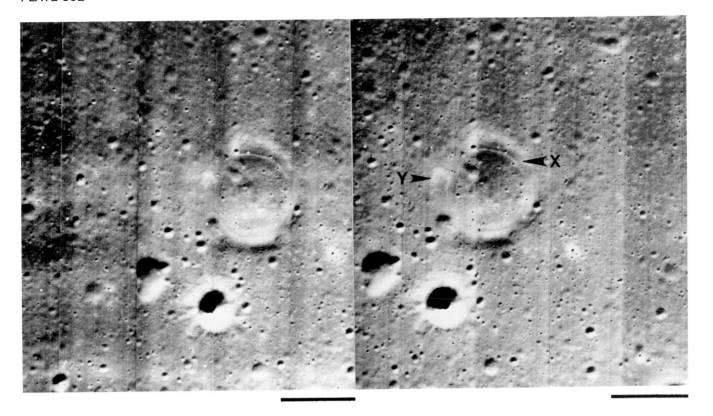

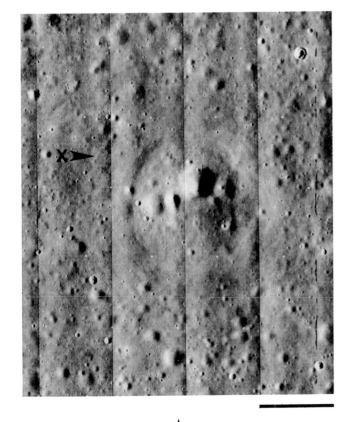

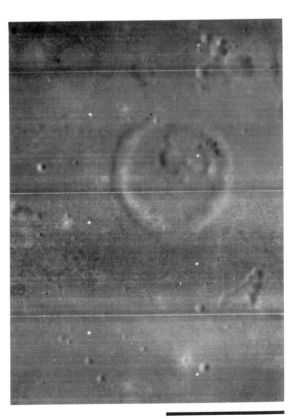

a. III-144-M   1.8 km

b. III-150-M   1.5 km

c. V-159-H3   0.64 km

d. IV-072-H2   12 km

Ring Structures: Medium-Scale Rings

Plate 153 shows large-scale (10 km-25 km) arcuate ridges or broken rings. The first example is Maskelyne F, in eastern Mare Tranquilitatis (see Plate 124a, arrow Z). Similar arcs and insular forms are common in the area, most of which are domelike or are subdued and have irregular plans.

The harsh lighting gives the ridge shown in Plate 153a a crested appearance, although the actual profile is more subdued. The ridge is bordered by an apron, which makes a relatively sharp contact with the mare surface. Two sets of narrowly spaced lineations, trending NW (arrow X) and NE (arrow Y), cross the ridge but apparently are not expressed across the apron. Several craters along the inner base have been partly buried by ridge material, and three craters (arrows Z, XX, YY) have been inundated almost completely. The extrapolated rims of these craters extend beyond the relatively narrow apron. Consequently, the apron material cannot be considered the only mobile component of the ridge.

The example shown in Plate 153a probably represents an inundated and breached crater, similar to the more-complex arcs considered in the chapter "Crater Rims" (Plates 91–92, and 109–111). This interpretation requires considerable degradation of the original crater because of the relatively symmetrical profile it now displays. If this smoothing occurred prior to the epoch of inundation by mare material, then it is not clear how the inner scarp could have been reduced in slope without significant reduction in slope of the outer rim as well. More likely, the symmetry was attained after this epoch, and the encroachment into small craters at the base of the remnant inner scarp is evidence for the rapidity of the process. Such rapidity, however, seems incompatible with the preservation of some premare structures, unless the original crater rim was relatively uncon-solidated or friable. For example, either the original impact may have occurred on a largely unconsolidated surface, or the rim was constructed volcanically by pyroclastics.

The symmetric profile also may be enhanced, in part, by collapse of the rim region along a graben encircling the inner rim. This collapse is suggested by inundated craters having platform-like rim profiles and by the moatlike region adjacent to the crested rim segment of Doppelmayer (Plate 191b, arrow YY). The concentric graben is thought to have developed along concentric weaknesses developed during crater formation. Several recently formed impact craters exhibit morphologic evidence for such weaknesses, and in several cases a synclinal region borders the inner rim (see Plates 93 and 101a).

Numerous circular or arcuate ridges are identified in Plate 153b, which covers an area north of Sinus Iridum. The local invasion of mare material was apparently insufficient to destroy the crater rims, and several craters along the border of the highlands to the south have been breached, with their floors partly flooded.

In addition to the encroachment of crater floors by mare units, the domes or peaks that interrupt the outline of the ghost-like features (arrow X) suggest that constructional processes may have occurred. Possibly such features are volcanic in origin and were triggered by the events that induced mare flooding.

As suggested for Maskelyne F, the arcuate ridges shown in Plate 153b also may represent inundated tephra rings or cones. The credibility of this suggestion is strengthened by the breached high-rimmed crater (arrow Y), which appears to be very similar to terrestrial cones. Many of the rings shown in Plate 153b, however, are remnants of secondary craters associated with the formation of the crater that is now Sinus Iridum.

PLATE 153

a. II-005-M   1.7 km

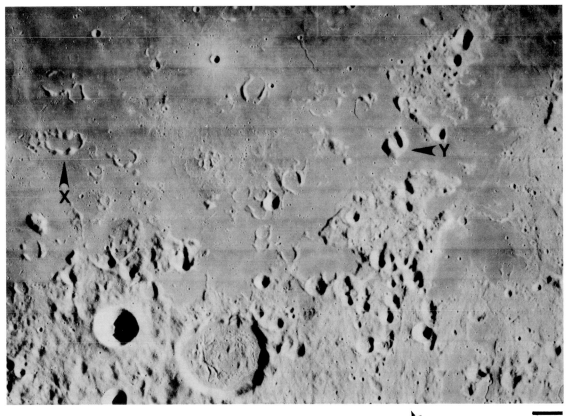

b. IV-145-H3   12 km

Ring Structures: Large-Scale Rings

The first Orbiter series yielded photographs of the Flamsteed region that prompted a controversy concerning the origin of the ring structure, long regarded as merely exposed portions of a partly buried crater. Plate 154a shows Flamsteed (the large "fresh" crater to the south) and the ring system, about 100 km in diameter. At this resolution, one can determine the following:

1. The outer ring is composed of islandlike relief features (arrow X).
2. An inner ring is formed by a wrinkle-ridge system (arrow Y).
3. Several insular features composing the outer ring are linked by wrinkle ridges (arrow Z).
4. The inner wrinkle-ridge system connects with an exterior system (arrow XX) through a breach of the ring on the north (easily identified on the original photograph).
5. There is an absence of the NNW-trending wrinkle-ridge system (noted exterior to the ring) within the outer ring.

Plate 154, b and c (stereo), reveals in more detail the major features of the Flamsteed Ring and the crux of the controversy. The basal terraces or aprons, which have already been illustrated with mamelonlike domes, are impressively developed along the base of the ring. The apron could be attributed to mass wastage, such as creep, which postdates mare flooding. In general, deposits formed along the base of terrestrial topographic features can be grouped as either talus or landslides. Arguments against this interpretation for the aprons shown here have been presented by O'Keefe et al. (1966):

1. Talus formation:
   a. Slopes are not steep enough to conform to terrestrial analogs.
   b. Talus slopes are typically concave in cross section, whereas the aprons considered here are clearly convex.
2. Rock creep and landslides:
   a. There is not enough weathering to produce talus slopes; it is unlikely, therefore, that creep and landslides would accompany talus formation.
   b. Rock creep would be expected to produce concave slopes, but concavity is not observed.

In defense of mass wasting, the foregoing arguments do not allow for an ancient crater wall that has been weathered by bombardment of meteoroids (or other processes) before mare flooding. That is, O'Keefe et al. appear to assume that the preexisting relief was an unweathered and competent block. Perhaps talus formation, landslides, and rock creep could be eliminated owing to the relatively gentle slopes (about 16°) and profiles, but stronger arguments are necessary to eliminate other forms of slow mass movement involving finer material, especially in a lunar environment.

A more formidable objection to mass wasting might be the existence of the apron around relatively low relief features. In particular, the width of the apron does not appear to be dependent strongly on the height of the relief. Consequently, the apron may not be composed of material derived from the upper slopes (also see Plates 60b and 147, a and b, which show other sections of the Flamsteed Ring).

The alternative to mass wasting suggested by O'Keefe et al. was that the aprons represent termini of viscous lava extruded from ring fissures, which are now the Flamsteed Ring. In support of this suggestion, it has been shown that wrinkle ridges commonly are linked to the mamelonlike domes (see Plate 129). If the ridges are in part extrusive, it might seem reasonable (but not convincing) to draw a similar conclusion about the insular features. Note the nearby ring (Plate 154a, arrow YY; also see Plate 148d), which was probably formed by volcanic processes, and the volcanic cones (Plates 154a, arrow ZZ) having relief comparable to that of the domes forming the Flamsteed Ring. The actual debate is much more encompassing than the formation of the Flamsteed Ring, for the existence of such ringed extrusions would open a large number of possibilities concerning crater formation in general. Several comments can be made in regard to an extrusive origin:

1. The proposed lava termini (the aprons) are surprisingly regular and nonlobate.
2. The apron appears to be continuous for considerable distances, a circumstance that would seem to require a remarkable simultaneity of extrusions.
3. Even at medium resolution, a change of texture is apparent between the apron and the upper slopes.

Perhaps some of the objections to an extrusive origin could be eliminated by not requiring that the basal aprons be lava termini. The individual hummocks comprising the ring could be analogous to terrestrial cumulo domes or plugs, which are highly friable. The apron is then envisioned as mass-wasted debris during construction, yet the ring remains as a postmare volcanic structure.

A third set of possible mechanisms arises from the effects of mare flooding. For instance, when the mare units encompassed the pre-existing relief, the weak contact between the lava and the relief may have resulted in frothing or spattering. Another suggestion is that the aprons are high lava marks that resulted after settling or migration of the underlying molten lava (see Plate 128). Such possibilities explain the usual absence of aprons in nonmare contact regions. Contacts without these phenomena could be rationalized as buried under talus or as lacking the proper conditions.

Arrow X (Plate 154, b and c) locates the region shown at higher resolution in Plate 155. Arrow Y identifies surface texturing, which is characteristic of several regions within the Flamsteed Ring and which is thought to represent preserved flow patterns from the last period of mare emplacement (see Plates 192–193).

PLATE 154

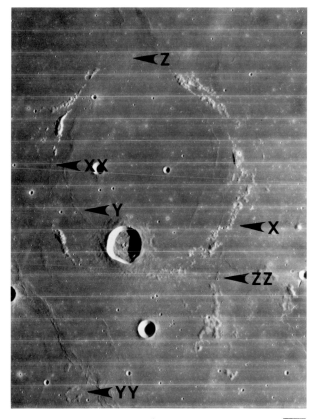

a. IV-143-H3   12 km

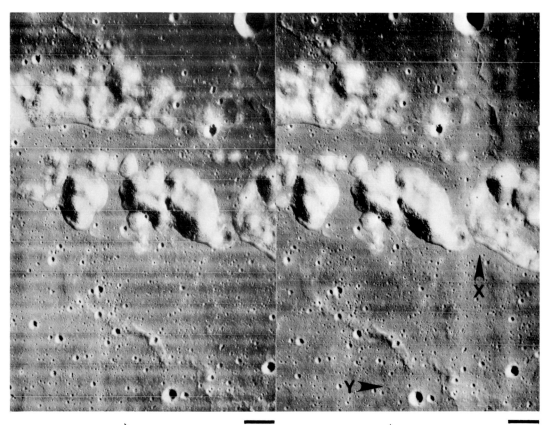

b. III-181-M   1.8 km   c. III-184-M   1.8 km

Ring Structures: Large-Scale Rings

Plate 155 shows a detailed view of the Flamsteed Ring seen in Plate 154, *b* and *c*. The mare-relief contact is traced approximately 9.5 km, and the apron (arrow X) is distinguished readily by its change in slope and finer corrugated texture. The apron is not composed of large blocks, which are found more typically farther upslope or along boundaries of large subdued craters (center).

Movement of the apron is indicated by the partial encroachment on the small crater (arrow Y); however, preferential filling of the crater floor has not occurred. Movement also is inferred by the reduction in crater counts on the relief features relative to the density on the adjacent mare surface. In particular, large craters and very small craters appear to remain, whereas those of intermediate size are either extremely subdued or nonexistent. Just such a size distribution would be expected on a mobile surface, for smaller craters are frequently replenished and larger craters require considerable time to become obliterated. The apparent advancement onto the mare surface is, of course, not diagnostic for an extrusive origin of the ridge but only indicates movement, which could be the result of mass wasting and slow creep.

More-rapid movement appears to be indicated by the parallel flowlike texturing of the surface extending southeastward more than 3 km from the base of the hills at left (arrow Z). The surface exhibits a low density of craters 50 m–100 m in diameter, and most of the larger examples are crossed by parallel lineations, in contrast to the tree-bark texture that typically parallels local contours. An unusual feature on this surface is a ring of boulders (arrow XX) having a diameter of about 0.8 km. Within the blocky ring, the texturing is less marked. The origin of this unit could be a landslide or an extrusive flow that apparently was derived from the valley to the north and perhaps ultimately from a cascade of mare material originating on the other side of the hills. Although the present level of the mare units does not seem to support the latter suggestion, Plate 156, *c* and *d*

(arrow Z) shows evidence for previously higher levels. Alternatively, the textured unit corresponds to an extrusive unit related to a poorly defined northern extension of a wrinkle ridge visible in Plate 154, *b* and *c*. The ringed structure mentioned above could represent the inundated source of this unit.

Despite the evidence for mobile surfaces, there is a set of NE-trending lineations that crosses the upper slopes of the hills, the apron, and the flow-textured unit. These lineations are not expressed clearly, however, on the mare surface, except for isolated textured patches. Their location on surfaces supposedly mobile enough to produce a decrease in crater density suggests that they either must be very recent or are effects of lighting. With respect to the first possibility, their poor expression on the mare surface is enigmatic, for such a relatively level surface should be more stable than sloped surfaces.

The dilemma might be resolved if the textured patches and hills are underlain by competent units that responded to stresses as closely spaced joints. The mare surface, however, probably is interbedded with competent and incompetent units, with an overlying regolith. The irregular crater plans in the area suggest this. Consequently, if the system of joints is regional, it will be expressed poorly on the mare and difficult to identify without favorable lighting. The point remains, however, that under this model the lineations across the apron must indicate relatively recent release of stress or that the subdued craters on the mare are endogenetic and cannot be used to estimate relative erosion rates.

It is interesting to note that, besides the obvious differences in texture between the hills and mare, there is also a greater number of bright-haloed craters on the hills. This is especially apparent around the large subdued crater (arrow YY), where the sun angle is not very different from that on the mare surface. Presumably, this is an indicator of differences in the target, either physical or mineralogical.

PLATE 155

III-183-H3; III-184-H3   0.21 km

Ring Structures: Large-Scale Rings

Plate 156 displays in stereo segments of the eastern portion of the Flamsteed Ring. Both stereo pairs also include the inner concentric wrinkle ridges. The apron is prominent (arrows X, Y, and Z) and is nearly continuous along the inner base of the insular structures seen in Plate 156, a and b. Its width is relatively independent of relief of the over-all insular structure, as noted previously for the western portion of the ring (Plate 154, b and c). It becomes barely recognizable on the sunward side of the relief (arrow XX).

Unlike the mamelonlike relief composing most of the ring, the relief shown in Plate 156, a and b, has a well-defined W-facing scarp (accentuated by the sun's being only about 19° above the horizon). The stereo view reveals that the E-facing slope is broken by a series of smaller scarps (arrow YY). Consequently, the feature appears to be a tilted block with terraces. More detailed coverage of this area is shown in Plates 157–158.

To the south of the above area, Plate 156, c and d, shows positive-relief structures similar to those shown in Plate 154, b and c, but, in contrast, they appear more interconnected and less well defined. The mare material displays more variations in texture than is noted in other areas around the ring. The aprons are not everywhere clearly identifiable at the base of relief features. Around such areas, there is typically a zone of increased texturing (arrow X). These surfaces are marelike with respect to albedo but display fewer craters and have rougher topographies with 0.1 km wavelengths. Note the apron (arrow Y) that crosses this textured area and encircles a closed depression, which contains marelike material. In this case, the apron has not developed just at a mare-relief contact. In addition, the apron marked by arrow Z has a ridgelike profile and is separated from the relief by a moat. These observations are consistent with the interpretation that these aprons are high-level marks of mare units or, at least, were formed in association with mare emplacement (see Plate 128).

PLATE 156

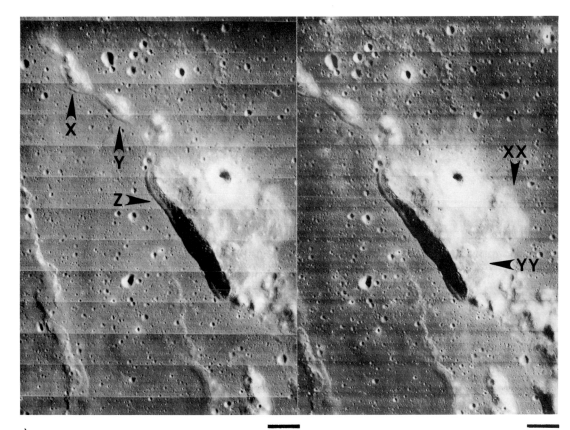

a. III-197-M 1.6 km (*top*) 1.9 km (*bottom*)   b. III-200-M 1.6 km (*top*) 1.9 km (*bottom*)

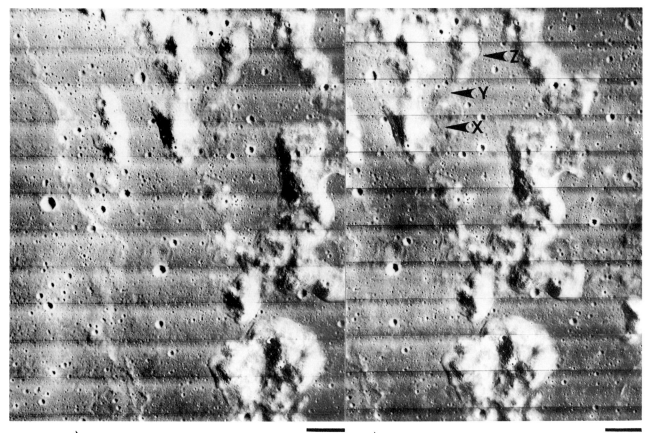

c. III-199-M 1.6 km (*top*) 1.9 km (*bottom*)   d. III-200-M 1.6 km (*top*) 1.9 km (*bottom*)

Ring Structures: Large-Scale Rings

The major relief feature shown in Plate 156, *a* and *b*, is isolated in Plate 157. Immediately obvious is the absence of craters with diameters exceeding 0.3 km on the upland surface. Such a size distribution might be expected for a mobile (through creep) terrain subject to relatively high destruction rates.

Outcrops and boulder fields are scattered across the relief feature. Some of these are associated with abrupt changes in relief or crests of local swales (arrows X and Y) but do not appear to be part of a continuous stratification. Two very prominent steeplelike outcrops are also present: one to the northwest of the funnellike crater (arrow Z), the other along the SW-facing scarp (arrow XX). The former displays a very high albedo, as is characteristic of many of the blocks, and might be associated with the funnellike crater. The latter outcrop is shown in Plate 158*b* and discussed in more detail.

The 1.5-km-diameter funnellike crater shown in Plate 157 (arrow YY) also is shown in Plate 85*b*, and, in the accompanying discussion, it is pointed out that mass transfer must have been quite rapid if an impact origin is assumed. This is necessary because of the asymmetric funnellike profile, with preservation of blocks along the rim and diffuse bright rays overlying the adjacent mare. Consequently, this portion of the Flamsteed Ring is inferred to be very unconsolidated. This crater may have been formed endogenetically, with only minor ejecta.

In Plate 157, the apron is clearly visible on the W-facing slope (arrow ZZ) but is indistinct on the west slope (arrow XY). The nearly absent detectibility of an apron on this side is probably due to unfavorable lighting conditions since the apron is continuous in other sections of the ring. Plate 158 shows selected enlargements of the apron.

PLATE 157

Ring Structures: Large-Scale Rings

The series of photographs in Plate 158 shows selected enlargements along the western face of the relief shown in Plate 157. Plate 158a reveals numerous boulders lining the upper slopes (arrow X) along the northern part of the feature. The blocks are concentrated near the upper break in slope, and such a disposition has been noted previously along the rims and upper walls of extremely subdued craters. This is an expected placement for objects exposed through erosion.

The apron-mare boundary intersects two subdued craters (arrows Y and Z), and, although no preferential filling is apparent, indentations in the apron occur directly above each. It appears that the apron is advancing onto the mare. Close examination does not reveal well-defined termini of the mare material against the apron; however, the mare appears wrinkled at several locations (arrows XX and YY) just before the break in slope.

The upper slopes have a coarse tree-bark pattern (arrow ZZ), which approximately parallels the apron-mare contact (see also Plate 155), but, because this texture is seen under such extreme lighting conditions, caution must be exercised in assigning significance to it. In contrast, the apron (arrow XY) displays a finer parallel design.

Plate 158b shows the continuation to the southeast of the area shown in Plate 158a. A prominent feature is the cluster of steeplelike outcrops (arrow X). The extreme lighting conditions make a description difficult, and the sharp-rimmed craters illus-

trate the exaggeration produced by the lighting. The steeplelike outcrops are also exaggerated; nevertheless, the boulders may be the result of accumulation of material upslope from the steeple or of debris resulting from the erosion of the parent body. Numerous boulders are strewn downslope (arrow Y) and probably originated from the outcrop. The remnant steeple complex may be due to preferential removal of more friable and mobile material.

The apron-mare contact displays more structure in Plate 158c. Of particular interest is the partial overlap by the two craters in the lower right (arrows X and Y). Again, preferential filling of their floors apparently has not taken place. The upper slopes are much more littered with debris (arrows Z and XX) with no accountable source. Note the prominent parallel lineations (arrow YY) that disappear at the apron-relief contact. A relatively well defined boundary marks this contact between the apron and upper slopes (arrow ZZ). A similar boundary is found on other relief features (see Plate 128).

Plate 158d covers a region immediately to the east of that shown in Plate 158c. Two subdued craters (arrows X and Y) appear to be partly filled with apron material. An extension of apronlike material into the higher relief is identified (arrow Z). The block-strewn slopes are very prominent in this region (arrow XX).

PLATE 158

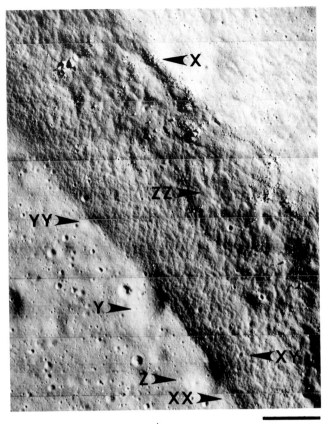

⬑ a. III-199-H3   0.23 km

⬑ b. III-199-H3   0.23 km

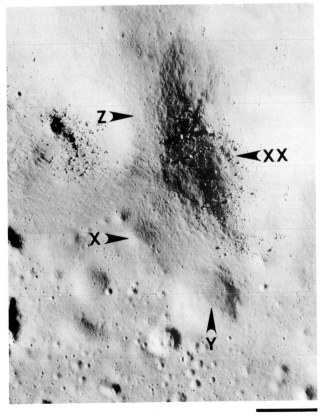

⬑ c. III-200-H2   0.24 km

⬑ d. III-200-H2   0.24 km

DISCUSSION AND SUMMARY

Isolated positive-relief features are broadly classified by their profiles and include flat-topped forms (platforms), convex forms (domes), cones, and crested features. Both constructional and relict features are represented within each class, and this genetic distinction is qualitatively supported by specific morphologies and analogies to terrestrial forms. Such positive-relief features are distributed widely across the Moon but are obvious on the maria or marelike surfaces, where they appear as islands.

Platforms are described as either polygonal or irregular in plan. As the plan approaches circularity, the classifications of platforms and low-relief domes may merge, but a distinction is maintained by the abruptness of the bordering scarp relative to the platform's over-all linear extent.

The absence of erosion and transport by water on the Moon requires that lunar platforms be primarily volcanic constructions, structurally raised blocks (horsts), or relict forms left in relief through mass wasting of surrounding terrain or drainage of encircling lava. The distinction in geometry generally helps in selecting from these possibilities. The polygonal plan commonly indicates a structural influence, but the relative importance of upthrown blocks to downthrown borders is difficult to assign. The typically small size (less than 25 km) of such platforms implies that they do not reflect major upheavals; rather, they may be local effects of such processes as mare flooding. In general, most polygonal platforms appear to be relict structures enhanced in relief by mass wasting or stoping and engulfment of surrounding topography along structural weaknesses. This statement obviously must be tempered, for the plan may be of secondary importance with respect to specific morphologies.

Irregular platforms are more common than are polygonal ones. They also are found near the borders of maria or as members of a cluster of volcanolike domes, such as the Marius Hills, and typically are situated on marelike surfaces and the higher-albedo highland plains-forming units. In addition, irregular platforms may comprise the rim/wall of large (greater than 25 km) ring structures such as Pitatus (Plate 42, a and b), Regiomontanus (Plate 47, a and b) and the Crisium ba-

sin (Plate 113). As illustrated in the chapter "Crater Floors" (Plate 46d), they also may be the central protuberances of craters.

The examples shown in Plates 123 and 125, a and b, appear to have been formed or influenced by volcanic processes. Irregular platforms less distinctly textured and extrusive appearing are illustrated in Plate 124 and can be attributed to a variety of possible volcanic or nonvolcanic origins. Nonvolcanic processes are similar to those proposed for polygonal platforms, except that irregular platforms develop without control by structural weaknesses. For irregular platforms forming a ringed plain, it is suggested (see text accompanying Plate 113) that they might be brought into relief by relatively rapid isostatic adjustment of a large crater. Their platformlike appearance is accentuated by the development of moat-like depressions and the accompanying emplacement of marelike units to the interior and exterior of the original crater rim. The platforms illustrated in Plate 123, b and c, probably represent a large-scale analog of this process because this particular region corresponds to one of the mountainous rims encircling the Nectaris basin. A major difference, however, is the volcanism that appears to have modified the region, perhaps including the emplacement of plains units that cap the remnant platforms.

Several irregular platforms appear to be thick extrusive units (often multiple) without well-defined vents. Their surfaces are typically pocked by the presence of more numerous medium-size (0.4 km–1.0 km in diameter) craters than are on the surrounding mare, and this anomalous crater density may represent an endogenetic component. Their ages are difficult to assign, but they are probably contemporaneous with the emplacement of the maria. Relative to the mare basalts, they are a minor contributor to surface extrusions. Their importance lies in the evolution of the magma chambers that resulted in the mare-flooding epochs; they could, in fact, represent late stages of those epochs.

Although numerous irregular platforms were originally formed by volcanic extrusions, their present relief is believed to have been produced, or at least enhanced, by the subsidence or removal of the surrounding mare basalts. Such a process occurs around terrestrial vents and is suggested in the lunar case by high-level marks collaring pre-existing relief. In terrestrial volcanoes, this removal results from drainage of highly fluid lava into adjacent topographic lows and, in some cases, back into the original vent. Similar episodes are suggested to have taken place on the Moon. The widespread evidence for previous lava levels, however, may require additional mechanisms, such as a reduction in lava volume during cooling, filling of pre-existing craters that act as local drains, and perhaps ancient Earth-produced tides.

Domelike features are subclassified and illustrated according to their height-to-diameter ratios. For the purpose of this study, the ratio is a qualitative estimate, and numerical values might prove useful in proper assignments to a subclass. Such a refinement might overwork the classification scheme, but it could be valuable for comparisons to terrestrial domes and for a better understanding of the degradational history of individual examples.

Low-relief domes have a flattened lens profile and typically display a rimless summit pit, which may have a circular, dumbbell, elongate, or rillelike plan. They are relatively small features—less than 5 km—and commonly are found in clusters on mare or marelike surfaces. Arthur (1962) noted from visual observations that such domes are found in the shallow regions of mare fill. In general, the contact between the dome and mare is ill defined. Plate 125c, however, shows multiple lobate extensions, and there appears to be little albedo difference between dome and mare. This similarity between dome and mare surfaces is probably more distinctive than the dome profile.

There can be little doubt that these domes are volcanic in origin, but uncertainty arises in designating the appropriate terrestrial analog. Structural doming may be indicated by the ill-defined dome-mare contact. Terrestrial laccolithic intrusions generally display the same areal extent. The summit depressions could thus be calderas or extensional fractures. Most terrestrial laccoliths are intermediate in composition, but a similar conclusion for the intrusions responsible for the lunar domes is premature.

An alternative to intrusive laccolithic doming is small-scale shieldlike construction. The general absence of well-defined termini could be rationalized by the probability of highly mobile metal-rich lavas (indicated by the Apollo samples). Consequently, most of the flows would be thin, with termini easily masked by later degradation, blanketing, or poor resolution. The extrusive origin is preferred for multiple coalescing mare domes with a summit fissure (Plate 206, a and b) and for those adjacent to highland relief (see Plate 134, a and b). It is plausible that many of these domes are relatively unobtrusive late-stage expressions of vents once responsible for a respectable fraction of the mare basalts.

Very small scale domes (less than 500 m in diameter) having surfaces similar to the surrounding mare also have been found. Several examples are surrounded by moats (see Plate 15, b and c, arrow ZZ; Plate 125e), a feature commonly recognized around domes within the large pristine craters (Plate 227, b and c). They are interpreted as remnants of the last mare-flooding epoch and place an upper limit on the degradational processes in their respective regions.

Mamelonlike domes generally have greater height-to-diameter ratios than do domes in the foregoing subclass, but six more distinctive characteristics are the following:

1. Basal aprons that are recognized by a change in slope and texture at the base of the dome
2. Absence of summit pits
3. Nonmarelike surfaces having considerable small-scale texturing, commonly narrow-spaced lineations, and relatively low crater counts
4. Higher albedo than the maria
5. Scattered blocky outcrops along the upper slopes, without visible accumulations at the base
6. Convex and rounded profiles

These forms also occur in groups, and two or more may coalesce to form more complex plans. Ring structures commonly are composed of aligned and interconnected domes of this type. They also occur on the floors of mare-filled craters and may comprise their central peaks.

Although mamelonlike domes are most obvious on the maria, they are not easily recognized on rougher

355

terrains as a result of typically poorly developed basal aprons. Since the basal apron also is to be seen around positive-relief features other than these domes, its importance as a criterion in a classification scheme may be in question. In fact, several mamelonlike domes are surrounded by a narrow moat rather than an apron. As discussed in reference to specific examples, basal aprons and moats may be largely a contact phenomenon between mare and highland relief. Until such an origin is established, however, these features aid in defining a particular morphologic class of domes.

Mamelonlike domes are interpreted as premare relict-relief forms or volcanic constructions. As nonvolcanic relict forms, they may represent remnants of hillocks, such as those found in the zone between the rings of multiringed basins (see Plate 207). These mounds could be fall-back debris from the major basin-forming event; alternatively, they represent extensive volcanic constructions. The latter suggestion is prompted by the similarity of several hillocks with summit pits to subdued tephra cones. The northern shores of Mare Serenitatis (bottom of Plate 66, a and b) exhibit numerous examples of such features. Consequently, mamelonlike domes also can be interpreted as relict volcanic forms.

Further credibility of a volcanic origin, whether pre- or postmare in formation, results from comparison to similar-appearing domes within calderalike depressions (see Plates 24–25). These are interpreted as volcanic domes analogous to viscous extrusive features found on Earth. In addition, several domes and cones with summit pits on the floor of Copernicus (Plates 226–227) are believed to be volcanic forms and could with time be degraded into domal profiles.

As noted, mamelonlike domes are not easily recognized in rougher terrains; however, Plate 38, a and b (arrow ZZ), and Plate 250 (arrows ZX, ZY) show very similar features occurring on the fractured floors of large craters. If the apron and moat surrounding such features indicate engulfment by flow units, then their existence suggests that such units have subsequently receded.

It is highly probable that an attempt to assign a common origin to all mamelonlike domes will be fruitless. Moreover, the apron and moat may indicate a variety of processes; for example, they could represent mass wasting of an overlying mantle of ashlike material overlying a more competent surface (see Plate 166c). Detailed studies are needed in order to reveal which formative or modifying process is applicable to specific examples.

The third dome subclass generally displays higher relief than do mamelonlike domes and exhibits a coarse furrow and ridge texture that extends radially from the summit. The Gruithuisen domes are used as illustrations of relatively symmetrical examples, but there are also more extended irregular forms that approach a flat-topped profile with similar texturing. At least one member of this subclass exhibits a much higher albedo than does its surrounding terrain (Plate 190d).

It is suggested that these domes are extrusive features, and the Gruithuisen domes clearly superpose the more subdued hills of the highlands but apparently were embayed by the last local extrusions of mare basalts. The ridge and furrow topography can be interpreted as results of multiple lava flows or rapid mass wasting during dome construction. In the latter hypothesis, the domes are visualized as expanding from the central buildup of viscous extrusions.

Lunar cones are distinct forms, owing to their pyramidlike profiles and summit pits. As the summit pit increases in size, the classifications of lunar cones and tephra rings merge; an example of the latter is shown in Plate 130f (arrow Z). Cones found on the maria or marelike surfaces may have lower albedos than do their surroundings, whereas examples found on rough volcanic floors of large craters (such as Copernicus: Plates 226–227) have relatively high albedos in comparison to that of their regional setting. Cones found on the maria are shown to occur both in groups and as isolated forms.

The size and morphologic appearance of lunar cones are analogous to those of terrestrial cinder cones. Several lunar examples also exhibit breached flanks (Plate 153b) and elongate plans (Plate 130f), and they indicate multiphase or fissure eruptions. None of the cones examined showed clearly associated flows, although tephra-ringlike forms within Vitello (Plate 251a) and Gassendi (Plate 86b) have related flowlike features.

The destruction of the summit pits should have been relatively rapid, owing to their placement and pyroclastic development. This implies that the present recognizable number density of cones may be only a small representative of this type of activity. In particular, many of the domelike forms possibly are remnant cones.

The occurrence of volcanolike cones on the Moon requires low-velocity and high-trajectory ejecta; otherwise, it is unlikely that the large rims could have been constructed (see Discussion and Summary at the end of Chapter 4). If these cones represent spatter cones, which are less violent eruptions than ash-built cones, then many of the lunar cones are much larger than the terrestrial examples. In addition, the lunar analogs are clearly not the last escape routes for cooling mare units below, for such cones have typically been surrounded by later flows of mare units. I feel that several irregular cones and ridges are more likely examples of lunar spatter cones (see Plates 131–132d), whereas the larger and more symmetric cones (Plates 130, e and f, and 190a) represent sites of more significant pyroclastic eruptions. A large fraction of the erupted material, originally expressed as a much broader rim region, has been buried by the last fluid flows of the mare basalts.

Thus, isolated positive-relief forms exhibit a variety that could not have been appreciated until Lunar Orbiter photographic quality became available. Many of the examples on the maria appear to represent a late or final stage of extrusive activity, some of which appears to postdate considerably the last local mare-flooding epoch (see Plate 60a, arrow Z). Their differences in surface expression reflect differences either in viscosities or in total extruded volume. Other examples on the maria appear to be relict or remnant features, but to say this is largely to describe rather than to propose a statement of origin. The distinction between relict volcanic features and relict nonvolcanic topographic highs is not always obvious. The coalescence or arrangement of several isolated forms can yield a variety of extended and composite forms, a topic that is the basis of the next discussion.

Ridges and composite positive-relief features are con-tinuous, in contrast to the isolated forms considered in the foregoing subclass. In many instances, the distinction between isolated and continuous features is admittedly artificial, for the component members are merely aligned isolated forms. Strictly as a broad morphologic subclass based on geometry, however, the classification remains distinct and provides a noninterpretive categorization, as well as continuity, for descriptions of typically more complex histories.

The first example of this subclass is irregular ridges that appear to have been constructed volcanically. Their trends commonly reflect other alignments defined by rilles, crater chains, or other ridges. Although many have low albedos or are surrounded by diffuse low-albedo deposits, there are examples that clearly do not have such an association. These ridges typically break into discontinuous alignments of irregular platforms, cones, or domes with summit pits. Although not a steadfast rule, such ridges commonly occur in the supposed "shallow" areas of mare flooding, i.e., where relict-relief structures also appear abundant.

In general, irregular ridges are analogous to certain isolated forms but have developed along fissures, which reflect large-scale trends. Hence, they appear to be related to major stresses. Their location in the apparently thin sequence of mare units adds further credibility to the suggestion that such areas are not simply the result of overflow from the main mare basin. Rather, nearby sources were responsible for the inundation of these regions. Several ridges may represent vestiges of the last gasps along such vents.

As in the case of isolated forms, the differences in surface expression appear to be the result of differences in lava viscosities and eruptive mechanics. The ill-defined dark haloes indicate that most of the eruptions produced pyroclastic material, although the ridge itself may not be composed of tephra. Subject to lower gravity, lunar spatter cones should develop to greater heights than the terrestrially observed 15 m maximum. Since these features are constructions of molten lava rather than tephra, they should be more durable to the weathering processes that occur on the Moon; therefore, their approximate form should be preserved longer than that of a tephra cone. This is the suggested analog for the examples shown in Plate 132, *a, b, c,* and *d*.

Septa include any straight ridge that is slightly crested. Not all septa are associated with a basin-forming event, but these are the most impressive representatives. In the text, two general interpretations of septa are suggested: They are surficial phenomena, i.e., the result of deposition or erosion; or they are internal phenomena, i.e., structurally or volcanically formed. All these processes can produce septa having similar morphologic appearances. Erosion and deposition of massive ejecta flows (see "Crater Rims") were responsible for many of the smaller corrugations. The channeling of this ashlike flow through radial grabens could have produced levee-built rims that remained as septa after partial mare flooding. Other septa, however, appear to be remnants of rims of elongate craters, some of which are interpreted as volcano-tectonic depressions.

Mountainlike masses are termed "sharp-crested massifs" and are illustrated as composite structures having multiple peaks. The most impressive examples (2-km elevation differences) are remnants of the multiple rings around enormous basins. The detailed events during the catastrophic moments of basin formation are impossible to reconstruct confidently from photography. The construction of the broken and isolated portions of the rings could have occurred during the initial basin formation or developed through later adjustments. The resulting massifs do not typically resemble tilted blocks or upturned rims but seem more appropriately interpreted as horsts.

Most crested massifs are not volcanolike but are shown to be associated with volcanolike features, such as low-relief mare domes. If these massifs represent upheaved blocks, then their borders reflect faults that may have acted as feeders. Mare flooding appears to have been initiated along faults expressed by the interior-facing scarp of the mountainous rings (see Appendix A: Mare Orientale, Plates 204–207), and the mare domes at the base of the remnants of the rings are suggested to be indicators of this activity.

Wrinkle ridges can be examined individually for peculiar characteristics, in association with other features, and in association with other ridges. They are classified broadly into two types: marelike and nonmarelike. A system in Oceanus Procellarum illustrates the marelike ridges. They exhibit broad elevated marelike surfaces bounded by the tree-bark pattern. There is typically only one well-defined scarp, but, along the length of the same ridge, this face may change sides. A few ridges in the Littrow system illustrate the nonmarelike ridges, which commonly have a more convex profile. Their "sculptured" or knobby surfaces have blocky exposures and lower crater count. In addition, such ridges may be bordered by a low-relief moat.

Narrow ropelike ridges are associated with both Oceanus Procellarum- and Littrow-type wrinkle ridges and enhance the twisted or "wrinkle" appearance of the ridge complex. Ropy ridges have a braided appearance produced by elongate component hills adjacent to one another but offset. Their plans may reflect, cross, or disregard the plan of the broader component. They commonly occur on the crests of the broad marelike ridges. The ropy ridges in the Littrow system appear to be connected to the larger nonmarelike ridges and could be contiguous features at smaller scales.

To seek a universal origin of wrinkle ridges is not a reasonable goal, for the variety of forms and associations inevitably leads to contradiction. Clues to origin are provided by individual morphologic features or interaction with other ridges. The limited high-resolution coverage by Orbiter photography precludes a detailed classification that might relate these two indicators.

The structural influence of many of the ridges is illustrated repeatedly by *en échelon* arrangements and their reflection of regional trends. The marelike surface of the Oceanus Procellarum wrinkle ridge is consistent with the interpretation that they are elevated folds or horsts that preserve the original mare surface. On the other hand, the close spatial association of several ridges to volcanolike features suggests an eruptive history.

A wide range in the ages of wrinkle ridges is inferred. Ridges overlapping Copernicus ejecta (Plate 96, *a* and *b*, arrow Z) demonstrate that relatively recent periods of ridge formation have occurred. Furthermore, an example is shown in the following chapter (Plate 177*b*) in which a rille cuts a broad wrinkle ridge. The dissection of such relief by surface channels requires long-term erosion, structural control, or peculiar circumstances. Some rilles can be explained as extensonal fractures, rather that surface channels, but the rille shown in Plate 177b is connected to irregular depressions analogous to subsided lava lakes. As a result, slow buckling of the mare surface after rille formation is a more reasonable explanation.

Relative uplift of the mare surface can be expressions of either vertical stresses, resulting in elongate doming, or horizontal stresses, producing compressional forms. The arrangement of some marelike ridges favors the latter alternative. Specifically, several plates shown in this chapter illustrate the exchange of the more abrupt scarps to opposite sides of the ridge. The scarps could represent exposed faults where stresses exceeded the elastic limit of the medium. Which of the meeting masses will thrust over the other probably is determined by inhomogeneities in local stratigraphy, such as the depth of inundation by the mare units. In at least one example (Plate 148f), a pre-existing crater was severed during ridge formation, and remnants of the downthrown crater rim are absent. This hanging crater suggests either that subsequent mare basalts engulfed a portion of the crater or that the ridge represents a very low angle thrust in which one mare surface slid over the adjacent surface. The broad convex surface on the upthrown side of numerous wrinkle ridges having a single abrupt scarp suggests associated surface buckling.

Baldwin (1963) proposed that wrinkle ridges are compressional folds that developed as the domed circular maria subsided during withdrawal of underlying magma or from the weight of the overlying lava. The more recent discovery of positive gravity anomalies (mass concentrations or "mascons") beneath the maria is consistent with the latter idea that these areas are overburdened and should subside. Although the mare-filled basins are topographically low, they are not concave in profile but convex, because of the curvature of the Moon. Thus, previous doming of the maria is unnecessary; merely subsidence over a large area is sufficient.

Baldwin also postulated that tension should develop along the mare shores, whereas compression should dominate the interior. Consequently, peripheral rilles on the mare surface and interior wrinkle ridges are expressions of these stresses. There are, however, peripheral wrinkle ridges, such as the Littrow system, which is found on the eastern edge of Mare Serenitatis near the base of a scarp marking the mare boundary. Similar associations are visible in Maria Crisium and Nectaris. Thus, a slight amendment to Baldwin's theory is proposed in which the margins of some of the maria are subjected to considerable horizontal compressional stresses near the surface. In eastern Mare Serenitatis, folds result where the more deeply rooted mare shores meet the mare units. Narrow rilles and small-scale fractures occur peripheral to these ridges and are interpreted as tension features associated with broader but still-local bowing of the surface in response to the adjacent compression. The peripheral tension-produced rilles as envisioned by Baldwin are believed to be represented beyond the mare shores as grabenlike rilles crossing the shelf units and highlands. Further complexity in this idealized picture can occur if the subsidence results in separate crustal plates.

Within Mare Imbrium—and at numerous other locations—wrinkle ridges are related spatially to sharp-crested massifs. Baldwin suggests that ridges form in regions of shallow mare fill where buried structures have altered a uniform response to the stress field. Thus, the relationship with massifs, which can be viewed as exposed portions of otherwise buried mountainous rings, might seem reasonable. The ridges extend, however, as basal collars around parts of the massifs. If these peaks are stable with respect to the settling mare crust, then compressional folds should develop on the basin side of the peaks. In general, this is not the case. In many instances, massifs simply interrupt the ridges without evidence for extensive buckling. Furthermore, the trends of the ridges do not always parallel the axes of the massifs. Very likely, settling of the maria did not occur as a unit but as broken plates. Such an event might explain the rectilinear ridge systems as well as the more complicated associations with massifs.

Ridge systems are not restricted to mare surfaces, as illustrated in Plates 109, c and d, 145c, 146, and 249 (arrow XXY), and many appear to have been unrelated to the subsidence of the maria. Highland extensions of ridges commonly are expressed as relatively subtle vertical offsets, narrow rilles, or ropy ridges. Although wrinkle ridges appear to be more common on the maria, this impression is largely the result of easier identification on these flat plains. Where such ridges cross both mare and highland terrains, their expression on the maria generally is more complex. This is suggested to be the result of a greater likelihood for folding of the mare units relative to the highlands. Whereas much of the highlands contains a jumbled stratigraphy owing to eons of impacts, the mare units were probably emplaced by a series of flows with interbedded layers of incompetent regolith. Furthermore, the base of the mare units generally marks an abrupt transition in rock strength, which is a likely zone for shearing if subjected to compressive stresses. Thus, the mare is more likely to develop a complex region of folds and low-angle thrust faults under laterally directed stresses.

The other mode of elevation is by vertically directed stresses. This may develop by intrusions following fracture systems (see Plate 147, c and d). Thus, structural control is preserved, but the stresses responsible for fracturing and the resulting arrangement are varied. Examples shown in Plate 119, a and b (arrow XY) and Plate 173 are radial to the Imbrium basin and are interpreted as surface expressions of dikes. The ridges seen in Plate 173 appear to be associated with low-albedo strata, perhaps indicating related extrusions. The radial ridges from the ringed wrinkle-ridge system Lamont in Mare Tranquilitatis also suggest a tension-produced dike system associated with regional uplift. This system, however, crosses mare units of different colors, believed to be related to different stages of mare emplacement; therefore, the origin of the colored units and that of the ridges probably represent separate events. Another radial-ridge system appears to be associated with the crater Taruntius in Mare Fecunditatis.

Thus we are led to the possibility that some wrinkle ridges might be formed volcanically. Fielder and Fielder (1968) proposed that wrinkle ridges represent remnants of fissure eruptions responsible for much of mare flooding. A plausible example is in Mare Imbrium, where well-defined flow units (Plates 192–193) extend for great distances and may have been related to a large annular system of wrinkle ridges. Limited mare units in the Orientale basin indicate that the vents in multiringed basins are associated with faults corresponding to the enormous multiringed scarps. Thus, the annular wrinkle ridges in Mare Imbrium may be expressions of these vents. The relation of ridges to sharp-crested massifs can be explained, therefore, as combined intrusive doming and limited extrusive flows along reactivated faults connected to the massifs. This theory cannot be the entire story, however, for several wrinkle ridges in Mare Imbrium cross separate flow units and do not appear to have been the conduits responsible for these particular flows.

Evidence for the extrusive origin of wrinkle ridges is shown in Plate 181c. Offset elongate depressions clearly are related genetically to a ridge, and there appears to be a related flow unit (arrow ZZ). This particular example may represent a fissure or a partly collapsed lava tube. In addition, Plate 182, a and b, shows several rilles that are believed to originate along the crests of wrinkle ridges, and several ridges are accompanied by endogenetically produced craters (Plates 141, 148d, and 150a) that are interpreted as volcanic vents.

Narrow ropy ridges that are superposed on many of the broad marelike ridges can be interpreted as limited extrusions through axial tension fractures developed by doming of the mare surface. If such a process occurred while the mare was cooling, then the preservation of these narrow ridges indicates that they have not been as vulnerable to smoothing processes as have craters comparable in size. This constraint on smoothing processes is not damaging to the hypothesis of the extrusive origin of these ridges, because of the evidence for other small-scale features preserved since the last stage of mare flooding (Plates 16, c and d, 78d, 87, 125e, 128, 132d, 151, 152, a, b, and c, and 156, c and d).

The preservation of features related to mare flooding prompts another plausible origin for some wrinkle ridges. Plate 145a illustrates the collaring by a ridge around a premare dome, and the morphology of this association suggests that the ridge corresponds to ramparts of flows or crusted lava surfaces that subsided prior to complete cooling of the mare basalts (also see Plate 203, arrow XZ). Three mechanisms are proposed. First, crusted lava was driven by movement of still-molten lava beneath the surface. Second, cooling of the mare basalts resulted in considerable subsidence but left behind high-level marks. Third, Earth-produced tides were much greater four billion years ago and resulted in periodic flooding and ebbing during the epochs of mare emplacement. This, in turn, suggests an additional mechanism for ridge formation in which surface displacement is the result of subsidence over pre-existing structures. For example, buried craters might be exposed as ringed ridge systems, such as those seen in Plate 150, a and b. Ridges formed by this process will probably not display the narrow ropy component; rather, they will have broad marelike surfaces with relatively subdued boundary scarps (Plate 143a).

I believe that each of these various mechanisms—thrust faulting, folding, igneous intrusions, extrusions, uncollapsed lava tubes, and previous lava levels—was responsible for the formation of certain lunar wrinkle ridges. The question is not "which process was responsible for all wrinkle ridges," but "which process was responsible for which wrinkle ridge."

The last series of illustrations shows ring structures that range in size from 1 km to greater than 100 km. "Ring structures" is another catch-all label that includes several examples discussed in previous chapters on craters. As noted at the outset, discussion at this juncture is necessary in order to preserve nongenetic implications of structures that have debatable interpretations.

Again, the smallest ring forms can be interpreted as either relict or constructional features, but either alternative leaves difficulties. As relict structures, their preservation seems incompatible with the highly cratered mare surface. As constructional forms, they are very strange protrusions. The first suggestion is more reasonable and implies that many of the subdued craters on the maria are formed endogenetically.

At larger scales, broken rings commonly occur on the maria. Such features as Maskelyne F are most probably relict forms. But their being relict relief does not require

that they are exogenetic in origin, and many examples are interpreted as remnants of volcanic forms, such as tephra rings.

At the largest scales, rings occur as wrinkle-ridge associations or as discontinuous arrangements of mamelonlike domes. The Flamsteed Ring is examined in detail, but reference also should be made to other complexes examined in "Crater Rims" (Plates 110–111). The mamelonlike domes comprising portions of the Flamsteed Ring appear analogous to mamelonlike domes examined under the subsection "Isolated Positive-Relief Forms" and are comparable to domes on crater floors. The existence of similar-appearing domes in other volcanic settings supports the plausibility of internal origin for segments of these rings. In general, however, the rings of mamelonlike domes represent remnants of crater rims, which were heavily modified during the stages of mare emplacement. Rim modification includes the structural development of multiringed plans and extrusions along rim faults.

In summary, many positive-relief features provide evidence that the Moon has yielded volcanic eruptions of considerable variety. Although the maria comprise the most obvious evidence for extrusives, there are also volcanic cones, domes, fissures, and shields that identify some of the sources for the mare units and supply information on the geochemical evolution of the Moon. The process of mare flooding cannot be viewed as solely overspill from lava-filled basins. This is illustrated by unbreached mare-filled craters and by numerous volcanic forms in the "shallow" zones. Guest (1971) provides further review of volcanically formed positive-relief features and their relation to mare flooding. In addition, the lunar highlands exhibit evidence for volcanically constructed forms. This has been pointed out by Milton (1968), Wilhelms and McCauley (1969), and Wilhelms (1970, 1971).

7. Negative-Relief Features

INTRODUCTION

The major negative-relief features on the lunar surface include craters, rilles, and irregular depressions. Crater characteristics are treated in the first five chapters of this book. In the present chapter, the morphologic appearances and genesis of rilles are presented. Since rilles include any linear depression, the term is ambiguous. Curvilinear and rectilinear rilles are two main classes that can be distinguished generally by their plans. The former class is subdivided further into nonsinuous and sinuous categories. Nonsinuous rilles may be either arcuate or relatively straight and presumably are formed in response to stresses within the lunar crust. It has been proposed that sinuous rilles are structural features, collapsed lava tubes, or surface channels.

In addition to the rille plan, the profile provides a clue about rille origin. For many rilles, a flat-floor profile suggests a grabenlike structure in which the floor represents a dropped surface. Such a profile is not fully diagnostic of this type of event, because a similar appearance can evolve from extensive blanketing and subsequent lava flooding. The **V**-shaped profile results from the meeting of facing scree slopes, and continued degradation of this or the flat-floor profile may produce a **U**-shaped profile. In some examples, the **U**-shaped profile is thought to represent relatively slow subsidence that did not yield exposed fault scarps. As on Earth, such tectonic events are associated with, or result from, igneous activity, which is revealed by related topography, regional location, or albedo contrasts.

Examples of curvilinear nonsinuous rilles are shown in Plates 159–165*b*, and the accompanying discussions describe the rilles and associated features that might provide information about their origin. Nonsinuous rilles typically occur with other nonsinuous rilles, resulting in a variety of arrangements, including branching, *en échelon*, parallel, and intersecting rilles. Groups of these arrangements result in large-scale associations termed rille systems. Plate 165–170 illustrate both orders of complexity.

Sinuous rilles represent the other class of curvilinear rilles. They are subdivided into more sinuous (meandering) and less sinuous (nonmeandering) subclasses. A more sinuous rille exhibits repeated curvilinear changes

in its path over relatively short distances, whereas a less sinuous rille exhibits fewer and less abrupt directional changes. Typically, the most sinuous rilles are nested within large, less sinuous rilles (see Schröter's Valley, Plates 171–172) or closed depressions (Plate 177*a*), but portions of larger rilles also may exhibit very tight meanders. Examples of more sinuous curvilinear rilles are shown in Plates 171–173 and of less sinuous rilles in Plates 174–176. These examples also illustrate the variety of profiles, their enormous linear extents, the effect of pre-existing topography, locations, and selected details that might reveal origin or later modifications. Most sinuous rilles have a crater or elongate depression at one end, which is typically at the highest elevation. Surface flows of lava (channels), subsurface flows of lava (tubes), and structural events are believed to be formative processes.

Rectilinear rilles (Plates 178–180) show abrupt angular changes of path and are either fractures or structurally controlled surface channels. Numerous rectilinear rilles also have craters or elongate depressions at one or both ends that commonly have low-albedo ejecta, in contrast to those craters associated with sinuous rilles.

The preceding discussion outlines the sequence of plates for this chapter and contains a scheme that provides a guide, or "check list," of characteristic features according to their general descriptions and interrelationships. The description of rille features, such as the profile and morphologic appearance, can be considered a "zero-th" order of inspection, whereas the description of the rille plan and its continuity can be viewed as first order, i.e., characteristic form at a next-higher order of size scale. Association of two rilles forms a rille arrangement, which comprises a second-order description, and, finally, the association of two or more rille arrangements is termed a rille system, which comprises a third-order description. Thus, the purpose of the following outline is not to describe the order of subsequent plates in this chapter but to supply a systematic method for examination of rilles and to reveal their variety.

I. Local rille features
  A. Profile
    1. Flat floor
    2. **V** shape
    3. **U** shape
    4. Raised rim
    5. Asymmetric
  B. General description
    1. Freshness
      a. Sharp breaks in slope
      b. Subdued slopes
    2. Albedo contrasts
    3. Floor compared with surrounding region
    4. Regional location
      a. On dome or depression
      b. On maria or highlands
      c. Restricted to crater floor
      d. On crater rim
    5. Associated topography
II. Rille plan and continuity
  A. Plan
    1. Curvilinear
      a. Nonsinuous
        (1) Straight
        (2) Arcuate
      b. Sinuous
        (1) Meandering
        (2) Nonmeandering
    2. Rectilinear
  B. Continuity
    1. Continuous
    2. Discontinuous—rille interrupted by:
      a. Negative relief
        (1) Irregular depression
        (2) "Pinched" crater
        (3) Symmetric crater
      b. No relief
      c. Positive relief
      d. Filled with subsequent units
III. Rille arrangements
  A. Branching rilles
    1. Single offshoot
      a. Right angle
      b. Acute angle

2. Multiple offshoots
  B. *En échelon* arrangement
  C. Parallel rilles
  D. Intersecting rilles
    1. Right angle
    2. Acute angle
IV. Rille systems
  A. Parallel systems
  B. Concentric systems
  C. Radial systems
  D. Intersecting systems
  E. Complex systems

Nonsinuous Rilles: Flat-Floored Profiles

The Hyginus rille is shown in an oblique view (Plate 159a) and in stereo (Plate 159, b and c). It illustrates a curvilinear nonsinuous rille and is near the center of the lunar Earthside. Two major trends intersect at the crater Hyginus: the NW trend aligns with the Imbrium lineations, and the WNW trend aligns with the Ariadaeus rille system to the east. The latter system extends west of Hyginus in an en échelon arrangement.

Certain features of the rille and Hyginus, shown on the three succeeding plates (Plates 160–162), are examined in detail. The rille is flat floored, or grabenlike, and contains several crater groups that form contiguous chains. The coincidence of such craters with the rille favors a genetic relation to the system rather than fortuitous impacts. Fielder (1965) suggested from telescopic observations that the rille system, including the Ariadaeus rilles, is the result of first-order and second-order releases of compressional stress with strike-slip motion. If this is correct, craters along the rille were formed after the displacement, but there is relatively little topographic evidence for such a large strike-slip component.

Hyginus and its rille illustrate a typical arrangement found on the Moon in which a dark-haloed depression is at the intersection of two major structural features. In addition, Hyginus occurs at the bottom of a broad saucerlike depression about 100 km in diameter and 1.5 km in depth (from LAC 59). Low-relief ridges surrounding this region are interpreted as compressional forms due to either subsidence or viscous extrusions that followed weaknesses created by dilation. Eruptive removal of the magma from a chamber and removal of subsurface pressure from volatiles perhaps triggered this subsidence.

Hyginus is centered in a diffusely bordered low-albedo region, which is recognized more easily under full solar illumination through Earth-based telescopes. It suggests pyroclastic eruptions that may account, in part, for the absence of well-defined flow units. Regional blanketing also is inferred from the large number of shallow-floored craters, many of which have dimpled floors (Plate 159, b and c, arrow X). Recall that dimple-floored craters commonly accompany regions of suspected volcanism and are interpreted as maarlike eruptions or blanketed craters (see Plates 20 and 123a).

Several craters surrounding Hyginus rest on low-relief mounds (arrows Y, Z). Arrows XX and YY identify two additional examples that have sharp wall-rim transitions. Such craters are inconsistent with an interpretation that they reflect long-term degradation of impact craters and are believed to represent peripheral eruptive vents.

Very low relief ridges and rilles radiate from Hyginus. The most prominent example is located by arrow ZZ, and a crater along this lineation is elongate in the same direction. Such radial systems probably correspond to tension fractures developed during an earlier stage of dilation centered in the crater Hyginus.

Volcanic activity along the grabenlike rille is also indicated. As mentioned above, the crater chains superposed on the rille are interpreted as endogenetic features. Note that the chain along the NW-rille extension follows the rille plan in detail. Such craters illustrate the difficulty in distinguishing between endogenetic and exogenetic craters, for, if the crater members were separated and removed from their association with the rille, the interpretation that they were formed by internal processes could be seriously questioned.

Arrow XY identifies the northern border of a low-relief circular platform (5 km in diameter) that is bisected by the eastern branch of the rille. The interior of the rille at this location exhibits a positive-relief form (arrow XZ) and an adjacent crater that is placed approximately at the center of the circular platform. The platform is interpreted as the result of intrusive and extrusive events prior to the formation of this segment of the rille. The interior crater represents postrille collapse during a period of renewed volcanic activity.

Note that a narrow ridge (arrow YX) borders the rille at several locations. This ridge is thought to represent lava or ash erupted along faults defined by the graben prior to or during its collapse. "Ghost" rings, moat-surrounded relief, subtle irregular platforms, low-relief domes, and irregular ridges indicate a complex geologic history.

Another example of endogenetic activity accompanying rille intersections is seen in Plate 131, which shows the Fra Mauro region. There, grabenlike rilles with raised rims and a local low-albedo ridge are easily identified.

PLATE 159

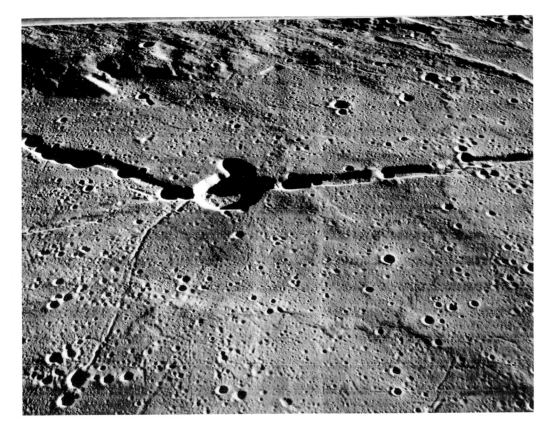

a. III-073-M   (oblique)

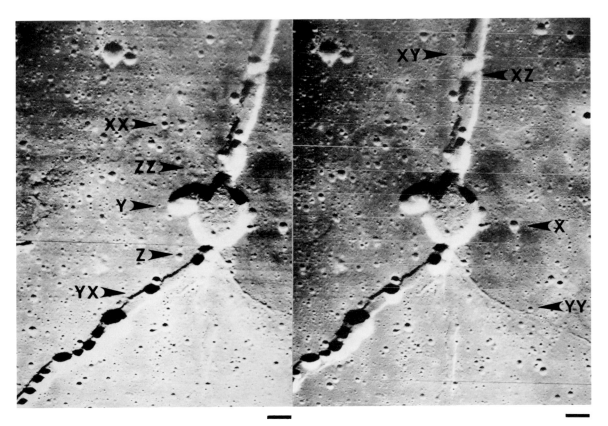

b. V-097-M   2.9 km (top)   4.0 km (bottom)   c. V-094-M   2.9 km (top)   4.0 km (bottom)

Nonsinuous Rilles: Flat-Floored Profiles

Plates 160–161 show detailed features in the crater (caldera?) Hyginus that can be compared with a segment of the graben-like rille (Plate 162).

The floor of Hyginus is similar to the surface of the surrounding country and exhibits a rolling but smoothly textured surface with numerous subdued craters. The domes evident in Plate 159, b and c, are visible in Plate 160 (for example, arrow X) and are enhanced by a tree-bark pattern, which typically accompanies change in relief. Other characteristic features on the floor are the highly reflective zones (shown in detail in Plate 161b) that commonly occur along the upper wall of subdued craters (arrow Y) or within irregular closed depressions (arrow Z).

The walls display numerous blocks (arrow XX) on the upper slopes, with an accumulation of debris upslope from several outcrops. In contrast, the lower part of the wall (arrow YY) appears to be composed of finer material having a gentler slope and is scarred by numerous boulder trails. At least three depressions (arrows ZZ and XY) along the rim of Hyginus have been severed by either the retreating wall of the crater or its initial collapse. One of these (arrow XY) appears to be related to a crater chain on the floor. The wall of a circular depression approximately 2.5 km in diameter is identified by arrow XZ on the floor of the northern extension of Hyginus. This nested depression is considered to be a site of further collapse.

PLATE 160

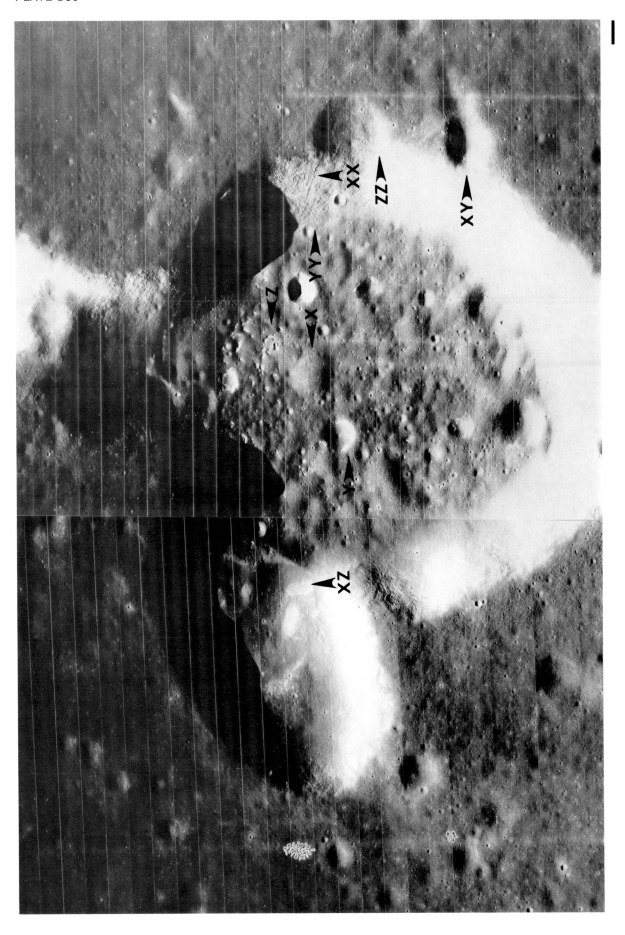

V-096-H1; V-095-H1   0.43 km

Nonsinuous Rilles: Flat-Floored Profiles

Plate 161a is an enlargement of the southern wall shown in Plate 160. At least fifteen boulder trails can be counted in this section, with varying degrees of freshness (see arrows X and Y), and such scars attest to the unconsolidated nature of the local regolith. There are also blocks on the floor without a trace of a scar; if scars existed, they may have been destroyed by erosion or covered during later floor eruptions. If isolated boulders had been hurled to their resting place, then they should be in association with craters, but very few, if any, are observed. The abundance of the trailless boulders near the base of the wall supports the idea that they were derived from the overhanging outcrops. The degree of preservation of boulder trails could provide insight as to the relative time of formation of other features on the floor of Hyginus. Alternatively, the presence of blocks and an absence of trails indicate a form of rock creep.

The large number of boulder trails in Hyginus, compared to those in other lunar craters, perhaps is due to the number and type of "outcrops." The ill-defined layering displayed on the Hyginus walls is in contrast, for instance, to the well-defined strata found in Dawes (Plate 63a), which exhibits few boulder trails. Outcrops found in Hyginus are exposed in only limited regions and are strewn down the wall. The relatively smooth surface outside the crater commonly can be followed onto the crater wall, and the result is a rounded rim. The formation of Hyginus probably was accompanied by eruptions of ash that obliterated or covered portions of the original blocky wall.

Some blocks may have been dislodged by major movement along the wall. This is suggested by the subdued lineation (arrow Z), which is interpreted as a fault. Subtle extensions of the fault are recognized on the floor and rim of Hyginus. In addition, numerous trails occur on the wall of the first elongate depression of the eastern rille extension from Hyginus. As visible in Plate 159, b and c, the same feature on its southern wall has a small scarp, which is probably a slump feature.

The crater on the wall visible in Plate 161a (arrow XX) exhibits a well-preserved rim but does not have a blocky ejecta blanket. The absence of blocky outcrops in this region, however, may reflect blanketing associated with crater formation. The rim nearest the floor (arrow YY) is broader than the rim upslope, and this difference probably is the result of migration

of debris down the wall of Hyginus. Late-stage pyroclastic eruptions have occurred at the floor edge within the Hawaiian calderas (Jaggar, 1947), and the small crater in Hyginus may be an analogous construction.

The high-albedo zones within closed depressions noted already on the floor of Hyginus are included in Plate 161b. Several zones are crescentic in plan and follow the upper walls of otherwise subdued craters (arrows X and Y). Others are found at the bottom dimplelike depressions (arrow Z). These regions might represent:

1. Zones of avoidance of an extrusive flow, i.e., kipukas: Therefore, the highly reflective areas represent pre-existing surface material.
2. Collapse features: The higher-albedo zones are perhaps due to either exposures of underlying material or escaped volatiles that resulted in sublimate deposits or alteration of surface structure.

These zones commonly occur in topographic lows, but kipukas are expected to be around topographic highs. Consequently, the first suggestion is not favored. The second proposal includes collapse features whose formation may have been either relatively recent or at the time of floor formation. The preservation of these zones indicates that, if they were formed with the floor, then surface processes have been extremely slow and cannot be responsible for the subdued state of associated and nearby depressions. Possible terrestrial analogs of these zones are found in the Diamond Craters area, Harney County, Oregon (see Green and Short, 1971, Plate 122A). If their formation was relatively recent, then their association with subdued craterforms suggests relatively current endogenetic production of subdued depressions. This has considerable significance for subdued craters on the maria. Recent formation of the high-albedo zones may have been triggered by release of trapped volatiles in subsurface pockets or subsidence into structures during the last stage of floor emplacement. Similar zones occur in the maria near Arago (see Plate 194), near Secchi X (Plate 132b), and in the Harbinger Mountains region (Plate 261a, arrow ZZ).

PLATE 161

b. V-095-H1   0.43 km

a. V-095-H1   0.43 km

Nonsinuous Rilles: Flat-Floored Profiles

Plate 162 is a detailed view of the rille floor west of Hyginus and adjacent to the first member of a contiguous series of low-rimmed craters (bottom left, Plate 159, b and c). The surface features of the rille floor are similar to features on the surface outside the rille. In contrast to the boulder-strewn wall of the crater (arrow X), the rille wall is textured yet generally smooth, with only a few scattered boulders. This difference probably reflects differences in rapidity of formation, slopes of their initial scarps, history of later depositions, states of degradation, or their respective compositions.

The following additional observations can be made:

1. Moatlike depressions parallel the floor-wall contact of the rille (arrow Y).
2. An apron occurs along the base of the west rille-wall contact (arrow Z).
3. The floor of the large crater is at a significantly lower level than the flat-floored rille and connects with the adjacent crater to the northwest without an abrupt change in slope.
4. The large crater has a slightly raised but smooth rim.
5. Boulder trails are found on the south walls of the large crater (shown in the original photograph).
6. Blocks on the crater wall typically have material accumulated on the upslope side (shown in the original prints).

The moatlike depression and apron are believed to indicate multiple periods of rille-floor subsidence. Such periods are clearly indicated along other portions of the rille not included in Plates 156–161. The crater chains were formed after rille formation, and their low-relief rims suggest limited erupted material. Because adjacent craters generally are not extensively filled (see Plate 159, b and c), it is likely that these craters were formed simultaneously. A possible exception is shown in Plate 162, in which adjacent craters have flat floors and are interconnected; these are thought to represent two of the earlier eruptions. Plate 159,b and c, also shows that despite the inferred simultaneity, intervening septa as illustrated by multiple impact craters (Plate 116b) did not form. As discussed previously (see Discussion and Summary at the end of Chapter 4), lunar pyroclastic eruptions have been predicted to disperse their ejecta over a wide area. This could account for the low-relief rims. More important, these craters probably resulted, for the most part, by collapse and did not have a long eruptive history —if any—after this event. The absence of low-albedo deposits surrounding the rille craters perhaps supports this hypothesis, but it is more likely that later secondary craterings, which has been extensive in this region, has lightened the original albedo. The low-albedo surrounding Hyginus suggests much later—and probably the last—eruptions from smaller vents on the floor and around the rim of Hyginus.

PLATE 162

V-097-H2   0.46 km

Nonsinuous Rilles: Flat-Floored Profiles

The Sirsalis rille, shown in Plate 163, extends more than 300 km on the outer fringe of the Orientale ejecta and provides a highland example of a curvilinear nonsinuous rille with a flat-floored profile. Although offset near the center of the photograph (arrow X), the rille is remarkably continuous. It is grabenlike, and elements of local topography from the Orientale event can be identified on its floor. Widening of the rille occurs where it intersects topographic highs (arrow Y). This indicates that its boundary faults dip inward, provided that the extrapolated intersection of these faults occurs at a depth that is relatively independent of surface topography. McGill (1971) has compared such changes in width with changes in elevation for ten nonsinuous rilles and found that their faults typically dip 60° inward. He interprets this value as evidence for vertical gravity-induced maximum principal stress.

The northernmost extension of the Sirsalis rille ends abruptly at a small crater on the border of Oceanus Procellarum. This abrupt termination suggests that it disappears beneath the last unit of this mare and, therefore, that the Orientale event also must predate mare emplacement.

To the south, the eastern scarp of the rille becomes significantly higher than the western scarp, and the rille makes an abrupt change in direction along a SE-trending graben (arrow Z). In addition, numerous minor scarps lie within the southern extension (arrow XX).

There are other rilles in the area, some of which reflect the NE trend exhibited by the Sirsalis rille, whereas others have WNW alignments. The most obvious intersections are identified in the lower left of the photograph.

Evidence for extensive volcanism surrounding the Sirsalis rille includes:

1. Subsided lava lakes (arrow YY)
2. Hummocky-floored craters (arrow ZZ; also see Plate 28f)
3. Breached tephra-ringlike forms (arrow XY)
4. Rimmed linear depression that connects two craters (arrow XZ)

PLATE 163

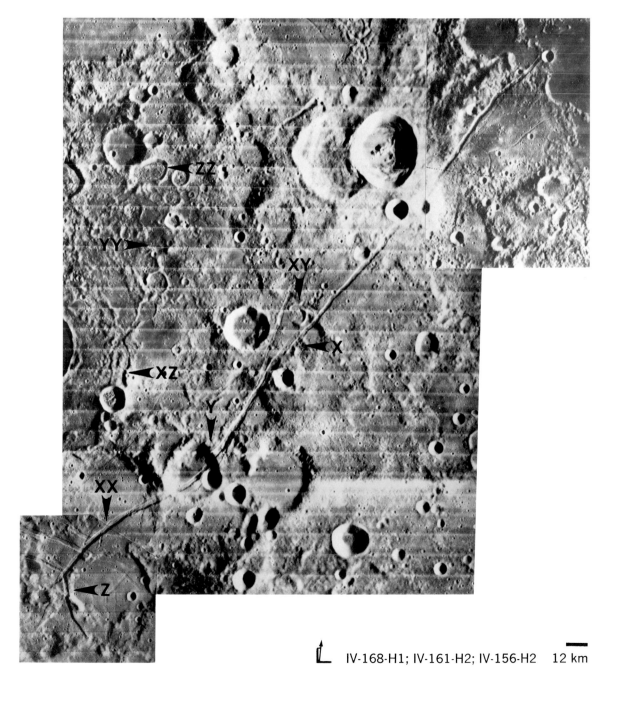

IV-168-H1; IV-161-H2; IV-156-H2   12 km

Nonsinuous Rilles: V-Shaped, U-Shaped Profiles

The Hevelius rille system (Plate 164) includes several nonsinuous rilles and illustrates the transition from flat-floored (grabenlike) to U- and V-shaped profiles. As noted in the preceding discussion, the width of a grabenlike depression commonly increases where it crosses features at higher elevations; consequently, the marginal faults are thought to dip toward each other. On the crater floor (Plate 164a, arrow X) and mare (arrow Y), where the rille cuts the lowest levels, the rille profile is V shaped. This simplified interpretation does not take into consideration changes in depth and intensity of the zone of weakness, nonparallelism in strike of bounding faults, and effects of degradation on different rock types.

As in the Hyginus rille, volcanic activity has occurred along one of the Hevelius rilles (Plate 164a, arrow Z; Plate 164, b and c, arrow X). Noteworthy features include three "pinched" craters on its western extension (Plate 164, c and d, arrow Y); a transition into a normal fault (arrow Z); an extrusive sheet extending northward (arrow X); and generally low albedo along the length. In addition, three prominent dark-haloed elongate craters to the right (arrow XX) reflect the trend of this part of the rille system. They are clearly extrusive centers that postdate the stippled zone (arrow YY).

PLATE 164

a. IV-169-H1; IV-162-H1   12 km

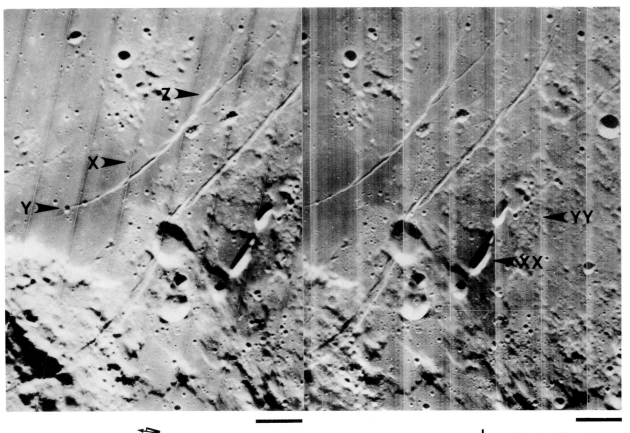

b. IV-162-H1   12 km

c. IV-161-H3   12 km

Nonsinuous Rilles: Arrangements

Plates 165–170 illustrate associations of rilles as outlined in the introduction to this chapter. Plates 165–167b are primarily concerned with the arrangements of rilles, whereas Plates 167c–170 include the larger-scale ordering of rilles—rille systems. The following discussions identify the types of arrangements or systems and include other details that distinguish one rille from another, either in form or origin.

Plate 165, a and b, shows in stereo curvilinear nonsinuous branching rilles in western Mare Humorum (also note eastern Mare Humorum in Plate 167c). The rilles are apparently related to the ghostlike crater (arrow X) since they appear to intersect at this location and since two rille branches terminate at its floor-wall contact. The rille walls are scalloped along several portions as if they once were a contiguous crater chain (arrow Y); in fact, the southern extension shows such related crater groups (arrow Z). To the south of the converging rilles, there is a definite raised rim (arrow XX).

The generally low albedo of this region and raised rims of the rille give evidence for a fissuretype eruption. Telescopic investigations in the infrared have shown this to be a region of anomalously high thermal emission, an effect that apparently cannot be explained by differences in solar heating due to the local low albedo or regolithic structure (Hunt et al., 1968).

Plate 165c shows an arrangement of rilles with a 90° join. The rilles are largely restricted to the interior of a flat-floored crater that has been highly modified by the Imbrium sculpture. The rilles obviously postdate both the stippled material to the southwest, which has overridden the crater, and the very dark mare material that fills the craterforms and postdates the Imbrium sculpture. The region west of the rille is topographically lower than is the surface to the east, and the 90° join of two major rilles suggests localized upward-directed stress. Near the southern extension of the main rille is an elongate blisterlike feature (arrow X), which probably was an extrusive center. In addition, near the northern end, a NW-trending ridge (arrow Y) meets the main rille and also may be the result of igneous processes, intrusive or extrusive. Thus, this rille pattern indicates faulting and fracturing that were associated with volcanic activity.

Plate 165d shows an area adjacent to the eastern border of the northern Caucasus Mountains in northwestern Mare Serenitatis. The rille pattern is similar to that of the foregoing example but includes not only right-angle joins but also acute-angle offshoots. One branch is composed of connecting pinched craters (arrow X). In addition, the V-shaped profile of the western extensions grades into flat-floored profiles to the east (arrow Y). In discussion accompanying Plate 164a, this transition is attributed to changes in elevation of the terrain that the grabenlike rille crosses. In this example, it is interpreted as the result of partial filling of the fracture. Regions of low albedo are also present along the rille, and close inspection of the original photograph reveals sheetlike flows to either side of the depression. Also note:

1. The apparent transition of the southern extension into a ridge (arrow Z)
2. Smoothing and sculpturing performed by ejecta from the crater Aristillus to the ENE (adjacent to arrow Z)
3. The "cat's-paw" crater (an ejecta product shown below arrow Y) along the upper branch, which postdates the rille but is subdued compared to the rille
4. The almost complete isolation of a portion of the mare by the branching and rejoining of the rille (arrow XX)
5. The raised lip exhibited by the major rille and especially evident near the 90° join

PLATE 165

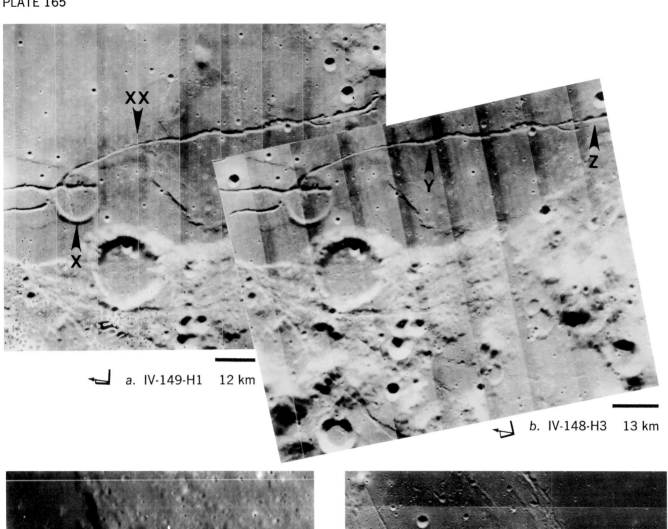

a. IV-149-H1   12 km

b. IV-148-H3   13 km

c. IV-097-H2   12 km

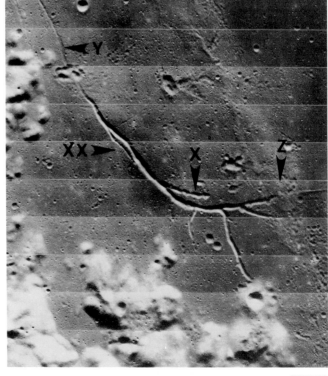

d. V-086-M   5.7 km

377

Nonsinuous Rilles: Arrangements

The Sulpicius Gallus rilles border southwestern Mare Serenitatis and are nonsinuous arcuate rilles. They illustrate the second-order arrangement of multiple offshoots, which resembles a bird's foot. Such an arrangement is not unique and can be identified down to small features (0.05 km in width) in other regions. The three rilles have been labeled (top to bottom in Plate 166, a and b) Rima Sulpicius Gallus I, II, and III. Sulpicius Gallus I exhibits a flat floor, part of which has resulted from mare flooding. Near the join of the rilles is a possible flow front within Rima I (arrow X). Sulpicius Gallus I and II display V- and U-shaped profiles.

The characteristic plan of these rilles warrants their consideration as a class, and their setting provides clues to their interpretation and history. They are on a low-albedo shelf that is darker and has fewer craters than the adjacent mare. Similar low-albedo units occur adjacent to other maria (Plate 188) and appear to be associated with extrusive activity (Carr, 1966). The low-albedo deposits are not restricted to the shelf but extend westward onto the peaks of the Apennines. This is revealed clearly under full solar illumination in telescopic views. In addition, there are numerous low-rimmed, flat-floored craters with small dimplelike depressions in the center (arrow Y). These are suggested to be possible extrusive centers or products of catastrophic blanketing (Plate 20).

Other interesting morphologic features in the overview provided by Plate 166, a and b, are:

1. NE-trending "pinched" craters (arrow Z)
2. Domal feature on the mare (arrow XX)
3. The rille that bisects a dome and appears to be the "head crater" of a partially collapsed rille (arrow YY)
4. Broadened but hummocky floor of the eastern extension of Rima I
5. Numerous high-albedo specks on the peaks (arrow ZZ; see plate 185c for discussion)

Plate 166c permits closer examination of the center of the area shown in a and b. The subdued topography recognized at lower resolution is even more apparent here; furthermore, the marelike floor of Rima I makes a relatively well defined contact with the walls. The presence of several half-buried craters (arrow X) along the base of the rille walls suggests an advanced (or advancing) wall. In addition, there is a poorly defined apron along the mare contact, a feature that has been discussed in connection with a variety of positive-relief features at the contact with marelike surfaces. Subdued medium-size craters on the marelike rille floor are more numerous than those on the shelf regions. Plate 166c also shows narrow lineations (arrow Y)

that follow the subdued rilles. In addition, note the islandlike peak to the west of Rima I (arrow Z). It has a rougher texture than the surrounding low-albedo material and exhibits a relatively well defined contact with this material. A similar exposure on the other side of Rima I (arrow XX) may have been a related topographic high prior to rille formation.

If crater abundance alone is used as a measure of relative age, then the shelf seems to be much younger than the rille, but the rille is clearly embayed by mare units. This contradiction is resolvable if smoothing processes were much more effective on the shelf or if the subdued craters on the mare surface were formed endogenetically. As mentioned previously, low-albedo material mantles not only the shelf but also the more rugged terrain to the west. The inferred copious eruptions of ash would result in a deep layer of incompetent material that could not preserve craters, owing to its mobility in response to later impacts and Moonquakes. The incompetency of this layer, and perhaps the buried strata, is suggested by the absence of large blocks in the ejecta of bright-rayed craters. The mobility is indicated by the exposed high-albedo knobs (arrows Z and XX). The relatively well defined boundary surrounding these exposures suggests that mass movement by landslides and creep is more effective than is the dispersing and smoothing action by impacts. Note that, if the narrow lineations within the subdued rilles on the shelf represent reactivated faults, then the locally rapid degradation of surface features requires that this faulting continued until relatively recently.

At least some of the subdued craters on the mare units are believed to reflect remnants of internally produced craters rather than degraded impact craters. This belief is based on the evidence for small-scale features that appear to have been preserved since the last stages of mare flooding (see Plates 125e, 128, 181c).

The Sulpicius Gallus rilles are interpreted as tension fractures and boundary faults related to basin subsidence during the early stages of mare flooding. During the formation of these peripheral rilles, eruptions of ash mantled the borders of the Serenitatis basin. Early subsidence of the maria along these peripheral rilles left the margins of the blanketed plains in relief and formed abrupt scarps adjacent to the mare plains in several regions. Subsequent flows of mare basalts embayed the rilles and inundated the interior down-dropped shores. A similar sequence of events can be inferred for eastern Mare Serenitatis (shown in Plates 74a and 138). Thus, the present boundary of Serenitatis is pictured as resulting from a combination of volcano-tectonic subsidence and mare flooding (see Discussion and Summary in Chapter 9).

PLATE 166

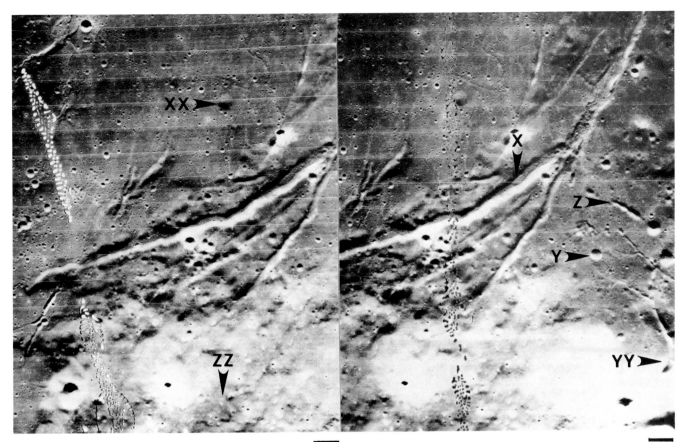

⟵◻ *a.* V-093-M   3.6 km (*top*)   4.2 km (*bottom*)   ⟵◻ *b.* V-091-M   3.6 km (*top*)   4.2 km (*bottom*)

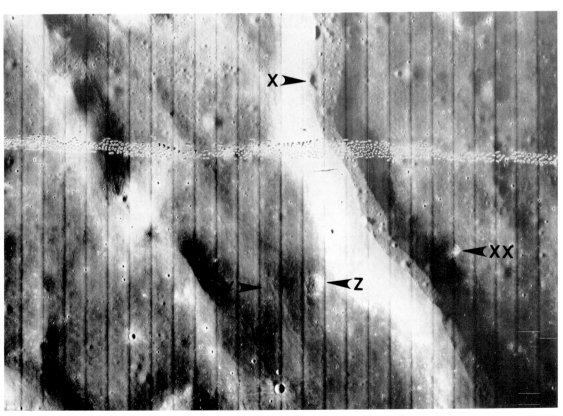

◻↑ *c.* V-092-H1   0.48 km

Nonsinuous Rilles: Arrangements and Systems

Plate 167 provides two examples of *en échelon* grabenlike rilles, Rima Bond I (Plate 167a) and Rima Ariadaeus (Plate 167b). Both show interior dropping of mare and highland surfaces, with widening of the rille generally occurring over the latter. The *en échelon* arrangement also was noted for the western extension of the Hyginus Rille (Plate 159). Such offsets are encountered frequently in terrestrial geology and are attributed to extensional or shear stresses.

Plate 167c introduces a series of examples of rille systems to be considered through Plate 170. The concentric system of grabenlike rilles shown in Plate 167c outlines eastern Mare Humorum. The rille system is more recent than the highlands and mare but is older than the bright-rayed craters that interrupt its path. Within the large flooded crater shown at left center, however, the graben appears to be overlapped by portions of hills (arrow X). This could be an illusion created by an offset, as is exhibited at other locations along the rilles.

The eastern border of Mare Humorum is irregular and has numerous inlets of mare material, whereas its western border is marked by an abrupt scarp (see Plate 168). The eastern concentric system of grabenlike rilles is not continued on the northern, western, or southern borders. Lighting may have some bearing on their identification, but if they are present (an unlikely possibility), they must not be connected to the N-S array and must lie exactly in an E-W direction. The northern and southern borders, however, do have other features not found on the eastern and western regions. The northern boundary of Mare Humorum connects with Oceanus Procellarum between islands of highland relief, with several N-trending sinuous rilles (Plate 191a). The southern border displays lava lake depressions and sheets of extrusions.

Quite clearly the eastern rille system shown in Plate 167c is genetically related to the Mare Humorum basin, but, if similar events occurred along other borders, they were later erased by subsequent lava flows, a plausible event because the mare plains in northern Mare Humorum have a lower albedo than does the mare where the concentric grabens are found. Subsid-

ence of the Mare Humorum basin with respect to the eastern boundaries has taken place, as is evident from the contour levels given on LAC 93. The annular grabens probably developed along extensional weaknesses created by this process, which may have occurred relatively early in the inundation process (see Discussion and Summary in Chapter 9). LAC 93 shows, however, that the general elevation does not increase at the western contact but continues to decrease.

If LAC 93 is correct, the rilles may indicate synthetic and antithetic faulting due to large-scale gravity sliding into Mare Humorum. This might be analogous to smaller-scale slump features along the northwestern coast of the Gulf of Mexico, where an underlying plastic layer of salt acts as a lubricant. Here, the lubricant could be an underlying magmalike layer. Such a slump may have been initiated by release of support at the toe, resulting from extrusion of the magma in Mare Humorum during late stages of mare flooding. This gravity sliding may have produced compressional folds corresponding to the wrinkle ridges in eastern Mare Humorum. It should be noted, however, that these wrinkle ridges extend around northern Mare Humorum, where annular grabens are either absent or have been removed by more recent mare flooding.

More likely, the arrangement of ridges and rilles in Mare Humorum reflects compressional ridges to the interior and tension fractures near the outer borders in response to regional subsidence (Ronca, 1965). Recent gravimetric maps by Sjogren et al. (1972) show a maximum gravity anomaly near the eastern shores of this mare, consistent with the development of compressional folds in the regions of maximum subsidence. It remains possible that the formation of the peripheral grabens and the interior wrinkle ridges represent separate events. As implied in foregoing discussion, the grabens may have been formed at an earlier stage of basin subsidence. If this sequence is correct, then perhaps the ridges formed as a result of subsidence or gravity sliding restricted to the interior of Mare Humorum.

PLATE 167

c. IV-132-H1   12 km

b. IV-079-H1   13 km

a. IV-090-H1   12 km

Nonsinuous Rilles: Systems

The major features of the western-border area of Mare Humorum are shown in Plate 168. In contrast to the eastern border (Plate 167c), the western border exhibits an abrupt scarp between the mare and highlands, and the region is crossed by two separate linear systems trending NW and NE (also see Plate 191a). The western boundary also includes numerous intersections of grabenlike rilles and a fissurelike complex in the mare (see Plate 165, a and b). The grabenlike features cross a variety of surfaces, including craters, mare surfaces, and highland relief. Note that two rille intersections occur near the center of mare-filled craters (see Plate 169).

The region shown in Plate 168 exhibits a large number of volcanolike features that are probably related to mare flooding and crustal weaknesses, indicated by the grabenlike rilles. These features include:

1. Low-relief domes (arrows X and Y)
2. Volcanolike cones (arrow Z)
3. Low-albedo deposits covering highland relief (arrow XX)
4. Calderalike structures (the 35-km crater [arrow YY] that rests on a broad domelike relief)

The region also illustrates a large parallel system of NE-trending rilles whose members include not only the parallel associations shown here but also the Sirsalis rille examined in Plate 163. This system is not concentric to Mare Humorum and is more extensive than the concentric system of rilles in eastern Mare Humorum (Plate 167c). Thus a larger-scale stress system is inferred.

PLATE 168

IV-149-H1   12 km

Nonsinuous Rilles: Systems

The crisscrossing grabenlike rilles shown in Plate 168 are shown also in Plate 169 but have been reoriented to achieve maximum stereo effect. Plate 169, *a* and *b*, shows that the northern wall of the crater predates the formation of the rille. The dome on the southern wall (arrow X) and the irregular relief form southeast of it (see Plate 168), however, could be interpreted as more recent structures. More likely, rapid mass wasting of these positive-relief features has masked surface displacement associated with the rille. It is not obvious which of the rilles is most recent, but the *en échelon* pattern on the rille floor at their intersection (arrow Y) suggests that the SE-trending rille is the older of the two. There apparently have been only vertical displacements, with little or no lateral offset. Note the continuation of the NW-trending rille (arrow Z) into the highland relief. Although close inspection reveals both rille walls, the feature is suggestive of collar-type features associated with wrinkle ridges (see Plate 144, *a* and *b*).

Plate 169, *c* and *d*, shows the intersection of four different rilles near the center of the crater, but it seems unusual that three of the four changed their trend before meeting. The two NW-trending rilles (arrows X and Y) merge toward the center of the crater and then continue SE to the crater pictured in Plate 169, *a* and *b*. The two NE-trending rilles (arrows Z and XX) also merge within the crater yet remain distinct farther west (arrow YY). Both systems again show the widening of these grabens as they cross high relief. Perhaps the intrusive doming centered in the craters shown in Plate 169 altered the local stress field and caused deflections of the rilles into the craters. Note the arc within the crater seen in Plate 169, *c* and *d* (arrow ZZ). This feature might represent the remnants of multiple floor eruptions prior to floor inundation by plains materials.

PLATE 169

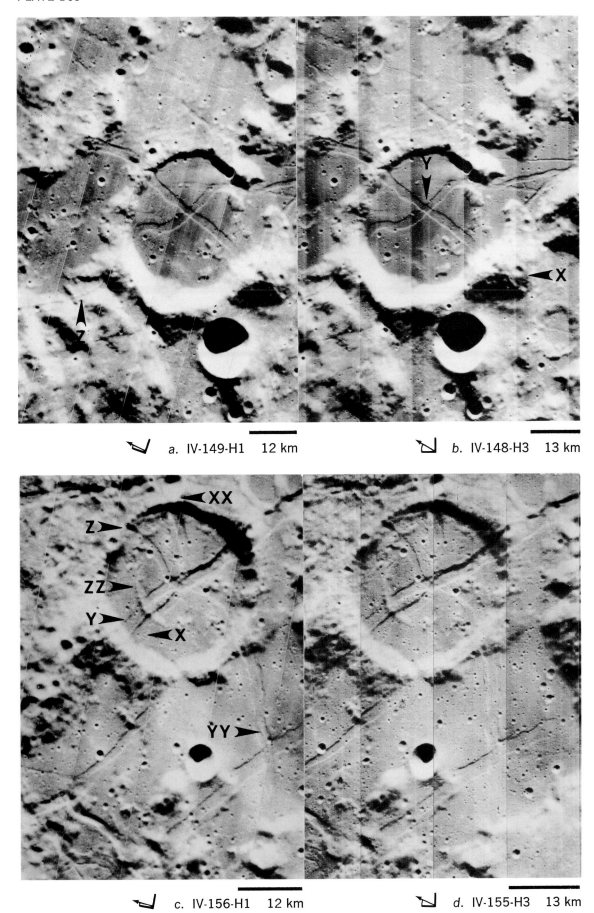

a. IV-149-H1   12 km

b. IV-148-H3   13 km

c. IV-156-H1   12 km

d. IV-155-H3   13 km

Nonsinuous Rilles: Systems

The Triesnecker rilles, shown in Plate 170, are southwest of Hyginus in Sinus Medii and compose a complex rille system, which has a general N trend. The crater Triesnecker, from which these rilles derive their title, is shown near the left center of Plate 170a and is obviously a more recent event than at least the nearest rille.

In contrast to the Hyginus rille, the Triesnecker rilles have segments that display V-shaped and concave profiles. Rima Triesnecker I is the rille extending SSE (arrow X). Rima Triesnecker II (arrow Y) extends northeastward and joins Triesnecker I at a 120° angle to the east of the crater Triesnecker. The shallower rille that extends northward from this same join is Rima Triesnecker VII (arrow Z), and the southward component is Rima Triesnecker V (arrow XX). Whereas the Hyginus rille is in general 3 km in width, these rilles range from 0.75 km to 1.5 km in width. In further contrast, few craters along the rille appear to be genetically related, as do the chains along Hyginus. A possible exception occurs along a branch from Rima Triesnecker I to the south, which continues into the highlands as a crater chain (not shown). There may be similarity in the roles of the crater Rhaeticus A (arrow YY) and Hyginus with respect to their positions along intersecting rilles, but such a comparison is restricted to their placement rather than morphologic similarities. In addition, there are no low-albedo deposits along the Triesnecker system. This last point may attest to an older age, a blanket from the Triesnecker ejecta, or the absence of associated extrusive activity.

At much higher resolution (Plate 170b), Rima Triesnecker I appears subdued. In this particular region of Rima Triesnecker I, a stippled texture of both the plains surface and the surface within the rille is prominent. The brightening of the eastern edge indicates a raised rim, but stereo is needed for confirmation. The profile is not simply concave but has a bulge (arrow X), which probably resulted from step faulting along the wall.

Before introducing the subject of sinuous rilles, it would seem helpful to refer at this point to the part of the outline given in the introduction that is concerned with curvilinear nonsinuous rilles, rille arrangements, and rille systems, and to review specific examples illustrated in Plates 159, 163–168, and 170.

II. Rille plan and continuity
  A. Plan
    1. Curvilinear
      a. Plate 163
      b. Plate 165, a, b, and d
      c. Plate 166
  B. Continuity
    1. Continuous
      a. Plate 163
      b. Plate 165c
    2. Discontinuous
      a. Negative relief
        (1) Irregular depression: Plate 165d
        (2) "Pinched" crater
          (a) Plate 165, a and b
          (b) Plate 165d
        (3) Symmetric crater
          (a) Plate 159
          (b) Plate 163
      b. No relief: Plate 165, a and b
      c. Positive relief: Plate 165d
III. Rille arrangements
  A. Branching rilles
    1. Single offshoot
      a. Right angle
        (1) Plate 165c
        (2) Plate 165d
      b. Acute angle
        (1) Plate 159
        (2) Plate 165, a and b
        (3) Plate 165d
    2. Multiple offshoots: Plate 166
  B. En échelon arrangements
    1. Plate 163
    2. Plate 167
  C. Parallel rilles: Plate 164
  D. Intersecting rilles: Plate 163
IV. Rille systems
  A. Parallel systems: Plate 168
  B. Concentric systems: Plate 167c
  C. Radial systems: Plates 1a, 42c, and 215
  D. Intersecting systems: Plate 168
  E. Complex systems: Plate 170a

PLATE 170

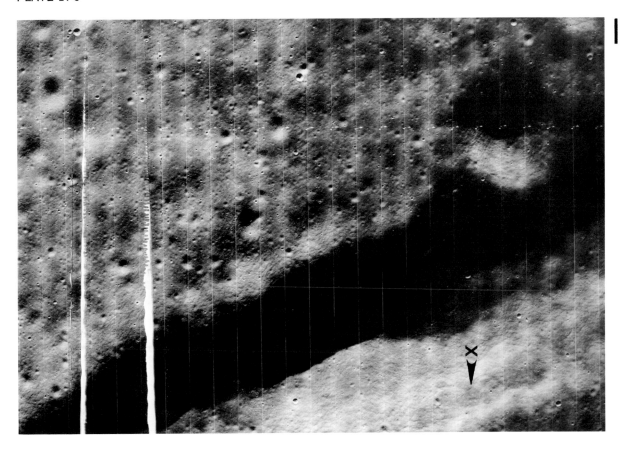

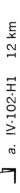

b. II-092-H1   0.18 km

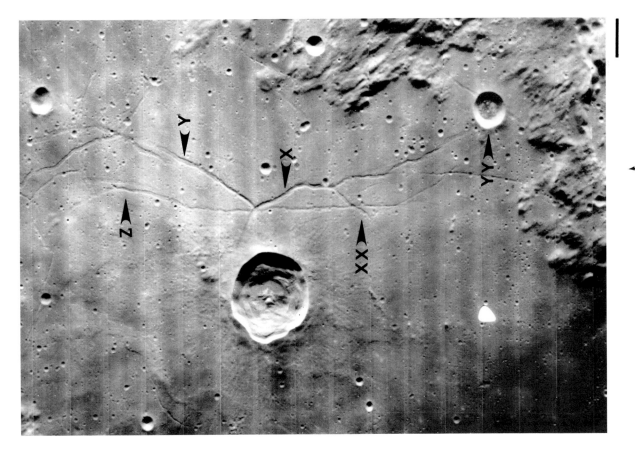

a. IV-102-H1   12 km

Sinuous Rilles: More Sinuous

Plate 171a shows Schröter's Valley (arrow X) and surrounding topographic features on the Aristarchus Plateau. The rilles to the east (arrow Y) are in the Harbinger Mountains, an area examined in Appendix A (Plate 255). In the immediate vicinity of the Aristarchus Plateau, there are at least six rilles varying in degree of sinuosity. One of the major characteristics of sinuous rilles is the presence of a crater at one end, which in Schröter's Valley is known as the Cobra Head (arrow Z). Such craters commonly are better designated as irregular depressions (see Plate 174a).

Sinuous rilles could be defined as rilles that display a certain meander wavelength over a designated distance. Such a subdivision might be useful in distinguishing between rilles, such as Schröter's Valley and the Hadley Rille (Plate 174a). A quantitative classification recently has been undertaken by Oberbeck et al. (1972) in which the rille is described in terms of the Fourier coefficients required to reproduce its plan. In the following discussion, however, such refinement is used only in the context of "more sinuous" and "less sinuous." Less sinuous examples typically display a marked curvilinear plan, but changes in their path occur abruptly or without repetition.

In Plate 171a, most of the rilles observable at this resolution are labeled "less sinuous." Plate 171, b and c, illustrates the distinction between "more" and "less" sinuous. On the floor of Schröter's Valley is a narrow and extremely sinuous inner rille. Even at this medium-resolution view, tight meander loops are evident (arrows X and Y). Portions of the inner rille disappear beneath the base of the wall defined by the outer rille (arrow Z), and, shown near the bottom of the photographs, the rille is traced adjacent to the wall (arrow XX). The "hugging-the-wall" phenomenon is characteristic of rilles, whether the relief is an interior rille wall or another form of positive relief. The disappearance of portions of the rille indicates advancing screelike slopes and provides a measure of how far this process has advanced.

The direction of decreasing elevation is of prime importance for an understanding of the nature of the sinuous rilles shown in Plate 171, b and c. LAC 39 shows that the elevation decreases from nearly 4.8 km at the Cobra Head to less than 0.9 km at the mare contact, a distance of more than 140 km. To judge from the over-all view in Plate 171a, the path of the major valley appears to be at least partly controlled by structural trends. In particular, the abrupt change from a NW to a NE direction, which continues to the mare plains, parallels the ridge to the NE (arrow XX) and may reflect a regional trend.

Ejecta from the crater Aristarchus on the rille floor clearly indicate that this crater postdates the formation of the rille (Plate 171, b and c, arrow YY). Besides the sculpture caused by the Aristarchus event, the surface just outside the valley is similar to the floor of the valley itself. Surrounding the local plains-forming units is a region that has many small hills (right and lower left in the photograph). Such terrain is typical of the Aristarchus Plateau.

Arrow YY in Plate 171a identifies a narrow rille extending from a breach in the wall of the crater Krieger. Similar arrangements can be noted in Plates 177a and 206d.

PLATE 171

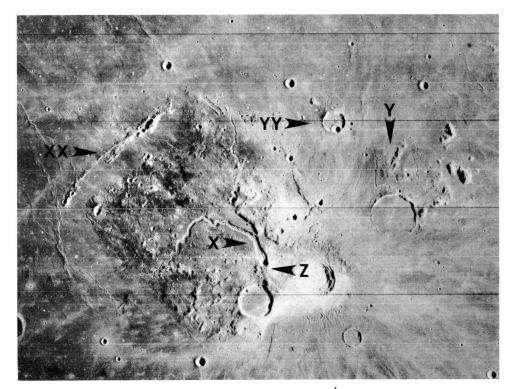

a. IV-157-M   (oblique)

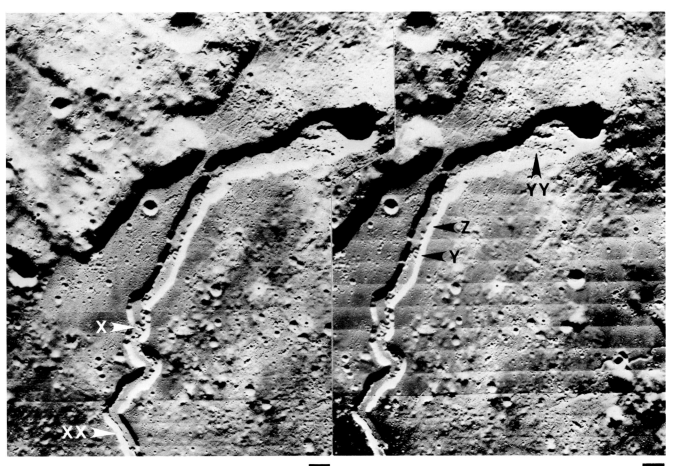

b. V-204-M   4.2 km (top) 4.5 km (bottom)   c. V-202-M   4.1 km (top) 4.5 km (bottom)

Sinuous Mare Rilles: More Sinuous

At the increased resolution of Plate 172, the meander pattern of the inner rille in Schröter's Valley (Plate 171, *b* and *c*, arrow Y) becomes clear. The meander loop shown in the center of the photograph has been cut off (arrow X). The bottom of the inner rille displays a convex profile (arrow Y), which may represent the last material to use the rille as a channel. Not all exposed segments exhibit such a profile, and U-shaped or flat-floored portions are also common along the rille. Although only a limited area of the rille is observable owing to shadows, it does appear that the convex profile is generally typical of tight meanders, whereas U-shaped or flat-floored profiles are more typical of less sinuous stretches. In addition, such profiles become V shaped as the distance from the Cobra Head increases.

Flat-floored craters in the region occur both on the floor of the large valley (arrow Z) and on the surfaces above it (arrow XX). Their presence implies that ashlike material (perhaps ejecta from Aristarchus) filled local depressions and perhaps was responsible for the partial filling of the rille. This does not explain, however, the preferential fill in tight meanders or the fact that similar flat-floored craters occur in the regions where the inner rille displays a V-shaped profile. Consequently, the material in the rille might be better explained as material that once followed the meandering channel. Such material may have been more viscous or of lesser volume, hence insufficient to traverse the entire rille length.

As mentioned, the walls of the outer valley appear to have encroached over the inner rille (arrow YY). Numerous blocks are shown on the wall, with an accumulation of larger ones at the base (arrow ZZ). Although numerous boulders are in place along the upper walls, there does not appear to be well-defined stratification. The boulder trails (for example, arrow XY) that cross both wall material and the floor of a partially visible inner rille indicate that both regions have a regolith.

The interpretation of sinuous rilles has spurred considerable controversy. Suggestions have been made that they are the result of fluvial processes. One view (Urey, 1967) supposed a lunar atmosphere in order to allow more direct comparison to terrestrial rivers. Another view, suggested by Lingenfelter et al. (1968), does not require a gaseous blanket. Rather, through calculations involving the triple point of water, the researchers concluded that water could exist in a liquid state as long as a thousand years. In this case, however, there is necessarily a coexistence of solid and gaseous phases, as well, and the actual process in creating the meander is not made clear. Interesting—and not necessarily contradictory—work has shown that, when water is released in a vacuum, it will immediately freeze without flowing in the sense needed for "lunar rivers" (Adler and Salisbury, 1969b). The scale of the model, however, was much smaller than that proposed on the Moon. The "head" crater in such a theory would be the source of the water, which was trapped in pockets of ice along the margins of the maria but released by an impacting body.

An alternative to the lunar-river model relies on volcanic processes, such as collapsed lava tubes or lava channels. Remarkable similarities can be presented between sinuous tubes in Craters of the Moon national monument, Idaho, and sinuous lunar rilles. In every case, the tube has been formed in pahoehoe basaltic flows. The primary attraction of such an analogy is that special circumstances are not necessary, such as those imposed by the hypotheses involving water. Lunar volcanism was a common occurrence, yet the preservation of pockets of ice beneath the lunar regolith with even a small thermal gradient seems doubtful. The problem with comparisons with terrestrial analogs, however, is that very few lava tubes or channels, if any, attain the extreme sinuosity exhibited by the inner rille shown in Plate 172 and other lunar examples. There are, however, many less sinuous rilles that could easily fit into this scheme.

Finally, a problem exists in the tremendous linear extent of the lunar rilles. The example shown in Plate 176 is a rille that extends for 700 km, with transitions from a flat-floor to a V-shaped profile. Possibly, the differences noted between lunar rilles and terrestrial lava tubes or channels are the result of differences in composition that reduce the viscosity of the rille-forming material even in a lunar environment. Samples brought back from the Apollo 11 mission exhibit extremely low viscosities when melted (Murase and McBirney, 1970), a property believed to be the result of the high metal content. Frothing of a lavalike melt at the surface upon exposure to a vacuum would be accompanied by a rapid increase in viscosity as the volatile content rapidly decreased, but the actual mobility would depend on such factors as composition, gas content, surface slope, and temperature. On Earth, a lava flow may continue great distances under the insulating layer of its cooled surface. Such a layer might provide protection for the lunar extrusion, as well, and, coupled with compositional anomalies, lava tubes or channels of considerable linear extent might be formed (Greeley, 1971a; Murray, 1971; Cruickshank and Wood, 1972). The almost complete absence of water-bearing minerals in the samples returned from the Apollo missions should be a deterrent to any hypothesis dependent on fluvial processes.

One further suggestion has been made that sinuous rilles were formed by ashtype flows (Cameron, 1964). The low viscosity is obviously attained, but the creation of a narrow meandering valley seems improbable.

The problem faced by all proposals is the degree of sinuosity attained in a few examples. Probably not all rilles were formed by the same process. In some instances, one could imagine the degree of sinuosity was created merely by a channellike flow following a path of least resistance. But in the case of the inner rille within Schröter's Valley, as well as in other examples, processes must have acted over a period of time rather than during a single on-rush of material in order to form the tight meander patterns.

PLATE 172

V-204-H3   0.58 km

Sinuous Mare Rilles: More Sinuous

Plate 173, *a* and *b*, presents in stereo the local relief around Rima Plato II (also refer to upper right of Plate 179*a*). The hummocky pattern is the outer zone of ejecta from Plato, the rim of which is less than 100 km to the east. The highest peaks shown of Montes Alpes attain elevations 2.8 km above the sinuous rille. The head of the rille is an irregular depression about 4 km in diameter (arrow X). Two separate rilles diverge from the crater but join 5 km to the northwest (arrow Y). The plan of the rille to the west of the latter point appears to be an extremely tight meander (arrow Z). Appoximately 45 km from the head crater, another rille joins Rima Plato II at nearly right angles (arrow XX), and the path continues through a variety of topographic features, which are partly marelike material and partly the presumed ejecta blanket from Plato. The plan remains roughly sinuous but not in the same sense as that exhibited by the inner rille of Schröter's Valley. The major difference is the frequent abrupt changes in direction taken by Rima Plato II, which produce a haphazard plan rather than a smooth sinusoid. The same portion of the rille displays a narrow inner rille (arrow YY), which continues to the abrupt kink (bottom center). Although not shown, the rille extends onto Mare Imbrium, where it loses much of its sinuosity (Plate 179*a*, arrow ZZ).

Plate 173*c* shows a portion of Rima Plato II (Plate 173, *a* and *b*, arrow Z) where the rille displays extremely "tight" meanders. The term *meander* is perhaps ill chosen, for the boundary of the rille appears to be more scalloped than sinuous. At several locations, the wall does not mirror the plan of the facing wall. Plate 173*c* clearly shows outcrops of a low-albedo competent stratum (arrow X) within the rille, within several craters, and along the edges of several platforms. The low reflectivity of this unit is not a shadowing effect, since it is identifiable on sun-facing walls. Two linear ridges, radial to Mare Imbrium, also exhibit these outcrops and are cut by the rille (Plate 173*c*, arrow Y; Plate 173, *a* and *b*, arrow ZZ). The low-albedo unit is not well exposed or present along the entire rille length but only in the approximate area shown in Plate 173*c*.

Portions of the rille shown in Plate 173*c* display raised rims that are platformlike rather than leveelike (arrow Z). The impression is that the rille crosses a series of interconnected platforms. Some form of blanketing is indicated along the rim by changes in texturing, low crater counts, and several partly buried craters. Furthermore, possible flow termini can be traced outside the rille (in the original photograph).

Several interpretations are possible from these observations. First, this portion of the rille may have been a fissure that subsequently collapsed in an offset arrangement, producing a pseudomeandering plan. Examples shown in Plate 181 may represent the initiation of this process. The fissure would be the source of the platformlike constructions and the low-albedo unit. Other contemporaneous fissure activity is reflected by the NE-trending ridges, which are interpreted as radially directed dikes from Mare Imbrium. Such an explanation is not applicable for the entire length of Rima Plato II but only for the area near the rille head, including the portion shown in Plate 173*c*. To the northwest, the rille probably developed as a surface channel.

Schumm (1970) reproduced similar meanderlike plans by jetting gas through various thicknesses of particulate material. Tilting of the model resulted in coalescence of the separate blow holes.

Perhaps the rille represents a combination of lava channel and tube. The portion of the rille shown in Plate 173*c* is interpreted as the collapsed lava-tube segment. The meanderlike plan again can be explained as offset collapses or, alternatively, as the reflection of actual flow. In the latter case, the arrangement of the facing walls requires that this flow continued over a long period of time, with erosion of the walls by a small interior channel. The incised Canadian River in northeastern New Mexico shows a similar plan, although the transporting medium was water rather than highly mobile lava. The platformlike rim area may represent uncollapsed portions of the roof of a tube or spillover from a channel. The high-level mark along the positive relief (arrow XX) suggests that the plains-forming unit shown in Plate 173*c* has subsided. Consequently, the local gradient was probably low and thus likely to encourage the development of a crust or roof and a meandering plan. Farther "downstream," the gradient increased, and the lava flow continued as a channel without the tight meander plan. The collapsed roof was removed by subsequent flows of lesser volume following the well-established channels. This interpretation can be modified slightly such that the entire length of the rille is a channel, in which case the platformlike rim is suggested to be spillover. In both interpretations, the ridges are viewed as contemporaneous constructions, since the path of flow was not altered and the ridges clearly do not cross the rille.

PLATE 173

a. V-132-M   8.4 km (top)   7.6 km (bottom)   b. V-129-M   8.0 km (top)   7.2 km (bottom)

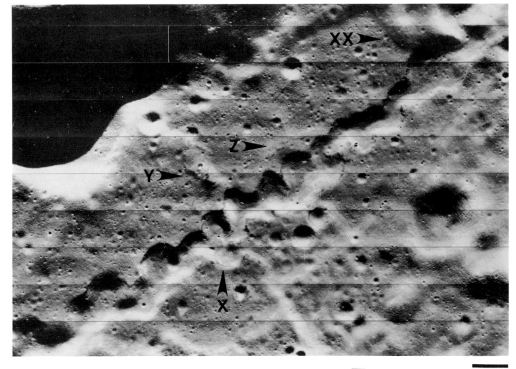

c. V-130-H1   1.0 km

Sinuous Mare Rilles: Less Sinuous

The Hadley rille, shown in Plate 174, is a "less sinuous," or nonmeandering, rille. This is in contrast to the plan illustrated by the inner rille of Schröter's Valley (Plates 171–172), although there is, perhaps, no distinction in origin. Plate 174a discloses the over-all path of the rille adjacent to the west-facing scarp of Montes Apennines. Some of these peaks rise more than 2.7 km above the mare floor called Palus Putredinus. Mt. Hadley, shown at the top of the plate, is about 3.5 km above the rille. The head of the rille is an irregular depression (arrow X) and is shown on LAC 41 to be at a higher elevation than the remaining portions of the rilles. The rille's path parallels the concentric cliffs and ridges of Mare Imbrium and is apparently controlled by these structural and topographic features.

The rille profile is, for the most part, V shaped. As is true for other examples, the rille goes around an obstruction rather than disappearing beneath it (arrow Y). Farther to the north on Plate 174a, the path diverges (arrow Z), with one branch continuing to the northwest, the other to the northeast. The northwest branch connects to a flat-floored depression (arrow XX). The northeast branch (Rima Fresnel II) appears to be controlled by structural weaknesses concentric to the Imbrium basin. This section of the rille is sufficiently different from the more curvilinear segment to represent a possible separate grabenlike rille. The apparent disappearance of the rille to the northwest of the contact between the rille and obstruction (arrow YY) is probably due to unfavorable lighting, and the high-resolution photo-

graph of the region shows continuation, although the floor is flat and the walls are subdued.

In addition to the Hadley rille, the Rima Fresnel system illustrates the much wider and flat-floored linear depressions that parallel the Apennine front, i.e., concentric to Mare Imbrium. Rima Bradley is similar and dissects similar relief to the southeast (see Plate 112).

Plate 174, b and c, shows a stereo pair of the Hadley rille head. The relief is exaggerated and permits extrapolation of the depression into the shadows (arrow X). Note that the fracturelike head crosses both the smooth mare floor and the hummocky highland relief. The northern extension (arrow Y) is double and is apparently not the result of slumping.

Observations of further interest are:

1. The northern extension of the head ends in a small craterlike depression (arrow Z).
2. Platformlike feature is crossed by the rille (arrow XX).
3. The edges of the rille are slightly raised.
4. Opposite sides of the rille appear to be at different elevations (confirmed by the Apollo 15 mission, 59 km to the northwest).
5. The contact between the highlands and mare, seen near the upper left edge of Plate 174, b and c, is marked by a terrace (arrow YY), probably indicating a previous level of mare flooding (easily identified in the original photograph).

PLATE 174

a. V-107-M   4.5 km   V-105-M   4.2 km

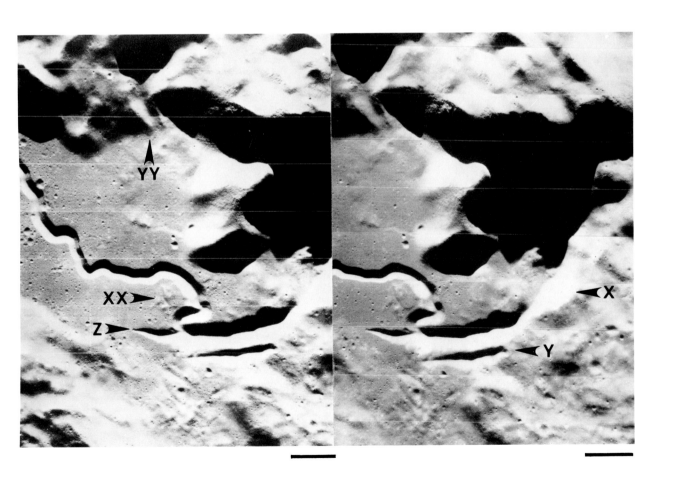

b. V-105-M   4.2 km

c. V-104-M   4.2 km

Sinuous Mare Rilles: Less Sinuous

Plate 175, a and b, permits examination of the northeastern section of the Hadley rille in stereo. Plate 175 includes selected high-resolution views and more detailed close-ups of the region to the south of the large crater (visible in Plate 175, a and b). Plate 174b shows the precise area covered by these detailed views.

The stereo pair reveals several interesting aspects of this type of rille. The **V**-shaped profile becomes apparent, and segments that seem to disappear are clearly the result of alignment with the direction of solar illumination. Along these parts of the rille, outcrops near the upper slopes of the walls are visible. The rille has a raised rim, which must be explained in any hypothesis for rille origin.

A slightly elongate crater 6 km in diameter (Hadley C) is visible to the right in Plate 175, a and b (arrow X). Its ejecta blanket covers the rille to the east; hence, it was formed after the rille. Such a time sequence indicates that degradation of the less consolidated ejecta blanket was much more rapid than that of the crater or rille. This is partly due to the screelike nature of the crater wall and rille. Note that the blanketing of ejecta did not produce a flat-floored profile.

As the rille is traced northward in Plate 175, a and b, it is seen to abut against two mountains. In a few areas the mountainous slopes have advanced into the rille, but in no instance does the rille appear to undercut the massifs significantly. The rille is not completely continuous but breaks into connecting elongate craterforms above the contact between the rille and mountain (Plate 175, a and b, arrows Y and Z). Just below the discontinuity marked by arrow XX, the rille deepens and widens at the juncture with another shallow rille, which extends from the left (above the mountain at left). The shallow rille changes into an east-facing scarp just before joining the Hadley rille (arrow YY). Farther to the north, the main rille is seen to mark the contact between the mountain and mare surface. Along this section are three short and shallow rilles perpendicular to the rille to the right (arrows ZZ, XY, XZ). The first (arrow ZZ) is another east-facing scarp.

Three origins for the Hadley rille are proposed. One, it resulted from the surface flow of material from the elongate craters at its head. This hypothesis would account for the construction of levees and the apparent obstruction by the Apennine Mountains. The discontinuity indentified by arrow XX and the coalescing craters, however, suggest that, at least along this portion of the rille, a roof formed. Consequently, a second suggestion (proposed by Greeley, 1971a) is that most of the rille may have been a lava tube. The absence of the collapsed roof on the rille floor may be evidence for the thinness or friability of the roof.

The third interpretation involves a structural origin, which is suggested by differential movement. In particular, facing rille rims exhibit differences in elevation, and two faultlike scarps connect with the rille. In this hypothesis, the short secondary rilles connecting at right angles can be interpreted as extensional fractures. In addition, the northern extension of the Hadley rille links with structurally produced grabenlike rilles that cross a variety of terrains. It is possible that the Hadley rille was initially formed as a fracture (Quaide, 1965) that provided a

path for lava. The scarps connected to the rille may reflect renewed displacements along this structural weakness.

Aside from the rille, there are other interesting features shown in Plate 175, a and b:

1. The faceted peak near the center (arrow YX)
2. The "blisterlike" swelling and summit crater chains (arrow YZ)
3. The dimplelike floor of the crater on the mare surface 5 km south of the rille seen at arrow XZ (commonly found in regions apparently once endogenetically active)
4. Terraces bordering the mountains (arrow ZX)

Plate 175c shows a segment of the rille to the south of the region shown in Plate 175, a and b. Plate 175, d and e, provides more detailed views of the floor and walls. Outcrops are easily identified along the upper portions of the walls but typically are seen below the break in slope from the surrounding mare surface. If the outcrops represent "bedrock," then the region is covered by a thick layer of regolith that may be due to repeated small-scale impacts, larger-scale ejecta, or ash. The outcrops, instead, could represent stratification produced by successive flow units.

The high-resolution photograph and its enlargements reveal three slope regimes: a gentle lower slope, a steeper slope near the upper wall (below the outcrops), and a gentler slope near the top, which merges uniformly with the flat mare surface at several locations. The over-all **V**-shaped profile appears to be formed by the meeting of two screelike slopes. Evidence of continued slope reduction is indicated by the relative absence of blocks resting at the bottom of the valley (covered by mutually advancing walls) and the severed irregular depressions and craters (Plate 175c, arrows X and Y) along portions of the rille rim (note two small subdued craters on the north rim [Plate 175e, arrow X]).

The present configuration is sufficiently stable, however, that several craters remain along the narrow floor and rille rim. At the former location (Plate 175d), relatively rapid degradation should be expected, since the rille bottom would act as a trap for ejecta impacts along the walls and outside the rille (Salisbury and Smalley, 1964). In addition, numerous small craters can be identified on the walls (Plate 175, d and e), although their abundance is much less than that of the surrounding mare surface. Slope stability is also indicated by the smooth transition between the subdued 0.8-km crater shown in Plate 175d and the rille wall. This transition suggests progressive smoothing by meteoroid bombardment, rather than reduction of slope of the screelike rille wall. Note the two boulder trails seen in Plate 175e (arrow Y).

Closer examination of the rille from these plates reveals continuing degradation but does not resolve the problem of its origin. Outcrops of strata can be rationalized in channel, tube, or structural origins. The problem is reconsidered in Discussion and Summary at the end of this chapter. For a closer look at the rille with Apollo 15 photography, and for further review of the regional geology, refer to Howard and Head (1972).

PLATE 175

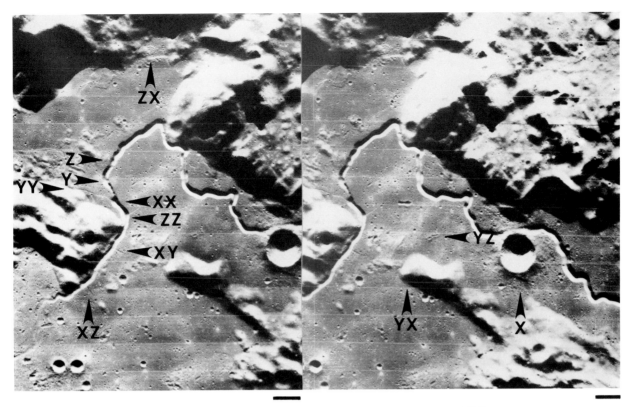

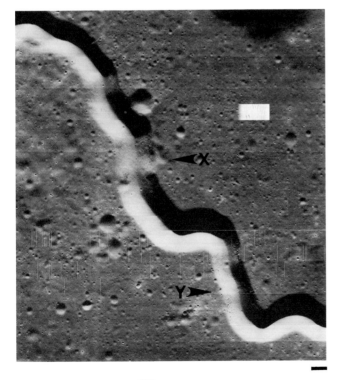

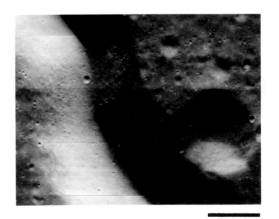

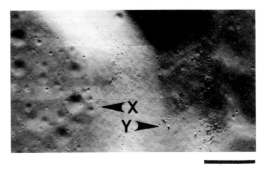

a. V-106-M   4.5 km

b. V-105-M   4.2 km

c. V-105-H2,H3   0.55 km

d. V-105-H2   0.55 km

e. V-105-H2   0.55 km

Sinuous Rilles: Less Sinuous

In a few regions, rilles are traced over enormous distances. A typical example is Rima Sharp I, which extends more than 360 km in the area shown in Plate 176 and continues almost the same distance beyond the area shown at the bottom edge of the photograph. As mentioned already, this extent is in sharp contrast to the very limited extent of collapsed lava tubes and channels on Earth. The elongate crater (arrow X) is apparently related to two rilles. Just to the east of this crater are several irregular depressions (arrow Y), which also may be genetically related to the rilles. The path of the rille is influenced by pre-existing structures, such as domes (arrow Z) or highland relief (arrow XX). The plan of Rima Sharp I is curvilinear but less sinuous than that of the inner rille of Schröter's Valley.

The profile of this rille changes from **V** shaped to flat floored. A transition zone appears near the highland peninsula (near arrow XX). South of this peninsula, an inner rille is recognized on the flat rille floor and continues to near the area shown at the bottom of Plate 176 (arrow YY), where the rille widens. Note the branching and rejoining of the rille near the center of the plate (near arrow XX) and along the northeastern extension (arrow ZZ).

The formation of both V-shaped and flat-bottom profiles along the same rille is evidence against the formation of the flat-bottom profile by erosion and deposition from meteoroid bombardment. Perhaps the profile change indicates that the mare region to the south of the peninsula has undergone at least one additional phase of mare-filling activity. Such a suggestion would appear to require a more recent formation of the adjacent rille shown near arrow Z. More likely, the transition reflects a change in regional gradient resulting in "downstream" pooling of lava from extrusions that originated in the vent at arrow X.

Note the highly subdued crater (arrow XY) that is in distinct contrast to the adjacent rille. If this is an impact crater subdued by meteoroid smoothing, then the rille must be a much more recent collapse structure; however, I interpret this feature as a preserved volcanic vent related to the last period of mare emplacement. This interpretation is consistent with the preservation of small-scale irregular platforms and terraces (arrow XZ), also thought to have been related to mare flooding.

PLATE 176

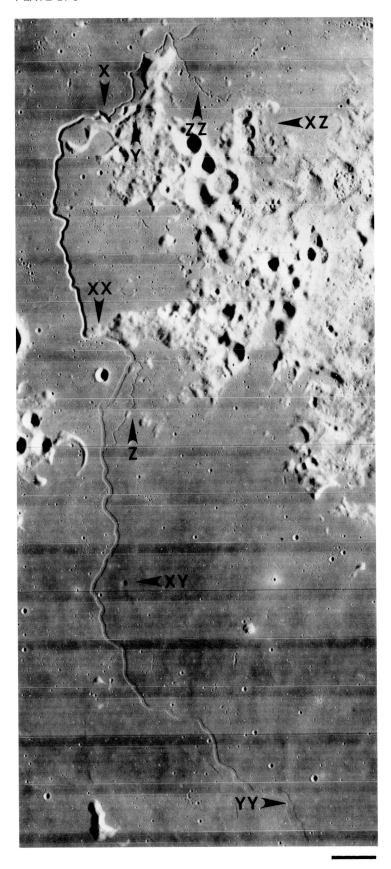

IV-158-H2   12 km

Sinuous Rilles: Location

Sinuous rilles typically occur on plains-forming units, including "pooled" regions within the highlands and the floors of large flooded craters, as shown in Plate 177a. The crater Posidonius (Plate 177a) is also discussed with Plate 2b for its subconcentric (eccentric) plan. On its floor is an extremely "sinuous" rille that extends between a narrow breach in the crater wall (arrow X) and an area adjacent to the northern wall (arrow Y). As observed for rilles on the maria, the rille path parallels the base of positive-relief forms. This "hugging" occurs on the southwestern wall (arrow Z) and along the insular massif (arrow XX).

The plan of the rille is very complex near the area marked by arrow YY, where it is composed of interconnected irregular and scalloped depressions. Several segments of the rille exhibit meanderlike patterns, with each side of the rille approximately mirroring the facing side (arrows ZZ and XY). In comparison to the plan of terrestrial rivers, this rille appears much wider relative to the meander wavelength. Under high magnification, large loops and cutoffs are recognized and have been confirmed by examination of Apollo 15 photography (AS15-91-12366).

The rille shown in Plate 177a is interpreted as a lava tube or channel that developed within a large lava pool represented by the plains-forming floor unit. The extreme "sinuosity" is attributed to the relatively low gradient established within this closed depression and to successive periods of stagnant and mobile flow (see Discussion and Summary at the end of this chapter). It is not clear whether the source of the flow was from the narrow breach of the crater wall (arrow X) or the region near arrow Y. I do not believe this rille can be interpreted completely as coalesced craters, owing to the remarkable offset arrangement required to reproduce segments of the meanderlike rille plan. By the same token, it cannot be understood as completely a surface channel, because of the complex plan developed along other segments (arrow YY). A few rilles in the "open" maria exhibit similar tight "meanders," but rilles typically are not so sinuous as these examples. As in the case of Schröter's Valley, a low surface gradient may have been established within an essentially closed basin, a situation that should encourage the development of flow meanders.

Plate 177b shows part of a large linear massif to the northwest of the Aristarchus Plateau (see upper left of Plate 171a). The plateau exhibits prominent NW and NE structural trends; the latter trend is paralleled by ridges, such as the major feature shown in Plate 177b. The primary interest here is with the sinuous rilles parallel to, and on either side of, the NE-trending massif.

The rille (arrow X) to the northern (left) side of the linear massif is connected to a complex of flat-floored depressions (arrow Y) whose interior surfaces are marelike except for textures approximately reflecting local boundaries. In particular, note the depression seen at upper left (arrow Z). These structures are interpreted as subsided lava lakes, and it is proposed that the rille was a path for drainage. The rille appears to be more deeply entrenched where it crosses the wrinkle ridge (arrow XX). This suggests that the rille formed during or after ridge formation rather than being a preserved surface feature on a later fold. If it formed as a channel later than the ridge, then continuous erosion is required. More likely, the rille cut the ridge while it was being formed by either surface folding or the subsidence of the surrounding mare level. Later ridge formation, probably by folding, is indicated at arrow YY where it crosses the rille.

The rille to the south of the linear massif is shown only in part in Plate 177b. From LAC 39, the local elevation is highest along the ESE-trending wrinkle ridge (arrow ZZ) and decreases to either side. This might indicate that the apparently continuous rille is actually two separate rilles extending downslope on either side of the wrinkle ridge. If this was the case, then two craters forming the heads of the rilles should be found at the higher elevations adjacent to the wrinkle ridge. Unfortunately, ejecta from Aristarchus have effectively erased possible head craters as well as any continuous trace of the rille across the ridge. The similarity between portions of the rilles on either side supports the interpretation that they were once continuous. In addition, there is an excellent candidate for the head crater at the northeastern end of the rille (not shown). Thus, it is suggested that the rille was once continuous and followed an original decrease in elevation from northeast to southwest. Later the region was uplifted during the formation of the wrinkle ridges. This interpretation is supported by the obvious vertical displacement along the same ridge to the southeast of the area shown in Plate 177b but included in Plate 177, c and d (arrow YY).

Plate 177, c and d, shows a stereo pair of a sinuous rille less than 30 km southeast of the area shown in Plate 177b. The rille is at the northern edge of the Aristarchus Plateau, shown in Plate 171a. It originates in a "pool" of marelike material on the plateau, but, in its path to the mare to the left, the rille cuts the nonmarelike plateau unit. This is one of several rilles that cross positive-relief forms of highlandlike terrain. Consequently, to interpret the rille's entire length as a lava tube is not reasonable. In addition, it demonstrates that any proposed medium responsible for forming the rille must have caused significant erosion. Note that the rille has meander loops (arrow X) and a possible cutoff (arrow Y) along its path in the pooled area but becomes less sinuous, more narrow, and more shallow as it cuts the plateau (arrow Z) and reaches the mare (arrow XX).

PLATE 177

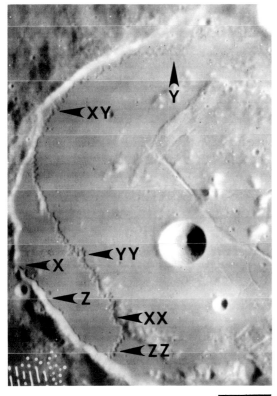

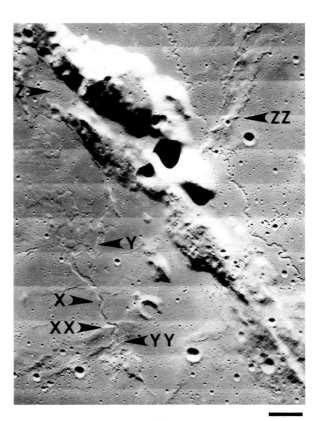

a. IV-086-H1   13 km

b. V-209-M   4.7 km

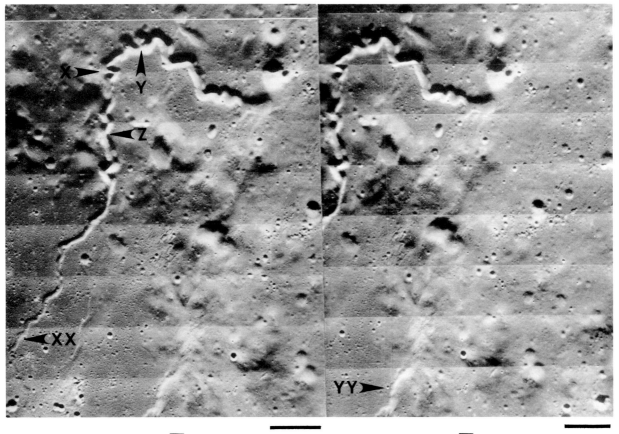

c. V-209-M   4.7 km

d. V-206-M   4.7 km

Rectilinear Rilles

Plate 178 includes examples of rilles with craterforms at both ends and rectilinear rilles. Plate 178a shows a "two-headed" rille west of the crater Birt (center). Both "head" craters (arrows X and Y) are elongate in the direction of the over-all rille trend, which is parallel to the well-known scarp called the Straight Wall (arrow Z). This nonsinuous rille exhibits offsets (arrow XX) and a relatively angular change in its plan (arrow YY). At both locations, the rille is composed of interconnecting craters. The structural influence—if not origin—of the rille is suggested by the offsets and its alignment with the Straight Wall. The craters at either end are interpreted as related volcanic vents.

Plate 178b shows a rectilinear rille in an inlet of mare material between the craters Doppelmayer and Vitello (see Plate 249). It is controlled by a prominent NE trend (arrow X) with a minor W trend (arrow Y). Structural origin for the rille is indicated by the absence of a head crater, the angular changes of plan (arrow Z), and the crossing of the elevated unit shown near the top of the photograph without deepening or change of direction. The change in rille profile from its **V** shape on the plains-forming unit to a **U** or a subdued **V** shape on the elevated region is interpreted as a change in competence of local rock type.

The oblique view in Plate 178c shows an area in western Mare Fecunditatis and includes several rilles. Arrow X identifies an incipient rectilinear rille, composed of numerous connecting and contiguous craters (see Plate 117, a and b). It appears to extend northward and is tangent to the large flat-floored crater, which is about 10 km in diameter (arrow Y). Examination of Apollo 11 photographs (AS 11-42-6310) suggests that it continues north-westward (arrow Z) and connects (arrow XX) with a rille that breaches a dark-haloed volcanic crater (arrow YY).

If the rille segments listed are (or were) part of a continuous rille, then their relationship with other topographic forms indicates a structural origin for the rille. In particular, the grabenlike rille (arrow ZZ) is shown in Apollo photographs to predate the formation of the rille segment marked by arrow Z. The formation of the rille (arrow Z) as a surface flow from the breached crater is inconsistent with this time sequence. In addition, the tangency with the crater (arrow Y) and the abrupt directional change (arrow XY) at the rim of the grabenlike rille are inconsistent with surface flows. These relationships also seem inconsistent with structurally controlled surface or subsurface flows because they require such unusual circumstances.

Plate 178d shows a region adjacent to Rima Bradley (arrow X), which is adjacent to the Apennine Mountains. Three elongate dark-haloed craters (arrows Y, Z, XX) form end members of interconnected rectilinear rilles. The over-all trend of the rille and the elongate craters are approximately on-axis with a flat-floored rille (arrow YY) that connects (connection not shown) with Rima Bradley.

If this rille complex was formed by structurally controlled flows of extrusions, then it appears that the material flowed from one crater to another and eventually drained. More likely, the rilles attained morphologic expression as structural features, although the dark-haloed craters represent volcanic vents through which ash had been erupted.

PLATE 178

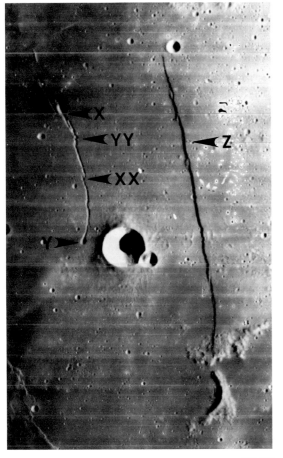

a. IV-113-H2,H1   12 km

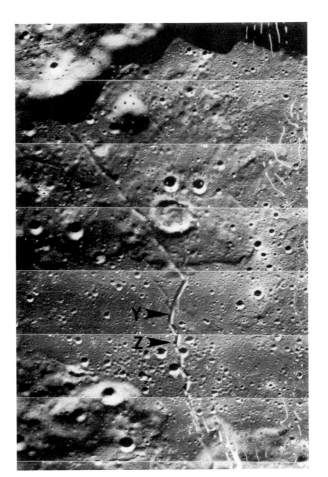

b. V-168-M   5.7 km

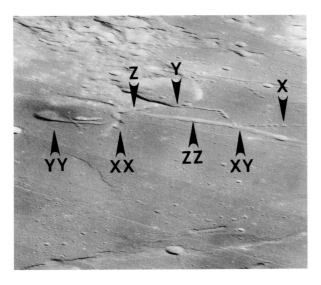

c. V-042-M   (oblique)

d. IV-109-H3   12 km

Rectilinear Rilles

The Alpine Valley, shown in Plate 179a, extends approximately 190 km across a hummocky region separating Mare Imbrium from Mare Frigoris. Along the southwestern section it divides a mountainous region, the Alps, shown in stereo in Plate 179, b and c. The valley floor is cut by a rille that extends most of its length and has a roughly rectilinear plan. As is commonly known, terrestrial fault lines rarely follow a straight line, and as an analogy one need only to trace a crack in cement.

The Alpine Valley type of structure is not unique. In Plate 2a, a similar feature is shown that extends from a large concentric crater near the south pole. It also appears comparable to a raised-rimmed valley—Bouvard—extending from the Orientale basin; a short rillelike structure on the valley floor adds to this comparison (see Plate 209). The obvious difference is that the Alpine Valley has been filled with plains-forming materials.

The raised rim displayed by such broad valleys possibly indicates that the entire region was raised as a blister. Along the Alpine Valley, the rim appears to be composed of individual or coalescing mountainous blocks. Along Bouvard, it appears more like a levee, and the floor is striated by an inferred on-rush of material associated with the formation of the Orientale basin. Close examination of the northeastern end (arrow X) of the Alpine Valley reveals similar subdued lineations. Consequently, it is suggested that the large structures represented by Bouvard and the Alpine Valley are grabens that formed as a result of tension developed by the initial doming of these enormous basins during their formation and that acted as channels for enormous ejec-

ta flows. Hartmann (1963, 1964) reviewed pre-Orbiter observations and interpretations and arrived at a similar conclusion.

The Alpine Valley is crossed by several NNW-trending features and is surrounded by volcanolike structures. In particular, a NNW-trending depression (arrow Y) near left center of Plate 179a can be traced across the floor of the valley and farther to the right (arrow Z). In the same region, a wrinkle ridge (arrow XX) extends across the floor and across the NW highlands (possibly associated with the dark-haloed crater [arrow YY] about 12 km from the NW wall). At this location of crossing features the rille exhibits a "kink." Thus, such regions have been affected by later structural trends and associated volcanic activity.

Plate 179, b and c, is a stereo pair showing the southwestern end of the valley. The fracturelike structure just to the northeast of the gap (arrow X) may represent the head crater for the rille, and a flowlike feature (arrow Y) extends from the gap into the adjacent depression. The stereo view should be compared to Plates 179a and 180. In particular, note the crater-topped peaks (arrow Z and adjacent to X) and the tephra-ringlike structure (arrow XX).

The rectilinear rille on the floor of the Alpine Valley may represent a tension fracture related to continued uplift. The rille on the hummocky floor of Bouvard supports this interpretation. However, the inundation of the floor of the Alpine Valley by a marelike unit and the existence of a craterform near one end suggest an origin comparable to that for sinuous rilles. Further discussion of these alternatives accompanies Plate 180.

PLATE 179

a. V-102-M (oblique)

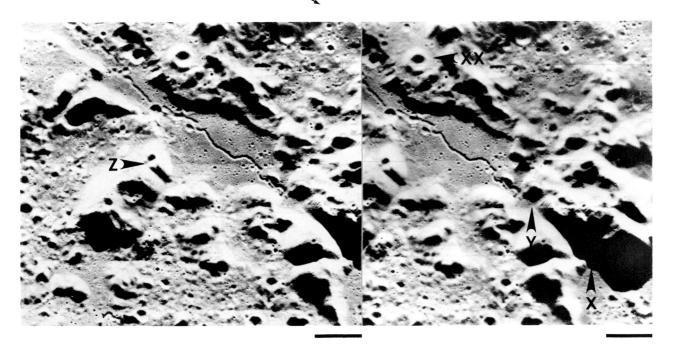

b. V-131-M   8.2 km

c. V-129-M   8.0 km

Rectilinear Rilles

The oblique view of the Alpine Valley provided by Plate 179a is shown at higher resolution in Plate 180a, with selected enlargements in Plate 180, b through e. Plate 180b shows in more detail the southwestern end of the Alpine Valley and the possible head crater (arrow X). The floor of the inner rille typically has an abrupt—and in some regions a moatlike—contact at the rille wall (Plate 180c, arrow X; Plate 180d, arrow X). Similar floor-wall transitions and floor material are identified within craters in the region (Plate 180c, arrow Y). The crater shown in Plate 180e (arrow X) is almost completely filled with material, and its floor is surrounded by a doughnutlike ring in contact with the wall (arrow Y). These craters typically display low rims and are surrounded by low-albedo areas relatively devoid of small craters. The surface is similar to the floor of Hyginus (Plate 160), and craters with similar floor-wall contacts are shown in Plates 16c and 152, a and b. Small craters on the floors of the rille are typically highly subdued and less numerous than craters outside the rille.

The rille- and crater-floor material is interpreted as the result of either catastrophic blanketing—for example, local eruptions of pyroclastics—or processes associated with the formation of the rille and craters. The sharp contacts at slopes seem inconsistent with secular erosional processes (see the discussion with Plates 16c and 152, a and b). The fact that both the rille and crater floors exhibit similar floor units seems to favor regional blanketing. However, the almost completely buried crater shown in Plate 180e (arrow X) requires either that such blanketing was very extensive in certain restricted localities or that this craterform was the result of collapse. The absence of similar fill material within some of the subdued craters seen near the center of Plate 180a suggests either that these craters were, in part, responsible for the inferred local blanketing or that they postdate this period.

Plate 180d isolates a tephra-ringlike feature (arrow Y) just above the valley, shown also in stereo in Plate 179, b and c. One could argue that it is the result of an ancient impact on top of a mound, but more likely it is the result of volcanic processes.

There appear to be a few subdued dark-haloed craters along the valley, but local changes in illumination may be misinterpreted as albedo differences. In particular, note the crater at the foot of the northwestern wall of the valley shown in Plate 180c (arrow Z; also 180a, arrow X). The albedo contrast in this case has been confirmed in other views of this region under different illuminations. The subdued textured topography and the low al-

bedo are similar in appearance, and perhaps origin, to the dark-haloed volcanic craters on the floor of Alphonsus (Plate 187). The edge of the valley floor is a likely site for such eruptions owing to the adjacent fault scarp and the discontinuity represented by this contact. The adjacent wall apparently has advanced over a portion of this deposit but diffuse mantling of the low-albedo material by talus apparently has not occurred. This suggests that the eruption was relatively recent (after mare emplacement) and that degradation of the scarp has occurred primarily by scree advancement rather than by degradation by meteoroid impacts, which spreads scarp material across the valley floor.

As the valley is traced to the northeast (downward in Plate 180a), sharp contacts are recognized between the confining walls and floor (arrow Y). At several locations, these contacts are marked by a moatlike depression (arrows Y and Z), which is separated from the floor by an apron with a convex profile. The apron around the edge of the floor of the Alpine Valley is interpreted as a previous level of the floor.

Thus, it seems that, after emplacement of the floor of the Alpine Valley, subsidence occurred. The inner rectilinear rille was formed during or soon after floor emplacement and was followed by cratering events—probably secondary impacts—that interrupt its plan near the center of Plate 180a. The rille is interpreted as a channel for the valley plains units that were derived from either Mare Imbrium, through a breach in the Alps Mountains, or craterforms near the southwestern end of the valley. The flowing molten material was structurally controlled by a tension fault developed during the formation of the Alpine Valley. It is plausible that some of this material drained into Mare Frigoris and that this resulted in floor subsidence. Pyroclastic eruptions occurred during the last stages of floor emplacement and were, in part, responsible for the mantle indicated on the floors of the rectilinear rille and shallow craters. Local eruptions of plains-forming units and pyroclastics are indicated outside the valley floor, as well (Plate 179a). More recent activity is suggested by transecting ridges (Plate 179a, arrow XX). It is believed that loosely consolidated material presently composes the floor of the rectilinear rille and nearby craters and that this material was derived from both the late-stage pyroclastic eruptions and the ejecta from nearby impacts. The sharp contact with the confining rille wall is perhaps a result of continued regional seismic activity that triggers mass wasting and possibly flowlike migration within the rille.

PLATE 180

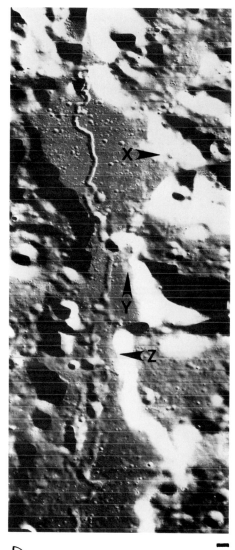

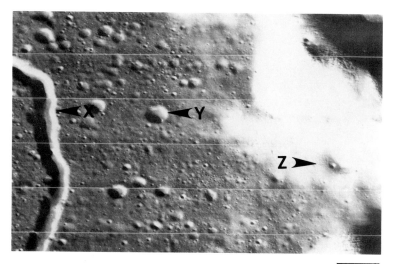

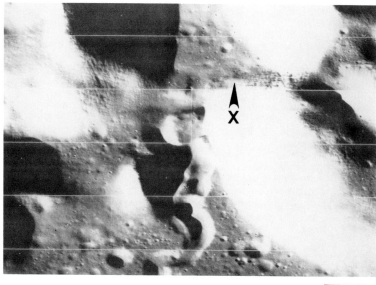

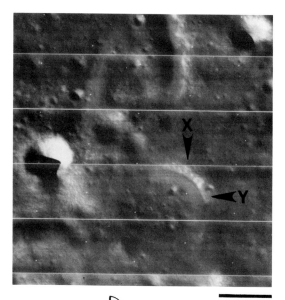

a. V-102-H2,H1   1.6 km (*top*)
1.3 km (*bottom*)

b. V-102-H2   1.3 km

c. V-102-H2   1.3 km

d. V-102-H2   1.2 km

e. V-102-H1   1.1 km

Sinuous Rilles and Mare Ridges

As has been shown in several photographs, many rilles have raised rims or levees. In addition, several display a plan that could be matched by coalescing depressions rather than by surficial flow of some unspecified material (see Plate 173c). This observation is also discussed with plates displaying mare ridges, where several superposed craters appear to be genetically related to the ridge-forming process. Plates 181–182, a and b, illustrate additional evidence for a genetic relationship between some rilles and wrinkle ridges.

Plate 181, a and b, shows in stereo an area near the crater Gruithuisen in which numerous elongate craters occur on the crests of ridge segments (arrow X, which corresponds to arrow X in Plate 181c). The ridge segments are separated by wider elongate craters that typically have narrow but well-defined rims parallel to their major axes (Plate 181c, arrows Y and Z). These craters clearly scallop the adjacent positive-relief forms (Plate 181c, arrow XX). Terraces are recognized within several craters (for example, Plate 181c, arrow YY).

A large crescent-shaped depression forms the northernmost extension of this complex (Plate 181, a and b, arrow Y). It exhibits a greater depth relative to the surrounding surfaces than that of the aligned craters to the south. Two small wrinkle ridges extend southwestward from the depression (arrows Z and XX). A shallow flat-floored depression (arrow YY) extends parallel to the eastern branch of the crescent-shaped depressions. Its south-facing wall is not easily recognized, owing to the direction of solar illumination, whereas its north-facing wall and adjacent rim are clearly identified. An irregular-shaped depression (arrow ZZ) marks an abrupt change in direction toward the south that is revealed by a rimmed arcuate depression (arrow XY).

The southern extremity of the complex appears to be a wrinkle ridge (Plate 181, a and b, arrow XZ). The interconnecting depressions (arrow YX) and the branching wrinkle ridge (Plate 181c, arrow ZZ) also are thought to be related structures.

These observations prompt several comparisons to sinuous rilles that are illustrated in this chapter. The large crescent-shaped depression seen to the north resembles in plan, profile, and location the depressions that occur at the high-elevation end of several less sinuous rilles—for example, the Hadley rille (Plate 174a, arrow X). The low-relief rims of the elongate craters

and the arcuate depression (Plate 181, a and b, arrow XY), are also similar to the rims of rilles. In addition, the broader flat-topped rims of several elongate craters resemble those along a segment of Rima Plato (Plate 173c).

Consequently, the example shown in Plate 181 can be interpreted as an incipient rille. The positive-relief forms that separate the elongate craters are believed to be remnants of once-continuous wrinkle ridges, parts of which remain relatively intact (Plate 181, a and b, arrow XZ; Plate 181c, arrow ZZ). This relationship is apparently not unique; in particular, note in Plate 119, a and b, the association between an elongate depression and a wrinkle ridge. The elongate craters indicate later collapse that at several locations has engulfed most of the pre-existing ridge. This incipient rille may represent partial collapse of a lava tube (suggested by Oberbeck et al., 1969). Note the fanlike feature (Plate 181c, arrow XY) that is characterized by elongate craters and furrows extending from the wrinkle ridge (arrow ZZ). This may represent a flow unit that flowed through the tube branch, the furrows representing small collapsed tubes. There are numerous sets of subdued small-scale (100-m widths) furrows in this region. Although some of the furrows can be attributed to secondary cratering, a large proportion is believed to be remnants of flow patterns. The association between rimmed pits and wrinkle ridges is also consistent with the interpretation of Fielder (1965) that wrinkle ridges are produced by fissuretype eruptions. The aligned depressions may be eruptive centers or collapse pits.

The region shown in Plate 181, a and b, displays a variety of morphologic features that are discussed in other chapters. In particular, refer to the Gruithuisen domes (Plate 130, a and b) and the calderalike depressions with multiple-domed floors (Plate 25). Each of these plates displays filled craters, craters with radial ridges, irregular depressions, high-relief domes, and rilles. Also note the small bulbous-rimmed pit seen in Plate 181c (arrow XZ) and the larger dark-haloed pit in Plate 181, a and b (arrow YZ), both of which are interpreted as volcanic vents. The ridge associated with the crater chain (Plate 181, a and b, arrow XZ) can be traced southward to a very complex region, including the Aristarchus Plateau and the Harbinger Mountains, both of which display numerous rilles.

408

PLATE 181

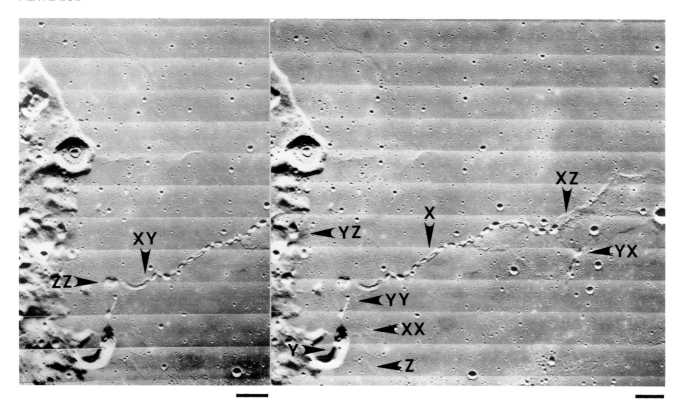

a. V-184-M   5.7 km

b. V-182-M   5.7 km

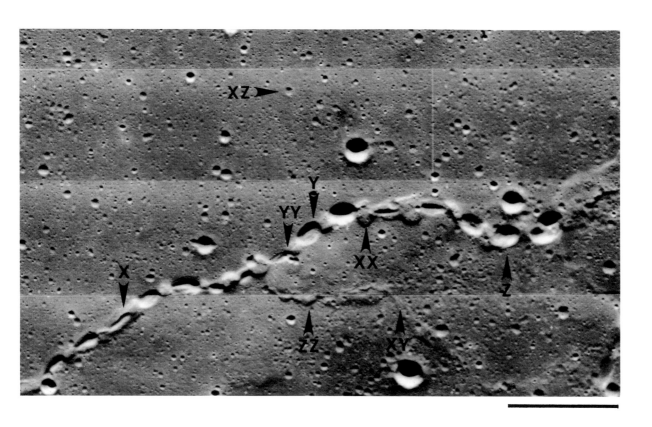

c. V-182-M   5.7 km

Sinuous Rilles and Mare Ridges; Incipient Rilles

Plate 182, a–c, shows further evidence for a genetic relationship between wrinkle ridges and sinuous rilles. The rilles shown in Plate 182a are northeast of Gassendi on the border of Mare Humorum and Oceanus Procellarum. These and several nearby rilles (not shown) trend between Oceanus Procellarum and Mare Humorum. It is difficult to determine how many separate rilles are shown in Plate 182a, but, if a head depression is used as a criterion, then there are five. Each head crater or depression is on a wrinkle ridge, and most appear to be associated with the narrow ropy-ridge component. The elongate depressions in the upper right (arrows X and Y) parallel the wrinkle-ridge trend, whereas that in the upper left (arrow Z) is perpendicular to the ridge trend.

The location of the "head" craters on the wrinkle ridges is consistent with the interpretation that these wrinkle ridges represent fissurelike vents and that sinuous rilles are lava channels or tubes. The same conclusion was reached by Young (1972) from examination of Apollo 16 photography. Although several segments of these rilles appear to be structurally controlled (arrow XX), the meanderlike patterns (for example, arrow YY) and the absence of offsets suggest that the theory of solely a structural origin is unreasonable.

Plate 182, b and c, shows a rille with a raised rim (arrow X) that makes a transition into a mare ridge (arrow Y). The region shown is in northeast Mare Serenitatis to the northwest of Littrow. Although the photographic quality is poor, the transition from rille into ridge apparently occurs without a crater chain. The rille shown in Plate 182, b and c, might represent a later stage in development (either active or arrested) than the feature shown in Plate 181, a and b. Alternatively, they are end products of entirely different sequences of events. For instance, the rille shown in Plate 182, b and c, may have undergone complete collapse before drainage of the lava tube had terminated, or it developed a roof only along the downstream segments.

Plate 182d shows a nonsinuous branching rille near the Sulpicius Gallus rilles (beyond left edge of area shown in Plate 166a). The deep portions of the rille follow shallow U-shaped rilles that cross the platformlike feature near the top. The rille to the left extends onto the mare within a projection of the platform (arrow X). To the right, the shallow rille crosses the ridge (arrow Y), and another extension is traced due south (arrow Z). The latter extension may be related to small-scale parallel lineations (visible in the original photographs) that approximately follow valleys into the highland peaks. In addition, note the rooflike remnant within the rille where it bends toward the south (arrow XX).

The rilles seen in Plate 182d are given three possible interpretations:

1. The shallow rilles represent uncollapsed portions of lava tubes, which extend uncollapsed to a source in the highlands.
2. The shallow rilles are filled segments of a channel or structurally formed crevice.
3. The deep portions of the rille are late collapse forms along pre-existing structural trends represented by the shallow rilles. These also may have been centers for extrusions.

Origin as a collapsed tube is supported by the lobate extensions of the platform along the branches of the rille. But there remain several objections. First, the remnants of the collapsed roofs are absent on the floor of the deep segments. This might be rationalized if the deep segments were unroofed channels. Second, the very subdued SE extension crosses the ridge at right (arrow Y) and may connect either with Rima Sulpicius Gallus I and/or another system farther to the southeast on the mare. These connections, although obscure, suggest a structural origin. Third, the borders of the platformlike feature do not appear to be flow termini. Mare material appears to have spilled over the platform borders (arrow YY) in the upper right (visible on the original photograph), resulting in the erasure of this portion of the scarp.

The region surrounding the Sulpicius Gallus rilles shows numerous collapselike features, including incipient rilles (see Plate 166, a and b). One rille has a head crater similar to that of the Hadley rille. Consequently, one's elimination of the rille shown in Plate 182d as a structurally controlled collapsed lava tube or channel should be delayed until better photography is available.

Note the almost totally buried crater (arrow ZZ) at the base of the ridge at the right of Plate 182d. As shown in Plate 166, this is a common occurrence along the contact between mare and the low-albedo shelf in this area. It is an indication of the unconsolidated and mobile nature of the material.

Plate 182e shows a rille complex in Oceanus Procellarum to the southeast of the crater Marius. The rille identified by arrow X is continuous and connects with the elongate depression, which has a small interior crater (arrow Y). Another rille extends to the north of this depression (arrow Z), and a complex array of short rilles occurs farther to the north, with an over-all N trend (arrow XX). The northernmost extension of the disconnected rilles appears to form a continuous sinuous rille (arrow YY). The Marius Hills region exhibits another possible incipient rille (Plate 265, a and b, arrow ZZ), and further evaluation of such features as collapsed lava tubes has been made by Greeley (1971b) through comparisons with terrestrial analogs.

The N-trending association of short rilles is interpreted as an incipient rille related to the elongate depression to the south. It does not appear to be associated with any positive-relief forms. If one attempts to interconnect these rille segments, he must either construct a very sinuous plan or concede more than one rille.

410

PLATE 182

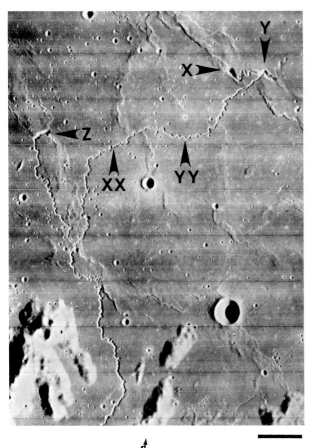

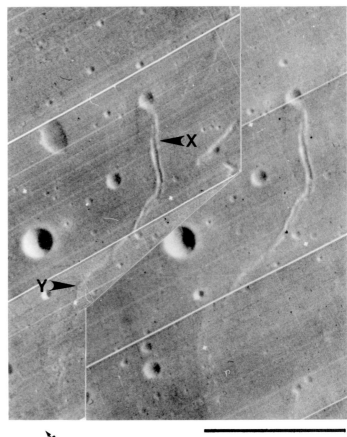

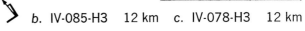

a. IV-137-H2   12 km

b. IV-085-H3   12 km   *c.* IV-078-H3   12 km

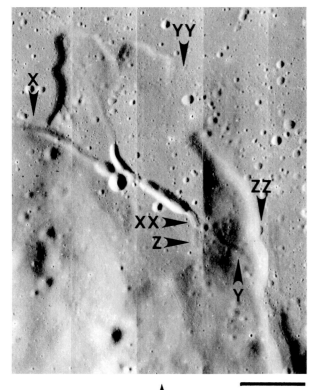

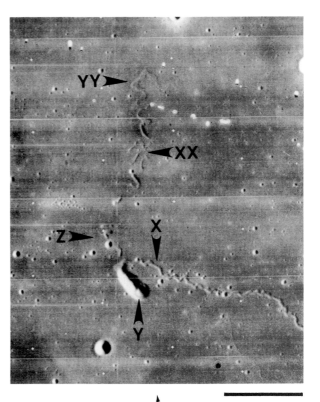

d. V-092-M   3.9 km

e. IV-150-H2   12 km

DISCUSSION AND SUMMARY

Rille is a nongenetic term encompassing a variety of linear negative-relief forms that are categorized by their plan, continuity, profile, and morphologic features, which include crispness of detail and associated features.

Most curvilinear nonsinuous rilles are fractures of the lunar crust, and many have flat floors that represent dropped surfaces. Alignments and groupings of non-sinuous rilles over large areas reflect large-scale release of stresses. For example, Mare Humorum and Mare Serenitatis display concentric systems of rilles that generally fall into two classes:

1. Large arcuate grabenlike rilles that cross both mare and highland terrains along or beyond the periphery of the maria (see Plate 167c)
2. Small isolated rilles that are typically inside the mare-highland contact, display **V**- or **U**-shaped profiles, and commonly have low-albedo deposits (for example, Plates 138a, 165, a, b, and d, and 166)

The first class is well developed around Mare Humorum, whereas the second is best displayed in Mare Serenitatis.

Grabenlike rille systems are associated with other mare-filled basins, but the rilles with **V**-shaped profiles are commonly absent. In general, the irregular-bordered maria, such as Maria Procellarum, Frigoris, and Vaporum, do not have large concentric systems, although smaller interior rilles are recognized. Mare Crisium also appears to be without such systems. The following list summarizes the annular systems around other maria:

1. Mare Tranquilitatis exhibits an annular grabenlike rille system on its western margin, but the component rilles generally do not extend into the highlands. Smaller peripheral rilles are not well developed.
2. Mare Fecunditatis and Mare Nectaris have an extensive set of grabenlike rilles that cross highland terrain between them. The poor quality of available photography prevents a statement about the smaller rilles with **V**-shaped profiles.
3. Mare Imbrium has several annular grabenlike rilles, but they are generally restricted to the interior of the outer mountainous ring. Many have been partly inundated by later mare flooding. Annular nonsinuous rilles with **V**-shaped profiles are not abundant.
4. Mare Nubium has several rilles that appear to be part of an annular system.
5. Mare Orientale has an annular system of grabenlike rilles that cross the crater Riccioli. Within the inner mountain ring, there is also a complex array of narrow rilles, some of which are part of an annular system.

The concentric arrangement of nonsinuous rilles with respect to mare-filled regions indicates subsidence. Hartmann and Kuiper (1962), Baldwin (1963), Quaide (1965), and Ronca (1965) have proposed that gravity settling of the maria produced tension fractures along the hinge zones of crustal subsidence. The preceding observations, however, suggest that there were several separate zones of tension. The grabenlike systems outside the mare-highland contact reflect an outer hinge line of gradual down warping, whereas the systems of rilles having **V**-shaped profiles within this contact correspond to hinge zones related to more rapid subsidence of mare plates. In Mare Serenitatis and Mare Smythii, a period of rapid subsidence is believed to correspond to volcano-tectonic collapse along boundary faults. The branching rilles in the shelf regions of Mare Serenitatis (Plate 166) are interpreted as remnants of this phase during which erupted low-albedo ash mantled the mare borders. Subsequent lava flows have hidden, in part, the total displacement associated with this collapse. Such a collapse history is responsible for the circularity of several old basins. Later fracturing produced **V**-shaped rille profiles on the maria exterior to the compressionally formed wrinkle ridges. These fractures also were frequently accompanied by volcanic eruptions through linkage to the cooling magma chambers responsible for mare flooding (for example, Mare Humorum: Plate 191).

In general, the older irregular and less circular basins are regions of relatively shallow mare flooding (see page 462). The mare units are suggested to have been extruded through a network of crustal fractures in a sequence similar to that for smaller-scale floor-fractured craters. Early periods of upwelling and subsidence produced annular grabens, followed by gradual isostatic adjustment producing additional systems of postmare grabens. These ancient basins, therefore, exhibit small mass concentrations and extensive systems of annular grabens.

As suggested previously, however, several old basins underwent volcano-tectonic collapse. Prior to this event, a system of annular grabens was produced by crustal displacements related to periods of mare eruptions, but basin collapse and subsequent eruptions largely erased this earlier history. In addition, subsequent partial isostatic adjustment was confined, for the most part, to the interior mare plate, and peripheral grabens failed to develop. These rejuvenated and expanded basins exhibit large mascons as a result of increased volumes of erupted mare basalts filling the voids and perhaps as a result of a direct link to the lunar mantle, as suggested by Wise and Yates (1970). This appears to be analogous to the large positive gravity anomalies associated with terrestrial calderas that have erupted large volumes of basalt (Williams and McBirney, 1968).

Provided that volcano-tectonic collapse does not occur, the older less circular mare-filled basins will display both small mascons and systems of grabenlike rilles. Basin collapse, however, results in large mascons—regardless of basin age; furthermore, it may erase previous graben systems and perhaps prevent their later formation.

Not all systems of nonsinuous rilles are associated with mare-filled basins. Plates 163–164 illustrate a large parallel system that postdates the formation of the Orientale basin but predates mare flooding. Another parallel system occurs near the southeastern limb of the lunar nearside and is believed to be related to large-scale weaknesses perhaps associated with a pan-lunar belt of maria (see Discussion and Summary at the end of Chapter 9). Some of these systems may reflect tidal stresses or the redistribution of mass expressed by mare flooding.

Numerous nonsinuous rilles were associated with volcanic activity, some of which postdate local mare flooding. Examples are:

1. Rima Hyginus (Plates 159–161): calderalike depression at the intersection of two rilles
2. Rima Parry V (Plate 131): volcanic constructional form parallel to rille
3. Rima Hevelius (Plate 164): fissure eruption with associated flow unit, pyroclastics, and maarlike craters
4. Rimae Sulpicius Gallus (Plate 166): pyroclastic eruptions

Most of these volcanic centers indicate eruptions of ash, which implies gas-rich magmas. This is probably related to evolution of the magma chambers, perhaps on a Moon-wide scale.

Sinuous rilles are divided into two subclasses, more sinuous (meandering) and less sinuous (nonmeandering). These rilles typically occur on mare surfaces, although several cross highlandlike terrain. Schubert et al. (1970) and Murray (1971) have shown that they are distributed along the margins of mare-filled basins.

General origins of sinuous rilles have been proposed by various investigators. Lingenfelter et al. (1968) and Urey (1967) suggest erosion by water. Cameron (1964) credits erosion by volcanic ash, whereas Strom (1966), Oberbeck et al. (1969), Greeley (1971a), Murray (1971), and Cruikshank and Wood (1972) believe fluid movement of lavas have produced large-scale lava channels and/or lava tubes. Schumm (1970) prefers outgassing along fractures, and Fielder (1965) and Quaide et al. (1965) propose collapse associated with intrusives along structural weaknesses.

As in the dilemma on the origin of wrinkle ridges, the attempt to assign a single process for the origin of all sinuous rilles does not appear to be realistic. Such an approach, however, does provide a better understanding of the problem. Several selected features observed in the preceding plates of this chapter are consistent with rilles produced solely by structural weaknesses:

1. Crossing of positive-relief forms and nonmarelike surfaces
2. Alignment of the general rille plan with regional structural trends
3. Extension of rille into "pinched" craters and crater chains

413

4. **V**-shaped or flat-floored profiles
5. Meanderlike pattern created by offset collapses and crater coalescence
6. Apparent genetic relationship to wrinkle ridges, such as those near Gruithuisen (Plate 181)
7. Difference in elevation of one side of a rille with respect to the other

A solely structural origin, however, is inconsistent with other features of sinuous rilles:

1. Most sinuous rilles have craters or elongate depressions at one end that represent the widest and deepest portion of the rilles and typically, if not always, occur at the highest relative elevation.
2. Typically, sinuous rilles have rims.
3. Many sinuous rilles appear to be obstructed by positive-relief forms.
4. Many sinuous rilles extend between a crater or depression in the highlands and the maria.
5. Most sinuous rilles extend for enormous distances without offsets, typical of some structural features.
6. At least one sinuous rille is connected with irregular depressions that appear to be subsided pools, rather than structurally dropped blocks (see Plate 177b).
7. Facing walls of sinuous rilles typically are not mirror images but have differences in curvature; for example, a concave wall element commonly faces a spurlike projection of the opposite wall.

The interpretation of sinuous rilles as collapsed lava tubes is consistent with several observations. "Head" craters and elongate depressions are interpreted as sources of lava. Incomplete collapse is perhaps represented by such examples as the offset pits along the wrinkle ridge shown in Plate 181. Plate 119, a and b, also shows an elongate depression connected to a wrinkle ridge similar to the relationship between "head" crater and rille. Consequently, it is plausible that some wrinkle ridges are uncollapsed lava tubes, and the rims observed along numerous rilles can be interpreted as the remnants of the roof. Further evidence for incomplete collapse is the existence of discontinuities in the path of rilles, such as the Hadley rille, shown in Plate 175, a and b. Similar examples are found in the Har-

binger Mountains region (Plate 256), near the Marius Hills (Plate 182e), and near the Sulpicius Gallus rilles (Plate 166b, arrow YY).

As in the exercise of claiming solely a structural origin for sinuous rilles, one encounters inconsistencies in the interpretation of all sinuous rilles as lava tubes. The most inflexibile inconsistency is the existence of sinuous rilles that cross highland topography and positive-relief structures. Lava tubes develop within flow units and do not cut subsurface channels through other terrains. Thus, such an interpretation is clearly not applicable in all cases. A less severe objection is the apparent structural control of several rilles. This has been observed in the formation of terrestrial lava tubes (Harter and Harter, 1971).

Although several rilles shown have not completely collapsed, most rilles are remarkably continuous over large distances, and there is little evidence for remnants of roofs on rille floors. If sinuous mare rilles are collapsed tubes, mechanisms were required, therefore, to initiate collapse and obscure or remove the remnants. Six possibilities for initiation of collapse are:

1. Collapse prior to the last stages of flow owing to insufficient support
2. Collapse due to internal spallation of the roof (Oberbeck et al., 1969)
3. Collapse due to erosion by subsequent flows
4. Collapse triggered by large seismic events
5. Collapse due to overburden from later blanketing by ash or thin lava units.

The first suggestion engenders arguments in definitions: whether such rilles should be termed "surface channels" or "subsurface tubes." It and the second suggestion raise the question, however, on the requirements for roof suport. Oberbeck et al. (1969) considered earlier work by Nichols (1946) and concluded that, under the reduced lunar gravity, a maximum 500-m-wide vesicular roof could be supported if its thickness equals $\frac{1}{8}$ the inner tube width. The parameters, however, are near the limits for those found for terrestrial lava tubes, and the formation of rilles 1 km in width requires significant postcollapse widening through erosion. Their calculations used thick tube roofs, which are possible under proper cooling conditions. On the Moon, however, cooling should be very slow. As a result, the lava should retain its low viscosity, increasing the likelihood for rapid drainage from a tube and thereby encouraging the formation of thin roofs. These arguments suggest that such roofs probably were fragile and collapsed relatively soon after formation.

The absence of roof remnants on the floors of sinuous rilles could be due to three processes:

1. Erosion and deposition by meteoroid bombardment on the roof remnants and rille walls
2. Blanketing by regional deposits
3. Subsequent surface flows following the collapsed tube

The first suggestion is not always applicable. If sinuous rilles represent once-subsurface tubes, then, as surface expressions, they obviously are contemporaneous with, or are later than, the mare within which they formed. Since that time, surface processes have produced regolith depths on the order of 6 m in the maria. This does not seem sufficient to destroy large blocks of roof remnants. The floor of a rille should act as a depositional trap for ejecta from the small-but-continuous influx of meteoroids. If the variety of rille profiles is explained by such a process, then an enormous range of ages is implied by rilles having similar widths but different profiles, e.g., flat floored versus V shaped. An even greater contradiction results from the same rille's having a variety of profiles along its length. Thus, blanketing of the rille floor by small-scale meteoroid ejecta is not a reasonable explanation for the variety of profiles and the burial of roof remnants.

The various profiles could reflect differences in roof remnants and tube geometry. For example, flat-bottomed profiles result from grabenlike collapse of a relatively thick intact roof or from collapse into a wide tube. Sinuous rilles with V-shaped profiles result from collapse of thin roofs into deep tubes. Subsequent smoothing by meteoroid bombardment then might account for the destruction of remaining irregularities. Many sinuous rilles have facing-wall heights that decrease toward the termini, and typically the rille merges with the maria. From this reasoning, one might propose that the roof

becomes thicker or the tube becomes gradually elongate and more shallow "downstream," but I believe there are other more reasonable explanations, yet to be discussed.

The suggestion that regional blanketing of ash is responsible for burial of roof remnants, although not always applicable, is a plausible alternative. The numerous flat-floored craters along Schröter's Valley and in several regions having low albedo show that this configuration may result from such blanketing.

Numerous flat-floored rilles have narrow inner rilles that suggest resurgent activity. The fact that such narrow features have survived indicates either that they are very recent structures or that they have been relatively well preserved. The latter alternative is further evidence that flat-floored profiles are not the result of subsequent degradation. Resurgent activity also provides a mechanism that can remove or bury roof remnants; produce a variety of rille profiles, depending on the local flow velocity and supply of lava; and modify the rille plan. Multiphase development of tubes and channels, having been observed in terrestrial volcanic fields (Greeley, 1971c), is not an unfamiliar phenomenon.

The previous discussion leads to the suggestion that some sinuous rilles formed as surface channels for molten lava, and several features are consistent with this interpretation:

1. The absence of roof remnants is easily explained because they never existed.
2. Like lava tubes, the head craters or elongate depressions are sources for the flow material.
3. Rims of sinuous rilles are explained as levees constructed by spattering and spillover, as observed in terrestrial channels. Scott et al. (1971) have noted a good example near Lansberg (AS14-73-10121).
4. Rille profiles are interpreted as a function of local gradient and lava supply.
5. The relation of at least one rille to a lava-lake-like depression (Plate 177b) is readily understood as a drainage channel.

Lunar lava channels and tubes have common departures from terrestrial analogs. One of the most obvious is the enormous linear extent, but this departure can be attributed to the low viscosity of lunar basalts.

Furthermore, terrestrial channels and tubes generally are not regarded as the results of erosion; rather, they describe the path of most rapid movement within a larger flow unit. Consequently, their development is, for the most part, different from river channels that erode and deposit the surface on which they flow. This leads to the discussion of two properties of several rilles. First, several sinuous rilles develop plans comparable to meandering rivers on Earth. Second, several rilles do not reside within lava units.

Tight meanders, including ox-bows and meander cut-offs, occur in a few terrestrial lava channels and tubes (Greeley, 1971a). However, none of the terrestrial lava tubes and channels have meanders comparable to the inner rille of Schröter's Valley. Nor do they exhibit the width and relatively short meander wavelength that characterize Rima Prinz II (Plate 259). More typically, these terrestrial features are described as sinuous but nonmeandering, except for a few meander loops or cut-offs. Lunar conditions may favor development of meandering channels. The slow rate of cooling—surmised because of extremely low heat transfer—probably results in molten lava beneath an insulating layer of degassing froth, which is supported by the still-molten lava; consequently, the path of most rapid transfer of materials is a roofed tube but perhaps not in the usual sense. The channel or tube formed at this stage could readily adjust itself to a meandering plan, depending on the local gradient, by adjustments between the more viscous and less viscous fractions of the lava pool. The latter component is maintained by a continuous supply of lava from the source, presumed to be the head crater or depression. The process is remotely analogous to the development of meandering subglacial streams, the plans of which are revealed by the depositional forms— eskers. Initially, subsurface structures act as guides for later surges of lava that modify the original rille plan and that might eventually produce apparently deep-cut rilles and smaller interior rilles. It is suggested that the absence of roof remnants is the result of settling of the froth layers or of their destruction by these later surges or flows. Tightly meandering rilles are mostly shallow structures and have developed in the later-emplaced units, after the preceding units had reduced the gra-

dient. The absence of terminal deposits is due to burial by subsequent thin units and because the resulting canyon continuously developed, rather than being eroded by a single erosive surge. The origin of long sinuous rilles by this process requires that large portions of the maria be molten. This is consistent with the general absence of well-defined topographic expressions of flow units and the recognition of massive flows in Mare Imbrium (see Strom, 1965; and AS15-M3-1558).

The existence of rilles crossing nonmare terrains is possible evidence that the supposed transporting medium is capable of extensive and continued erosion. Typically, such rilles are sinuous but nonmeandering. The crossing of obstructions, such as wrinkle ridges, provides clues to rille or ridge formation. If the ridge is the pre-existing feature, then either the path of the rille was guided by a pre-existing weakness (perhaps a structurally formed rille) or the on-rush of material was sufficient to carve a channel. Gradual erosion of topography, however, requires slow uplift of structures such as wrinkle ridges. Alternatively, the general or local lava level represented by the present mare surface was once significantly higher, making possible the transection of features currently standing in relief. Terraces interpreted as high-level marks of the mare support this hypothesis. In particular, the rille shown in Plate 259 is thought to have crossed a large massif during such a high-level stage, and a similar situation may be applicable for some wrinkle ridges.

If the formation of more sinuous rilles in the mare plains is attributed to lava flows, then similar rilles in the high-albedo plains units in the highlands (for example, Rima Bode III in Plate 188a, arrow XX) may indicate an origin of some of these units by local extrusions of lava. Further discussion of this possibility is found at the end of Chapter 9.

From these discussions, it is apparent that more than one process has been responsible for sinuous rilles. Scott et al. (1971) suggest that those rilles with levees are the result of fluid flow—not necessarily lavas—and that those without are either subsurface tubes or flow channels. They also suggest that a rille may exhibit both collapsed-tube and surface-channel segments. I propose

416

in addition that highland sinuous rilles are indicative of either rapid surface flows or flows that followed structural weaknesses. Further, tube formation on the maria is inferred from identification of incipient rilles and perhaps from wrinkle ridges that have related head-crater-like forms. It is suggested that some mare rilles develop within large pools of lava capped by a layer of froth.

8. Albedo Contrasts

INTRODUCTION

Albedo is a measure of the reflectivity of a surface relative to a given standard, and various types of albedo result from altering either the standard or the spectral range of the radiation in use. The *Bond albedo* is defined (see de Vaucouleurs [1964] for definitions and discussions of photometric units) as the ratio of the total luminous flux radiated in all directions from a surface to the total flux incident on this surface. Observations rarely use the total luminous flux, owing to spectral limitations of the eye or other photometric instruments as well as to practical difficulties, such as atmospheric transmission. As a result, astronomers define a visual albedo that covers the visual region of the spectrum (0.48μ–0.65μ) and is centered at 0.555μ. Owing to the spectral response of the human eye and instruments constructed to reproduce this response, apparent visual albedo contrasts actually may be indicative of color contrasts.

The term *albedo* is not strictly applicable to light and dark areas on Lunar Orbiter photographs. A single photograph will not give the luminous flux in all directions, and the spectral response of the photographic system is not calibrated to astronomical photometric standards. "Albedo" contrast in the context of this book is strictly the relative *luminance* between two surfaces (see de Vaucouleurs, 1964). If the phase angle (angle between illuminating source and observer) and angle of incidence are small (full Moon view), then it may be safe to state that a "luminance" contrast approximately corresponds to an "albedo" contrast. Most Lunar Orbiter photographs were taken, however, with relatively low angles of solar illumination, and, as a result, sloped surfaces appear to result in possible "albedo" contrasts. The photometric function empirically relates the relative reflectivity to the angles of incidence, observations, and phase, and it is strongly dependent on the physical structure of the uppermost surface layer.

Hence, Lunar Orbiter photographs are inadequate for an extensive classification of surface features based on albedo contrasts, but reference to telescopic photographs under full illumination aids in the distinction between effects expressed by the photometric function and possible albedos. In the following discussions, the term *albedo* is used for general surface contrasts although it should be realized that its application is incorrect. It is used where extrapolation from relative luminances to relative albedos seems appropriate.

Albedo and color contrasts provide parameters to distinguish between different selenologic units that may or may not have topographic expressions. Extensive telescopic work by Peacock (1968) has shown that there appears to be no consistent correlation between color and visual albedo for particular lunar regions. The highlands, however, generally appear redder than the average lunar surface, and these regions also exhibit a higher visual albedo than do the maria. The crater ray patterns, which generally display higher reflectivities, are typically blue. Such color contrasts may indicate different mineralogic units, although the strongest signatures of composition reside in the infrared. In addition, Adams (1967) found that albedo is strongly affected by particle size. Transitions from one albedo or color to another can be a nontopographic indicator of different events, and the preservation or destruction of such contrasts is a possible criterion for the apparent age and surface processes.

Minnaert (1961) has given typical normal (zero angles of incidence and reflection) visual albedos for lunar features from telescopic observations:

1.	Plains (dark):	0.065
2.	Plains (bright):	0.091
3.	Mountains:	0.105
4.	Crater bottoms:	0.112
5.	Crater rays:	0.131
6.	Aristarchus:	0.176
7.	Darkest spot:	0.051

The total range in reflectivity is relatively small, and this should be kept in mind for the discussions and illustrations to follow.

Reiner γ Formation

Reiner γ (Plate 183) is a swirllike complex of high- and low-albedo units in western Oceanus Procellarum south of the Marius Hills (see Plate 263), which can be identified on the horizon in Plate 183a. The oblique view is complemented by a vertical view in Plate 183b. The only topographic associations appear to be a possible rise in elevation and wrinkle ridges, which extend into the Marius Hills, but it is clear from Plate 183b that the Reiner γ unit crosses these ridges. The *Consolidated Lunar Atlas* (Kuiper et al., 1967) provides full-illumination photographs of the region and verifies the impression given in Plate 183a that the higher-albedo deposit is outlined by a unit having a lower albedo than the surrounding mare surface. This outline can be traced into the Marius Hills and appears to be part of the Reiner γ complex.

Reiner γ apparently overlies ejecta blankets from the craters Reiner (Plate 183b, arrow X) and Cavalerius (arrow Y). The latter crater was formed after emplacement of the last local mare units, whereas the ejecta blanket of Reiner has been embayed on its southern edge by relatively recent mare units. Examination of adjacent Orbiter photographs (not shown) and Plate 183a reveals that it also blankets the volcanic complex called the Marius Hills. The variety of units mantled by this formation suggests that it was associated with the formation of a nearby impact crater, perhaps Cavalerius, which has a relatively well preserved ejecta blanket and a diffuse ray system. Alternatively, it might be associated with the bright-rayed crater Ölbers A, which is 400 km to the west and which exhibits an extensive plains unit overlapping the inner rim area (see Plate 105c).

Kuiper (1965) and Whitaker (1969) have discussed the possibility of lunar sublimates in reference to such features as Reiner γ. More recently, El-Baz (1972a) has suggested that such sublimates may have been released through fractures triggered by the formation of a nearby impact crater.

PLATE 183

a. II-215-M   (oblique)

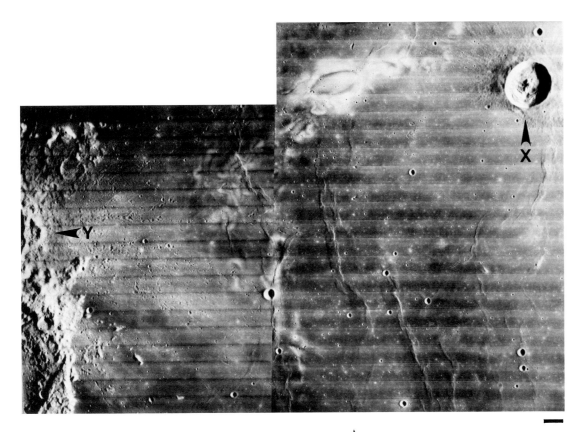

b. IV-162-H1; IV-157-H1   12 km

Reiner γ Formation

The Reiner γ type of feature is not unique and is illustrated further in Plate 184 by two examples on the lunar farside. Plate 184a shows swirls of high-albedo units on the floor of a large (300 km) crater partly inundated by mare units, Mare Ingenii. The albedo contrast is revealed clearly despite the very low sun angles in Plates 183–184. The pattern is complex but appears to have been blocked, in part, by the ringed plains. Although the association of the relief and the high-albedo swirl simply may be coincidental, the inferred topographic control suggests a near-surface flow feature. Note, however, that this unit overlies both mare units and the more hummocky terrain. It appears to extend to the northwest edge of the large crater and may be related genetically to the crater shown in Plate 88a (arrow X), which includes this region.

Plate 184b shows the flooded crater Goddard (arrow X) in northern Mare Marginis. Based on almost full illumination photographs from the Apollo 8 mission, this example was specifically referred to by Whitaker (1969) as a possible lunar sublimate. These photographs reveal that the majority of the high-albedo pattern is distributed over the highlandlike relief below Goddard, which is not apparent in the low sun-angle view of Plate 184b. I prefer, however, the interpretation that this pattern was produced by ejecta from the small circular crater, which is just below Goddard (arrow Y) and which is centrally located in the deposit (Apollo 8 frame AS8-12-2208). This preference is based on a close spatial relation between swirllike deposits and a recently formed crater shown in Plate 109 and the existence of a nearby bright-rayed crater wherever such deposits are found. The complex patterns may be produced by an unusual geometry in the impact (see Plate 109 and the discussion with Plate 105). This cannot be the only cause, however, for impacts at such locations commonly occur without unusual ejecta patterns. An alternative is that these craters represent small cometary impacts in which the encompassing cloud of volatiles interacted with the secondary ejecta from the impact of the nucleus.

PLATE 184

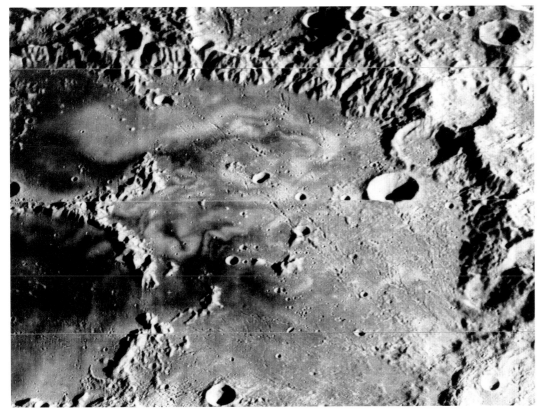

a. II-075-M   (oblique)

b. IV-165-H3   (oblique)

Crater Ejecta

Plate 185 illustrates the variation in albedo associated with craters. Plate 185a shows the bright-ray streamers extending from a small (about 70 m in diameter) crater. The occurrence of high-albedo rays ranges from large craters, such as Copernicus (Plate 94), down to microscopic examples found on lunar samples returned by the Apollo 11 mission. Numerous large pristine craters, such as Tycho, exhibit highly reflective ray systems, whereas others have systems that are more subdued (Copernicus) or are almost entirely absent (Eratosthenes, Plate 106). This conclusion led researchers (Shoemaker, 1962; Hapke and Van Horn, 1963) to suggest that bright-ray systems darken slowly with apparent crater age. Possible ray material of Copernicus returned by the Apollo 12 mission indicates an age of about 1.0 b.y. (Silver, 1971). Consequently, the darkening rate is extremely slow or is offset by bleaching processes.

Plate 185b shows the effect of the angle of incidence—the angle between the vertical and the sun—and the angle of observation—the angle between the vertical and the observer. The crater (arrow X) appears to have a dark halo, but note that this oblique view is in the direction of the illuminating source; that is, both the incidence and the observation angles are large. However, the same crater under a large incidence but small observing angle exhibits a bright halo. This effect is perhaps the result of lighting rather than peculiar ejecta composition. For example, the blocks ejected from the crater may be large, and their shadows control the photometric function under large incidence and observing angles. The relative deficiency of dark-haloed craters under such illuminations does not seem to be the result of an absence of recent craters, and craters with bright haloes under all lighting conditions simply may indicate fine-grained ejecta.

Plate 185c illustrates the brightly reflecting specks on highland terrain (the Apennines) north of Sulpicius Gallus (Plate 166). Close inspection reveals that many of these areas are directly associated with small (less than 20 m) craters (arrow X) and, hence, may indicate their ejecta blankets. Other areas are perhaps either the result of craters too small to be resolved or the result of other processes (arrow Y). Small bright specks blend at lower resolutions to produce local regions of high albedo. Furthermore, these regions typically are restricted to topographic highs, the upper walls of subdued rimless craters, and the rims of subdued dimple craters. The local angle of illumination is an important parameter, but in most examples the high albedo not only is on the side directly facing the sun but also extends around the topographic feature. These specks are interpreted as clusters of high-albedo blocks that have been exposed through the migration of overlying debris. Low-albedo ash is believed to have mantled this particular region (see Plate 166), perhaps during the volcano-tectonic collapse of the Serenitatis basin. Such a blanket provides a distinct contrast to the underlying higher-albedo ejecta from the Imbrium basin.

Plate 185d shows a selected region of Mare Crisium, which includes the craters Picard (arrow X) and Lick D (arrow Y), as well as the ringed plains Lick (arrow Z) and Yerkes (arrow XX). These features display low-albedo haloes and/or ejecta blankets that become much more accentuated under full illumination.

Similar halos are observed around a variety of craters, for example, Dawes (Plate 62a), Reiner (Plate 183b), and Gambart (Plate 91). The low-albedo deposits around such craters appear to be relatively diffuse in contrast to the filamentary structure exhibited by the high-albedo rays. An exception is the dark-ray filaments around Dionysius (Plates 26 and 89, a and b).

The low albedo accompanying otherwise recent appearing craters perhaps indicates compositional differences at depth revealed by impact excavation. This appears to be particularly applicable to the dark-ray system of Dionysius. The diffuse blanket around such craters as Picard, however, possibly represents deposits from an ash flow or fall. This implies a multi-phased history, which might be inferred from the narrow terrace surrounding the floor of Picard. As illustrated in the chapter "Crater Walls" (Plate 60), such terraces are distinct from the larger slump terraces typical of large craters and are suggested to mark previous levels of the floor. The craters Aristillus (Plate 99) and Copernicus (Plate 94) also display irregular low-albedo patches that apparently overlie the bright-ray systems and that could be the result of ejecta melt or later extrusions.

Many bright-rayed craters are dark-haloed craters when examined under full illumination. This is clearly displayed by Tycho and Aristarchus, in which the low-albedo deposit approximately coincides with Zone II of the ejecta blanket (see Plates 101a and 236a). The halo is suggested to be the combined effects of composition and particle size resulting from the impact and its subsequent stratigraphic inversion of pre-existing subsurface material. Eratosthenes (Plate 106a) is a more highly subdued crater without an extensive bright-ray system but with a low-albedo halo. If it originally exhibited a bright-ray system, then the dark halo apparently survived.

Plate 185e shows the crater Lichtenberg in northwestern Oceanus Procellarum. As shown in Plates 85e, 107, and 253, ray patterns around several craters are distributed asymmetrically as a result of oblique impacts or multiple simultaneous impacts. The ray pattern of Lichtenberg has a prominent asymmetry in which the bright rays extend to the northwest within a 130° sector. In this case, however, the inner ejecta blanket on the southeast side of the crater contains plains units that separate portions of the hummocky blanket and that appear to merge with the surrounding mare plains (arrow X). The albedo of this portion of the inner blanket is lower than that of the blanket to the northeast. This low-albedo sector appears to correspond to a dark mare unit extending to the northeast.

These observations suggest that a recent mare unit has buried the bright rays and embayed the inner ejecta blanket. The low albedo of the ejecta blanket is believed to represent mantling by pyroclastics erupted during this emplacement, a phenomenon that is not unique (see Plate 191b, arrow Z). If this interpretation is correct, then either the darkening rate of high-albedo rays is very low or the overlapping mare unit represents a very recent period of volcanism (perhaps less than 0.5 billion years ago). The large bright-rayed crater Manilius (IV-97-H2) also predates the last stages of mare flooding in Mare Vaporum.

PLATE 185

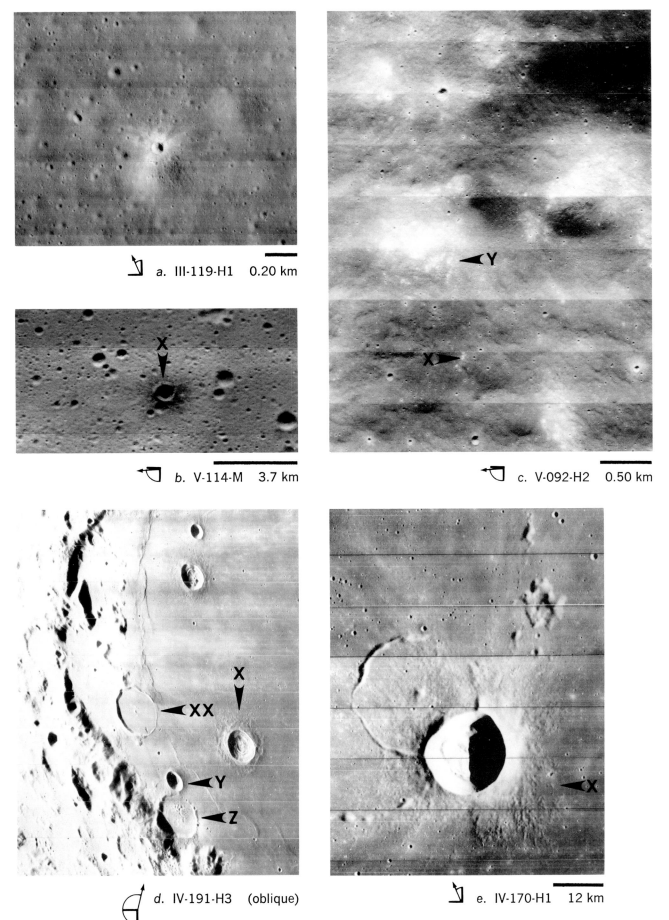

a. III-119-H1   0.20 km

b. V-114-M   3.7 km

c. V-092-H2   0.50 km

d. IV-191-H3   (oblique)

e. IV-170-H1   12 km

Crater Ejecta

Whereas low-albedo deposits around craters shown in Plate 185d could be interpreted as impact or volcanic ejecta, the low-albedo blanket surrounding craters along rilles shown in Plate 186 is most easily understood if the material was derived internally. Plate 186a gives an impressive oblique view of the large backside crater Schrödinger (also see Plate 2a), with an interior mountainous ring, numerous fractures, and a prominent low-albedo crater (arrow X). This dark-haloed crater interrupts a fracture, an association that is typical of terrestrial volcanoes.

Plate 186, b and c, is a stereo pair of the floor of Alphonsus (also see Plates 44 and 72) that shows four dark-haloed craters (arrows X, Y, Z, and XX), which are also associated with fractures. In addition, the craters display a **D**-shaped plan, in which the straight-wall segment aligns with the adjacent fracture. Ranger IX transmitted numerous photographs of other dark-haloed craters along fractures in Alphonsus. Kuiper (1966b) suggested that they are analogous to maartype craters resulting from explosive releases of gas. Rough calculations of the volume of ejecta substantiated this interpretation. Hartmann (1967) examined in more detail the amount of material required to fill adjacent craters 600 m in diameter, which are absent within the low-albedo ejecta blanket. He found that the ejecta volume must have been five times that of the volume represented by the crater and concluded that the craters were a continual source of material. The craters shown in Plate 186, b and c, appear to rest on top of mounds of material that were deposited during the eruption.

PLATE 186

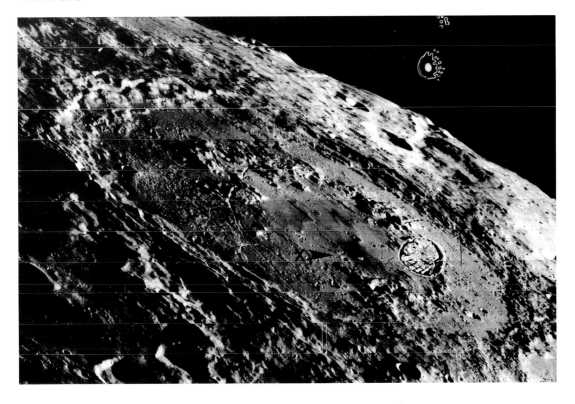

a. V-021-H2   (oblique)

b. V-118-M   4.2 km (top) 3.4 km (bottom)   c. V-116-M   4.3 km (top)   3.4 km (bottom)

Crater Ejecta

Plate 187 provides a high-resolution view of the dark-haloed craters in Alphonsus shown in Plate 186, b and c. Their appearance is subdued, although several blocks and their trails can be identified on the walls in the original photograph. Boundaries for the low-albedo deposits are not well defined; hence, the ejecta blanket was either largely pyroclastic, a thin flow unit, or a flow unit whose borders have since been eroded. Because the rim deposits appear to thin gradually with distance from the crater, the blanket most likely was the result of pyroclastic eruptions.

Perhaps the darkening of bright rays associated with fresh craters involves progressive mixing by meteoroid bombardment with the surrounding and underlying terrain. It is noted (in discussion with Plate 106a) that morphologic features remain even though the albedo contrast has disappeared. Thus, under this hypothesis it seems that the highly reflective component must be surficial. Moreover, the preservation of the low albedo around the subdued dark-haloed craters in Alphonsus warrants the following possibilities:

1. If the craters once displayed much more structure but have since been subdued by meteoroid weathering, then either the low-albedo blanket must be much thicker than the bright counterpart found around fresh impact craters, or it has not been altered by the influx of meteoroids to match the higher albedo surroudings.
2. The dark-haloed craters were formed relatively recently and therefore are well preserved.
3. The mechanisms responsible for lowering the albedo of bright crater rays with time are not the same as those for raising the albedo of dark deposits.

PLATE 187

V-118-H2   0.49 km

Low-Albedo Hills

Bordering eastern Sinus Aestuum is a low-albedo shelf shown in Plate 188a (arrow X). Plate 188b is an oblique view of the region under low solar illumination, and the low-albedo character is largely masked by shadowing and photometric properties of the material. The appearance is similar to that of highland topography, exhibiting the familiar tree-bark pattern and numerous subdued craters.

The Apollo 12 mission brought back oblique views (AS12-52-7733) of the same area under a much higher sun angle. The contact between the low-albedo shelf and the mare is photometrically and topographically well defined. The shelf has numerous much lower albedo patches that typically correspond to subdued dimple craters having bulbous (Plate 188b, arrows X and Y) or low-relief rims (arrow Z). Dark material also surrounds subdued rille systems (arrow XX) and elongate craters. Many subdued craters, about 1 km in diameter, have relatively high albedo walls with dark streaks extending from near their rims (best seen in Apollo 12 photographs of the region). Within a few craters, these dark streaks appear to correspond to outcropping ridges.

Several craters along the contact between the hills and the mare appear to be partly buried by advancing shelf material (arrows YY and ZZ). Unfortunately, these craters are partly in shadow, but Orbiter IV and Apollo 12 photogaphs show that the highland material encroached into the craters. The mare contact typically is accompanied by a "wrinkling" of the surface (visi-ble in the original photograph and similar to that shown in Plates 195–196, a, b, and c). If the shelf represents deposits of pyroclastics, then the well-defined contact is consistent with its emplacement before the mare units. Furthermore, the encroached craters suggest shelf material that advances by creep rather than by widely redistributed material from meteoroid impacts. The wrinkled pattern, however, is thought to be associated with the emplacement of the mare units, and its preservation along several contacts indicates that the shelf material has not advanced appreciably. Therefore, albedo contrasts can place constraints on the type and extent of postmare degradational processes.

The region near Rima Bode (Plate 188a, arrow Y) which is northeast of the area shown in Plate 188b, appears to be part of the same low-albedo complex. It also displays numerous indications of extrusive activity (see Plate 119). Another, and apparently related, region is identified by arrow Z in Plate 188a, and is examined in Plate 189. The low-albedo shelf areas are thought to have been mantled by pyroclastic eruptions during the early stages of mare flooding. As discussed at the end of Chapter 7, some of these eruptions may have been triggered by volcano-tectonic collapse of the mare basins.

Arrow XX (Plate 188a) identifies Rima Bode III, a sinuous rille that crosses a high-albedo plains unit (see Discussion and Summary at the end of Chapter 7).

PLATE 188

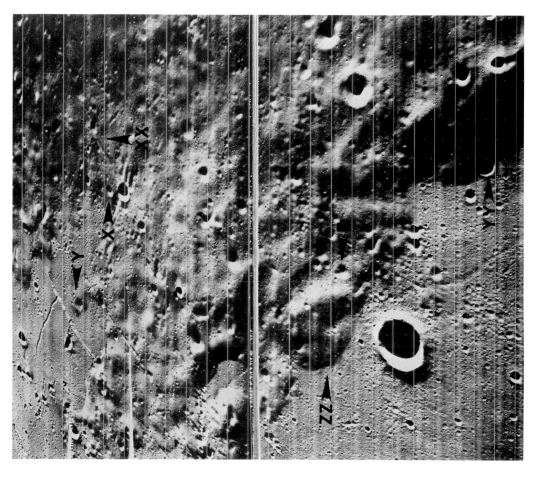

b. III-102-M (oblique)

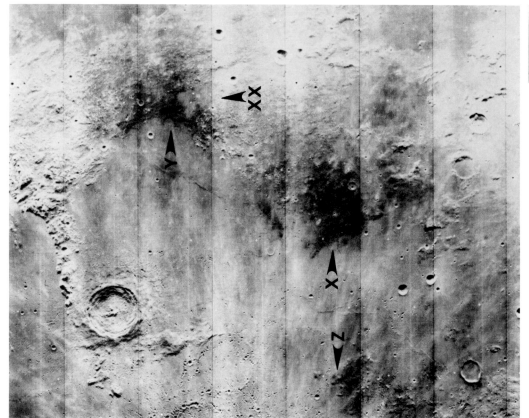
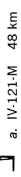

48 km

a. IV-121-M

Low-Albedo Hills

Plate 189a shows the general region southeast of Copernicus and identified in Plate 188a (arrow Z). As viewed from terrestrially based telescopes, low-albedo units appear to overlie parts of the Copernicus ejecta blanket. This impression is confirmed in Plate 189.

Plate 189b provides an enlarged view of an elongate depression with the herringbone pattern (arrow X) that is part of the outer rim sequence of Copernicus. The western border of this secondary impact is overlapped by a smooth low-albedo unit (arrow Y). Plate 189d (arrow X) shows additional evidence for this unit.

The region shown in Plate 189 shows several features suggesting an interesting local geologic history:

1. Numerous flat-floored craters indicate regional blanketing or internal origin (Plate 189b, arrow Z; Plate 189d, arrow Y).
2. The flat-floored crater with an interior dimple implies blanketing, stoping, or multiphased activity (Plate 189d, arrow Z).
3. The subdued interior terrace of the crater shown in Plate 189d (arrow YY) is evidence for differential floor subsidence.
4. A larger crater overlapping a smaller crater (Plate 189d, arrow ZZ) without extensive obliteration of the smaller suggests a relatively small amount of ejecta with collapse.
5. Ropy ridges may be viscous extrusions along fractures (Plate 189d, arrow XX).
6. Low-rimmed elongate depressions (Plate 189a, arrow X).

In addition, note the similarity of the subdued craters having low, wide rims (Plate 189b, arrow XX; Plate 189c, arrow X) to the craters seen in Plate 188b. These are possible sources for the ashlike blanket of this region. Plate 189c (arrow Y) shows a dark-haloed crater with high-albedo outcrops, and this is also similar to features shown in Plate 188b. High-albedo outcrops occur near the crest of several hills (Plate 189c, arrows Z and XX) and are also recognized near the Sulpicius Gallus rilles (Plate 166).

The low-albedo hills of this region are believed to indicate mantling by pyroclastics, most of which were erupted during the epochs of mare emplacement. Mass movement of this overlay exposed underlying high-albedo material on the hills and on the walls of craters. The dark-rayed crater seen in Plate 189d (arrow XY) suggests that Copernicus ejecta overlie local low-albedo units, but the embayment of Copernicus ejecta elsewhere suggests rejuvenated volcanism. Portions of the low-albedo plains units, however, may represent Copernicus ejecta. Similar units appear to have been associated with Aristarchus (Plate 258) and possibly indicate molten ejecta that acted as flow units after impact.

PLATE 189

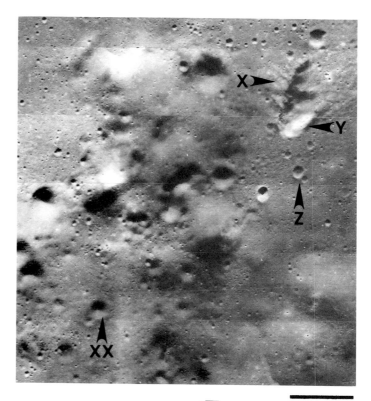

a. V-137-M 3.0 km *(top)* 3.6 km *(bottom)*

b. V-137-M   3.1 km

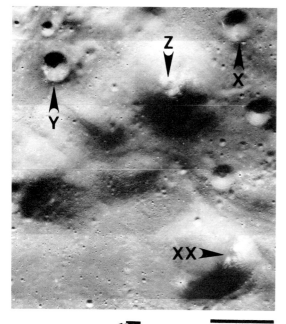

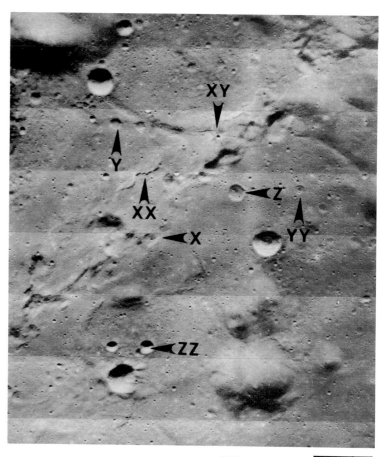

c. V-137-M   3.3 km

d. V-137-M   3.6 km

Low-Albedo Units and
High-Albedo Relief Features

Plate 190 illustrates albedo contrasts created by mare flooding and viscous volcanic extrusions. Plate 190a shows the contact of two plains-forming units. The higher albedo unit (arrow X) exhibits numerous subdued craters, whereas the lower albedo unit displays fewer craters but several positive-relief features that were probably volcanically formed (arrows Y, Z, XX, and YY). Note the irregular scarp (arrow ZZ) that suggests a previous lava level. The area under consideration is on the Orientale ejecta blanket. The higher albedo unit is typical of topographically low regions, but the darker unit is relatively unique to the surrounding area.

Tsiolkovsky, shown in Plate 190b under full illumination, can be contrasted to views shown under lower solar elevations in Plates 11a and 40b. The low-albedo floor material is one of the darkest units on the Moon. In contrast, note the high-albedo peak and hummocky floor area bordering the mare fill. These probably represent more highly silicic impact melt and extrusions predating mare flooding.

Plate 190c includes a high-albedo zone near Gassendi. This is presumably an islandlike remnant of a lighter plains-forming unit and complex hills.

Plate 190d shows a high-albedo irregular dome (arrow X) near the craters Billy (upper left) and Hansteen (lower right). The dome has surface textures similar to those of the Gruithuisen domes (Plate 130, a and b) and, it is suggested, has a similar origin. Note that nearby hummocky terrains have a much lower albedo, comparable to that of the mare plains. Thus, albedo contrasts may result from viscous extrusions of higher albedo material, rather than from emplacement of surrounding low-albedo mare material.

PLATE 190

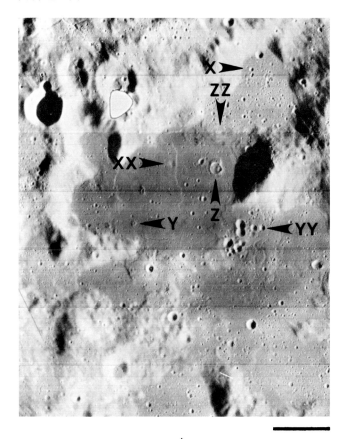

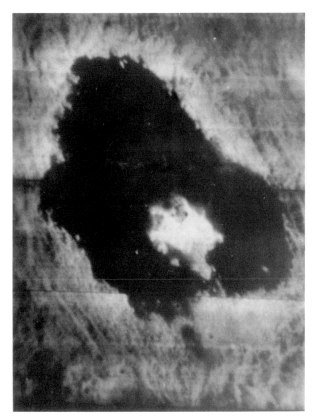

_a._ IV-160-H3   13 km

_b._ I-102-M   (oblique)

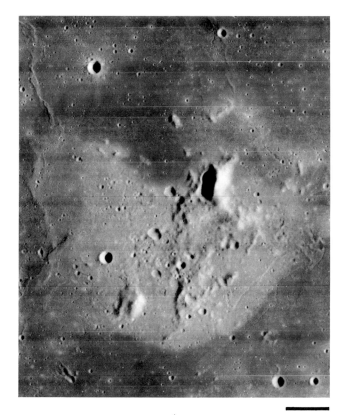

_c._ IV-132-H2   12 km

_d._ IV-149-H2   12 km

DISCUSSION AND SUMMARY

The relatively small range in lunar-surface albedo is evidence that some process (or processes) alters surface reflectivity. Prior to the Surveyor landing missions, it was generally believed that surface reflectivity is lowered with time, but the subsurface material thrown out by the Surveyor footpads was shown to be darker than the surface layer. Various mechanisms alter the surface albedo in a lunar environment:

1. Bleaching by ultraviolet irradiation (Cohen and Hapke, 1968; Hapke et al., 1970)
2. Alteration of the surface porosity—a decrease producing a higher albedo (Hapke, 1963; Hapke et al., 1970)
3. Darkening of materials with high absorption coefficients by reduction in particle size (Filice, 1967; Adams, 1967; Minnaert, 1969; Salisbury, 1970)
4. Darkening by a "varnish" emplaced by the solar wind (Shoemaker and Morris, 1969; Borg et al., 1971)
5. Darkening by proton bombardment (Hapke, 1965), although a relatively small amount is indicated in experimental results (Dollfus et al., 1971; Nash and Conel, 1971)
6. Impact-induced mixing and vitrification that reduce the albedo by increasing the percentage of dark Ti- and Fe-rich glasses relative to crystalline material, believed to compose the bright-ray material (Adams and McCord, 1971)

These mechanisms apparently alter the albedo by varying degrees, but the dominant process and rates remain in considerable debate. In addition, the local albedo can be changed through either vertical mixing due to stratigraphic inversion by nearby impacts or lateral mixing due to material transported considerable distances from much larger impacts.

It is reiterated that surface coverage from Lunar Orbital photography is limited with respect to different solar illuminations. Consequently, "albedo" contrast derived solely from such photography is actually a "luminance" contrast. The significance of this contrast depends not only on surface composition but also on surface structure, which, in turn, is commonly a function of slope.

The following examples, shown in the cited plates, do not comprise a complete tabulation, but they do illustrate typical albedo contrasts that may be associated with:

1. Isolated swirllike deposits (Plates 183–184)
2. Ejecta sequences (also see "Crater Rims")
   a. Bright-rayed craters (Plate 185a)
   b. Dark-rayed craters (Plates 89, 143a, and 189d)
   c. Dark-haloed craters (Plates 185, b and d,–187 and 202)
3. Crater-wall features (see "Crater Walls")
   a. Outcrops (Plates 64 and 173c)
   b. Wall debris (Plates 57 and 64b)
4. Rilles
   a. Wall outcrops (Plate 173c)
   b. Surrounding units (Plates 119, 159, and 164–165, a and b)
5. Highland "specks"
   a. Associated with craters (Plate 185c)
   b. Associated with outcrops (Plates 140c, 166, a and b, and 189c)
6. Positive-relief features (Plates 154, 166c, and 188–189)
   a. Volcanolike ridges (Plates 131–132, a and d)
   b. Septa (Plate 133)
   c. Cones (Plates 60a and 130, e and f)
   d. Irregular platforms (Plate 123a)
   e. Low-relief crater-topped domes (Plate 125d)
   f. Mamelon domes (Plate 126)
   g. High-relief domes (Plates 130, a and b, and 190d)
7. Plains-forming units (Plates 103 and 190)

The bright-ray systems associated with relatively recent craters represent one of the most obvious albedo contrasts. As suggested in Chapter 4, this high albedo can be attributed to (1) a decrease in the surface porosity from the impact and scouring by a molten and low-density ejecta cloud; (2) mineralogic composition and physical structure of deposited material from the primary impact (secondary debris); (3) the properties and composition of subregolith layers excavated by the impact of secondary debris; and (4) the deposition of local regolith and subregolith materials physically altered by the secondary impact. The first contribution to the albedo of crater rays is inferred from high-albedo patches that show evidence for surface scouring without well-defined secondary craters. The second contribution is indicated by Apollo 16 surface photography of rayed regions that exhibit numerous blocks resting on the surface. The third contribution seems to be confirmed by dark-rayed craters that have penetrated an overlying blanket of high albedo and have excavated a low-albedo layer known to be present. Last, the fourth mechanism is suggested by small bright-rayed craters occurring in regions believed to be heavily mantled by low-albedo material.

Each of these four mechanisms is believed to have been operative to a greater or lesser degree, dependent on the size of the primary impact. The first mechanism —scouring of the surface—is believed to characterize the rays of craters 1 km–15 km in diameter. This contribution should be erased rapidly by the "gardening" of the regolith by small impacts, and such an erasure seems consistent with preserved irregular-bordered secondary patches not having craters and not having a notable albedo contrast. The second mechanism—deposition of material excavated by the primary impact—is very important around smaller craters (less than I km in diameter). Around larger craters, this material combines with locally derived debris ejected from complex secondary impacts. The latter process is illustrated in certain regions by the "disappearance" of bright rays where they cross low-albedo terrain. This disappearance can be attributed, in part, to locally excavated low-albedo material. In general, the secondary craters responsible for the high albedo of these rays are highly subdued in morphology—a result of interacting ejecta. Closer to the primary crater, deposition of secondary debris—not ejecta from the impact by this debris—is believed to determine the albedo. There is evidence, however, that impact melt from the primary crater may alter the local albedo at greater distances. Deposition of materials physically altered by the impact event dilute the other three mechanisms to an unknown degree.

Subsequent alteration of the albedos determined by the last three mechanisms should be slow relative to the alteration of scoured surfaces. It is noted that the inferred changes in the albedo associated with both large secondary complexes and the inner ejecta blanket surrounding the primary crater support a physical or chemical alteration of the upper surface, rather than mixing with subsurface layers. The process of vertical mixing, however, must act to some degree and is believed to mask more rapid albedo changes performed by other processes.

These four plausible contributions to the albedo of crater rays coupled to six possible mechanisms for subsequent alterations illustrate the difficulty in assigning crater ages only on the inferred disappearance of these rays. Moreover, not all craters originally displayed high-albedo rays. The absence of bright rays around Tsiolkovsky (Plate 190b) and similar large craters or basins is probably determined by the initial event and not subsequent processes. The ray systems of Aristarchus (Plates 171a and 236) exhibit a distinctly lower albedo than do those of Kepler or Copernicus, but its morphologic features clearly indicate relatively recent formation.

Albedo contrasts can be used, however, to determine the effects of surface processes through the preservation of contacts between contrasting units. The topographic contact between the low-albedo shelf material and the mare typically corresponds to the albedo contact. If extensive degradation and mixing had occurred, then such contacts should be diffuse. This is particularly significant if such craters as Secchi X (Plate 57) are postulated to have resembled originally Mösting C (Plate 80). Such a transition requires considerable degradation of the ejecta blanket, but this is inconsistent with the preservation of the albedo contact between mare and shelf units over approximately the same time interval.

Adjacent units of contrasting albedo are also useful in the identification of otherwise unimpressive features. Many craters with low-albedo haloes are thought to be volcanic in origin, but without their haloes they are morphologically similar to nearby craters. More detailed studies might determine the degree of similarity and perhaps yield reliable parameters for determining differences in crater origin.

437

9. Maria and Other Plains-Forming Units

INTRODUCTION

The lunar maria are the most obvious surface features visible to the unaided eye from Earth and characterize the lunar Earthside. Telescopically, they appear smoother and display a low albedo (0.06) relative to that of the lunar highlands (0.10) and highland plains-forming units (0.09). The material comprising the mare units fills not only the enormous basins, such as Imbrium, but also the floors of many craters. Thus, mare filling represents an important epoch or epochs in lunar history and is the primary subject of this chapter. Although not all low-albedo plains-forming units necessarily represent the same general extrusive type, this is certainly a reasonable hypothesis, based on element analyses from the Surveyor missions and the samples returned from Apollo and Lunokod (the unmanned Russian Mooncraft).

Craters and major basins may be inundated by extrusions external or internal to their topographic boundaries. In the former event, mare material surrounds the crater or basin until the walls are overrun or breached. Thus, the depression does not supply the material for its inundation. The other mode of inundation occurs when mare material is extruded through feeders within the crater or basin.

Marelike units do not represent the only lunar extrusive form. A wide range of flow viscosities is indicated, and marelike units appear to be one of the least viscous. In addition, there are plains-forming units that exhibit higher reflectivities and higher crater densities than do mare surfaces. These are also considered in this chapter. The term *plains* in the present usage includes both limited flat-surfaced areas and larger expanses. More viscous extrusions and different types of eruptions appear to have been responsible for a variety of positive-relief features; illustrations and discussions of such forms are found in preceding chapters. Variations in viscosities indicate a range of chemical compositions and temperatures. In the lunar environment, departures from processes and compositions found in terrestrial extrusives might be expected. For example, terrestrial *nuées ardentes* are typically rhyolitic in composition, but sudden injection and disruption of a basaltic lava into a lunar vacuum could result in a basaltic *nuée ardente.*

Mare Humorum

Mare Humorum (Plate 191a) illustrates features related to the inundation of a large lunar basin by mare material. It is an approximately circular basin and has only remnants of large peripheral scarps, which characterize Mare Orientale (Plate 201) and Mare Imbrium (Plate 192). In addition, this basin does not display an extensive system of radial lineations. Bordering Mare Humorum are several shallow craters with extensively fractured floors, such as Gassendi (Plate 191a, arrow X; also see Plate 36), Vitello (arrow Y; also see Plate 249), and Doppelmayer (arrow Z; also see Plate 191b). Mare Humorum shows considerable variety in its associated features, which are summarized below.

The northern and northeastern borders of Mare Humorum are characterized by irregular islands of positive-relief features (arrow XX). N-trending sinuous rilles extend across the embayed northern border and typically have "head" craters on NW-trending wrinkle ridges in Oceanus Procellarum (arrow YY; see Plate 182a). A N-trending wrinkle ridge (arrow ZZ) also crosses this region.

A prominent system of concentric grabens crosses the eastern and southeastern borders (arrow XY; see Plate 167c). Several embayed craters in this region are breached on their southwestern side, parallel to the border of Mare Humorum and the concentric grabens rather than basinward (arrow XZ; see Plate 167c). Another concentric pattern is formed by wrinkle ridges within Mare Humorum (arrow YX). Remnants of a concentric mountainous ring (arrow YZ) are expressed by platforms and crested-relief massifs. Note the NE-trending scarp (arrow ZX), which is tangential to the basin, and the NW-trending furrows that are approximately perpendicular to the scarp trace. A peripheral embayment of mare units, Palus Epidemiarum (arrow ZY), exhibits several volcanolike forms, such as that shown in Plate 3, c and d.

The southern and southwestern borders also exhibit embayments of mare units. Two wrinkle ridge systems cross this region. A NE-trending wrinkle ridge may be related to the NE-trending scarp (arrow ZX) and extends at least 200 km across the highlands (visible only at higher resolutions). In addition, NW-trending wrinkle ridges (arrow XXX; see Plate 191b) approximately parallel a system of grabens to the west (arrow XYY). A complex array of irregular lava lakes and flow termini (arrow YYY; see Plate 87a) occur at the intersection of these NW-trending wrinkle ridges and the southwestern extension of the concentric wrinkle ridges on the eastern border.

In contrast to the concentric grabens in eastern Mare Humorum, a system of linear NE-trending grabens on the western border is approximately tangent to the northwest mare edge (arrow ZZZ; see Plate 168). The well-defined scarp (arrow XXY; see Plate 168) along the western mare border appears to be part of this NE-trending rille system. Another system of grabens has a NW trend (arrow XYX), one of which intersects NE-trending grabens near the center of two craters (see Plate 169). Note the branching narrow rilles with low-albedo deposits (arrow XYY; see Plate 165, a and b).

Plate 191b includes the region around Doppelmayer (arrow X) on the southern "shores" of Mare Humorum. As mentioned (also see Plates 34–38), craters on the margins of large mare-filled basins typically display shallow flat floors with complex fracturing. The breached wall of Doppelmayer on the mare side and partial flooding of its floor suggest an inward tilting of the basin. In contrast, the crater Puiseaux has only a narrow breach (arrow Y) but an almost completely inundated floor. The crater Gassendi (Plate 191a, arrow X; see Plate 36) also has a possible breach that has been camouflaged by mass movement.

Although these observations suggest that inundation of this basin occurred simply through overspill from a central source, such an interpretation is oversimplified. The inundation of Puiseaux requires an extremely fluid lava, which gained access to the crater interior through a narrow breach of its wall. More likely, its floor was filled with mare units through vents on the floor. This view is supported by the identification of proposed vents within Gassendi and by the existence of other unbreached craters and closed basins with mare-filled floors.

The shallow fractured floors of Gassendi and similar craters along mare borders suggest three alternative processes:

1. The floors of old impact craters are being raised through rising magma chambers related to the maria.
2. The floors of old impact craters are rising through rapid isostatic adjustment.
3. The floors indicate caldera formation.

Each suggestion has significance with respect to mare flooding. The first and third proposals indicate that these peripheral craters are potential extrusive centers, whereas the second proposal requires that a molten layer extended beneath the highland borders of the basin. Extrusive activity in Doppelmayer is indicated by the blanket of low-albedo material over portions of the hummocky floor unit (Plate 191b, arrow Z). This material probably represents pyroclastics erupted from sources on the crater floor. These interpretations are not consistent with a mare-filled basin that simply overflowed its borders.

The mechanism responsible for breached craters is suggested to be related primarily to structural events, such as crustal foundering associated with mare emplacement, rather than overspill and removal or burial of wall segments. Evidence for this mechanism is listed below:

1. Craters are breached parallel to the concentric fracture system (Plate 191a, arrow XZ).
2. The craters shown above arrow XXY (Plate 191a) and Gassendi (arrow XXZ) have been severed by a boundary fault.
3. In general, breached crater walls do not reflect a regional slope.
4. The WNW-trending wrinkle ridges shown in Plate 191b (arrow XX) appear to be related to both a highland scarp (arrow YY), which marks the highland-mare contact, and the breached portion of Doppelmayer.

Inundation of Mare Humorum probably occurred through several vents. NE- and NW-trending systems of grabenlike rilles, subdued faults, and wrinkle ridges possibly reflect conduits. The dark-haloed branching rille (Plate 191a, arrow XYY) is interpreted as remnants of fissurelike vents that may be part of a larger concentric system. In addition, overspill from Oceanus Procellarum may have been responsible, in part, for the island-like province in northern Mare Humorum. This is consistent with the interpretation that the sinuous rilles in the region reflect the direction of lava flow from sources along the NW-trending system of wrinkle ridges in Oceanus Procellarum.

The preservation of the original mare surface and related events is indicated by the existence of moats, terraces, and low-relief wrinkled surfaces at the contact between the mare and highland terrain. These are shown in Plate 249. Arrow ZZ in Plate 191b identifies a complex terraced region corresponding to that marked by arrow YYY in Plate 191a and shown in detail in Plate 87a. Two probable volcanically formed craters are located in Plate 191b by arrow XY (see Plate 87b) and arrow XZ.

PLATE 191

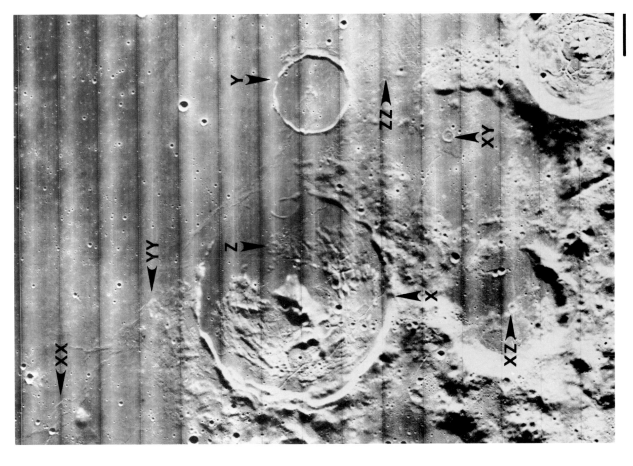

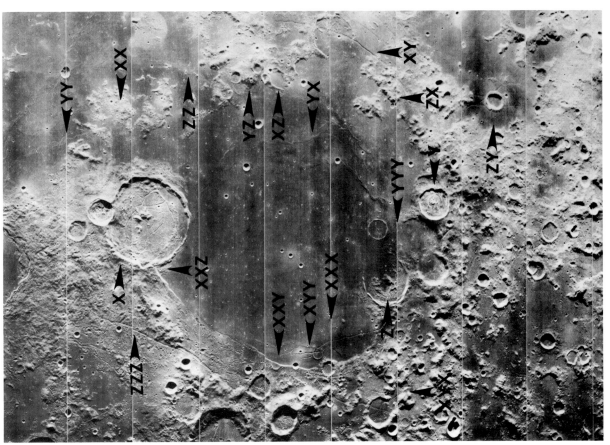

a. IV-143-M   (oblique)

b. IV-142-H3   14 km

Mare Imbrium

Mare Imbrium (Plate 192) is almost three times larger in diameter than Mare Humorum (Plate 191). It also displays several features that make it distinct from Mare Humorum:

1. A large arcuate scarp forms an abrupt eastern and southeastern boundary (Plate 192a, arrow X). Remnants of this scarp are traced to the south and northeast. The northern border (arrow Y) appears to be vestiges of an interior mountainous ring analogous to the Orientale basin concentric system.
2. Concentric grabenlike rilles occur interior to the outer eastern scarp (arrow Z), but they typically predate the last stages of mare flooding (see Plate 112). Recall that the grabenlike rilles associated with Mare Humorum are exterior to the basin and postdate local inundation by mare units.
3. Sinuous rilles occur along the periphery of Mare Imbrium. They typically have "head" craters within the mountainous borders and exhibit structural as well as topographic control in their paths to the mare surface. Several complexes of sinuous rilles occur near the interpolated western and southwestern borders. These include rilles associated with the Harbinger Mountains (arrow XX; see Plate 255).
4. Wrinkle ridges generally comprise a system concentric with the basin. The southwestern portion of this system, however, is composed of a NW-trending straight-ridge segment arrow YY). Wrinkle ridges typically interconnect insular sharp-crested massifs that presumably are remnants of an interior mountainous ring.
5. Low-relief mare domes with summit pits are common along and beyond the borders of the basin (see Plate 125). Low-relief mare domes with and without summit pits also accompany sharp-crested massifs (see Plate 134).
6. Volcanic cones, ridges, and rimless pits typically occur in the "shallow" areas of mare flooding (arrow ZZ; see Plate 132e).

In addition, Mare Imbrium displays a variety of crater appearances. Craters with shallow fractured floors, which characterize the borders of Mare Humorum, are typically absent; however, numerous craters exhibit marelike floor units. Shallow craters similar to Plato B (see Plate 28c) and Hortensius D (Plate 132e) are associated, in general, with volcanic complexes along the outer boundaries of Imbrium. Small craters with multiple-domed floors (Plate 25) and ringed plains (Plate 91) are also characteristic of these regions.

These observations suggest extensive volcanism, with considerable variety in expression along the borders of Mare Imbrium. The sites of these eruptions occurred:

1. Near the base of the scarps and remnant scarps
2. On the interior benches, such as the Archimedes bench and Montes Alpes near Plato
3. Beyond the border of the Imbrium basin in the "shallow" regions of mare flooding
4. In association with isolated complexes of positive-relief features

The interior portion of Mare Imbrium appears to lack evidence for such vents, with the possible exception of some wrinkle ridges, which may represent sites of fissure eruptions.

Telescopic observations from Earth reveal several flow fronts within Mare Imbrium that were mapped by Strom (1965) as six units having various colors. Two of these units were traced more than 200 km as single units. Plate 192, b and c, shows the lobate termini of one of these units that is located, but not identifiable, by arrow XY in Plate 192a. The surface of the flow exhibits shallow troughs (Plate 192, b and c, arrows X and Y) that are interpreted as collapsed lava tubes through which the flow advanced (Fielder and Fielder, 1968).

Statistical analyses of craters on the Imbrium flow units have been made from Orbiter photographs (Fielder and Fielder, 1968). The results indicated that craters of about 225 m in diameter are almost twice as abundant on the more recent unit as on the older surface. It was concluded that a large fraction of craters was produced by nonimpact processes. Although the authors attempted to exclude secondary craters from their statistics, elimination probably was incomplete owing to the inherent difficulty in differentiating secondary and primary craters. However, there are numerous endogenetic features preserved on these surfaces, such as subdued and rimless craters (arrow Z) as well as the flow borders. Arrow XX identifies a subdued ring that is interpreted as either an internally produced form or an inundated crater (see Plate 152c). There are at least four other similar rings in this region. Arrow YY marks a rimmed crater on the old surface that is partly encroached by the lobate flows. In addition, note the dark-haloed crater (arrow ZZ) that is subdued at higher resolutions and is interpreted as a nonimpact-produced crater, which was formed during the cooling of the flow. These observations put upper limits on the degradational rate and require relatively good preservation of features larger than 100 m.

Arrow XZ (Plate 192a) identifies the irregular mare, Mare Frigoris, discussed at the end of this chapter.

PLATE 192

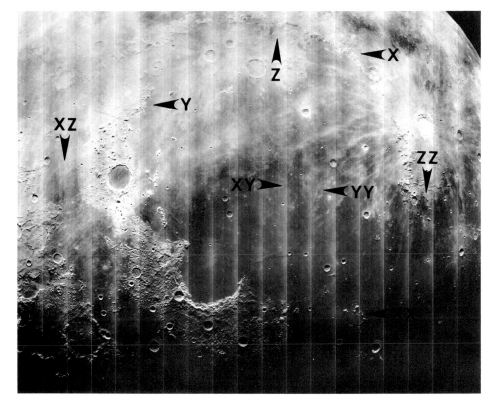

a. IV-134-M   (oblique)

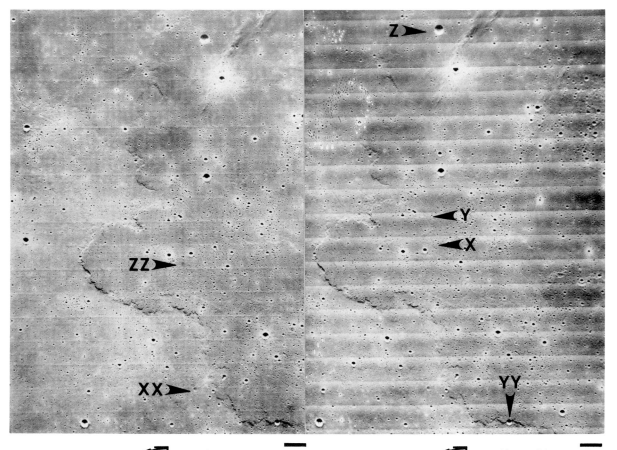

b. V-161-M   5.0 km   c. V-159-M   4.9 km

Mare Imbrium

Shown at higher resolution in Plate 193, the Imbrium flow surface loses much of the complexity that is anticipated from the resolution of Plate 192, b and c. The shallow troughs are subdued (arrow X), and a marked texture is found near the base of the termini from which minor flows extended (arrow Y). A few blocky outcrops can be identified along the lobate front (in the original photograph). The edge of the flow is characterized by the tree-bark pattern, which also is expressed within and around craters having broad subdued rims.

Estimating approximate thickness of the regolith is possible by estimating the smallest crater having a concentric plan (Quaide and Oberbeck, 1968). This gives a local depth of about 5 m, which is consistent with the Fielders' (1968) estimate. This should be contrasted to the maximum elevation difference of the flow front, which is calculated to be 25 m (Fielder and Fielder, 1968). Since the upper layer of the lava flow may froth during the release of volatiles, it is not clear what proportion of the regolith was developed by meteoroid bombardment.

A relatively well preserved surface is indicated by the following features:

1. Highly lobate termini (arrow Z)
2. Minor lobate extensions from beneath the flow (Note in particular the lobate feature associated with the subdued crater on top of the lobate extension identified by arrow XX.)
3. The boundary between the flow front and an encroached crater (arrow YY)

These features confirm observations noted in discussion of Plate 192 and comprise additional evidence that these surfaces are relatively well preserved. The subdued appearance of most of the craters is not the result of extensive degradation by meteoroid bombardment unless a large number of craters 200 m in diameter were formed immediately after emplacement of these flows. This inconsistency between preserved small-scale features and subdued 200-m craters can be understood if a significant fraction of such craters was formed endogenetically, in agreement with the conclusion of Fielder and Fielder (1968).

Aside from the more obvious subdued troughs on the most recent lava flow, the surfaces shown in Plate 193 exhibit a smaller-scale texture (arrow ZZ) that is interpreted as remnants of flow patterns. The mare plains within the Flamsteed Ring (Plates 147, a and b, and 154–157) and the region around the Gruithuisen domes (Plates 130, a and b, and 181) show similar textures as well as other endogenetic forms.

PLATE 193

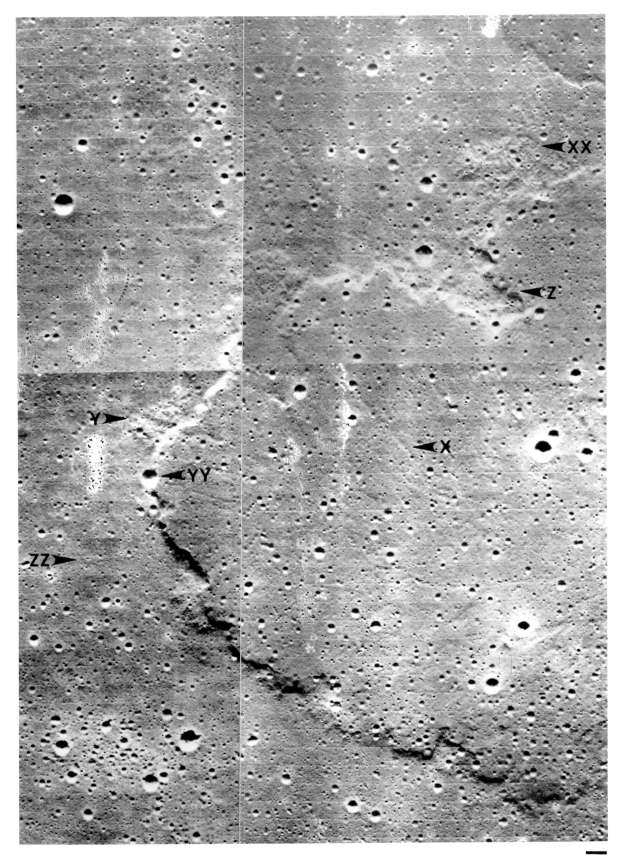

V-160-H2,H1; V-161-H2,H1   0.65 km

Miscellaneous Features

Plate 194, *a* and *b*, shows a region south of Arago in Mare Tranquilitatis that exhibits extensive tree-bark patterns (Plate 194*c*, arrow X). These patterns are not restricted to the walls of subdued craters but parallel the system of shallow arcuate rilles, which are approximately concentric to the "ghost" crater Lamont. A wrinkle ridge (Plate 194, *a* and *b*, arrow X) is crossed by these rilles and displays a variety of surface textures, which includes a jointlike system (see Plate 142).

Arrow Y (Plate 194, *a* and *b*) identifies a closed depression that is shown at higher resolutions in Plate 194, *c* (arrow Y) and *d* (arrow X). The floor of this depression has a greater texture (at the resolution shown in Plate 194*d*) than does the surrounding mare surface, and its borders are irregular or lobate in plan and abrupt in profile. A smaller closed depression is to the east (Plate 194*d*, arrow Y). Very similar features occur on the floor of Hyginus (Plate 161*b*) and near Secchi X (Plate 132*b*).

The interpretation of this closed depression is significant for the interpretation of both surface processes and, in turn, the evolution of the mare surface. If it represents a zone of avoidance—i.e., a kipuka—that developed during the emplacement of the mare, then the mare surface must be in a pristine state except for later impacts, which have not extensively degraded features larger than a few tens of meters. More likely, the depression formed in conjunction with the arcuate rilles, as it is situated along one of the subdued members of this system. If the depression is a collapse feature, then the sharply defined borders indicate that this event was rapid and recent. Analogous features on the floor of Hyginus commonly occur along the upper wall and rim areas of subdued craters. A similar occurrence is noted for a nearby crater to the east of the area shown in Plate 194 (see Plate 56*b*). If collapse occurred soon after mare emplacement analogous to the sinks in the Diamond Craters area of southeastern Oregon (see Green and Short, 1971, Plate 122A), then they indicate a well-preserved surface. If they represent recent formations, then they open the possibility for current production of nonimpact depressions. The latter interpretation is supported by the subdued crater chain shown in Plate 56*a* (upper left). The subdued appearance of the rilles might reflect the rate of formation rather than the amount of subsequent degradation. This is consistent with the observation that these rilles cross a wrinkle ridge that exhibits considerable small-scale detail.

As collapse features, these depressions comprise evidence for the contribution of nonimpact craters to the total crater density. Therefore, statistics of crater abundance and size might not be applicable to all mare or marelike surfaces. This is particularly important where photographic resolutions are not sufficient to permit examination of crater form.

PLATE 194

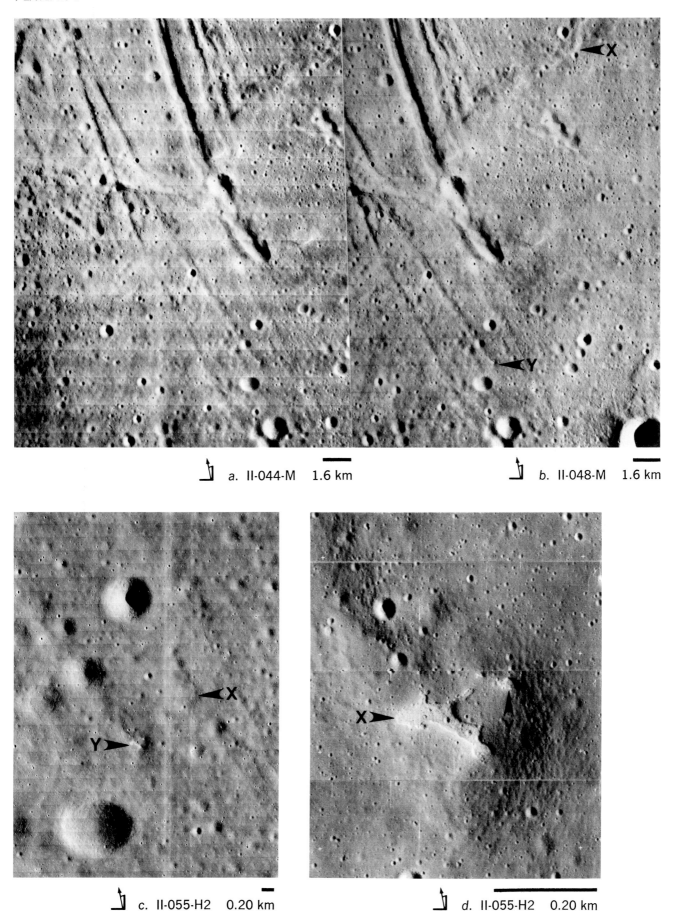

a. II-044-M   1.6 km

b. II-048-M   1.6 km

c. II-055-H2   0.20 km

d. II-055-H2   0.20 km

Miscellaneous Features

The effects of mare flooding on pre-existing topographic features provide additional information on the age and genesis of mare units. Plate 195a shows the large breached crater Maestlin R in Oceanus Procellarum. Another photograph of the same region is shown in Plate 195b (also see Plate 136a) and is an oblique view in the direction of arrow X in Plate 195a.

These views show the geologic setting of the two craters illustrated in Plate 195, c and d (Plate 195a, arrow Y; Plate 195b, arrow X). Plate 195b (arrow Y) reveals a sharp contact between the mare surface and a highly cratered terrain, which presumably is a remnant of a partly inundated crater floor. This photograph also shows several small ring structures (arrows Z, XX, and YY) that are interpreted as craters inundated by mare units (see Plate 151). The region is crossed by NE-trending ejecta (Plate 195a, arrow Z; Plate 195, c and d, arrow X) from Kepler and by NW-trending ejecta (Plate 195, c and d, arrow Y) perhaps from Aristarchus. These and other secondary impacts produce a mottled surface visible in Plate 195, c and d, that is characterized by relatively smooth patches having both a low albedo and numerous flat-floored craters (arrow Z) and by rougher patches having a higher albedo (arrow XX).

The adjacent craters shown in Plate 195, c and d, illustrate the preservation of surface features that are interpreted to be associated with mare emplacement. A wrinkle pattern (arrow YY; also see Plate 196a) or a narrow ridge (arrow ZZ; see Plate 196b) typically accompanies the contact between the crater rims and the mare surface. Plate 195a shows that these are not shallow craters and, therefore, were not inundated significantly by the mare units. In addition, it appears that the floors of these craters were not linked to the mare-flooding source.

The significance of features marked by arrows XY and XZ is discussed with Plate 196.

PLATE 195

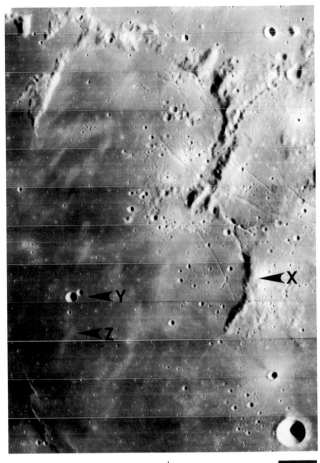

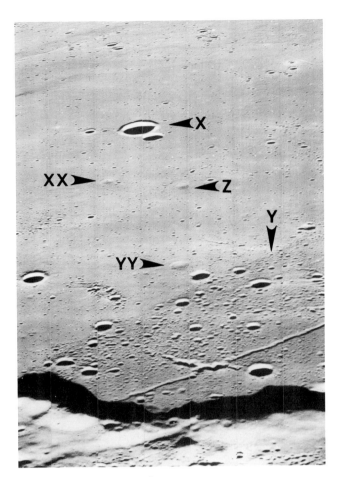

a. IV-144-H1 11 km

b. III-161-M (oblique)

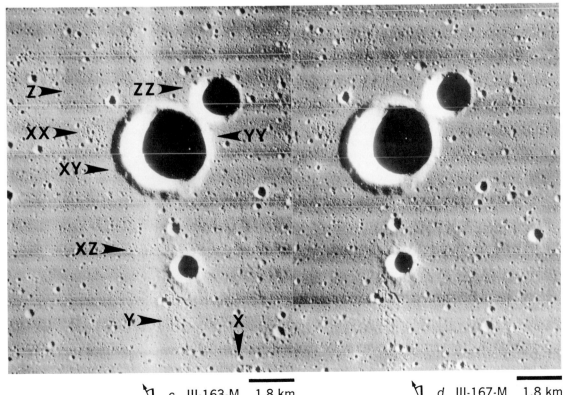

c. III-163-M 1.8 km

d. III-167-M 1.8 km

Miscellaneous Features

Plate 196, *a*, *b*, and *c*, shows detailed views of the contact between the mare surface and crater rims. The wrinklelike surface feature is identified by arrow X in Plate 196*a*, and the ridge by arrow X in Plate 196, *b* and *c*. These correspond, respectively, to arrows YY, ZZ, and XY in Plate 195, *c* and *d*. The moatlike depression between the textured surface and the rim (Plate 196*a*, arrow Y) suggests a genetic relationship to the mare unit rather than mass wasting from the crater rim. This relationship is more convincing for the ridges shown in Plate 196, *b* and *c*, since they cross the mare surface and are removed from the crater rim.

These contact features are interpreted as pressure ridges and termini that were formed at the time of mare emplacement. The preservation of such features is inconsistent with the interpretation that the high density of subdued craters represents degraded impact craters. The mottled appearance of the surface noted in Plate 195, *c* and *d*, probably was created by two generations of crater formation. The numerous subdued craters on the low-albedo surface may have been endogenetically formed, whereas the higher albedo and rougher patches are the result of secondary ejecta. Although a regolith has developed through meteoroid bombardment, it is thought that degradation has not been responsible for the larger subdued craterforms.

Plate 196, *d* and *e*, shows stereoscopically matched enlargements of an area identified by arrow XZ in Plate 195, *c* and *d*. The crater identified by arrow X in Plate 196, *d* and *e*, has no raised rim but exhibits numerous blocks (arrow Y) and an abrupt wall-rim transition. In addition, a flow unit appears to extend into the adjacent subdued crater (arrow Z). The rimless crater is suggested to be a volcanic feature and is further evidence for either continued local volcanic activity or well-preserved mare surfaces.

The inferred contact phenomena illustrated in Plate 196, *a*, *b*, and *c*, are not unique. A significant proportion of the aprons surrounding mamelonlike domes (see Plate 128) is attributed to the emplacement of the mare basalts. Narrow terraces within closed or partly closed depressions (Plates 87, arrow Y; 156, *c* and *d*, arrows Y and Z) as well as broader terraces bordering the maria at several locations (Plates 130, *a* and *b*, arrow XX; 173*c*, arrow Z; 174, *b* and *c*, arrow YY; and 175, arrow ZX) are thought to represent previous levels of mare flooding, thus preserved since that time. Such features add credibility to my belief that, despite a long history of primary and secondary cratering, evidence of processes associated with mare emplacement remains at numerous locations.

PLATE 196

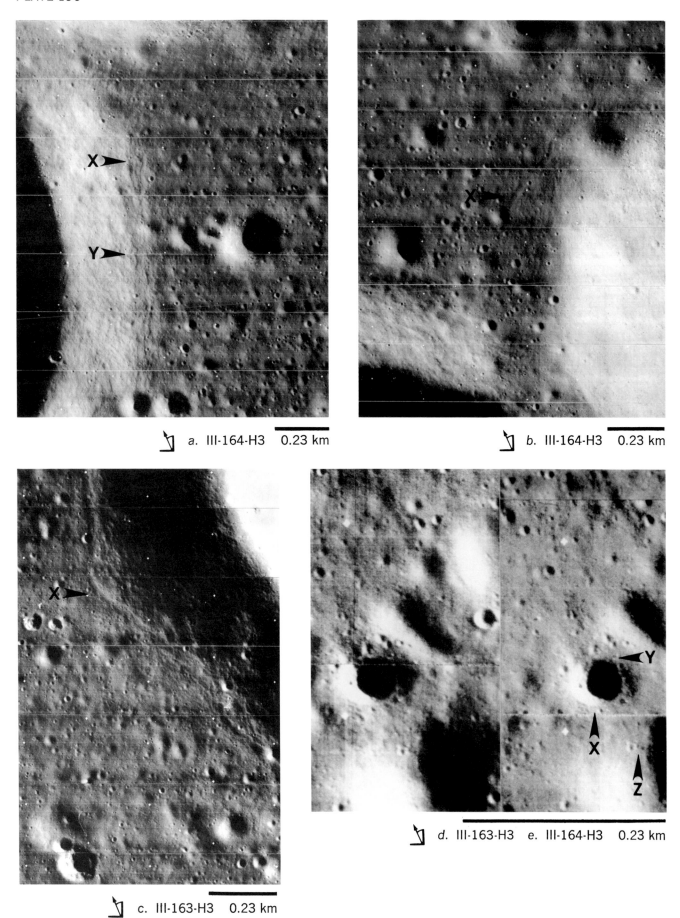

a. III-164-H3   0.23 km

b. III-164-H3   0.23 km

c. III-163-H3   0.23 km

d. III-163-H3   e. III-164-H3   0.23 km

Mare Australe

Mare Australe, shown in Plate 197a (arrow X), is distinct from the maria considered thus far. Whereas Mare Imbrium is bordered by relatively well defined arcuate scarps and is inundated almost completely by mare units, Mare Australe is identified by numerous partly filled craters in a roughly circular area, whose boundary is marked by the absence of similarly mare filled craters. It is a topographically low region that is not outlined by mountainous scarps, such as those around Mare Imbrium or Mare Orientale, although isolated massifs occur on the southeastern border (Plate 197a, arrow Y). Radial lineations, which are pronounced around Maria Imbrium and Orientale, are virtually absent.

Plate 197, b and c, shows a selected area of Mare Australe (Plate 197a, arrow Z). Mare flooding clearly did not issue from a single source, with breaching and subsequent filling of crater interiors. Rather, mare material probably was extruded within each crater from separate feeders, which were linked to a large magma reservoir. Possible flow features are noted by arrow X. Similar mare-floored craters west of Mare Smythii (Lunar Orbiter I-25, 26, and 27-M) show evidence for "bathtub" rings produced by previous levels of floor material.

Stuart-Alexander and Howard (1970) have noted that the mare plains in Mare Australe have a higher albedo than that of Mare Serenitatis, for example. This difference in surface reflectivity can be attributed either to differences in chemical composition of the extrusions at the time of emplacement or to later processes, such as blanketing by ejecta derived from impacts in the surrounding highlands over a long period of time.

If blanketing is the primary mechanism for long-term brightening of originally dark mare units, then higher albedo units must have been emplaced during a time when the cratering rate was much higher. This is inferred from the preservation of small isolated "pools" of dark mare plains found within the highlands. Such plains do not exhibit a bright halo near the highland-mare contact, as might be expected from ejecta transported from the surrounding highland terrains. If these deposits represent the most recent extrusions, but still older than three billion years, then the primary and secondary cratering flux must have decreased very rapidly. Such a decrease has been suggested from crater statistics (Hartmann, 1966).

If the higher albedo indicates a difference in the geochemical evolution of the magma chamber, then the later mare basalts appear to have been enriched in the mafic minerals, and the period of emplacement may have been relatively short. If the albedo is attributed to blanketing, then a long eruptive period is inferred, perhaps extending back to the time when the last major impact basins were formed. Thus, Mare Australe possibly was one of the earlier basins to be filled, or at least one of the better preserved regions reflecting this early history. The last period of inundation in Mare Australe postdates the formation of the large craters Humboldt (Plate 11b) and Schrödinger (Plates 2a and 49b), both of whose ejecta have been embayed by portions of Mare Australe, and apparently was contemporaneous with the formation of the mare-filled crater Jenner (Plate 46a), whose ejecta overlie portions of Mare Australe. Inundated basins, such as Mare Serenitatis and Mare Humorum, may have originally resembled Mare Australe but have undergone volcano-tectonic collapse with subsequent mare flooding, resulting in the burial of exposed craters characteristic of Mare Australe.

In addition to the mare plains shown in Plate 197, b and c, plains-forming units of even higher albedo have inundated several craters (arrows Y, Z, and XX) and local depressions (arrow YY). Such units are discussed in detail with Plates 198–200 and at the end of this chapter.

PLATE 197

a. IV-130-M   (oblique)

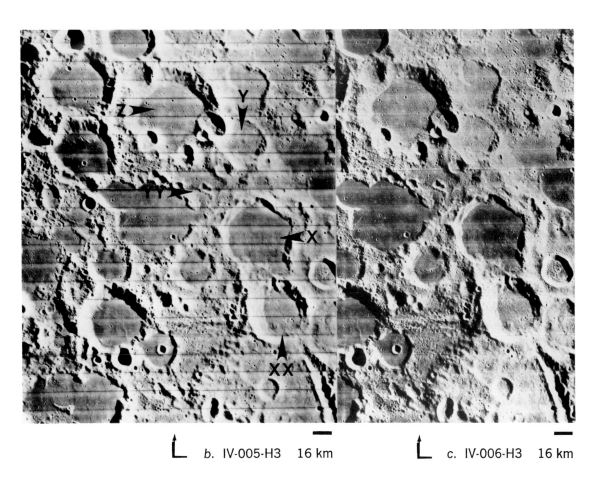

b. IV-005-H3   16 km   *c.* IV-006-H3   16 km

453

Highland Flow Units

Plains-forming flow units as well defined as the Imbrium flows on the maria (Plates 192–193) also occur in the highland areas but typically have higher albedos and numerous craters. Several units appear to be associated with the Orientale basin-forming event and are illustrated in Plates 198–199.

Plate 198a shows a region about 400 km north of the outer scarp of Orientale. The highly lineated flow features, interpreted as part of a massive ejecta flow from Orientale, are identified near the bottom of this plate (see Plate 210 for additional examples). To the north are numerous plains-forming units that suggest a range of flow viscosities.

The area shown in Plate 198b is included in that shown in Plate 198a (arrow X) and reveals a thick unit (arrow X) that apparently was derived from the breached crater to the north (arrow Y). Plate 198c shows other well-defined units (Plate 198a, arrow Y) whose surfaces display numerous subdued craters. A large contribution to the subdued crater density by nonimpact events is inferred by the preservation of the lobate termini (Plate 198c, arrow X). The crater doublet seen in Plate 198d (see Plate 198a, arrow Z) contains a relatively smooth unit with ridgelike borders (arrow X). This unit and the radial floor fracturing indicate floor resurgence. In contrast to these examples, arrow X in Plate 198e (see Plate 198a, arrow XX) shows relatively smooth units on the outer rim of Einstein A (Plate 198a, arrow YY) that apparently lack termini. These are similar in appearance, and perhaps origin, to the pools that commonly accompany large craters (see Plate 93). Note the narrow wall terrace in the adjacent crater (Plate 198e, arrow Y) which suggests floor collapse.

There are two general interpretations for the sources of these flow units: The units are derived from nearby volcanic vents or they are related to the Orientale event. The units shown in Plate 198, b and d, comprise evidence for the first suggestion. The second interpretation may be applicable for the units seen in Plate 198c, and it implies further that such units represent:

1. Multiple stages of the flowlike ejecta
2. Less viscous components of the lineated ejecta-flow deposit that were released at the termini
3. And/or terminal deposition of the ejecta flow

As terminal deposits, the ejecta-flow material lost much of its horizontal velocity but retained its relatively low viscosity. Consequently, such units behaved as lava. It is plausible that late-arriving secondaries from Orientale or a swarm of projectiles that accompanied the basin-producing projectile(s) penetrated the still-molten ejecta-flow deposits and produced volcanolike craters.

Since the ejecta flow is inferred to be a complex of hot volatile and liquid phases, it is reasonable to surmise that the terminal flow units contain a large proportion of volatiles. Such units may be conducive to the development of numerous collapse craters. Note that preservation of the lobate termini associated with these units also raises the problem of the degree of degradation of other unrelated postbasin features.

PLATE 198

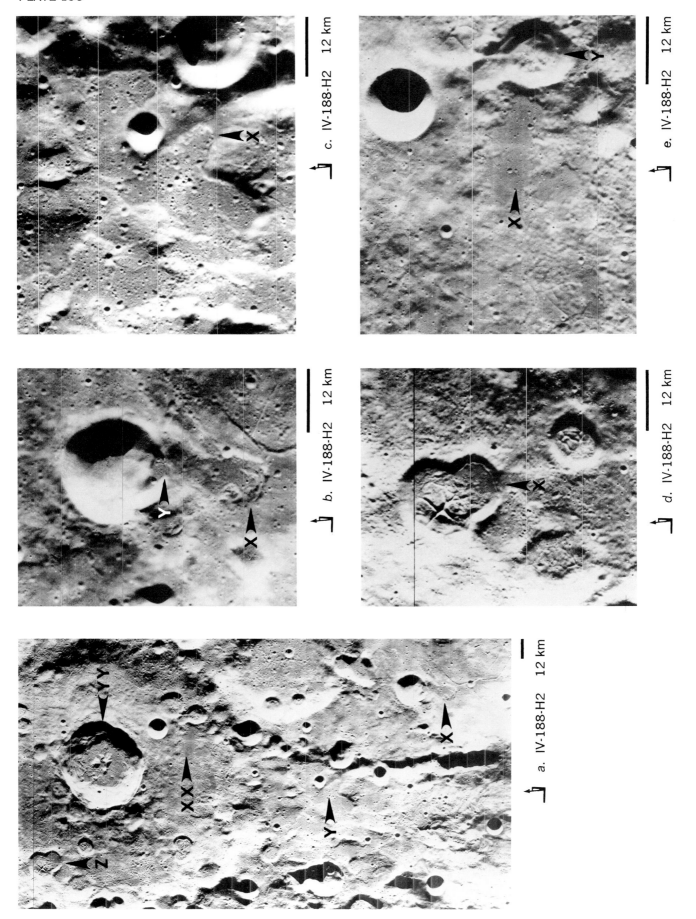

a. IV-188-H2   12 km

b. IV-188-H2   12 km

c. IV-188-H2   12 km

d. IV-188-H2   12 km

e. IV-188-H2   12 km

Highland Flow Units

Plate 199, *a* and *b*, shows in stereo plains-forming units that are associated with the Orientale basin (see Plates 201 and 211). The unit identified by arrow X exhibits a steep lobate front. Its surface has small SE-trending ridges and furrows parallel to the probable direction of movement. This flow overlies hummocky and striated units that are interpreted as part of the Orientale ejecta blanket. The borders of the unit marked by arrow Y are also lobate but exhibit numerous minor extensions (arrow Z). In addition, the NE-facing borders of this unit extend adjacent to the striated Orientale ejecta (arrow XX). Its surface exhibits furrows and ridges to the northwest (arrow YY) that are similar to the bulk ejecta from Orientale, but these lineations are less pronounced to the southeast (arrow ZZ). Plains-forming units without lobate borders typically are found in local depressions (arrow XY) and postdate the formation of the lineations radial to Orientale. The crater Wargentin (upper right; also see Plate 41, *c* and *d*) illustrates the inundation of the floor and northwestern rim by this type of unit.

Plate 199c reveals another steep-fronted flow (arrow X) ENE of the Orientale basin and adjacent to Grimaldi, whose floor is inundated by low-albedo mare units (right edge of the plate). This flow appears generally similar in albedo, crater density, and surface features to the flows shown in Plate 199, *a* and *b*. The flow termini are well defined away from Orientale but are poorly defined toward Orientale, where the unit appears to merge with the striated Orientale ejecta deposits. The flow also overlies the lineated and hummocky terrain (arrow Y), which is related to the Orientale event.

Grabenlike rilles (arrow Z) cross the flow unit (arrow X) and the "sculptured" terrain (arrow Y) but do not cross the mare units of Grimaldi. The rilles appear to be better expressed on the flow unit (arrow Z) than on the surrounding terrain (arrow XX). This difference in expression probably reflects a greater competence of the lobate flow unit.

The units shown in Plates 198–199 clearly postdate deposition of Orientale ejecta. The poorly defined and complex contacts between several of these plains-forming units with the lineated deposits from Orientale suggest a genetic relationship, which may be analogous to the pools and flows around smaller craters (see Plates 105 and 245).

Units without well-defined termini that occur within topographic lows, including craters, are similar in appearance to units displaying large flow fronts. This might imply a similar composition and age but not necessarily a similar mechanism of emplacement. It is possible that these units represent deposition of a *nuée ardente* type of flow. Mackin (1969) suggested such a mechanism for the inundation of crater floors by mare material. Although this proposal does not appear to be applicable for mare filling of craters, it may be applicable for filling by these higher albedo plains-forming units. There are several difficulties, however, in the Moon-wide application of such a proposal, and they are considered at the end of this chapter.

456

PLATE 199

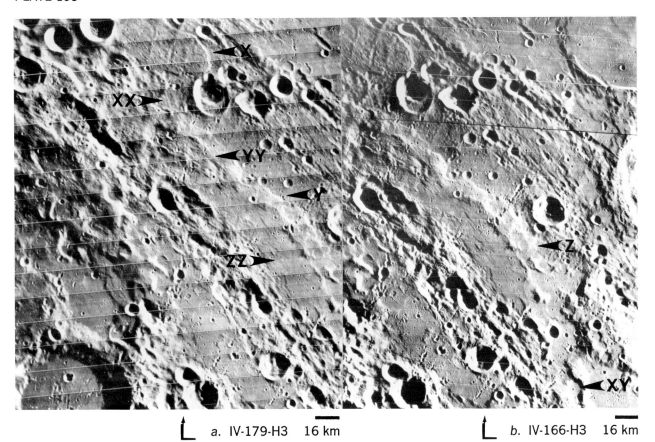

a. IV-179-H3   16 km

b. IV-166-H3   16 km

c. IV-168-H3   12 km

Highland Units

A major plains-forming unit of the highlands is the Cayley Formation (Wilhelms, 1964), which is characterized by a slightly
higher albedo than that of the maria and a higher density of
small craters (less than 2 km in diameter). At resolutions provided by terrestrially based telescopes, this unit appears widespread on crater floors and in local depressions. It postdates
formation of the Imbrium basin but predates the last stages of
mare emplacement (see Plate 43). Plate 200a illustrates its
occurrence west of the crater Delambre in the Southern Highlands within the breached crater Lade D (arrow X) and other
depressions (arrows Y and Z).

The southeastern portion of Lade D is shown in the upper
left of the stereo pair in Plate 200, c and d. This higher resolution view and an enlargement of the area shown in Plate 200b
reveal that the plains-forming unit is hilly at approximate wavelengths of 0.5 km–1 km. Its relatively rough appearance is the
result of several separate and overlapping topographic forms:

1. Numerous subdued craters that have broad bulbous rims
   (Plate 200b, arrow X).
2. Subdued furrows (Plate 200b, arrow Y), some of which break
   into crater chains
3. Low-relief mounds (Plate 200, c and d, arrow X)

The subdued crater with a central mound (Plate 200, c and d,
arrow Y) and relatively crater free patches (Plate 200b, arrow
Z) also appear to be generally characteristic of the Cayley unit
(see Plates 18c and 43). In addition, the tonguelike extension
of material from crater rim to floor (Plate 200, c and d, arrow
Z) is commonplace on the floor of Hipparchus, which is also
inundated by the Cayley Formation.

Arrow XX (Plate 200, c and d) identifies a small scarp that
borders at least the southeastern floor of Lade D. The disappearance of this scarp along the southern contact is probably
the result of unfavorable lighting conditions. Whereas such
floor boundaries do not appear to be characteristic in the limited
high-resolution coverage of this type of unit, other scarps, narrow clefts, and ridges are typical. A notable example is shown
in Plate 143b.

The scarp shown in Plate 200 (b, c, and d) may represent the
boundary of a flow unit. Subsequent subsidence could be responsible for its apparent traverse uphill. Such an interpretation requires a well-preserved surface and, in turn, requires
either a nonimpact origin of many craters or pre-existing craters
that were buried by the flow but revealed by later compaction
of the surface. Alternatively, the scarp reflects a fault zone that
might be relatively recent and that puts few constraints on the
origin of nearby craters. Note, however, that the floor was apparently raised relative to the crater boundary and that the fault
trace is continuous, without offsets and secondary faults.

At resolutions provided by Lunar Orbiter photographs, the
"Cayley Formation" appears to be a catch-all term for a variety
of plains-forming units, which include those shown in Plates
198–199. Consequently, high resolution and wide coverage of
the highland regions are required for meaningful statements of
genesis.

PLATE 200

a. IV-089-H3   12 km

b. III-082-M   1.5 km

c. III-081-M   1.5 km

d. III-082-M   1.5 km

DISCUSSION AND SUMMARY

A detailed discussion of each mare-filled basin on the Moon is beyond the scope of this work (see Stuart-Alexander and Howard [1970] for an extensive review of lunar maria and circular basins). Several general observations and their implications, however, can be made. Maria fall into two classes according to their shapes in plan—circular and irregular. Circular maria are further divided into the following subclasses:

1. Circular maria within well-defined mountain rings, such as:
   a. Mare Orientale (Plate 201)
   b. Mare Imbrium (Plate 192)
   c. Mare Crisium (Plate 113)
2. Circular maria within poorly defined or nonexistent mountain rings, such as:
   a. Mare Smythii (Plate 109b)
   b. Mare Nectaris (Plate 114)
   c. Mare Humboldtianum (Plate 1b)
   d. Mare Humorum (Plate 191)
   e. Mare Serenitatis (IV-097-M)
   f. Mare Tranquilitatis (IV-097-M)
   g. Mare Australe (Plate 197)

Irregular maria include Maria Procellarum (upper right, Plate 201), Fecunditatis (IV-084-M), and Marginis (IV-146-M). Irregular maria commonly occur in depressions beyond the outer scarp of circular mare-filled basins. For example, Maria Frigoris (Plate 192a) and Vaporum (IV-102-H2) are concentric with Mare Imbrium. Smaller units occur around Mare Crisium (Plate 113), and the partly inundated craters Grimaldi (Plate 199c) and Riccioli (Plate 213), it is suggested, represent analogous regions around Mare Orientale.

It is generally believed that most large circular basins represent impacts by enormous projectiles. The initial crater was probably enlarged considerably by slumping at grand scales along concentric rings of weakness or by shock-induced ripples of the lunar crust (see Plates 1 and 112). Subsequent inundation by mare material was a temporally separate event. This is indicated by basins having approximately comparable sizes, such as Orientale and Imbrium, but having distinctly different degrees of inundation. It is also suggested by the relatively recent inundation of basins that have only vestiges of

mountainous rings and ejecta blankets, which are assumed to be characteristic features of their initial appearances. Hartmann and Kuiper (1962) considered the morphologic evidence for such separate epochs of basin formation and inundation. Hartmann (1968), Stuart-Alexander and Howard (1970), and Hartmann and Wood (1971) later developed a time sequence of basin formation based on morphologic appearances and the crater density on their ejecta. As discussed below, however, basins may be structurally modified by processes related to the inundation by mare material.

Samples brought back by Apollo missions 11, 12, and 15 confirmed earlier indications that maria represent extrusions of basaltlike material. Relatively few flow termini, however, have been identified, with the exceptions of the large Imbrium flow sheets (Plate 192) and other relatively minor flows (Plate 164, b and c). Earth-based views of the lunar surface under low-angle solar illuminations reveal features only tens of meters high. The relative absence of flow termini, despite this broad telescopic coverage as well as the restricted but high-resolution coverage from the Orbiter missions, has several possible rationalizations:

1. Mare units were emplaced as numerous thin (less than 10 m thick) lavalike units that did not develop large fronts and that would be eroded with time.
2. The large Imbrium flows indicate that relatively thick flow units extend hundreds of kilometers without marked changes in thickness (see Strom, 1965). Previous extrusions may have been much greater in volume and extended to the highland contacts. Features are identified at such contacts in high-resolution photography that suggest preservation of termini or forms related to mare emplacement (see Plates 151–152, 177b, 193, 196, and 249).
3. Emplacement of mare units was accompanied by events not typical of terrestrial flows. For example, the upper flow surface and termini may have been highly vesicular and, consequently, vulnerable to lunar erosional processes.
4. Several "wrinkle" ridges (for an example, see Plate 145a) may represent flow fronts. Hence, the absence of termini may, in part, be the result of misinterpret-

ing such ridges as compressional or extrusively constructed features.

The major vents for mare flooding of circular basins were probably along ring fractures, some of which represent pre-existing weaknesses related to basin formation. This is indicated by comparisons of partly inundated basins, such as Orientale, and more completely inundated basins, such as Imbrium. Other evidence for these sites of eruption includes the following observations and interpretations:

1. Sinuous rilles typically occur along the margins of the maria. The "head" craters commonly are within or near the adjacent highlands. If sinuous rilles are volcanic in origin, then their vents—i.e., the head craters—reflect eruptive sites.
2. Low-relief mare domes commonly accompany insular massifs, which are interpreted as remnants of the concentric mountainous rings.
3. Wrinkle ridges typically interconnect insular massifs and are interpreted as stressed zones that could have been feeders for fissuretype eruptions.
4. Fissurelike rilles with low-albedo deposits occur on The borders of several maria. At least one of these rilles has a clearly identified flow extending from it (see Plate 164, b and c).

In addition, mare material is erupted through calderas and pre-existing craters that are linked to the magma chambers responsible for the mare basalts.

Inundation of less circular and irregular basins may occur, in part, through overspill from adjacent maria. This is suggested by sinuous rilles that link adjacent maria (see Plate 182a). Other important contributors to the volume of mare units are separate vents indicated by mare domes (Plate 125), volcanic cones (Plate 130, e and f), and volcanic ridges (Plates 131–132). Some wrinkle-ridge systems also may contribute through fissurelike eruptions. The following observations suggest, however, that the magma chambers were more widely distributed:

1. Fractured floors of craters occur on the borders of numerous irregular maria (see Plates 35 and 109b) and within isolated basins without mare fill (see Plate 34d). Such floors are thought to indicate a link with underlying or nearby magma chambers independent of basin formation.
2. Mare Australe and similar maria are expressed by craters that have mare units on their floors. These maria are far removed from other mare-filled basins.

It is possible that the maria are, in effect, windows to a once-molten lunar mantle. The largest basins have the deepest roots and consequently are most likely to tap this layer (see Wood et al. [1970] and Wise and Yates [1970] for proposed geophysical models). Both the wide range in mare fill for different basins and the existence of mare units within extremely old basins might suggest that the basin-forming events did not immediately tap a mantle layer. Subsequent degradation and later mare emplacement, however, could mask evidence for such tapping, and perhaps the extensive flows surrounding Orientale (Plates 198 and 199) represent ejected remnants of a once-molten mantle.

The concentration of maria on the lunar Earthside suggests a mantle layer that is either asymmetrically distributed or shallower on the Earthside. In addition, there appears to be a more subtle concentration of mare material along a great circle (Lipskiy et al., 1966; Stuart-Alexander and Howard, 1970), which is inclined approximately 50° to the lunar equator. The following major maria have geometric centers within 25° of this great circle: Imbrium, Serenitatis, Tranquilitatis, Fecunditatis, Nectaris, Crisium, Australe, and the Ingenii-Apollo complex. One important exception to this distribution is Oceanus Procellarum, a large irregular mare that is peripheral and perhaps related to Mare Imbrium (discussed below).

Two preliminary interpretations can be made. First, this apparent distribution of mare material may reflect a concentration of major lunar basins and, therefore, a preferred zone of early basin-forming events. The geometric centers of approximately seventeen of the thirty-two lunar basins are found within 25° either side of a great circle that is inclined N50W and crosses the lunar equator at longitude 45° E. The thirty-two basins include those recognized by Hartmann and Wood (1971) and in addition include Maria Australe, Serenitatis, Tranquilitatis, and Nectaris. This collection also includes the large complex around Apollo, which is interpreted as the remnants of an enormous basin rivaling Imbrium in size (see Plate 135). The 50°-wide belt comprises 42 percent of the total surface area of the Moon; consequently, seventeen out of thirty-two basins is not a significant statistic. But, these seventeen basins represent about 74 percent of the total basin surface area, if the most prominent ring (typically the third or second ring) is considered as the basin diameter. In fact, six of the seventeen basins having their geometric centers within this belt are greater than 600 km in diameter, in contrast to two of fifteen basins outside the belt. Thus, there is a marginal possibility that the major basins were formed preferentially along a great circle.

An alternative explanation is that mare inundation was "localized" through a major crustal fracture described by this great circle. Mare Fecunditatis displays a much greater surface area than the adjacent mare, Nectaris, yet Nectaris appears to be one of the larger basins (840 km in diameter, [Hartmann and Wood, 1971]). This seems inconsistent with the hypothesis that all concentric basins are punctures of the lunar crust. In addition, it could be argued that Mare Tranquilitatis (and perhaps even Mare Serenitatis) does not represent an old ringed basin. Extrusive centers for this mare appear to be centered near Jansen and Jansen B, which are northwest of the geometric center. This is in contrast to the typically peripheral vents of other maria. Therefore, Maria Tranquilitatis and Fecunditatis might represent volcano-tectonic basins formed along this belt.

These two alternative hypotheses depend on the interpretation of Maria Tranquilitatis, Serenitatis, Australe, and the Apollo complex. If these large basins are vestiges of impact scars, then perhaps there was a nonrandom cratering history. If they are not the result of impacts, then their localization of mare units indicates basin subsidence along a large crustal fracture.

El-Baz (1972b) has recently identified a large basin outside this belt of mare. It is an unusual multiringed basin in that the ratio of the exterior ring to the next interior ring departs from the value $\sqrt{2}$, being approximately 2, in contrast to other lunar basins (see Plate 1 and the end of Chapter 1). Furthermore, this ancient basin is adjacent to another ancient basin, Mare Smythii, as well as smaller mare-filled craters, yet it does not contain a significant unit of mare material. Such an exclusion of mare units, including those on the floors of craters, seems to require effective "plugging" of possible conduits along crustal weaknesses originally created by the basin-forming impact. This "plugging" may have been the result of its proposed formation early in lunar history; however, it should be noted that weaknesses within other very ancient basins have not been plugged, as demonstrated by their mare fill or mare-filled craters within their borders—even on the lunar farside. If the identification of this basin is correct, then the significance of the preceding statistics is lessened, and mare flooding seems to have been controlled by a major crustal weakness or fortuitous alignment of the largest basins.

Several significant features that accompany mare emplacement and their interpretations follow:

1. Commonly, segments of crater walls appear to have been removed by structural dropping, rather than tilting of the basin, with subsequent inundation. This may indicate foundering of the crust. In some examples, premare craters surrounded by mare units are bordered by outward-facing scarps, which suggest a form of mass wasting associated with mare emplacement (see Plate 110, a and c).

2. Irregular and polygonal platforms are produced along the periphery of maria. Several such platforms appear to be relict relief (see Plates 113 and 122) left after partial withdrawal or subsidence of mare units.

3. Several old basin structures, which are without well-preserved mountainous rings, have well-defined border scarps that indicate renewed basin subsidence. In particular, note western Mare Humorum (Plate 168), and Maria Serenitatis (Plate 166), Smythii (Plate 109b), and Nectaris (Plate 114). Subsidence also is indicated by concentric grabenlike rilles and some forms of wrinkle ridges.

4. Irregular terraced depressions in several maria are interpreted as receded lava lakes (see Plates 87a and

462

177*b*). Holcomb (1971) came to a similar but independent study of similar features.

5. Terraces and other indications of high-level marks commonly occur near the base of the highlands that border the maria (see Plates 174, 249, and 256, *c* and *d*).

6. Low-albedo shelves around Mare Serenitatis (Plates 138 and 166) and adjacent to Sinus Aestuum (Plate 188*a*) suggest that extensive pyroclastic eruptions predated the last stages of mare emplacement. Several—but in general, isolated—cinder cones and volcanic ridges postdate the mare units and indicate that such activity continued after the last major epoch of mare flooding.

As mentioned in the Discussion and Conclusion section of the chapter "Negative-Relief Features," mascons (mass concentrations) are associated with numerous circular basins. Hartmann and Wood (1971) point out that there is a good correlation between the total mass of a circular basin and its estimated volume of mare material. They also state that there is no correlation between the basin age and mascon mass; they conclude, therefore, that the mascons are the result of the mare material and not a buried meteoroid. It should be noted, however, that they did not include the irregular and less circular maria. As a result, Mare Serenitatis was the only large "old" basin considered. Since Hartmann and Wood (1971) used the total mascon mass, they integrated mass points over the surface area covered by the maria; therefore, mare-filled basins with the largest surface area generally will have the largest mascon mass, regardless of any other morphological criterion.

The following list provides a comparison between a mare-filled basin and the associated maximum mass point, which was taken from Sjogren et al. (1971). The surface mass is expressed in $10^5 g/km^2$. Starred values indicate that the data may not be reliable, owing to the nearness of the mare region to the terrestrially visible limb. The list is subjectively arranged so that the most circular mare in each group is at the top. The number in parentheses corresponds to the relative age of the basins (the youngest being number one) as proposed by Stuart-Alexander and Howard (1970).

1. Circular maria within well-defined mountain rings
   - a. Crisium   11   (3)
   - b. Imbrium   9   (2)
   - c. Orientale   7*   (1)
2. Circular maria within poorly defined mountain rings
   - a. Serenitatis   11   (6)
   - b. Humorum   8   (4)
   - c. Nectaris   11   (5)
   - d. Humboldtianum   5*   (7)
   - e. Smythii   5*   (8)
   - f. Tranquilitatis   4   (10)
   - g. Australe   4*   (11)
3. Irregular maria
   - a. Marginis   3*   (—)
   - b. Fecunditatis   2   (9)
   - c. Procellarum   3   (—)

The list reveals that in general the most recently formed basins have the greatest central surface mass, a correlation not revealed when the total mascon mass is considered. Furthermore, there appears to be a slightly better correlation between the central surface mass and the circularity of the maria, the irregular maria having the least central mass.

The maximum surface mass may indicate a buried Fe-Ni projectile, but unaltered lunar craters—i.e., not filled with mare units and without floor fractures—almost always display negative anomalies (Sjogren et al., 1972). Consequently, it seems more likely that the region of greatest surface mass corresponds to the region of greatest depth of the mare units, and this depth seems to correlate with the circularity of the mare plan. A reasonable model is that the most circular maria reflect a region of volcano-tectonic collapse similar to the idea presented by Quaide (1965). Such an event has been suggested previously to account for the low-albedo shelf regions (see Plates 166 and 188–189) and the mantling of the peripheral highland terrains through eruptions of low-albedo pyroclastics. As noted in Chapter 7, fault scarps border several maria and are responsible, in part, for the circularity and abruptness of the mare-highland contact. These scarps commonly are not directly related to the peripheral mountain rings. Volcano-tectonic collapse would permit greater volumes of subsequent mare units to be emplaced; furthermore, these maria may represent windows to a lunar mantle as theorized by Wise and Yates (1970).

Rather than proposing buried impact basins for newly discovered mascons, one can envision collapse and mare filling unassociated with a major impact structure. For example, Sinus Aestuum (Plate 188) contains a positive gravity anomaly and is bordered at several locations by low-albedo shelves.

The general correlation between basin age and central surface mass is thought to indicate a gradual "freezing" of faults associated with the oldest basins. This process was probably accomplished through slow isostatic adjustment of the basin prior to the epoch of mare emplacement. Impact basins formed immediately preceding or during the development of the lunar mantle would be drastically out of isostatic equilibrium. Rapid adjustment of the basin floor would produce extensive faulting and thus potential conduits for extrusions. Copious eruptions of mare units on top of the original basin floor eventually caused foundering and collapse, a sequence common to terrestrial calderas and apparently to some lunar craters (see Plate 40). The older basins developed less extensive fracture systems, and the relatively limited extrusions of mare units helped to compensate for a previous gravity low. In some regions, overcompensation resulted in a small gravity high. For example, the slightly overburdened crust in the region of Mare Tranquilitatis responded by the development of concentric grabens, which are still preserved on the maria. In contrast, the old basin floor within Mare Serenitatis foundered, and subsequent eruptions of mare units destroyed portions of a similar graben system related to this early period of crustal sagging. It is possible, however, that collapse was initiated in a small area of Mare Tranquilitatis corresponding to the "ghost" ring Lamont.

Although Stuart-Alexander and Howard (1970) interpret the Serenitatis basin as a relatively recent structure, Hartmann and Wood (1971) believe it to be very old, older than the Smythii basin. On the basis of the preservation of the surrounding mountain rings, the age determined by Hartmann and Wood (1971) from rela-

tive crater densities seems more reasonable. This age appears to be completely out of place in the preceding list, but its placement is understandable in terms of a model in which the morphologic appearance—in particular the circularity of the maria and the abruptness of the mare-highland contact—has been rejuvenated by volcano-tectonic collapse.

Many of the irregular maria are peripheral to the major mare-filled basins. For example, Oceanus Procellarum, Mare Frigoris, and Mare Vaporum are peripheral to Mare Imbrium; Palus Epidemiarum is peripheral to Mare Crisium; and Lacus Somniorum, to Mare Serenitatis. Concentric grabens around the Orientale basin intersect Grimaldi, Riccioli, and Rocca A, all of which have mare units on their floors. In addition, other volcanic features in this zone (see Plate 163, arrows YY, ZZ) suggest that the Orientale basin would have developed comparable peripheral irregular mare plains if greater volumes of magma had been available. This annular zone of weakness was probably produced during the impact event responsible for the Orientale basin, perhaps where stresses were not great enough to form another mountain ring but sufficient to induce large crustal weaknesses.

There seems to be a greater likelihood of the formation of an irregular maria where this inferred annular zone intersects the inferred annular zone of an adjacent basin. For example, Mare Fecunditatis lies between Maria Smythii, Nectaris, and Crisium; Mare Spumans between Maria Smythii and Crisium. It is also possible that the extensive mare plains of Oceanus Procellarum is related to the enhancement of crustal weaknesses between the Imbrium and Orientale basins, the two most recent major basins. Moreover, the massive outpourings of basalt within Mare Serenitatis could be a result of its location along the inferred annular zone associated with the Imbrium basin. Intersecting crustal fractures are likely regions for volcanism.

It should be noted that the mare plains are apparently not the only cause for positive gravity anomalies. Recent gravimetric maps by Sjogren et al. (1972) reveal a relatively large anomaly (about half of that associated with Mare Nectaris) near the crater Crüger (Plate 163) where comparatively limited mare units have been emplaced. In addition, a prominent positive anomaly has been noted west of Mare Nectaris in a region devoid of mare units. These identifications may indicate a relatively thin lunar crust overlying a dense mantle.

The surface mass distribution by Sjogren et al. (1971) reveals another significant feature related to mare emplacement. Contours of positive gravity anomalies in Oceanus Procellarum follow a NW trend that approximately corresponds to the wrinkle-ridge system in this region. Near the Marius Hills (a local positive gravity anomaly), the contours have a NNE trend, which corresponds to another wrinkle-ridge system between the Marius Hills and the Aristarchus Plateau. A similar correlation between topographic features and positive gravity anomalies occurs in northwestern Mare Tranquilitatis. The largest point-mass value approximately coincides with the "ghost" ring Lamont, which has a radial system of wrinkle ridges. A major wrinkle-ridge system extends northeastward and connects with a complex of volcanolike forms. This also corresponds to a trend of the gravity-anomaly contours. Such correlations are consistent with the interpretation that some wrinkle ridges are eruptive centers related to mare flooding (Guest and Fielder, 1968). The "ghost" ring Lamont may indicate the initiated but arrested collapse within Mare Tranquilitatis. This center of inferred volcanism perhaps corresponds to an old crater, originally formed by an impact but subjected to such modification that it now should be labeled a caldera.

These observations and interpretations yield the following generalized picture. A molten mantle developed beneath the lunar crust as a result of radiogenic or tidal heating. The asymmetry in the distribution of maria may reflect an asymmetry of this lunar mantle. If the mantle formed when the Moon was near its present orbit around Earth, then gravitational coupling later locked the Moon into such a configuration that a greater concentration of previous eruptions from this layer would face Earth. This means that the asymmetric distribution of the mantle was the cause for the present nearside concentration of maria, rather than the effect of migration of denser materials toward Earth. Reversal or modification of this cause and effect is possible only

if the Moon was very close to Earth during early lunar history, which is consistent with dynamical models. Smith et al. (1970) suggest that such gravitational interaction produced such a concentration of dense mantle material at shallower depths on the lunar nearside.

Enormous impacts may have immediately tapped this layer, but many developed fractures that penetrated deep into the Moon and later became paths for surface eruptions. Whereas a thin crust on the Earth is not prerequisite for surface eruptions, it is important on the Moon, where impact structures are roughly analogous to wells that extend to the mantle. In several instances, basins did not become eruptive centers, but a smaller crater or craters within them did (see Plates 34d and 74b).

Irregular depressions in the maria, some of which are interpreted as subsided lava lakes, and terraces on the mare shores imply that a considerable volume of material may have been erupted at one time. It is suggested that in some regions the basins became large molten pools and that the terraces may represent the results of subsidence or perhaps even ancient tides. Irregular and polygonal depressions within the Orientale basin (Plate 203) are evidence for crustal foundering into a subsurface magma chamber. The faultlike boundary scarps in Maria Humorum and Serenitatis perhaps reflect similar but larger scale events, and it is believed that several maria have undergone several stages of such foundering and enlargement as suggested by Quaide (1965). Present surface features, such as volcanic ridges (Plates 125 and 131–132) and volcanic cones (Plates 60a, arrow Z, and 130, e and f) are possibly the last expressions of relatively minor eruptions associated with this adjustment. Preservation of surface features that accompanied mare emplacement is one of the most reliable indicators of postmare surface processes and should be compared to craters of comparable size scales.

In summary, several inundated basins are viewed as enormous volcano-tectonic basins that were originally impact produced but were subsequently modified at a much grander scale than other crater floors. The hummocky ring zones within Orientale (Plates 205 and 207) and the low-albedo shelf on the margins of Mare Serenitatis provide clues to the preinundation surface and early eruptive history. It is reasonable to surmise that numerous volcanic structures, including large calderas, were formed during this period of extensive volcanism. Numerous crater-topped cones and domes border northern Mare Serenitatis (see the bottom of Plate 66, a and b) and may be remnants of extensive peripheral eruptions prior to mare inundation.

Several irregular maria are not simply regions of overspill from adjacent maria. Western Oceanus Procellarum is bordered by floor-fractured craters (see Plate 35) that suggest a subsurface link to magma chambers responsible for mare flooding. This linkage may be through crustal weaknesses created during crater formation and connected to deep-seated fractures that extend beneath the maria. Alternatively, the material represented by the maria extends beneath the highlands and is tapped directly by fractures beneath some highland craters. The latter suggestion is consistent with the maria interpreted as windows to an underlying mantle.

Vents for irregular maria may have been along weaknesses that now remain as wrinkle ridges. It is probable that large craters within the maria also became important eruptive centers. Volcanic complexes, such as the Marius Hills (Plates 263–264), are interpreted as sites of late stages of mare eruptions. Emplacement of the maria did not occur during one short epoch. Numerous craters in Oceanus Procellarum have inundated ejecta blankets. These craters are relatively small (1 km–25 km) and must have formed on a previous plains-forming unit. Otherwise, one would expect many more remnants of larger craters, whose rims represent the highest preinundated surfaces, than those observed. Partly inundated craters smaller than 1 km in diameter indicate that some individual mare units during the late stages of mare emplacement were thin and fluidlike.

Although the low-albedo maria are the most distinctive plains-forming units, higher albedo units compose the floors of numerous highland craters and local depressions. In general, their surfaces are plains at large scales (greater than 5-km wavelengths) but exhibit numerous craters, subdued depressions, furrows, ridges,

and low-relief hills at small scales (less than 1 km). These units are identified with and without well-defined lobate termini. Ring structures similar to "ghost" rings on the maria are commonly found (see the top of Plate 92,d and e). In addition, terraces (IV-089-H3) or narrow scarps (Plate 200) appear to border several of these plains units and may be analogous to the high-level marks characteristic of the mare borders. Without detailed studies with high-resolution photographs, it is premature to ascribe all these units to a single origin. The presence or absence of well-defined termini is no more reliable as a criterion for different geneses than it is for the maria, which typically do not have well-defined termini.

Whatever their origin, these highland units generally predate mare emplacement, but many units on the lunar Earthside postdate formation of the Imbrium basin. This constraint may provide information on early cratering rates during a relatively short period of time (approximately 0.5 billion years). Alternatively, the higher crater density and higher albedo than those of mare surfaces are clues for differences in unit type rather than cratering history. The latter alternative is plausible because of the following observations and comments:

1. Preservation of termini (see Plates 198c and 199c) seems inconsistent with a history of intense cratering by primary meteoroids. This might indicate a high density of secondary craters that are not indicative of smaller scale bombardment, or it might imply numerous non-impact-produced craters characteristic of these units.
2. Small-scale flow textures (500 m wavelengths) on Orientale-related units (Plate 199c) also indicate a relatively well preserved surface.
3. Preservation of narrow sinuous rilles on their surfaces (Plates 74c and 111) might place a constraint on degradational processes if the rilles are channels formed during the emplacement of these units.

Other characteristic features of several of these highland units include: small (less than 1 km in diameter) subdued craters and depressions having terraces and narrow upper-wall scarps; convex-floored craters; and small regions (typically less than 10 km² in area) that are relatively devoid of craters 0.1 km–0.5 km. If the terraces and narrow upper-wall scarps (see Plate 51) of small craters are interpreted as an indication of subsurface layering, then they are evidence for a relatively well ordered history at shallow depths rather than a chaotic jumble. Alternatively, they may represent recent slumps within a relatively homogeneous medium that were triggered by seismic events. It is noted that similar craters occur on the ejecta blankets of recently formed craters and on the ejecta blanket of their secondary craters. In addition, small convex-floored craters on these units (Plate 18) are typically 0.5 km–1 km in diameter, in contrast to similar mare craters, which are less than 0.4 km in diameter. According to conclusions reached by Oberbeck and Quaide (1968), such craters may indicate a subsurface competent layer. Such floor profiles might indicate either a subsurface layer or the convergence of debris slides from the wall that were triggered by seismic events.

The close spatial association of several lobate-bordered units with the Orientale ejecta blanket indicates a genesis related to basin-forming events. These units are thought to represent a part of basin ejecta that remained molten after deposition and separated from the bulk debris at their termini. The sheetlike flows around Aristarchus (Plate 245) are possible small-scale analogues.

These direct links to large basins suggest that similar units, including the Cayley, may be related to major basin-forming events. Wilhelms (1964) proposed, among several alternatives, that the Cayley unit might be the result of a *nuée ardente* type of flow, which travels over obstructions and is partially deposited in local depressions. Around the Imbrium basin, such deposits clearly postdate the Imbrium sculpture, which is a combination of bulk ejecta deposits and structural features, but they do not commonly occur close to the basin rim. Consequently, the origin of the unit can be envisioned as deposits from a rapidly expanding cloud-like ring. Near the basin rim, bulk ejecta bury any possible deposits from such a cloud, whereas farther from the rim, they are part of a matrix of fast-moving material that erodes and blankets pre-existing terrain. Fallout from this cloud fills topographic lows with plains-forming material. The relatively high crater density on this unit may be the result of late-arriving secondaries, endogenetic forms, or pre-existing features, revealed by compaction of the deposit. Similar units not associated with basin sculpture may be due to fallout from a temporary atmosphere produced by the enormous basin-forming event. On the basis of Apollo 16 and Lunar Orbiter photography, Eggleton and Schaber (1972) have attributed the Cayley Formation in general to be the result of such fluidized ejecta from major basins or tertiary ejecta from basin-produced secondaries. Differential compaction of this fluidized ejecta could reveal buried craters as highly subdued or "ghost" craters and leave high-level marks on the walls of craters inundated by this volatile-rich banket.

It is difficult to imagine the magnitude of the events that accompanied the formation of enormous lunar basins and their effect on pre-existing topography on a Moon-wide scale. Aristarchus shows evidence for possible blanketing of the mare plains by fluidized ejecta as far as four crater radii from its rim (see Plate 258), a process that might be comparable to the emplacement of the Cayley Formation but at considerably smaller scales. This also has been noted by Head (1972) for small-scale units associated with Aristarchus secondary complexes.

The depth of penetration by an impact responsible for large multiringed basins suggests that a portion of the fluidized ejecta may have been derived from released magmas in a manner similar to that envisioned by Mackin (1969). The proportion of ballistic and erupted debris probably would depend on the size of the basin and the time of formation. Note that the multiringed crater Schrödinger (Plate 49b, arrow Y) exhibits extensive plains units overlapping its western rim. As illustrated in Plate 105, plains units on the inner rim zone of large craters are common; however, the units surrounding Schrödinger are much more extensive. In contrast, similar units associated with the nearby but smaller crater Antoniadi (Plate 49b, arrow X) appear to be absent.

Several problems remain if all highland plains units are attributed to ejecta flows from large basins. Plate 41a shows that the ejecta blankets of small craters have been embayed by such units, yet the crater interiors have been relatively unaffected. This relation indicates that the smaller craters were formed prior to the emplacement of the plains material and that the crater interiors were not inundated during this emplacement. Such a sequence seems accountable under the following hypotheses:

1. The small craters were formed at the same time as the plains units, perhaps by secondary impacts.
2. The plains units were derived from ejecta melt that impacted adjacent terrain and that then embayed these crater rims by topographically controlled flows.
3. The plains units represent local extrusively derived lavas.

This specific problem leads to the more general one in which certain depressions have been filled by such units, whereas adjacent depressions remain relatively unfilled. For example, the crater Fabry (see Plate 49a) exhibits an extensive smooth-surface floor that has embayed the ejecta blankets of several small craters. The crater is relatively old or extensively modified, perhaps from the formation of the Humboldtianum multiringed basin. If the floor unit was derived from this basin, then the absence of smiliar units in the surrounding terrains seems unusual.

Another enigma is the wide extent of the inferred blanketing. Plains units occur at considerable distances from a major multiringed basin, yet the inner rim zone of the Imbrium basin (Plate 112) apparently has not been blanketed by the last major multiringed basin, the Orientale basin. However, regions far removed from major basins are not far removed from smaller, but still enormous, craters that may contribute to the mantling of nearby depressions over long periods of time.

The existence of sinuous rilles in the high-albedo plains prompts the analogy to similar rilles in the maria. If these rilles indicate fluid flow from a single vent, then portions of these units can be interpreted as results of volcanic eruptions. However, such rilles also may indicate fluid flow after emplacement of an ejecta melt. Although a structural origin cannot be ruled out, sinuous rilles in the highland units commonly are isolated from other structural features and appear to be topographically controlled. A notable example is Rima

Bode III, which is included in Plate 188a (also see IV-109-H1,H2). This sinuous rille extends from an elongate "head" crater through a high-albedo plains unit, which is bordered by terraces, and across a lower albedo mare-like shelf. It is interpreted as a lava channel or collapsed lava tube and is believed to have been related to the emplacement of the high-albedo plains unit.

The range in albedos within a single mare and the existence of relatively high albedo maria, such as Mare Australe and units associated with the Apollo-Ingenii region, suggest that some extruded lavas may not have been as dark as those usually associated with the mare plains. This interpretation must be held as a viable alternative to Moon-wide blanketing by impact basins. Isolated craters on the lunar farside show evidence for internal modification of their floors (Plate 29, a and b) without emplacement of mare units. Consequently, mare units should not be considered the only possible lunar lava. As remarked in Chapter 8, however, high-albedo plains units also might indicate surficial blanketing by ejecta derived from impacts in the adjacent highlands. Thus, many of the highland plains simply may represent a combination of early extrusive inundation and subsequent mantling by a veneer of impact ejecta.

As demonstrated throughout this book, a universal origin is not a reasonable goal for lunar features any more than it is for terrestrial features, and the origin of the high-albedo plains units is no exception.

General Observations

The floors, walls, and rims of craters typically possess different histories of modification:

1. Floors of numerous craters greater than 15 km in diameter preserve a record of lunar thermal history.
2. Walls reflect the degradational history and modifications of the floors.
3. Rims, through their appearance in plan and profile, are potential indicators of crater origin. They also may be used to measure crater age, if the original appearance can be established.

The large number of possible permutations of the various appearances of the floor, wall, and rim zones for each crater size leads to a prematurely detailed classification of crater types. For useful classification at the present stage of lunar morphology, we must determine which zone or zones are characteristic of a large number of craters. In a preliminary classification, I delineated more than forty different crater types, and these do not include probable degraded versions. This profusion indicates the difficulty in such a compilation. Nevertheless, classifications are instructive, for it is believed that they mirror common origins or histories of craters in the same class. Future quantitative studies should aid in selecting the parameters that will either reduce the number of principal crater types or add to their significance.

Several general observations result from this morphologic study of lunar craters:

1. Many small (less than 1 km in diameter) subdued craters on mare and marelike surfaces are thought to be nonimpact produced. The occurrence of other endogenetically produced features smaller than 0.2 km in diameter—such as collapse pits, irregular depressions, small-scale domes (tumuli?), and possible flow textures—indicates relatively recent formation or very well preserved surfaces. Degradation of the surface by meteoroid bombardment, it is suggested, affect most noticeably craters with diameters less than 100 m. This is consistent with the quantitative study of small-scale cratering by Chapman et al. (1970).
2. A significant number of craters 0.5 km–30 km in di-

ameter appear to be volcanic in origin because of the following perceptible distinctions:
   a. Association with or similarity to endogenetic-appearing chains or clusters
   b. Inconsistencies in reasonable models of degradation of impact craters
   c. Association with endogenetic forms
   d. Unusual rim profiles and floor features
3. Periods of volcanic activity modify the floors of large craters and are expressed by:
   a. Narrow interior terraces on crater walls that are suggested to be previous floor levels
   b. Highly fractured floors of craters along the borders of the maria and within some closed basins, which are partly inundated by mare units
   c. Volcanic-appearing features, such as nested craters, flow features, domes, and conelike forms
   d. Central peaks of several craters that have been volcanically formed or at least volcanically modified. (Although collapse is indicated by central peaks having summit pits and by annular peak complexes, this collapse is believed to be related in most—but definitely not all—cases to the crater-forming process or events soon after crater formation.)
   e. Inundated or partly inundated floors that clearly were formed at an epoch much later than that of crater formation
4. The rim areas of some large craters also indicate postformation modifications, such as flow features and irregular collapse forms.

Impacts, however, represent one of the most devastating modifiers of the lunar surface. The formation of a crater larger than 25 km in diameter produces ejecta that crater, blanket, and erode the surrounding terrain. Widely distributed secondary craters associated with such impacts generally mask the long-term contribution to the density of small craters by small primary impacts. The initially subdued appearance of these secondary craters camouflages their importance by making them almost indistinguishable from much older degraded craters and possible endogenetic craters. The relative contribution of small-scale secondary impact craters, primary impact craters, and endogenetic craters to crater statistics is as yet unknown. This problem is further complicated by the much greater rate of crater destruction in regions having deep layers of incompetent material. A proper solution requires the recognition of pristine morphologies in areas of widely differing surface and subsurface properties. Logical sites for such a study would be the different zones of large and recently formed impact craters.

The variety of positive-relief features indicates a complex history of the lunar crust. Structural uplift and subsidence accompanied the inundation of the maria and resulted in isolated polygonal and irregular platforms. In several mare regions, irregular platforms were left in relief after withdrawal of surrounding mare basalts. Volcanism is expressed by cinder cones, tephra-ringlike forms, spatter cones, volcanic ridges, marelike domes, mamelon domes, and large furrowed domes. Such features reflect the wide range of viscosities of extrusions, and their spatial distribution across the lunar globe provides data for the late-stage thermal history of the crust or mantle.

Some wrinkle ridges are also volcanic in origin and appear to be the result of extrusions along fractures. Their preservation indicates either very slow surface processes at scales smaller than 100 m or recent emplacement. Several wrinkle ridges are suggested to be flow fronts or ramparts, which were left in relief after removal of underlying lava. This removal may have been in response to ancient Earth-produced tides. Other wrinkle ridges are interpreted as compressional folds due to subsidence of the mare basins or crustal thrust faults. Structural influence is indicated by topographic displacements across the ridges, by en échelon displacements, and by the extensions of these ridges across highland terrain.

Negative-relief features include irregular depressions that resemble lava lakes, polygonal depressions that are structurally dropped plates, and rilles that reflect numerous different processes. The discontinuity of several sinuous rilles suggests the formation of lava tubes or large fractures. The existence of sinuous rilles that cross highland terrain, however, indicates erosive prop-

erties or structural origins (or, at the least, structural control) for some rilles. If sinuous rilles and lava-lakelike depressions are features that were formed during the emplacement of mare units, then they may be reliable indicators of subsequent surface processes.

The inundation of the large basins occurs through extrusions along the base of the boundary scarps and through numerous peripheral feeders, which include rejuvenated impact craters and calderas. Several mare-filled basins are thought to have undergone multiple stages of collapse, followed by extensive outpourings of mare basalts and a rejuvenation of the basin appearance. Prior to this series of events, extensive plains-forming units were emplaced in the highlands by both local extrusions and widespread blanketing by ejecta and eruptions associated with the formation of the major impact basins. The high albedo and large number of craters on some of these units may indicate short-term events associated with their emplacement—such as endogenetic processes and secondary cratering—rather than long-term exposure to primary cratering.

The post-Apollo period of absorbing and synthesizing the cornucopia of new data will result in constraints on the various hypotheses reviewed throughout this book. These new data must be treated, however, with the realization that planetary bodies may have undergone complex histories, which can easily be buried in sweeping generalizations.

APPENDICES

A. Selected Regions of the Lunar Surface
B. Data for the Plates

INTRODUCTION

In several instances, photographic coverage of particular lunar features by the Lunar Orbiters was very complete. Consequently, extensive use of photographs of these features into the main body of the volume was deferred, except where necessity demanded, in order to present these features intact. They provide an opportunity for detailed comparisons of floor, wall, and rim areas of craters ranging in size from the Orientale basin (930 km in diameter with respect to its outer major scarp) down to Messier (about 8 km by 15 km in size). Between these extremes one can examine:

1. Petavius: a large, but moderately subdued crater (about 175 km in diameter) with floor fracturing and central peaks
2. Copernicus: a slightly smaller scale (almost 100 km in diameter) crater that displays numerous relatively well preserved units
3. Aristarchus: a crater (36 km in diameter) of relatively recent formation that displays both similarities and differences with respect to the larger crater Copernicus
4. Vitello: a crater approximately the same size as Aristarchus but with marked differences in appearance as a result of differences in formation, degradation, or volcanic rejuvenation

In addition, two extremely fascinating mare areas are included: the Harbinger Mountains and the Marius Hills. The Harbinger Mountains region exhibits numerous sinuous rilles. The Marius Hills is a region of extensive volcanism, resulting in the formation of volcanolike hills, domes, ridges, rimless pits, and rilles.

The accompanying descriptions are primarily limited to the identification of features of interest. In the preceding pages more extensive discussions can be found that are concerned with a particular lunar form.

Orientale Basin

The Orientale basin, shown in Plate 201, is repeatedly referenced in preceding discussions. The two outer scarps were identified from telescopic observations and named the Cordillera (arrow X) and Rook Mountains (arrow Y). Three major units are recognizable:

1. Mare material that fills the interior basin; several areas between the scarps; and isolated zones within the Oriental ejecta sequence
2. Hummocky terrain between the scarps
3. Striated ejecta sequence oriented approximately radially to the Orientale basin

Comparison of Orientale with the Imbrium basin, which is inundated almost completely by subsequent flows of mare units, suggests that wrinkle ridges reflect buried inner mountainous rings. It is also possible that such ridges represent feeders following deep-seated fractures exposed in Orientale as concentric scarps. The exterior scarp is analogous to the Apennines bordering eastern Mare Imbrium, whereas the hummocky terrain between the Cordillera and Rook Mountains is similar to the region surrounding the Alps, which border northern Mare Imbrium. Thus, it appears that the Alps are members of the interior mountainous ring, and the Apennines represent the outer scarp, which is absent north of Mare Imbrium. Note the leveed valley to the south (arrow Z), Bouvard, which is analogous to the Alpine Valley.

Several large (greater than 40 km in diameter) craters were formed on the Orientale complex and appear to be more abundant than similar craters on an equivalent area of Mare Imbrium. Hartmann and Wood (1971) considered the size-frequency distribution for craters less than 16 km in diameter. Their results indicate that, although the Orientale basin appears to be the youngest of the concentric-ring basins, it formed prior to the period of mare formation. This finding is confirmed by the fact that mare material from at least one period of Oceanus Procellarum flooding overlies Orientale ejecta (arrow XX; also see Plate 163). It is possible, however, that some of the craters counted were formed by internal mechanisms.

PLATE 201

IV-187-M   (oblique)

Orientale Basin

Plate 202 permits closer inspection of the interior basin. Of particular interest are:

1. The inner scarp (arrow X), which is not continuous around the basin (see Plate 201) and appears to merge to the east with the next mountainous ring
2. The larger exterior scarp, which has sharp angular corners and which is crossed by numerous valleys with related fan-like texturing along its base (arrow Y)
3. The recent crater 60 km in diameter, displaying an appearance similar to that of Copernicus
4. The dark-haloed craters (arrow Z) on the ejecta blanket of the large crater
5. The Gambart-type crater (arrow XX), having a relatively symmetric rim in profile and an inner terrace
6. The mare ridge (arrow YY) with such related structures as the insular relief and the ringed plain north of the sharp-rimmed impact crater
7. The fracture pattern to the lower right (arrow ZZ), which is on top of a low-relief mare dome
8. The NE-trending rilles which cross the hummocky terrain (arrow XY) and which are approximately parallel to scarps of the mountainous ring
9. The scarplike borders of the insular massif (arrow XZ) and the high-albedo terrain to the lower left (arrow YX)

PLATE 202

IV-195-H2   12 km

Orientale Basin

The southern border of the central Orientale basin, shown in Plate 203, is extensively fractured. One set of fractures is approximately concentric to the basin but typically diverges to meet several peninsulas (arrow X) that extend into the mare surface. Another set of fractures (arrow Y) trends subradial to the basin but cannot be extrapolated to a common focus. To the right is an intersection of another subradial set with the concentric set (arrow Z). The intersecting fracture patterns perhaps reflect tension due to either intrusions or subsidence of a once-molten surface over underlying stable relief.

Numerous minor fracture patterns are related to local forms. Several elongate domes on the textured and hummocky terrain exhibit summit axial fractures (arrow XX). These hummocks are possibly extrusive forms rather than rubbled remains of the Orientale event. Alternatively, they reflect intrusives that bow and crack this distinctive basin unit.

Extensive and complex subsidence is also indicated in this region. Of particular interest is the flat basin (arrow YY) that has a ridgelike rim and is overlapped by a polygonal platform. A narrow hogbacklike ridge is tangent to the southeastern border of this basin and presumably resulted from extrusions or intrusions along a fracture. A similar ridge can be identified across the larger depression to the left (arrow ZZ), and one should note the transition of this ridge into a rille. The irregular platforms typically have marelike surfaces with a central array of peaks (arrow XY).

The wrinkle ridge (arrow XZ) forms an approximate boundary for the mare plains in this region. This is consistent with the interpretations that such ridges are compressional forms, such as pressure ridges, or complex flow termini preserved on bordering shelves.

PLATE 203

IV-195-H1   12 km

Orientale Basin

Plate 204 covers a region east of that shown in Plate 203. It includes three distinct physiographic units that characterize the interior of the Orientale basin. Hummocky, highly fractured terrain separates the inner mare-floored basin from the mare-floored ring interior to the Rook Mountains (upper right). Portions of the mare-floored ring are shown in greater detail in Plate 206. Fractures on the hummocky terrain are composed of trends concentric and subradial to the basin. Minor trends and fractures associated with positive-relief features add further topographic complexity. The disappearance of these fractures beneath the mare material indicates that the release of stress in this region was complete prior to or during the emplacement of mare units.

The hummocky terrain exhibits numerous mounds, some of which have fractures on their crests. These mounds are part of an inner blocky ring encircling the basin at the same radius (see Plate 201). Arrow X (Plate 204) identifies an isolated peak surrounded by or protruding through this unit. In addition, note the fingerlike mounds (arrow Y) that extend from the basin. The elongate mound west of arrow Y has a fracture along its crest.

The origin of the hummocky zone could be fall-back rubble, fall-back melt, or extrusions related to the basin-forming event. The mounds might reflect buried peaks that were brought into relief by protruding through this hummocky unit as it subsided. The extensive fracturing and the curvilinear interior scarp (arrow Z) indicate a complex history of adjustment.

Several distinctive crater types are shown in Plate 204. The large Gambart-like crater (arrow XX) has a diffuse ejecta blanket and fractured marelike floor. The absence of dunelike forms in the ejecta sequence indicates a long erosional history or an ejecta type different from that of the large crater shown in Plate 202. The first conclusion may not be consistent with the preservation of small topographic features that apparently originated soon after basin formation. A smaller Gambart-like crater is shown near the bottom of Plate 204 (arrow YY). Arrow ZZ identifies a crater with a complex floor that is common around the Orientale basin. These craters are possibly volcanic in origin.

PLATE 204

IV-187-H2   12 km

Orientale Basin

Plate 205 encompasses a region adjacent and to the east of that included in Plate 204. The two outer scarps (arrows X and Y) are shown with the associated partial flooding by mare material within the interior basins. The plains-forming unit along the outermost basin (arrow Z) displays a higher albedo than that of the marelike fill material to the interior (lower left). The higher albedo unit, it is suggested, represents a lower proportion of iron-bearing minerals rather than significant differences in erosional histories. A low-albedo pool, however, was emplaced in this zone (arrow XX) and is associated with a sinuous rille (arrow YY) that extends from the northeast and perhaps is related to the large crater. This pool does not have well-defined borders, and several small domes and surrounding relief appear to exhibit a relatively low reflectivity as well. Consequently, eruptions of ash are thought to have accompanied emplacement of the marelike pool, or the lava level subsided after covering the features.

The hummocky zone between the major scarps does not exhibit the extensive fracturing shown in Plate 204. Complex patterns are expressed by ridges and furrows, some of which are radial to central Orientale (arrow ZZ), whereas others are concentric to it (arrow XY). Arrow X identifies a radial arrangement that overlies the outer scarp. These may represent either flow patterns from the surge of material from the basin-forming event or later surges.

Note the large number of craters that display raised but subdued rims and the absence of identifiable ejecta sequences. In addition, there is a crater 8 km in diameter (arrow XZ) whose floor has been raised to near rim level. Also note that the inner scarp is not continuous but is composed of separated blocky masses probably analogous to the Teneriffe Mountains in northern Mare Imbrium. Of further interest is the sinuous rille that crosses the outer scarp (arrow YX).

PLATE 205

IV-181-H2   12 km

Orientale Basin

Selected close-ups of features shown in Plates 204–205 reveal several interesting ones associated with the mare units. Plate 206, *a* and *b*, is a stereo pair that reveals several low-relief domes and platforms composed primarily of mare material (see lower right of Plate 204). If these were extrusively built, then at least seven separate units are displayed. Four of these have central peaks (arrows X, Y, Z, and XX) that typically are split by fractures. Such an arrangement prompts comparison to fissure eruptions. It is also possible that these domes and platforms are the result of intrusive doming. In general, the marelike material lacks flow termini (except possibly those mentioned above) and has invaded almost every inlet.

Plate 206c isolates a region that can be identified also in Plate 204 (center right). Of interest is the sinuous rille that extends into the rugged topography to the north (arrow X). Just to the south of this point, the rille has a tightly meandering plan (arrow Y) but straightens to the south of the domelike relief (arrow Z). Note the dark-haloed crater near the south end of the rille (arrow XX). Several other dark-haloed craters are recognized in this plate (arrows YY and ZZ) and are suggested to be maars.

A Gambart-like crater shown in Plate 206d can be identified near the bottom of Plate 207d. The wall appears to be breached to the east, and a sinuous rille is traced from this point (arrow X). A similar arrangement is exhibited by Krieger, to the northeast of Aristarchus (Plate 171a) and suggests that such craters were sources of lava. In addition, two wider grabenlike rilles meet at the southern wall of the crater (arrow Y).

PLATE 206

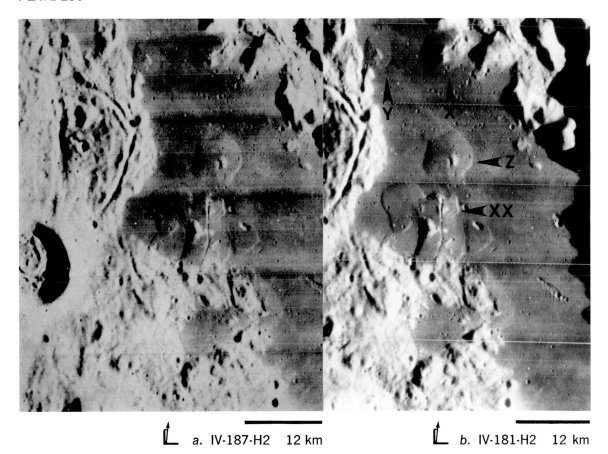

a. IV-187-H2   12 km

b. IV-181-H2   12 km

c. IV-187-H2   12 km

d. IV-187-H3   12 km

Orientale Basin

The northern outer scarp of the Orientale complex is shown in Plate 207 (arrow X). The transition from the hummocky terrain to the ribbon structure is abrupt and irregular in this region (arrow Y). As shown in Plate 205, the outer scarp marked this transition, but here the hummocky terrain continues beyond the scarp. This zone is characterized by small hummocks, corrugations, and domes. The domes may be analogous to the mamelon domes illustrated in Plate 126.

The relatively abrupt transition from the hummocky zone to the ribbonlike flow patterns suggests distinctly different origins. If the ribbonlike patterns resulted from an initial surge of ejecta during basin formation, then the hummocky zone appears to postdate this event. If the hummocky zone is viewed as another type of ejecta fall-back, then it is not apparent why the transition to the outer flow patterns is so abrupt. In addition, the hummocky zone near this transition appears to be topographically lower. Consequently, the hummocky terrain, it is proposed, represents a once-molten unit that was emplaced after basin formation. The melt was perhaps derived from melting of the parent rock or from deep-seated sources released through fractures developed during the catastrophe. It inundated not only the inner ejecta sequence but also the outer scarp shown in Plate 207. Volcanolike forms include cones with summit craters (arrow Z), nested cone (arrow XX), and steep-fronted flow termini (arrow YY). This zone is similar to the hummocky units on the floor of Tsiolkovsky (Plate 40b).

Several medium-size craters (around 10 km–20 km in diameter) occur on the hummocky terrain and on the ribbon-structured rim sequence (see Plates 201 and 211). The rims of these craters are relatively low and their ejecta sequences are very limited, either as a result of degradation or of having only a meager display at the outset. In general, the wall-rim transitions are abrupt, and in several cases, smaller craters on these contacts have been severed (arrow ZZ) by either a retreating wall or the initial crater-forming event (the latter requires relatively quiescent formation). Note the crater near the top (arrow XY), which can be viewed in stereo in Plate 61, a and b. Also note the NW-trending crater chain (arrow XZ).

PLATE 207

IV-187-H3   12 km

Orientale Basin

Plate 208 provides dramatic stereo views of several craters found along the outer scarps. Plate 208, *a* and *b*, shows a region along the outermost scarp to the southeast (refer to Plate 201) of the central basin. The three-dimensional view exaggerates a subdued ringed depression (arrow X), whose floor appears to be at approximately the same level as the hummocky basin to the northwest. A mountainlike mass (arrow Y), perhaps a bulk ejecta deposit, extends from the south into the depression.

The floor of the large crater to the southeast (arrow Z) appears to be at a depth similar to that of the subdued craterform, but its appearance is clearly different. If this crater was solely the result of an impact that produced a bulk-ejecta deposit similar to Aristarchus (Plate 236), then it seems unusual that this sequence could have been destroyed without destroying the many smaller scale structures in the region, noted later. The crater also displays a narrow interior terrace (arrow XX) that may be either a narrow slump or a pre-existing level of floor material (see Plates 60–61).

Other features of interest are the crater-topped peaks (arrows YY and ZZ) and the irregularly bordered mound (arrow XY). These are likely volcanic in origin. Numerous rimless craters in the region are also possible volcanic forms (for example, arrow XZ). Also note that the scarp (arrow YX) does not appear to be a tilted block but a ridge. Part of the effect may be enhanced by both the lighting and the exaggerated vertical scale.

Plate 208, *c* and *d*, covers a region to the northwest of the foregoing stereo pair and reveals two craters along the next interior scarp. Again, the mountainous ring does not appear to be a tilted block except for the portion shown near the bottom of the photograph. The significance of this profile with respect to origin of the multiple rings is uncertain, since the Orientale event must have been extremely complex. The mountainous ring appears to be composed of irregular platforms, ridges, and angular peaks. Note the smaller domelike (arrow X) and conelike (arrow Y) features that parallel the scarp to the north. At least four display summit craters. A system of fractures (arrow Z) is adjacent to these forms and crosses the rougher topography but not the marelike surface near the top.

The two large craters are highly scalloped and display limited rim sequences. Note the lobate flow front at the base of the scarp adjacent to the lower crater (arrow XX), which may have been its source.

PLATE 208

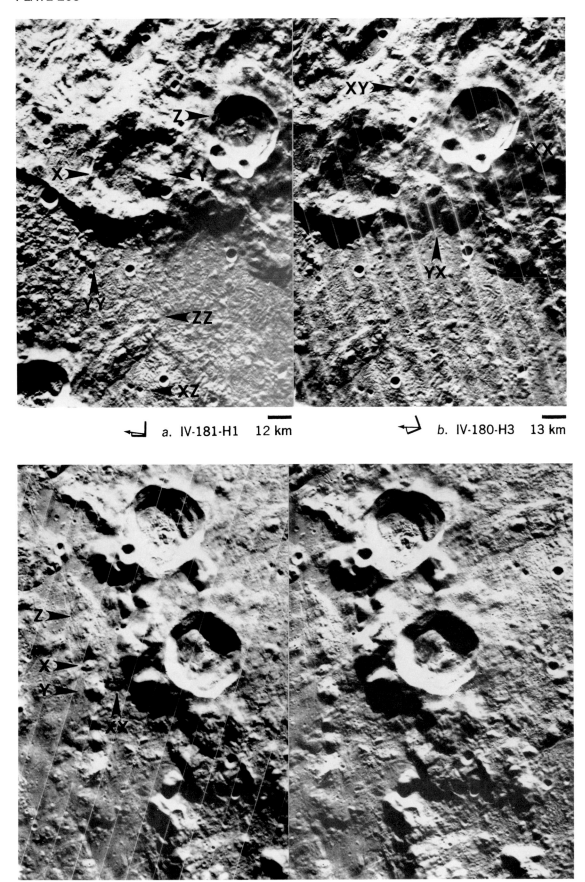

a. IV-181-H1   12 km

b. IV-180-H3   13 km

c. IV-187-H1   12 km

d. IV-186-H3   13 km

Orientale Basin

Radial lineations from Orientale are clearly shown in Plate 209, taken from the south of the basin. Bouvard, the large raised-rim valley, and a similar feature near the terminator are identified by arrows X and Y, respectively. Subsequent photographs are selected views of this complex zone, parts of which resemble the ejecta blanket of Copernicus on a grand scale. As illustrated previously (see Plate 199, a and b), some of the ejecta sequence involves flow of bulk material. The inner zone suggests massive movements of material, with later modification by impacts and/or volcanism. The outer zones reveal vestiges of pre-existing topography, which have been highly modified, as well as later units, such as smooth plains-forming material and local viscous extrusions.

PLATE 209

IV-193-M   (oblique)

Orientale Basin

Plate 210 provides a comparison between the inner zones of the Orientale rim sequence. The radial pattern displayed in Plate 210a, a contrast to the complex swirls shown in Plate 210b, is largely the result of differences in solar illumination for these two views. Plate 210a encompasses the northern rim area, where the radial pattern is accentuated by the eastern illumination. Plate 210b, on the other hand, covers the eastern rim, where identification of the radial pattern is difficult because of its alignment with the direction of the sun; furthermore, any transverse pattern is enhanced. Both components can be identified in both photographs.

The inner rim area shown in Plate 210a reveals little evidence of pre-existing topography, whereas, farther from the central basin, a large depression (arrow X) apparently controlled the flow (also see Plate 213). Similar control, or "focusing," was noted around large pristine craters in the Zone II region (Plate 83) and indicates a ground phenomenon similar to the base surge or a *nuée ardente.*

Plate 210b shows accumulations of material within pre-existing craters on the side farthest from the basin (arrows X and Y; also Plate 212, c and d). This is also analogous to phenomena observed around smaller (100 km in diameter) craters. Plate 210b includes several other interesting features:

1. A small crater with the contiguous crater chain extending to the south (arrow Z)
2. The floor-fractured crater (arrow XX)
3. The corrugated appearance of the relatively flat areas (arrow YY), which may represent later and slower flow movement

PLATE 210

b. IV-173-H2   12 km

a. IV-188-H1   12 km

Orientale Basin

Plate 211 covers an area southeast of Orientale, and, for reference, an edge of the "filled-crater," Wargentin, is seen at the lower right (see Plate 41, *c* and *d*). Both radial and transverse patterns are revealed. Of particular interest are the lineations near the center (arrow X). It is anticipated that a flow unit extending from Orientale might display arcuate transverse ridges, with the convex side directed away from the source. Perhaps the flow on the floor of the elongate valley had ceased further advancement, at which time a surge of material extended either a related or a separate unit into the crater (arrow Y). The renewed movement along the margins produced patterns with the convex side directed toward the source. This explanation implies multiphased deposition of ejecta, although the time intervals might be very short. Alternatively, flow of ejecta deposits was reactivated by large seismic waves associated with the formation of the Orientale basin. A stereo view is provided in Plate 212, *a* and *b*.

The gashlike furrow (arrow Z) is part of Bouvard. This may have been a channel for the units shown in Plate 199, one of which is indicated by arrow XX.

PLATE 211

IV-180-H2,H1   14 km

Orientale Basin

Plate 212, *a* and *b*, is a stereo pair of the complex indicated by arrow X in. Plate 211 and reveals a number of channellike features accompanying the lineations from Orientale (arrows X and Y). To the lower right (near arrow X), lineations from Orientale merge with plains-forming units, which might indicate later extrusions, a change in flow viscosity, or the termination of the Orientale blanket (refer to Plate 199, *a* and *b*). Note the numerous small craters and the system of crater chains extending from lower left to upper right (for example, arrow Z).

The second stereo pair (Plate 212, *c* and *d*) illustrates the accumulated material on the basin-facing wall of a pre-existing crater. The sharp break from the wall (arrow X) possibly developed during deposition or as the result of later slumping. The floor of the crater displays numerous shallow craters and crater-topped peaks (arrows Y and Z) that could indicate subsequent volcanolike activity.

PLATE 212

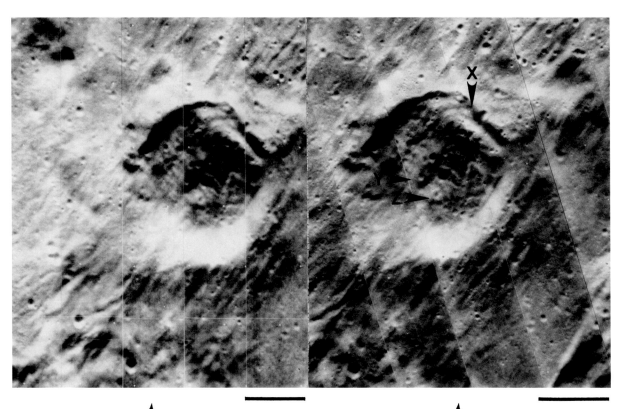

a. IV-180-H2   13 km

b. IV-172-H2   13 km

c. IV-173-H1   12 km

d. IV-172-H3   13 km

Orientale Basin

Both radial and transverse lineations are clearly exhibited in Plate 213, which primarily includes the outer blanket from Orientale. The large crater, Riccioli, has undergone subsequent flooding by mare material (arrow X). A more detailed examination can be made in Plate 214a, and it is clear that this mare-flooding epoch postdated not only blanketing but also the grabens concentric to the Orientale basin.

The low-albedo marelike unit in Riccioli is not the only plains-forming unit represented. Arrow Y identifies a highly pitted unit that has large lobate fronts. Similar units are recognized to the south of Riccioli (also see Plate 199c), and several have chaotic surface patterns (arrow Z). Note that this unit shows an absence of patterns where it surrounds the insular massif marked by arrow XX. A similar relation occurs on the floor of Copernicus (Plate 228b). This pattern in Plate 213 is interpreted as a zone of low-velocity flow and stagnation, caused by turbulence set up along the margins of the massif, but such an interpretation is probably inapplicable for the pattern in Copernicus. Extensive lineations from Orientale cross portions of these units but do not appear to be deposited on them.

The cratered plains-forming units may be a less viscous form of Orientale ejecta (see the discussion with Plates 198–199). It is also plausible that they were derived from the breached craters (arrows YY and ZZ).

Adjacent to the southwestern rim of Riccioli (arrow XY) is a classic example of the "focusing" effect created by a surge of material controlled by a pre-existing crater. Note, in addition, that despite the accumulation of a jumbled mass on the far wall, the striations across the floor of Riccioli continue.

PLATE 213

IV-173-H3   12 km

Orientale Basin

The northeastern quadrant of Riccioli is shown in Plate 214*a* and reveals a chain of irregular depressions (arrows X, Y, and Z) on the floor that is comparable to a similar complex to the west of the Aristarchus Plateau (see Plate 177*b*). These depressions are interpreted as subsided lava lakes. The mare material has invaded almost every accessible inlet, a saturation that requires a very low viscosity material. Note the disappearance of the grabenlike depression beneath the material (arrow XX).

Plate 214*b* shows a region to the northwest of Riccioli but not included in Plate 213. The grabenlike rille (arrow X) is part of the system that crosses Riccioli. Of interest is the sinuous rille (arrow Y) that appears structurally produced or at least controlled. The rille head could be interpreted as either the fracture on the scarp (arrow Z) or the depression to the right (arrow XX). Note that the rille crosses the corrugated relief pattern to the right (arrow XX) but appears to be blocked near arrow YY.

PLATE 214

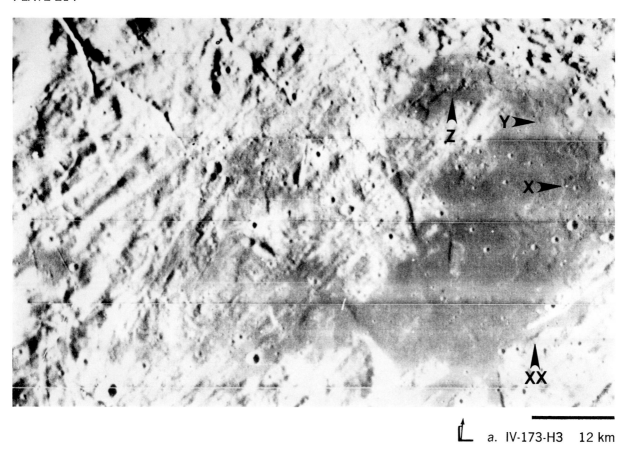

a. IV-173-H3   12 km

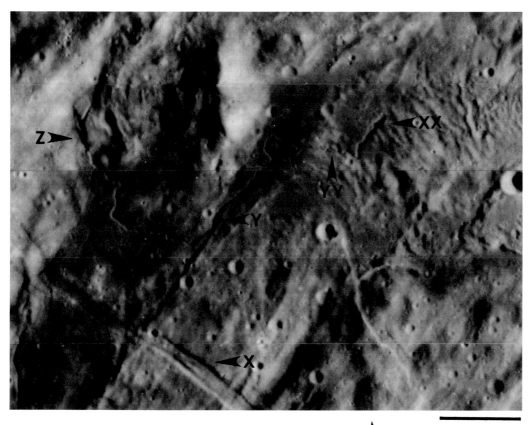

b. IV-174-H1   12 km

Petavius

Petavius is a large (about 175 km in diameter) crater on the southeastern limb of the terrestrially visible Moon. The following series of photographs provides comparison to the preceding mammoth Orientale basin and the following smaller, but still enormous crater, Copernicus. Plate 215a is an oblique view that reveals:

1. The "flat" floor crossed by several rilles, which are restricted to its interior
2. Central peaks, the largest of which rises more than 2.5 km above the adjacent floor, which is approximately 2.6 km below the western rim (LAC 98)
3. The central basin within the peak complex, which is interpreted as a collapse structure, a feature common to many lunar craters
4. The wide wall region, portions of which are rubbled (arrow X), terraced (arrow Y), and filled with plains-forming units (arrow Z)
5. A rim sequence similar to that found around Copernicus but predating the last stage of inundation by Mare Fecunditatis

Plate 215, b and c, is a stereo view of Petavius at higher resolution (note that it is rotated 180° with respect to Plate 215a). Three large grabenlike rilles extend radially from the central peaks. The most prominent NE-trending rille (arrow X) is perpendicular to a grabenlike rille on the southwestern wall of Petavius (arrow Y). The latter rille is part of a long grabenlike system that extends from Mare Fecunditatis to near Mare Australe, and it is modified in plan within Petavius to approximate the floor-wall boundary (see Plate 215a, arrows XX and YY). Portions of the peaks are dropped along the rilles (Plate 215, b and c, arrow Z). Less prominent rilles are also radial to the central peaks (arrows XX and YY) and are approximately mirrored by a rectilinear rille (arrow ZZ), which connects with a rille along the floor-wall contact (arrow XY). A smaller rille (arrow XZ) and wrinkle ridge (arrow YX) parallel the rille identified by arrow ZZ (also see Plate 215a).

The floor of Petavius resembles that of Gassendi in several respects (refer to Plates 36–38). Both display a hummocky terrain that is highly textured at higher resolutions (see Plates 216, d and e, and 217). Their floors are also crossed by numerous rilles, some of which have related volcanolike forms; for example, dark plains units occur on the northern floor edge around a floor rille (Plate 215, b and c, arrow YZ; see Plate 216a). Both floors exhibit relatively uncratered units (Plate 215, b and c, arrow ZX). At least one crater in Petavius (arrow ZY) exhibits an interior doughnutlike ring similar to those in several small craters in Gassendi. In obvious contrast to Gassendi, Petavius exhibits broad wall slumps and has no irregular depressions or a large moatlike zone between the floor and rim.

PLATE 215

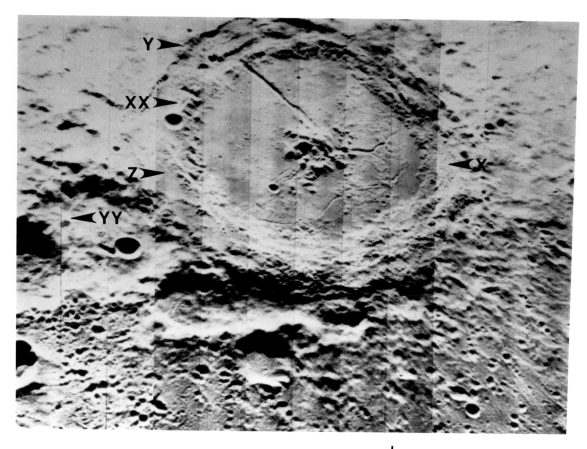

a. IV-184-H2 (oblique)

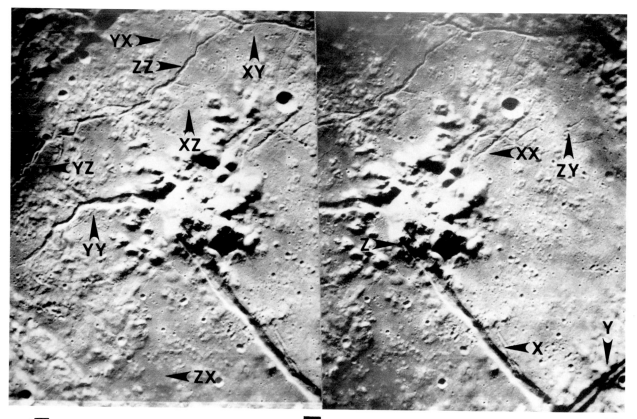

b. V-036-M   4.7 km (top) 4.4 km (bottom)   c. V-033-M   4.8 km (top) 4.5 km (bottom)

Petavius

Plate 216 shows closer views of the floor of Petavius. The rectilinear rille shown in Plate 216a (arrow X) can be identified to the left in Plate 215, b and c. Several insular features were formed within this rille (arrow Y). The rille profile is **V** shaped in the area near arrow X but flat bottomed in that near arrow Y. The rille floor at the latter location resembles the relatively low albedo unit, which it crosses, and the flat floors of several craters on this unit (arrow Z). The low-albedo plains-forming unit in this region is in distinct contrast with the hummocky floor (bottom right). Positive-relief features, including the wall seen at left, commonly exhibit aprons (arrow XX) where they meet the low-albedo unit. In addition, note the crater that is crossed by the rille (arrow YY) and the broad-rimmed crater (arrow ZZ).

Plate 216b shows an area east of the central-peak complex (Plate 215, a and b, below arrow Z). It reveals two contrasting terrains: topographically high hummocky units (arrow X) and topographically lower plains-forming units (arrow Y). The latter unit exhibits a higher albedo than that of the plains-forming unit seen in Plate 216a. Portions of this unit shown in Plate 216b exhibit a low crater density (arrow Z). Arrow XX identifies a bright-haloed crater with a central mound, a characteristic feature of craters on the Cayley Formation. Arrow YY identifies an elongate and polygonal crater that may be genetically related to the ropelike ridge that extends to the south (arrow ZZ). This configuration resembles features in Gassendi (Plate 86b) and Vitello (Plate 251a).

The region shown in Plate 216c is indicated by arrow XY in Plate 216b. It is a high-resolution view and contrasts the tree-bark surface texture of the hummocky floor unit (arrow X) with the plains-forming units (arrow Y), which appear to be relatively hummocky at this scale. The plains-forming unit is in a closed depression adjacent to a grabenlike rille. Note the relatively well defined horizontal layering (arrow Z) on the northwest-facing wall of the closed depression. This is not believed to be the grid pattern, which is rarely, if ever, horizontal. It also lacks the wrinkled appearance that characterizes the tree-bark pattern. Similar layering is identified within a rille in Gassendi (Plate 38, a and b, arrow XX) and suggests either a well-ordered history or high-level marks of the interior plains-forming unit. Note that both hummocky and plains-forming units exhibit similar abundances of small (less than 100 m) craters, but the latter unit displays a higher density of large (greater than 500 m) subdued craters.

Plate 216d isolates a small crater shown in Plate 216b (arrow XZ). It lacks structuring of bulk ejecta but exhibits several large blocks (Plate 216d, arrow X) with possible fillets and a cluster of blocks on the southeastern rim (arrow Y). The northeastern wall has two large outcrops of indeterminate structure (arrow Z).

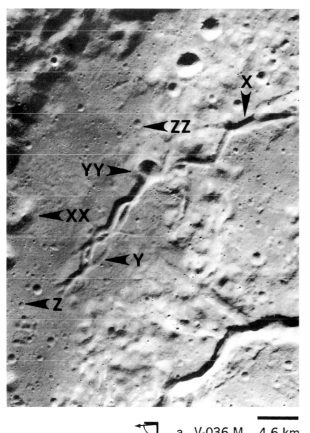

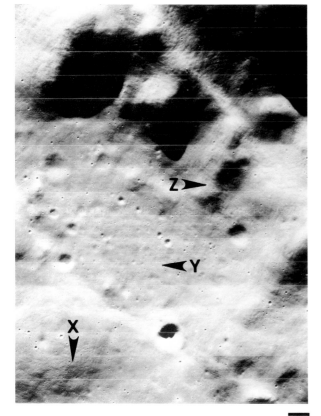

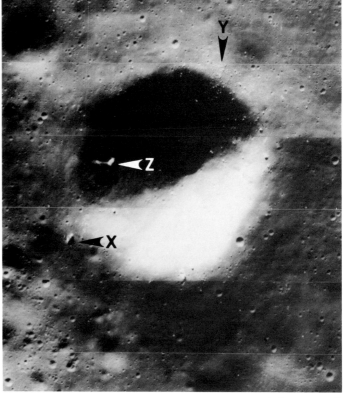

a. V-036-M   4.6 km

b. V-036-M   4.5 km

c. V-035-H3   0.59 km

d. V-036-H3   0.59 km

503

Petavius

Plate 217 isolates the eastern central-peak region (Plate 215, *b* and *c*, arrow Z) and shows the peak-floor contact (arrow X) and the large linear rille that crosses both topographies (arrow Y). Numerous outcrops line the upper wall, and several dislodged blocks produced trails as they rolled across the unconsolidated lower wall material (for example, arrow Z).

Note the curvilinear lineation crossing the large grabenlike rille (arrow XX). This can be traced to another fracture system, which is shown in Plate 215, *b* and *c*. Some ill-defined lineations that are aligned parallel to the large rille (arrow YY) probably represent partially buried scarps from multiple slump blocks. Other smaller scale lineations (arrow ZZ) may be layering.

The surface of the peaks is patterned, perhaps as a result of mass movement. In general, the contacts between the relief and floor are relatively ill defined.

PLATE 217

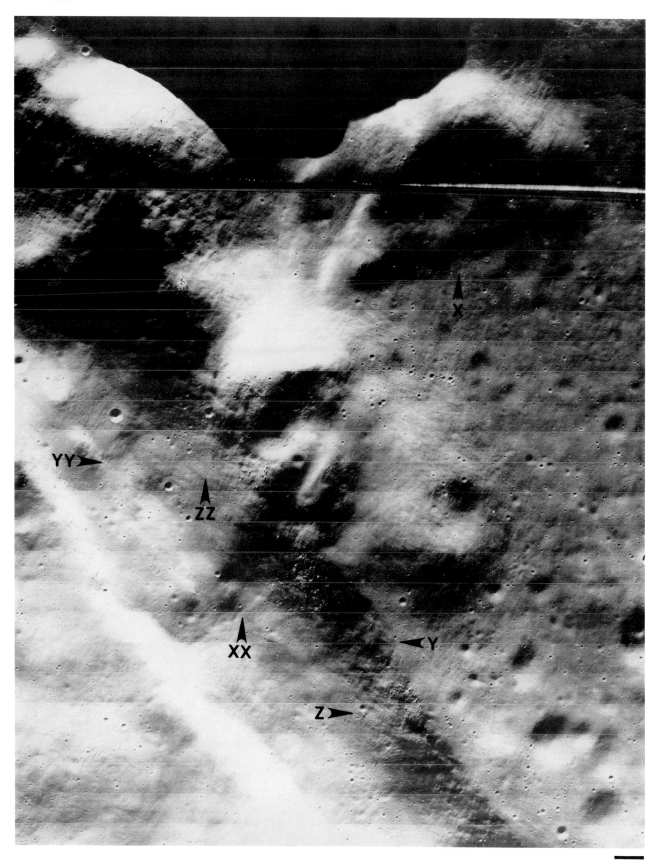

V-034-H3   0.60 km

Copernicus

The Orbiter II photographs of Copernicus were heralded as the "pictures of the century." They disclosed an unusual perspective of a large crater (96 km in diameter) familiar to Earth-based astronomers at considerably lower resolution. Plate 218a is the medium-resolution photograph of Copernicus and reveals its extensive ejecta blanket, terraced walls, central ridges, raised rim, and the Carpathian Mountains on the curved horizon. The corresponding high-resolution photograph, Plate 218b, shows the inner rim sequence in the foreground, details of the central peaks, and the precipitous scarps of the far wall. The near wall is not visible, because of the obliqueness of view. The elevation difference between rim and floor is approximately 2.7 km, whereas the peaks rise more than 0.5 km above the floor. For an over-all vertical view of the region, see Plate 94.

Several general features were immediately noticed and are illustrated in subsequent plates:

1. Outcrops and strata along the far scarps
2. Flow features extending down the wall, onto the floor
3. Outcrops on the central peaks
4. A generally flat floor at low resolution but hilly at higher resolution
5. The relatively smooth surface of the ejecta blanket, in contrast to terrestrially obtained photographs at lower resolution
6. Relatively gentle slopes of the mountain range in the background

Some volcanologists compared the photographs to similar views of terrestrial calderas and noted the volcanolike topography. The equally vociferous proponents of impact origin made comparisons to analogs of known impact craters and nuclear-explosion pits.

The following descriptions compare this oblique view to the vertical view transmitted by Orbiter V. Several of the latter examples can be examined in stereo but are best displayed with an apparent solar illumination from the top; consequently, their orientations will be with the east at the top.

PLATE 218

b. II-162-H3,H2   (oblique)

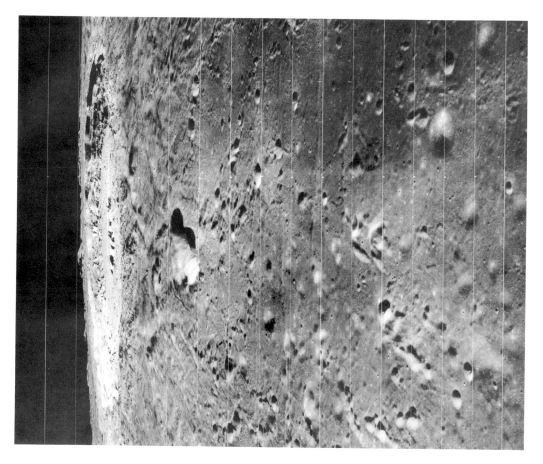

a. II-162-M   (oblique)

Copernicus

Plate 219 shows an over-all view of the floor, wall, and rim areas. The wall seen at left is that shown face-on in Plate 218. The following features are recognized and are illustrated in more detail in subsequent plates:

1. Floor
   a. NW-trending central-peak and ridge complex (arrow X)
   b. Two general morphologically distinct regions: hummocky topography to the east and south (arrow Y); smoother but textured plains area to the northwest (arrow Z)
   c. Numerous domes, some of which have summit craters or fractures (note the similarity at this resolution to the units in Orientale)
   d. Numerous craters having a variety of appearances
   e. Complex fracture systems
2. Wall
   a. Large scarp faces on the upper wall (arrow XX)
   b. Terraces at the base of the scarp (arrow YY) that are typically accompanied by pooled units having smooth surfaces (arrow ZZ)
   c. Jumbled units below the terraces (arrow XY) that also have small pools of material
   d. Numerous furrows that cross the lower wall region (arrow XZ)
3. Rim (also see Plates 94–95)
   a. Scalloped plan due to slumped masses now composing the wall
   b. Lip along rim—even along the scalloped region (arrow YX)
   c. Annular and radial sets of fractures or joints (arrow YZ)
   d. Radially directed furrows (arrow ZX)
   e. Pools of material that fill local depressions (arrow ZY)

PLATE 219

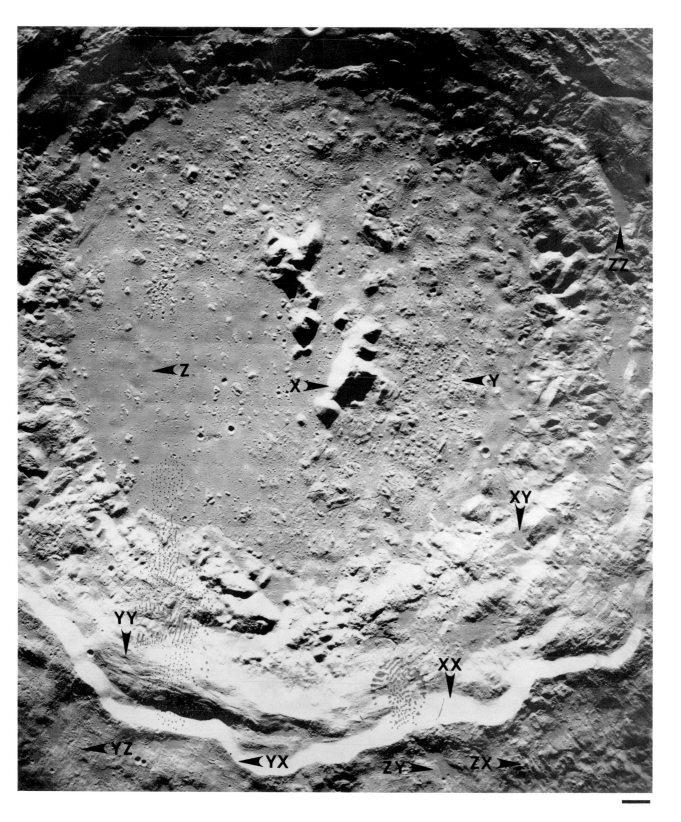

V-151-M   3.4 km (*top*)   3.2 km (*bottom*)

Copernicus

The next ten plates allow more detailed examination of floor features in Copernicus. Plate 220a is an enlarged section of Plate 218b and provides direct comparison to the stereo pair (Plate 220, b and c), especially with respect to vertical exaggeration. The oblique view is from the south of Copernicus; therefore, imagine the Orbiter II spacecraft to the right of the area shown in Plate 220, b and c.

The peaks mark an approximate boundary between the hillocky complexes (northeast, east, and south of the peaks) and plains units (northwest of the peaks). This is also revealed in the oblique view. The peaks display numerous outcrops that are examined in more detail in Plates 221–223. Several peaks appear to have summit pits or depressions; in particular, note the following examples:

1. The easternmost extension of northern peaks has channellike features extending to the north (arrow X).
2. The western extension of northern peaks has a summit pit with numerous blocks (arrow Y).
3. The highest peak of the southern peaks has two (or more) adjacent pits (arrow Z).

The validity of these observations requires further high-resolution coverage under different illumination. If these features are real, then they are evidence for volcanic processes. The first example (arrow X) might be analogous to flank eruptions observed on terrestrial volcanic peaks. If the peaks represent uplifted masses, however, then unloading and subsequent landslides could produce the channellike features; furthermore, the summit pits may be nonvolcanic collapse pits or later impact craters.

The stereo pair enhances the variety in detailed floor features:

1. Craters
   a. Rimless collapse pits (arrow XX)
   b. Broad-rimmed craters (arrow YY)
2. Positive-relief features
   a. Volcanolike domes with summit pits (arrows ZZ and XY)
   b. Low-relief ridges (arrow XZ)
3. Negative-relief features
   a. Moats around domes (arrow YX)
   b. Rille systems, many of which appear to be genetically related to the peaks and domes (arrow YZ)
   c. Fissurelike features (arrow ZX)

PLATE 220

a. II-162-H3 (oblique)

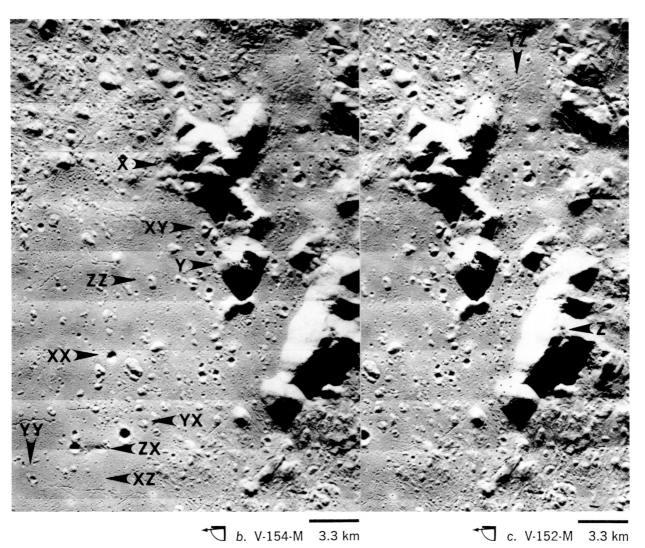

b. V-154-M   3.3 km

c. V-152-M   3.3 km

Copernicus

Plate 221 is a high-resolution view of the northern peak and ridge system (compare to Plate 220). It reveals numerous outcrops and blocks on the flanks of the peaks and the accumulation of debris at their bases. Blocks also occur on smaller mamelonlike domes (arrow X).

The floor to the north of the peaks (arrow Y) is characterized by a ropy appearance corresponding to ridge and furrow systems. At higher magnification, this surface is undulating or bumpy, with numerous small craters (but not nearly the apparent density of those of the maria). Between the peaks (arrow Z), the surface is markedly smooth at small scales, and the treebark pattern can be identified around the larger subdued craters. Note the numerous low-rimmed or rimless subdued pits in this area (arrow XX). Similar units are found around domes and rilles north of the peaks (see Plate 226a) and may be the result of local extrusions of ash.

PLATE 221

V-152-H1; V-151-H1   0.44 km

Copernicus

Plate 222, *a* and *b*, shows oblique and vertical views, respectively, of the western extension of the northern peaks. Several outcrops on the peaks shown in the oblique view in Plate 222*a* (arrow X) appear to be aligned with the lineation (arrow Y), but the vertical view (Plate 222*b*) demonstrates that this is an illusion due to perspective. The bulbous outcrops (Plate 222*b*, arrow X) are interpreted as viscous pluglike extrusions or dissected strata. The lineation (Plate 222*b*, arrow Y) is revealed to be composed of numerous blocks and is suggested to be a fault trace that parallels the general trend of the peak system. Displacement along this fault, which appears to dip to the west, may have been responsible for the outcrops (arrow X).

The screelike slopes of the peak display numerous streaks from mass movement, and debris typically has accumulated above the outcrops (arrow Z). The talus along the base is now obvious, and close inspection reveals that numerous blocks have split into several pieces and are surrounded by fillets (arrow XX). Block sizes are typically much larger than those found associated with the smaller domes.

The possible origin of the peaks is discussed in the chapter "Crater Floors," and the three general ideas that emerge are that they represent:

1. Rebounded masses from the impact, which is assumed to be responsible for Copernicus
2. Rubble representing the toes of concentric large-scale slumping and lateral plasticlike transfer of floor material
3. Volcanic peaks

With the respect to the last interpretation, recall that this peak was listed among those with summit craters (Plate 220). This feature can be identified in both Plates 222*a* (arrow Z) and 222*b* (arrow YY).

PLATE 222

a. II-162-H3   (oblique)

b. V-152-H1; V-151-H1   0.44 km

Copernicus

The peaks shown in Plate 223, *a* and *b*, are east of the peak shown in Plate 222, *a* and *b*. The relatively simple appearance in Plate 223a (the oblique view) is in contrast to the multiple crests shown in Plate 223b. In addition, the oblique view reveals the screelike slopes of the southern side, but the vertical view shows a much more complex character on the northern slopes, which appear to be crossed by numerous furrows (Plate 223b, arrows X, Y, and Z) and breached craterforms (arrow XX). One of these furrows can be identified also in Plate 223a (arrow X). In both views, transferred material is visible (Plate 223a, arrow Y; Plate 223b, arrow YY). These furrows are attributed to either avalanches (or similar mass transfer) or volcanic flank fissures.

An interpretation deduced from Plate 223b by itself could be deceiving, owing to one's inability to recognize equal elevations; therefore, reference should be made to both the oblique view and the stereo pair (Plate 220, *b* and *c*). For example, the ledge marked by arrow Z in Plate 223b appears to be well below the summit but is identified by arrow Z in Plate 223a to be near the peak crest. Similarly, note that the area near arrow ZZ in Plate 223b is also near the crest (Plate 223a, arrow XX) and appears to form a flat-topped summit. Such cross comparisons reveal possible stratification in the central peaks, which is consistent with their structural origin.

Several large blocks are identified at the bases of these peaks. The irregular block at arrow XY (Plate 223b) was apparently derived from near the summit and has left a deep scar down the slope and onto the floor (also see Plate 223a, arrow YY). An angular block with a trail is marked by arrow XZ (Plate 223b). Note the accumulation of smooth material in the valley (arrow YX).

PLATE 223

a. II-162-H3   (oblique)

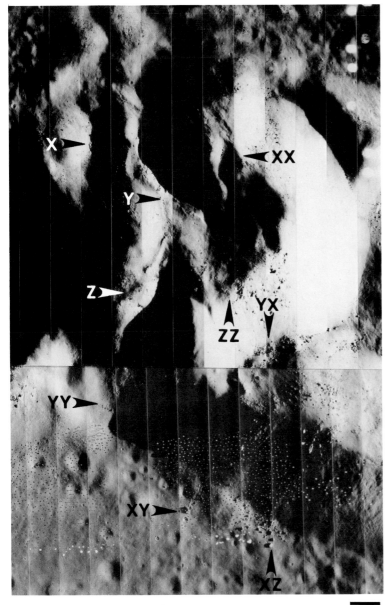

b. V-152-H1; V-151-H1   0.44 km

Copernicus

The "plains" area of the floor of Copernicus, shown in the upper and lower left of Plate 224, can be contrasted to the rougher topography shown in Plate 225 and to the right in Plate 224. It is north of the central ridges and can be identified also in Plate 220, b and c. Numerous mamelonlike domes are displayed, several of which have peripheral moats (arrows X and Y). The domes seen in the upper left corner (arrow Y) are shown in stereo in Plate 227, b and c. Domes with moats typically do not have the radiating fractures that characterize the hillocky complexes (arrow Z) and a few isolated domes (arrow XX). A smoother surface commonly surrounds domes having such fractures.

Perhaps the different appearances of domes resulted from different periods of dome construction, after which ropy floor units surrounded some of the domes and produced the moats. Alternatively, domes with moats indicate a slowly sinking mass at a time when the floor responded plastically, whereas domes with radiating fractures resulted from rapidly rising masses that produced tension cracks and limited extrusions of material. Tension cracks also could result from a subsiding molten unit draped over stable floor hummocks. Dome construction, however, was probably more complex and involved volcanic activity, as might be inferred from the numerous domes and cones with summit pits. For example, blocky domes are perhaps analogous to those in Mono Craters, California, which resulted from extremely viscous extrusions being forced through conduits, with rapid fragmentation from unloading. They have been interpreted as a last stage in the volcanic cycle in which first an explosion pit is formed, followed by the plug. E. Smith (1970) also has been intrigued by this possible terrestrial analog and recently completed a detailed study and comparison of such forms.

Small craters in this region show a wide range of morphologic forms, which include blocky rims (arrow YY), tufflike rings (arrow ZZ), and rimless pits (arrow XY). Note that relatively few craters display the flat floors, such as those found in other regions of possible extrusive activity. Craters in the dome and rille complex (arrows XZ and YX) have rims and walls that appear to be directly related to the surface texturing, as opposed to an impact, which spreads an ejecta blanket and would be expected to disrupt such continuity. In contrast, the crater identified by arrow YZ appears to interrupt the rille to such an extent relative to its size that it is interpreted as a volcanic feature.

PLATE 224

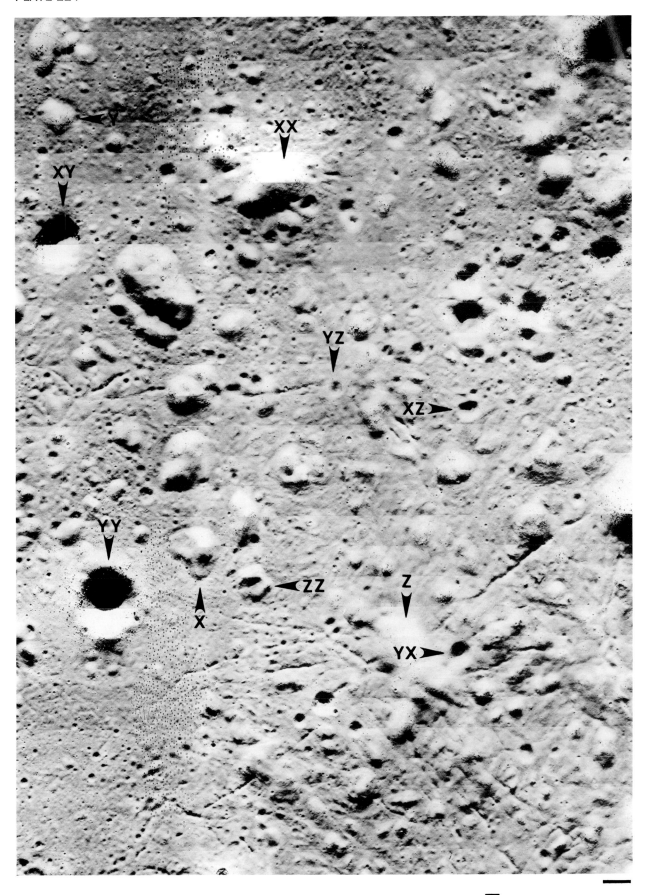

V-152-H2   0.43 km

Copernicus

In contrast to the surface examined in Plate 224, Plate 225 displays the more hummocky floor unit to the south of the southern central-ridge system, part of which extends into the area shown at upper left. It resembles a coalescence of many units similar to the dome and rille complex shown in Plate 224. The domes in Plate 225 typically display blocks across their surface (arrow X). The blocks are not limited to sudden changes of slope but are strewn with relative uniformity across the dome or are clustered around subdued crater pits (arrow Y), spinelike protrusions (arrow Z), or ridgelike outcrops (arrow XX). Between the domes, blocks are generally absent (arrow YY).

As in other areas of the floor, a variety of crater appearances is displayed. For example, the crater identified by arrow ZZ has a broad rim, which is breached. Also note the cluster of craters (arrow XY) that merge with the surrounding hummocky terrain.

Although well-defined flow units are not identifiable here, the original photograph shows an abundance of flowlike patterns.

PLATE 225

V-150-H2   0.43 km

Copernicus

Plates 226 and 227 illustrate in greater detail examples of fracture systems on the floor of Copernicus. The region shown in Plate 226a can be identified in the lower left corner of Plate 221 and beyond the peak shown in Plate 222a. A smooth-textured unit (arrow X) surrounds segments of the large fractures, whereas it is absent around the Y-shaped fracture to the northeast (arrow Y). The profile of fractures within the smooth areas is subdued, and it is suggested that eruptions of ash were responsible for the smooth units and this degradation.

The fractures are typically discontinuous, but their plans are reflected by elongate craters (arrow Z). In addition, the NW-trending fracture system (arrow Z) appears to extend SE as relatively subtle parallel lineations (arrow XX). Although outcrops typically are not recognized along these fractures, several do occur where these lineations cross the E-trending fracture and are probably the result of relatively recent displacements associated with the NW-trending system.

In addition to the fracture system, several other features are of interest:

1. Tephra-conelike form (arrow YY)
2. Tephra-ringlike feature (arrow ZZ)
3. Narrow basal aprons around several domes (arrow XY)
4. Moats around several domes (arrow XZ)
5. Central mounds in several small craters (arrow YX)
6. Spinelike protuberances (arrow YZ)
7. Block-lined collapse pit (arrow ZX)
8. Crater adjacent to the fracture (arrow ZY), a common configuration

Plate 226b shows another set of fractures near the northeastern edge of the floor. The smooth unit found around the fractures shown in Plate 226a is not so apparent here. Numerous outcrops along the crevice can be identified (arrow X), but the fractures are typically subdued and discontinuous. This suggests fracturing in a surface overlain by debris, local eruptions of ash, or subsequent degradation. Near the top of the photograph are several block-strewn mounds (arrows Y and Z) that may represent spatter cones; also note the blisterlike feature seen at the bottom right edge (arrow XX). The relatively large tuff cone (arrow YY) will be referred to later and can be compared to similar positive-relief forms shown in Plates 224, 226a, and 228c. Smaller examples occur along the fractures shown in this plate (arrows ZZ and XY).

The fracture system seen in Plate 226c is different from the foregoing examples in that it is associated with a foldlike form (arrow X) abutting against two dome complexes. The region shown is less than 4 km from the northern wall (refer to Plate 219).

PLATE 226

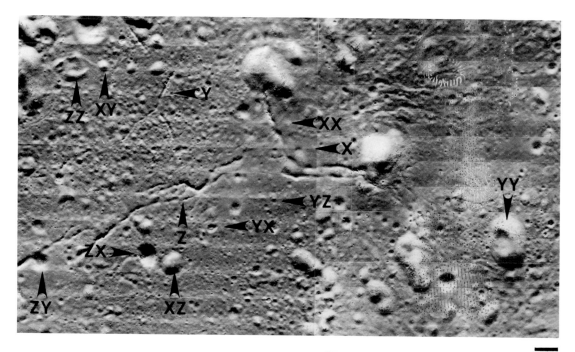

a. V-153-H1   V-152-H1   0.44 km

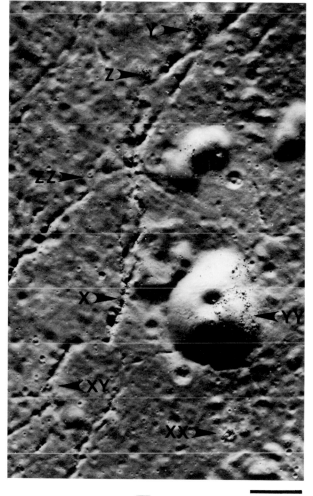

b. V-154-H1   0.44 km

c. V-154-H2   0.44 km

Copernicus

Plate 227a shows a fissurelike structure (arrow X) on the south-eastern edge of a subdued crater that is about 0.8 km in diameter (arrow Y) and adjacent to two domes. Identifiable is a flow unit (or units) extending to the north from the fissure (arrow Z), and several very subdued craters—presumably collapse forms (arrow XX)—can be recognized on its surface. The area shown in Plate 220b includes the enlarged area shown here.

Blocks ejected from the nearby crater to the north are strewn across the flow. The crater appears to lack the finer debris commonly exhibited by craters on the maria. This could be a result of a lower impact velocity—i.e., a secondary projectile—into a consolidated target. Blocks visible in the upper right originated from a similar crater just out of the field of view in this plate but shown in Plate 224 (arrow YY), as well as in Plate 220, b and c. These blocky craters should be compared to other craters and pits that are considered in Plates 228c–229d.

Positive-relief forms found on the floor of Copernicus are illustrated in Plates 227b–228b. The mamelonlike domes shown in the stereo pair (Plate 227, b and c) are bordered by moats, which closely outline the most irregular extensions of the domes. These domes commonly display blocks near their crests but have few blocks along their bases, an indication that finer debris has buried the blocks or that their location near the crest is relatively stable with respect to mass wasting.

Plate 227, b and c, also includes several other noteworthy features:

1. Lineations trending N30W cross the two large domes (arrows X and Y) and the surrounding area (arrow Z).
2. A circular structure (arrow XX), about 0.4 km in diameter, is revealed by a creaselike ring and slightly lower interior.
3. Crater with relatively large central mound (arrow YY) which is similar to craters on the floor of Flammarion (Plate 18c).
4. The apron around the small domes (arrow ZZ) exhibits a well-defined contact with the dome and is similar to structures on the maria (Plate 128) or in Vitello (Plate 250, arrows ZX and ZY).

PLATE 227

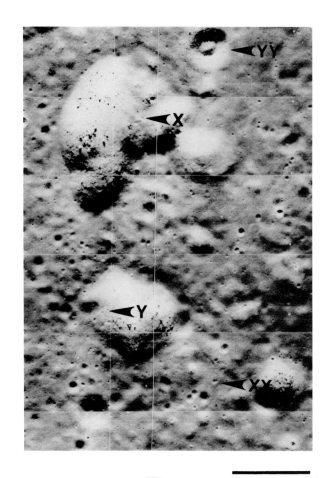

a. V-153-H2   0.43 km

b. V-153-H2   0.43 km

c. V-152-H2   0.43 km

Copernicus

Plate 228 illustrates additional examples of positive-relief features. The dome shown in Plate 228a (arrow X) displays a well-developed basal apron that is very similar to symmetric domes found on the maria (Plate 126). Other domes on the floor of Copernicus typically display narrower aprons. This particular example is near the terminus of a large flow unit extending onto the floor and shown in Plate 232.

Plate 228b shows an irregular platform (arrow X) without a basal apron or moat but with an encircling zone characterized by a concentric pattern (arrow Y) and by fewer subdued 200-m craters and hummocks. A large-scale example occurs near Grimaldi (see Plate 213, arrow XX).

Another positive-relief form that is characteristic of the floor is the cone shown in Plate 228c (arrow X). Other examples are shown in Plate 226, a (arrow YY) and b (arrow YY). The stereo view provided by Plate 228, c and d, shows that the cone is surrounded by a moat. Plate 228c also includes an elongate dome that appears to overlap a subdued crater (arrow Y). This is not a unique occurrence, as is illustrated by a dome shown in Plate 227a and probably reflects the incompetency of the domes.

The large pit (arrow Z), included in Plate 228, c and d, introduces a series of photographs (through Plate 229) that illustrates various craterforms. This example is a collapse feature and was influenced by a NW-trending zone of weakness (arrow XX). A similar pit is shown in Plate 226a (arrow ZX).

PLATE 228

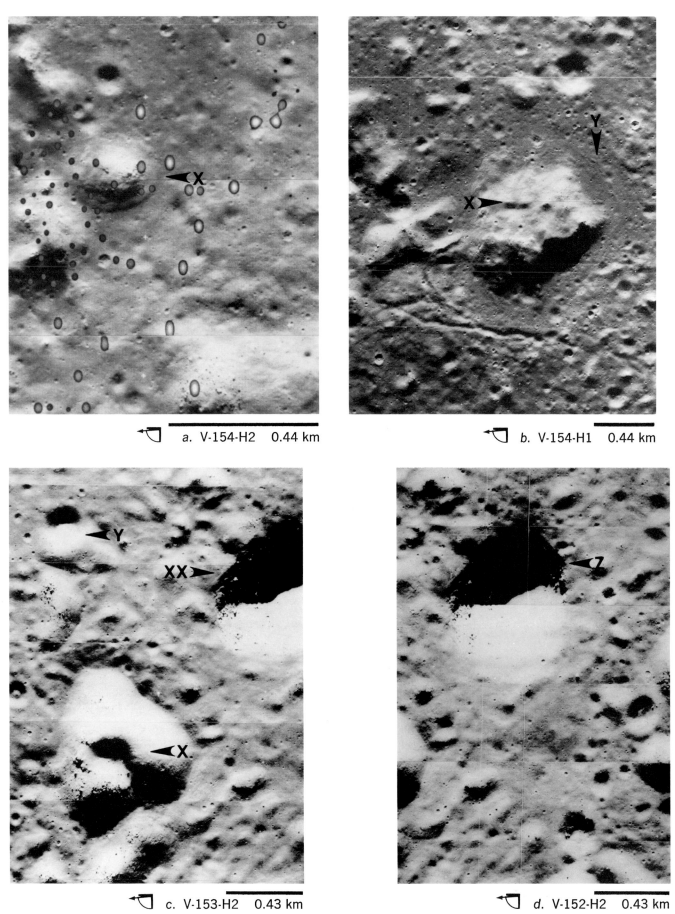

a. V-154-H2   0.44 km

b. V-154-H1   0.44 km

c. V-153-H2   0.43 km

d. V-152-H2   0.43 km

Copernicus

Additional examples of negative-relief features on the floor of Copernicus are presented in Plate 229. The elongate crater shown in Plate 229a is probably volcanic in origin. An impact with a low angle of incidence produces a saddle-shaped rim, with the minimum rim heights along the direction of the projectile trajectory (see Plate 253). In this example, however, the minimum rim heights are perpendicular to the major axis, i.e., the supposed projectile direction. In addition, the major axis appears to be an extension of a fracture (see Plate 220b, arrow YY), and the northwestern rim extends as a ridge.

The origin of the subdued crater shown in Plate 229b—as well as of similar craters abundant on the floor of Copernicus—is not so obvious. It might represent a volcanic structure or an impact that was subdued by later events, such as deposits of ash. Note that the northern rim is overlapped by a positive-relief feature.

Plate 229c shows an irregular crater with blocky ejecta and probably represents an impact into an inhomogeneous target. In contrast, Plate 229d reveals two blocky craters with more symmetric plans (arrows X and Y). The crater seen in the upper right resembles those near Maestlin R (Plate 76a). The example shown in the lower left is more subdued yet has numerous large blocks. The process responsible for the apparent smoothing of other floor features could have affected this crater also. Note that these and nearby craters (arrows Z and XX) display interior terraces that occur at different depths. If the terraces reflect outcrops of competent strata, then the local development of the floor was multiphased.

PLATE 229

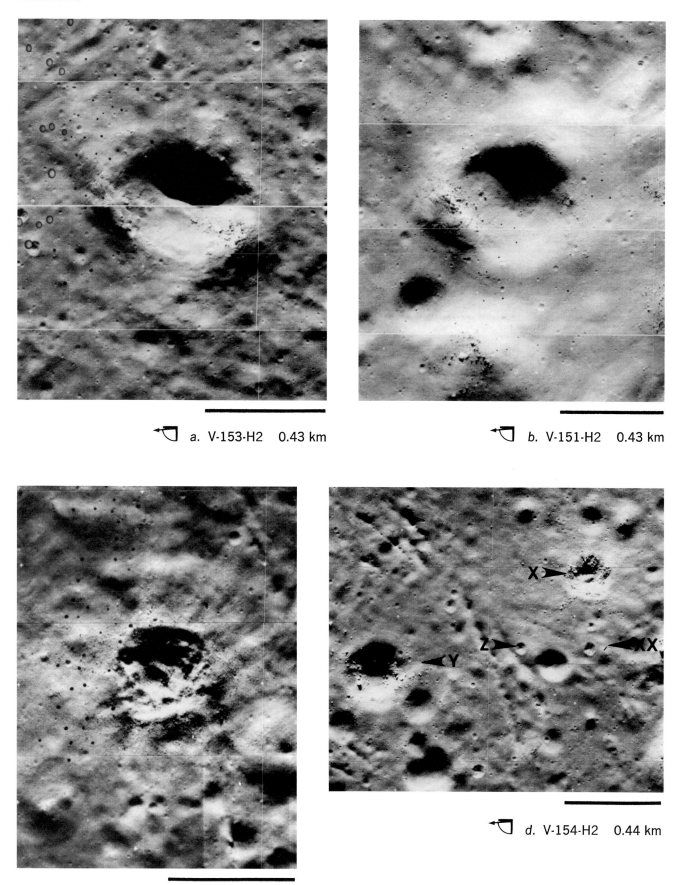

a. V-153-H2   0.43 km

b. V-151-H2   0.43 km

c. V-154-H1   0.44 km

d. V-154-H2   0.44 km

Copernicus

Plates 230–233 provide selected views of the floor-wall contact. The stereo pair (Plate 230, a and b) reveals the contact between the hummocky floor unit and the southeastern wall. The floor typically appears to overlap the wall, but arrow X identifies a lobate flow derived from the wall near the area marked by arrow Y. Numerous channellike features cross the lower wall (for example, arrow Z). The stereo pair also reveals an irregular depression (arrow XX) that is interpreted as a collapse form associated with the cooling floor.

Plate 230c shows the contact along the west-northwest wall and illustrates the highly fractured lower wall terrace (arrow X), which suggests floor subsidence. The floor area shown (top) is part of the "smoother" unit (identified on Plate 219), which also appears to separate the hummocky unit from the base of the wall. In contrast to the contact shown in Plate 230, a and b, a narrow moat borders the floor unit (arrow Y). Again, numerous leveed channels can be identified on the wall (arrows Z and XX), but none extend onto the floor. Therefore, it is suggested that the "plains" unit is representative of one of the last massive floor extrusions that postdate both the debris transferred to the floor from the wall and the period of floor subsidence.

PLATE 230

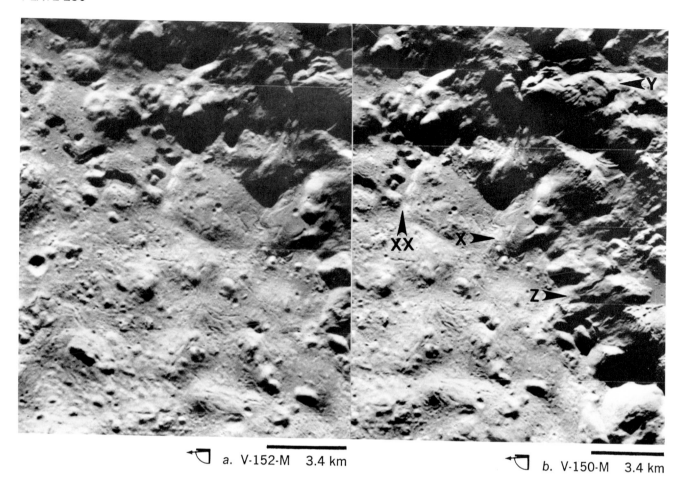

a. V-152-M 3.4 km

b. V-150-M 3.4 km

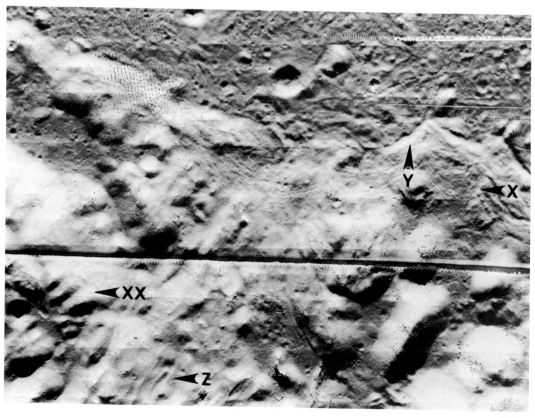

c. V-153-H3 0.43 km

Copernicus

Plate 231a is an enlargement of Plate 218b and shows the northern wall and prominent flow units extending onto the floor (arrows X, Y, and Z). Plate 231, b and c, provides a three-dimensional vertical view and includes the corresponding units (arrows X, Y, and Z, respectively) shown in Plate 231a, as well as a more massive unit to the east (arrow XX). These leveed and striated units are in contrast to the smoother unit shown in Plate 230, a and b (arrow X).

Several large **V**-shaped valleys and elongate craterforms cross the wall slumps. These are identified by arrows YY, ZZ, and XY in Plate 231, b and c; arrow XY corresponds to arrow XX in Plate 231a. Plate 234, a and b, also shows this and the surrounding region. The spatial relationship between the valleys and the large flow units suggests the following genetic relationships:

1. The valleys were carved by flows of material from the upper wall. The plains material on the upper terraces (see Plate 234, a and b) may be pooled units of molten rock that were responsible for the flows.
2. The valleys and elongate craterforms were eruptive centers for the flows.
3. The valleys and elongate craterforms are the remaining cavities of massive debris slides that produced the flows.

It is probable that molten material produced during the formation of Copernicus was injected or trapped beneath slumping wall debris. Such pockets could have acted as a lubricant for later slope failure, perhaps responsible for the cavities and flows illustrated in this plate.

PLATE 231

a. II-162-H3 (oblique)

b. V-157-M 3.3 km

c. V-152-M 3.3 km

Copernicus

The leveed flow units shown in Plate 231, *b* and *c* (arrows Y and Z), are shown in more detail in Plate 232. At this higher resolution, the units lose the crispness of detail suggested in Plate 231 and are pitted by numerous small subdued craters. The boundaries of the units become more difficult to define, and it appears that the flows either do not represent recent events or were rapidly degraded.

Shown in the lower left portion of Plate 232 is a collapse pit (arrow X), which unfortunately has been partially obliterated by photographic blemishes (see Plate 231a, arrow YY, and Plate 231, *b* and *c*, arrow XZ). The pit has removed a portion of the leveed channel (arrow Y) that extends from the valley (arrow Z) but may have been the source for another subdued channel that extends downslope (arrow XX). Possibly, the channel related to the pit was small and was guided by a pre-existing branch of the channel that extended from the V-shaped valley at the far left. The pit apparently formed after pooling of material on the terrace above it (arrow YY).

A large (2.5 km in diameter) subdued depression (arrow ZZ; also see Plate 231, *b* and *c*) has apparently controlled the course of the western branch of the two major diverging channels. Close inspection of the original photograph, however, shows that continued subsidence of this depression has severed the channel (arrow XY) that extends from the pit described above. Thus, the history of this wall region has been very complex.

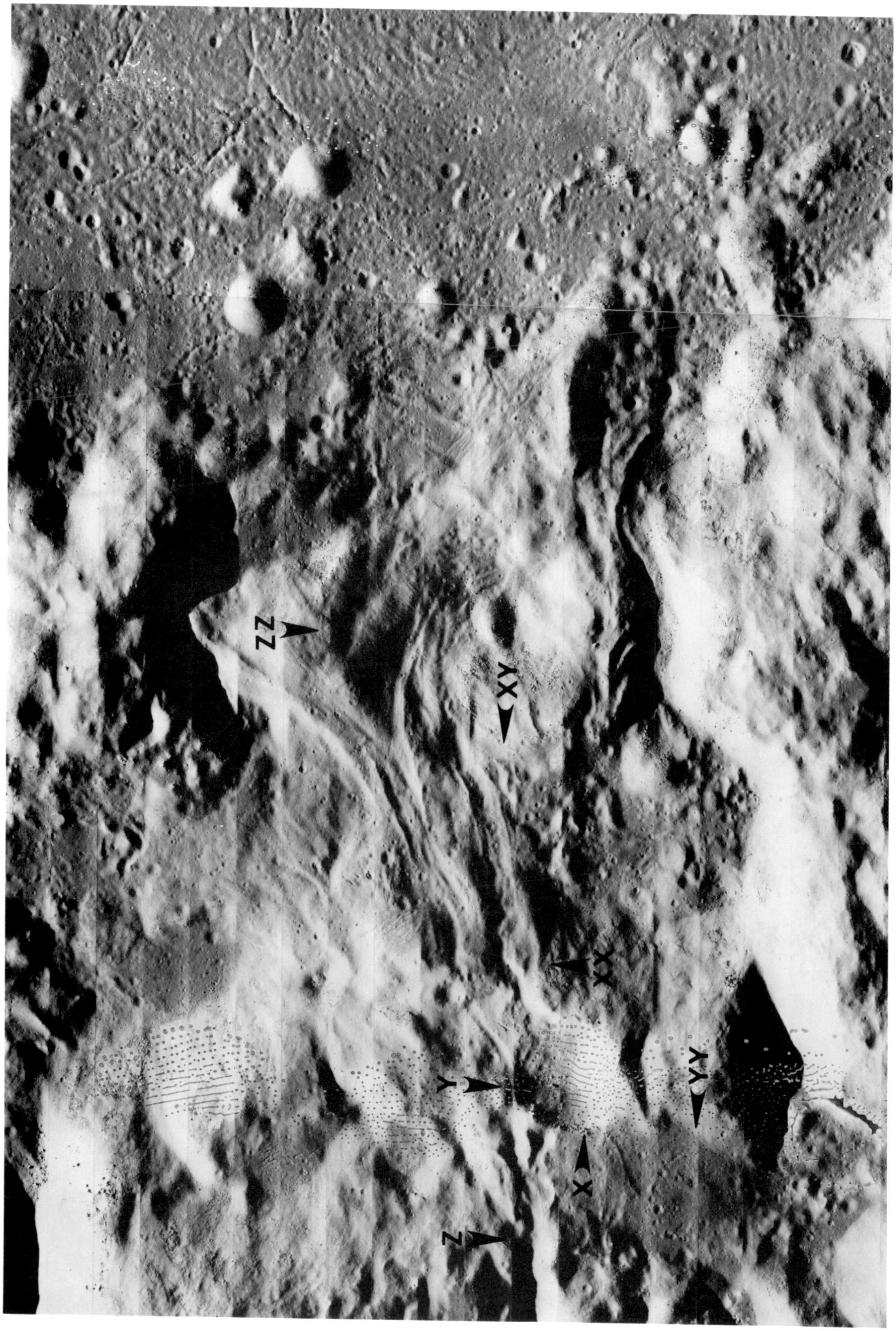

PLATE 232

V-155-H2; V-154-H2   0.44 km

SELECTED REGIONS OF THE LUNAR SURFACE

Copernicus

Plate 233 is a high-resolution photograph of the eastern flow unit shown in Plate 231, *b* and *c* (arrow XX). The portion of the unit shown is the terminus of material that was derived from near a highly dissected area close to the upper terrace and cascaded over an intermediate terrace before advancing onto the floor (see Plate 234, *a* and *b*).

In contrast to the units shown in Plate 232, the surface displays complex flow textures, which are enhanced near the termini (arrow X). The flow surface is pitted with craters, many of which are subdued. The concentric plan (arrow Y) of several craters indicates interbedding or the base of a regolith. Two different units occur at the base of the termini; a ropy complex floor unit (arrow Z) and a relatively smooth unit (arrow XX). The latter unit is shown in stereo in Plate 231, *b* and *c* (arrow YX). It is suggested that the smoother unit was derived from the last flows (shown in Plate 233) and that it was molten rather than debrislike.

536

PLATE 233

V-155-H1   0.44 km

Copernicus

The terraces of the wall of Copernicus are clearly revealed in Plate 234, *a* and *b*. In addition, the inner rim zones appear to be well preserved at this resolution (see Plates 94–97 for more detailed views of the rim regions). Pools of material occur on both the terraces and the rim area, and closer views of the former are shown in Plate 235, *a* and *b*.

As discussed in the chapter "Crater Walls" (Plates 66–67), the wall can be divided into three main regimes: an upper wall characterized by a large scarp, beneath which are numerous slump blocks; a lower wall characterized by hummocky and jumbled units (the toes of slumps and flow termini); and an intermediate area of less well defined slump blocks. The two lower zones are typically cut by leveed channels and valleys extending toward the floor.

Plates 234c and 234d are high-resolution views of the large scarps adjacent to the rim-wall contact (arrows X and Y, respectively, in Plate 234, *a* and *b*; arrow Z identifies the area shown in Plate 235). Plate 234c reveals large outcrops (arrow X), also shown in the oblique view of Plate 218. They are well below the crest and may be part of a stratified layer related to the outcrop shown in Plate 234d (arrow X). Near the top (Plate 234c, arrow Y) and base (arrow Z) of the scarp are levels of outcrops that may represent additional stratification. The sharp break of slope at the top of the scarp (arrow XX) indicates that rapid gravitative transfer of material continues to dominate any progressive smoothing. Note the cone of material at the base of the scarp (arrow YY) that was apparently derived from the furrow in the rim (arrow ZZ).

Plate 234d covers a region just to the west of that shown in Plate 234c. The top of the scarp is much more jagged (arrow Y), which could be the result of the event responsible for the relatively large crater on the rim. Beneath the crest is a layer (arrow Z) that can be traced more easily in an enlarged oblique view. Numerous blocks rest along the base of the scarp on a smooth-surfaced unit.

PLATE 234

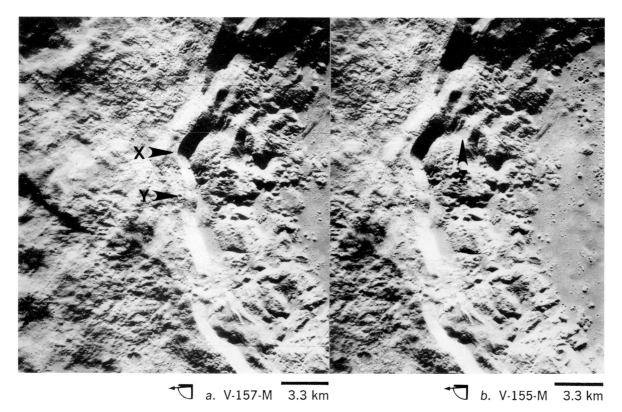

a. V-157-M   3.3 km   b. V-155-M   3.3 km

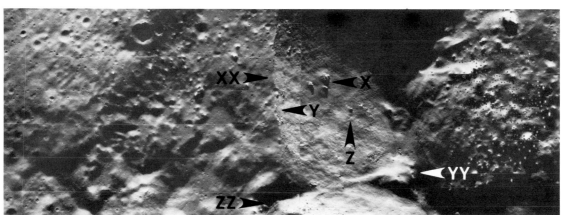

c. V-157-H1; V-156-H1   0.44 km

d. V-157-H1; V-156-H1   0.44 km

Copernicus

Plate 235a shows a smooth-surfaced unit included in Plate 234, a and b (arrow Z). The surface displays several small craters with irregular plans (arrows X, Y, and Z), which are similar to those near Maestlin R (Plate 76a). Such craters typically lack a symmetric ejecta blanket and have large diameter-to-depth ratios. Various stages of degradation seem to be represented. In addition, there are numerous subdued flat-floored (arrow XX) and mounded (arrow YY) craters. Thus, it appears that the material was deposited at a sufficiently early time to permit this degradation, but it is not clear what proportion of the subdued craterforms was endogenetic in origin. The pooled units have a regolith, which is indicated by boulder trails crossing a similar unit on the terrace adjacent to but beneath that in Plate 235a (not shown). Note that breakaway fractures have more recently cut the unit shown in Plate 235a.

Plate 235b includes a large slump block on the southeastern wall of Copernicus. Smooth-surfaced units can be identified on top of (arrow X) and at the base of (arrow Y) the block and were emplaced after its movement. The source for a few of these pools was apparently the inner zone of the rim, which exhibits abundant deposits of similar material. For example, one of the units on top of the slump block (arrow Z) was derived from channels extending back to the rim (arrow XX). Plate 235b also includes a highly textured pooled unit (arrow YY).

Plate 235c covers a small area on the northwestern wall and shows an overhanging lip (arrow X) with numerous dislodged blocks (arrow Y) and breakaway fractures (arrow Z). The terrace surface (to the left) is not a flat pool, such as that shown in Plate 235a, and does not resemble a single unit, because of its hummocky topography. The absence of slumped material at the base of the scarp is evidence that the overhang did not result from a sudden slide, but from progressive removal of underlying unconsolidated material.

PLATE 235

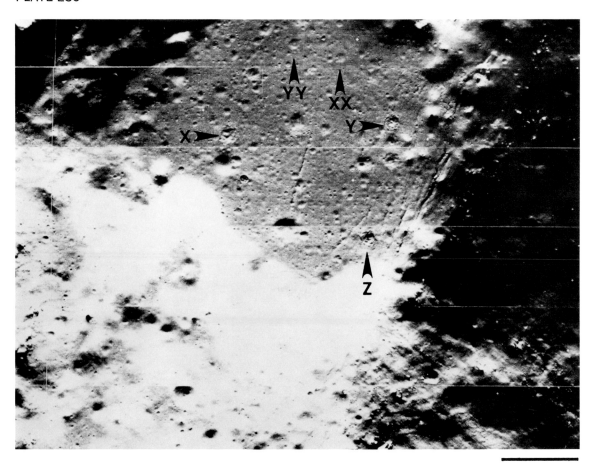

a. V-156-H1   0.44 km

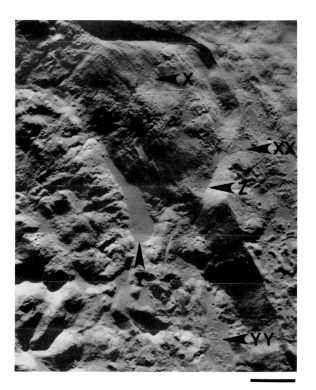

b. V-150-M   3.4 km

c. V-155-H3   0.44 km

Aristarchus

The crater Aristarchus is in Oceanus Procellarum at the edge of a large platform that is crossed by numerous sinuous rilles (see Plate 171a), most of which have "head" craters on this relief. The crater is about 36 km in diameter and resembles Copernicus, whose diameter is almost three times that of Aristarchus. Plate 236a shows Aristarchus and its extensive ejecta blanket, extensive wall slumps, "flat" floor, and central-peak complex.

Plate 236a suggests that the ejecta blanket has a relatively high albedo near the rim crest (arrow X), Zone I, but a lower albedo corresponding to the dunelike Zone II deposits of bulk ejecta (arrow Y). Plate 171a shows that ray patterns are extensive over the mare plains but are relatively mottled across the platform. In Earth-based telescopic views, the bright-rayed streamers are not so pronounced as they are in other similar craters, such as Kepler or Copernicus, under full solar illumination. The relatively recent formation of Aristarchus is obvious in this and following plates; therefore, the relatively low albedo of its ray pattern is interpreted as an intrinsic feature and not the result of degradational processes.

Plate 236, b and c, is a stereo pair and reveals that Aristarchus appears to rest on a dome whose boundaries (arrow X) approximately correspond to the transition from the Zone I to Zone II ejecta blanket. In the chapter "Crater Rims," this is noted as a common feature of large craters and is a pronounced feature for craters surrounded by later mare units. It may correspond to extensive bulk-ejecta deposits or structural doming associated with the impact event, which presumably formed Aristarchus. Note the markedly raised eastern rim (arrow Y).

The following six plates show certain selected regions in and around Aristarchus that are identified by the remaining arrows in Plate 236a.

PLATE 236

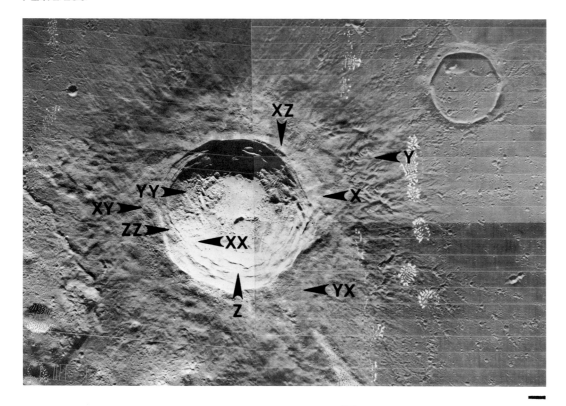

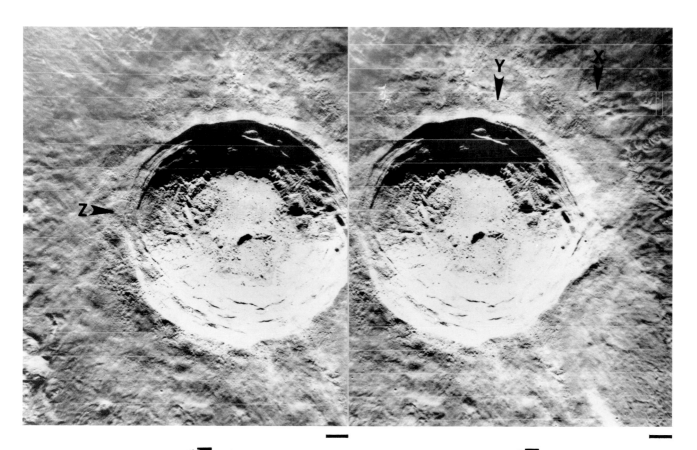

b. V-201-M   4.1 km

c. V-198-M   4.0 km

Aristarchus

Plate 237 shows the floor region of Aristarchus. Arrow X identifies the N-trending central-peak complex. The floor resembles those of Copernicus (Plates 219–230) and Tycho (Plates 31–33). In particular, the floor can be divided generally into two physiographic units: plains-forming units (arrow Y) and hummocky units (arrow Z). The plains-forming units exhibit ribbon-like patterns (arrow XX) and extensive fracturing (arrow YY). A prominent system of floor fractures approximately reflects the floor-wall contact and typically occurs on a convex apron (arrow ZZ; see Plate 238d). This system is attributed to peripheral tensional stresses developed during floor subsidence, and the convex apron represents a previous floor level, or "bathtub ring." The local stress field was altered around the southern extension of the central peaks (arrow XY). The floor generally subsided as a unit, with the wall and central-peak complex representing the major stable relief. This is suggested by the system of fractures that encircle the floor but not the hummocky areas.

Numerous positive-relief forms occur within the plains-forming unit. Several mamelonlike domes exhibit moatlike borders (for example, arrow XZ) that are analogous to features in Copernicus (Plate 227, b and c). In contrast, other mamelonlike domes exhibit basal aprons (arrow YX) that are characteristic of such domes on the maria (see Plate 126) and are also recognized in Copernicus (Plate 228a). Note the feature (arrow YZ) that displays a fractured apron resembling the floor-wall contact.

The hummocky area does not make an abrupt contact with the plains-forming unit and, in contrast to Copernicus, is not extensively fractured. This area appears to be composed of contiguous mamelonlike domes.

Plate 237 also includes portions of the lower wall (see Plates 239–242). It is crossed extensively by furrows that abruptly terminate at the floor-wall contact. In addition, "pools" of plains-forming material occur within local depressions on the wall (arrow ZX; see Plate 241c).

Arrows ZY, XXX, and YYY, shown on Plate 237, indicate areas examined in Plate 238.

PLATE 237

V-199-H2; V-198-H2   0.54 km

Aristarchus

Selected portions of the floor-wall contact are shown in Plate 238. Regions indicated by arrows ZY, XXX, YYY, and ZZ in Plate 237 correspond to the regions shown in Plate 238, *a*, *b*, *c*, and *d*, respectively. Fractures accompanying this contact and the relatively abrupt disappearance of the wall furrows are clearly illustrated in Plate 238, *a*, *b*, and *d*. Well-defined termini of the floor unit against the wall are absent.

Plate 238*c* shows an area to the north of the region shown in Plate 238*a*. It shows fractures that approximately parallel those at the floor-wall contact and cross several mounds (arrow X); this is also shown in Plate 238*d* (arrow X). Although ejecta from the subdued crater (upper left, Plate 238*d*) generally blanket the region and nearby fractures, a trace of a fracture is recognized on its wall (arrow Y). Note that the fractures along the floor-wall contact and those seen in Plate 238*c* are more jagged than the fractures identified by arrow X in Plate 238*b*. This difference may reflect different lengths of exposure to meteoroid bombardment. More likely, the subdued cracks developed while the floor was still relatively plastic and were accompanied by escaping material and volatiles that contributed to degradation, whereas the peripheral fractures were related to later and perhaps continuing floor subsidence.

Several features of interest are shown in Plate 238:

1. The breached tephra-ringlike feature shown in Plate 238*b* (arrow Y)
2. The pit seen in Plate 238*d* (arrow Z)
3. The numerous small (less than 30 m in diameter) subdued craters on the floor

PLATE 238

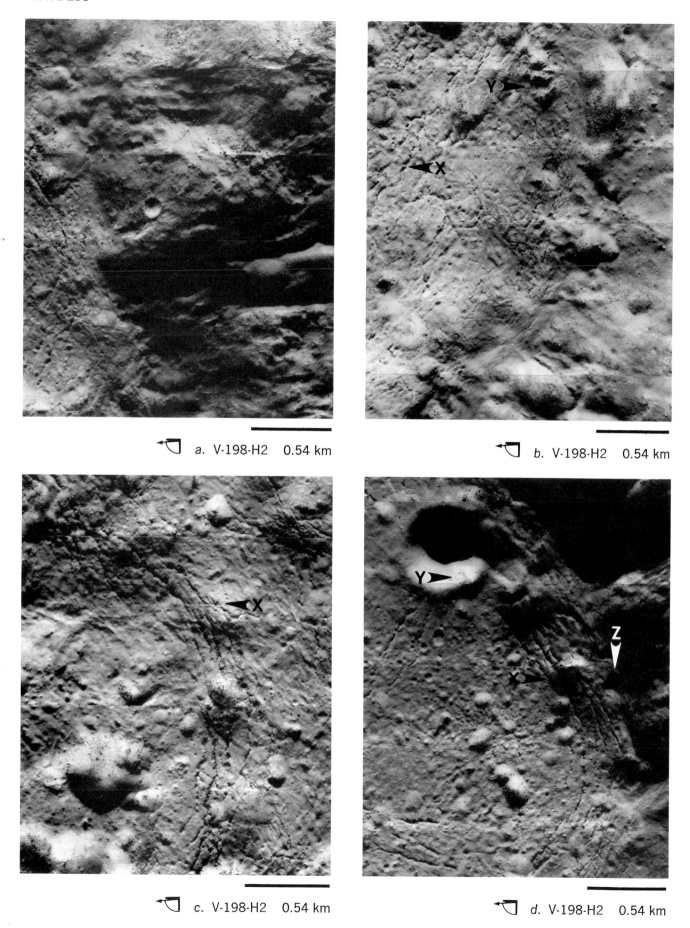

a. V-198-H2   0.54 km

b. V-198-H2   0.54 km

c. V-198-H2   0.54 km

d. V-198-H2   0.54 km

Aristarchus

Plate 239 shows the fully illuminated east-facing wall indicated by arrow Z in Plate 236a and reveals the results of extensive slumping. The slump blocks are typically narrow, and several upper terraces have trapped smooth plains-forming material (arrows X and Y; see Plates 241–242). Numerous rimmed furrows (arrows Z and XX) extend down the wall (from the bottom of the photograph to the top) and several can be traced to furrows on the rim (arrow YY). Blocks are strewn across the wall (arrow ZZ) but are not generally recognizable on the floor. High-albedo and low-albedo streaks (arrows XY and XZ, respectively) also extend down the wall.

The floor (seen at the top of the photograph) overlaps the lower wall and the surmised toes of slumps. Arrow YX identifies a relatively smooth unit that overlies the floor and that was derived from the wall.

PLATE 239

V-199-H3; V-198-H3   0.54 km

Aristarchus

Plate 240 shows a portion of the upper southeast-facing wall indicated by arrow XX in Plate 236a. The solar illumination of Plate 240 is more favorable for revealing the leveed furrows that extend across the wall from the rim. A notable example originates near the upper wall scarp (arrow X) and can be traced across the wall terraces to a clump of boulders (arrow Y). Note the large levees of this feature at arrow Z. Another example is clearly traced from the rim (arrow XX).

"Pools" of material occur in local depressions on the rim (arrow YY) and the wall (arrow ZZ). They commonly are extensively fractured. Note the blisterlike surface of one of these pools (arrow XY). This is not a unique feature (see Plate 242b). The pools commonly represent materials that were derived from sources near the rim, flowed to local topographic lows, and formed leveed channels. They are interpreted as once-molten material that was heaved as ejecta and/or extruded along fractures.

Note the relatively numerous subdued craters that are typically 0.3 km in diameter (for example, arrow XZ). These may represent either endogenetic features or late-arriving secondaries from Aristarchus.

PLATE 240

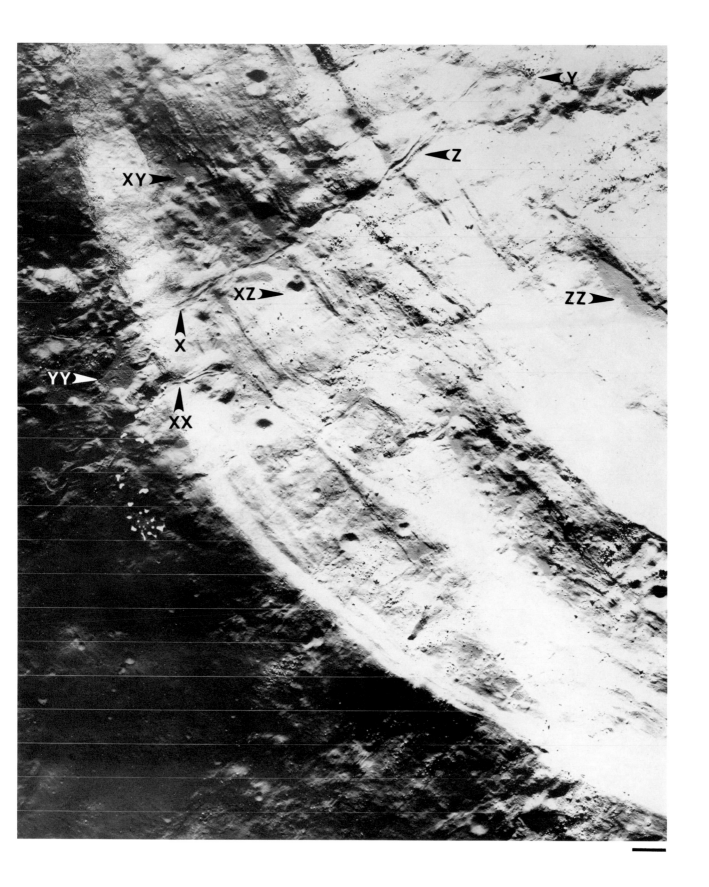

V-200-H3   0.55 km

Aristarchus

Plate 241 shows selected views of the "pools" on the walls of Aristarchus. The region shown in Plate 241a is included in that shown in Plate 239 (arrow X). The narrow terrace is capped by material derived, in part, from the upper wall (bottom of photograph). This observation is indicated by the conelike feature (Plate 241a, arrow X) and the termini (arrow Y) associated with the leveed channel (arrow Z). Fractures (arrow XX) near the edge of the terrace may reflect breakaway scarps associated with mass transfer; tension cracks produced by the addition of material from the upper wall; or tension cracks due to subsidence within the pool.

The pool (arrow X) seen in Plate 241b (see Plate 239, arrow Y) does not appear to be on a terrace, but this impression is largely the result of solar illumination. This small plains-forming unit does not exhibit well-defined termini and cannot be traced directly to channels from the upper wall (right side of photograph), although numerous furrows are identified (arrow Y). The sharp-rimmed, dark-haloed crater (arrow Z) is in contrast to the other pristine crater on the pool (arrow XX). Note the number of subdued craters on this surface that may have been formed endogenetically.

Plate 241c shows a pool on the southern wall (Plate 237, arrow ZX). It is crossed at several locations by fractures. Those shown near arrow X cross not only the pool material but also some of the hummocky wall material; therefore, they may be related to mass wasting rather than release of stresses associated with the pool. The trellis pattern (arrow Y), the concentric pattern (arrow Z), and the fractures confined to the more hummocky pool material (arrow XX) probably reflect stresses associated with the emplacement or cooling of the pool units. The corrugated surface pattern (arrow YY) is thought to represent transverse flow ridges due to the last surges of material from the upper wall (to the upper right).

Several types of craters are recognized on this surface. Concentric craters typically occur on such units (arrows ZZ and XY) and may reflect a thin regolith. In addition, note the irregular craterforms (arrow XZ and YX), which are interpreted as blisters, and the subdued rimless crater (arrow YZ), which is interpreted as a collapse pit.

PLATE 241

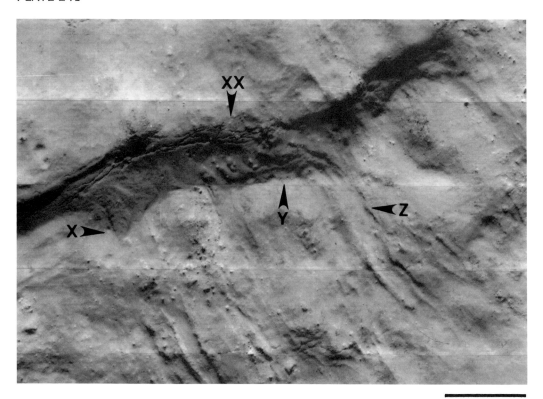

a. V-198-H3   0.54 km

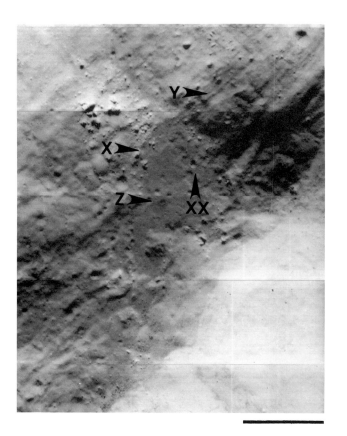

b. V-198-H3   0.54 km

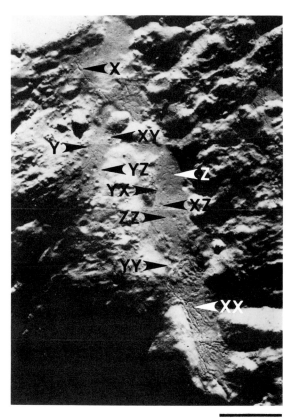

c. V-198-H2   0.54 km

Aristarchus

Arrow YY in Plate 236a identifies the portion of the wall shown in Plate 242a. The floor of the crater is to the lower right. The relatively smooth wall unit (arrow X) is in a narrow terrace and postdates the formation of the numerous furrows. The long fracture (arrow Y) crosses both the poollike unit and the wall hummocks. As suggested for the fractures shown in Plate 241c (arrow X), this probably reflects additional wall movement. The complex pattern (arrow Z), however, may be related to the removal of material into the furrow (arrow XX).

Plate 242b is also on the south-facing wall, indicated by arrow ZZ in Plate 236a. The pooled units in this plate have a blistered appearance (arrow X) and are similar to those recognized in Plate 240 (arrow XY) and in several other areas on the wall and rim. Their origin is not clear. They may represent trapped volatiles, buried blocks, or spatter remnants from molten secondaries.

PLATE 242

a. V-200-H2   0.55 km

b. V-200-H2   0.55 km

Aristarchus

This and the following five plates show the ejecta blanket of Aristarchus. Plate 243, *a* and *b*, is a stereo view of the southern rim area and reveals the high-albedo Zone I (arrow X) and low-albedo Zone II (arrow Y) regions of the bulk ejecta. Characteristics of these zones are noted in the chapter "Crater Rims" in reference to the crater Copernicus (Plates 94–97). Concentric corrugations (arrow Z) and radial lineations (arrows XX and YY; see Plate 248) are recognized in Zone I.

Zone II is extremely complex and clearly reflects the rapid outward movement of ejecta. Craterlike features (arrow ZZ) resemble dunes and are breached away from Aristarchus. Their floors commonly have relatively smooth units (arrow XY; see Plate 244). Bulk ejecta extensively blanket the region to the south of the dunelike features. Plate 236 shows that these craterforms are best developed on only the south rim of Aristarchus.

Plate 243, *c* and *d*, includes the transition from Zone II to Zone III of the ejecta sequence and shows the effects of ejecta on the pre-existing mare-filled crater Herodotus, shown also in the upper right corner of Plate 243, *a* and *b*. Several long chains of herringbonelike craters cross the region (arrows X and Y). Numerous secondary craters contain mounded floors (arrows Z and XX), and several have hummocky floors (for example, arrow YY). These features do not appear to be characteristic of a particular surface type; therefore, they probably reflect properties of the projectile and perhaps remnants of the secondary projectile itself. Note the irregular collapse depressions within Herodotus (arrows ZZ and XY).

PLATE 243

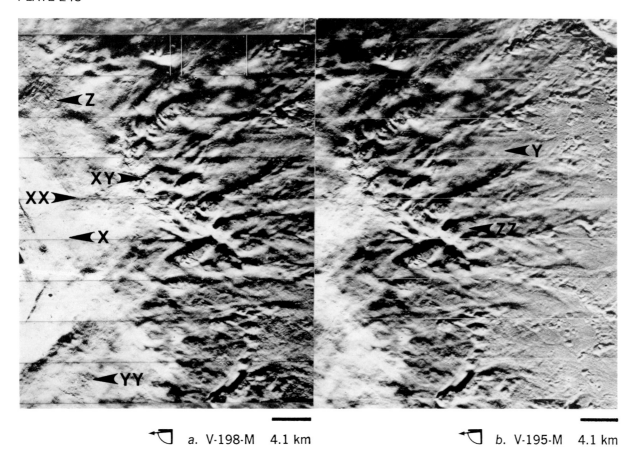

a. V-198-M   4.1 km

b. V-195-M   4.1 km

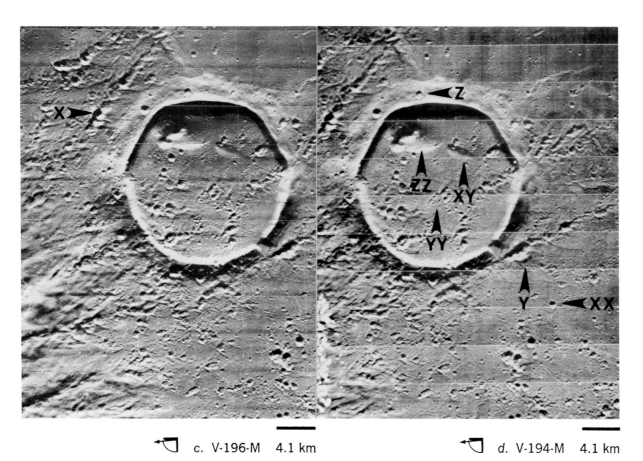

c. V-196-M   4.1 km

d. V-194-M   4.1 km

557

Aristarchus

Plate 244 is a high-resolution view of the ejecta-sequence transition from Zone I (left) to Zone II (right) (see Plate 243, *a* and *b*, arrow XY). Zone I in this region exhibits numerous radial lineations (arrow X). Blocks are scattered across the slightly corrugated surface (arrow Y) in the transition and Zone I regions but are comparatively absent on the dunelike topography to the right.

Plains-forming units occur in numerous, but not all, topographically low regions (arrows Z and XX) to the south of the striated portion of Zone I. Their surfaces are typically smooth, with few small craters and blocks. Exceptions are the textured flowlike units (arrows YY and ZZ), which are probably related to the relatively broad channellike features to the north (arrows XY and XZ, respectively). The unit identified by arrow XX is crossed by a fracture system that extends into the hummocky region to the south (also see arrow YX). These features may be related to settling and compaction of the bulk ejecta.

The location of numerous smooth-surface pools near the south end of the lineated zone deposit suggests a genetic relationship (see Plate 243, *a* and *b*), which is considered with Plate 245.

Craters are common in this region and display a wide variety in their crispness of form. They may represent late-arriving secondaries. Subdued craters perhaps indicate impacts that occurred while newly deposited bulk ejecta were still somewhat mobile. If craters without extensive ejecta blankets are the result of secular degradational processes, then the preservation of other topographic features at smaller scales indicates large differences in relative competency of the units. Three pristine craters (arrows YZ, ZX, and ZY) display blocky rays having low reflectivities at this angle of illumination.

PLATE 244

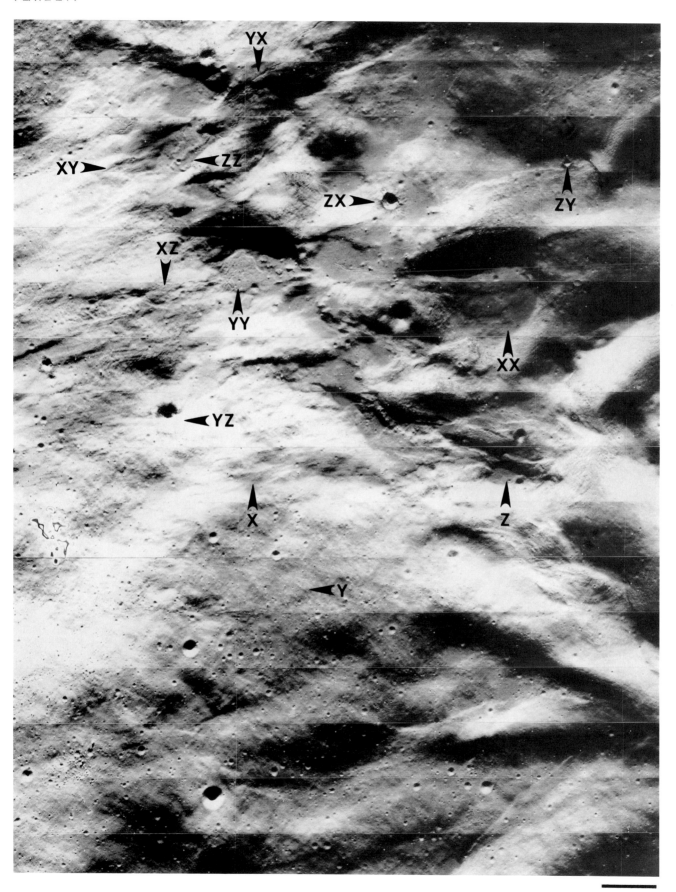

V-196-H2   0.53 km

Aristarchus

The region shown in Plate 245 and indicated by arrow XY in Plate 236a is on the northern rim of Aristarchus and includes also the transition between Zone I (right) and Zone II (left). Numerous prominent flow units are identified. Lobate flows (arrows X and Y), with related leveed channels, were derived from an ill-defined 1.5-km-diameter depression (arrow Z; also see Plate 236, b and c, arrow Z), which has a small pool of material on part of its floor (arrow XX). Leveed channels and flows to the west (arrow YY) also can be traced to craterlike sources (arrow ZZ). One of the leveed channels in this group extends down the wall of Aristarchus (arrow XY) rather than down the rim.

A second type of lobate-bordered flow unit on the rim of Aristarchus is thin and has a smooth surface (arrows XZ and YX). The examples shown in Plate 245 have blocks on their surfaces that may have been surrounded or carried by the flow. In addition, there are several subdued craters on the flows that are probably subsidence pits. These units are not clearly linked to craters or depressions and may be related to a third type of flow, discussed below. Other examples of such distinctive units are recognized to the west of the region included in Plate 245 (see V-201-H3) and cover broader areas than the limited units shown here.

The third general type of flow is much thicker and more jumbled in appearance (arrow YZ) than the two foregoing units. They typically exhibit highly irregular termini and may have one or more narrow extensions (arrow ZX). Their surfaces are commonly blocky, although portions appear to have been draped by the smoother units described above. At several locations, these hummocky units appear to merge at their termini with the smooth units.

The three flow types are perhaps units having different viscosities but a similar origin. For example, the thick, hummocky flows may be bulk ejecta originally composed of both molten and nonmolten debris. The thin, smooth units may represent a molten component that was initially entrapped in the hummocky units but subsequently separated and escaped because of its very low viscosity. The escape routes apparently were beneath the termini of several thick units. This phenomenon may be analogous to large-scale flows from Orientale (see Plate 199). The lobate units indicated by arrows X and Y are flows of intermediate viscosities.

The relation of several flow units to craters and depressions suggests two additional alternatives. Perhaps such units are the result of molten secondaries. However, the resulting craters typically do not exhibit an ejecta blanket, and the associated flows occur as units whose subsequent paths are largely controlled topographically rather than by the angle of impact. Therefore, the craters and depressions may represent internal sources of molten material, which was trapped by overlying ejecta or extruded through deep-seated fractures.

PLATE 245

V-201-H2   0.55 km

Aristarchus

The southeast rim area shown in Plate 246 is indicated by arrow XZ in Plate 236a. The position of the sun is favorable for enhancing the concentric (arrow X) and radial lineations (arrow Y). The concentric system is composed of relatively narrow ridges and furrows that commonly are continuous across the broader radial furrows. Several radial furrows have smooth-surfaced pools on their floors that clearly postdate the concentric system. In addition, several long, narrow radial furrows (arrows Z and XX) also appear to cut the concentric system. These particular furrows may be related genetically to the cluster of blocks near their termini (arrows YY and ZZ, respectively), although similar strings of blocks occur without an associated furrow.

The concentric lineations give the interior rim a corrugated appearance. They are interpreted as either concentric fractures or pressure ridges associated with movement of the interior bulk ejecta. Many of the narrow radial furrows appear to be produced by ejecta that clearly were among the last rim units emplaced. Ejecta also may have carved the larger radial furrows, but it is possible that some were structurally produced. As suggested in the discussion in the chapter "Crater Rims," the pools of material are either deposits of molten ejecta or extrusions along fractures.

PLATE 246

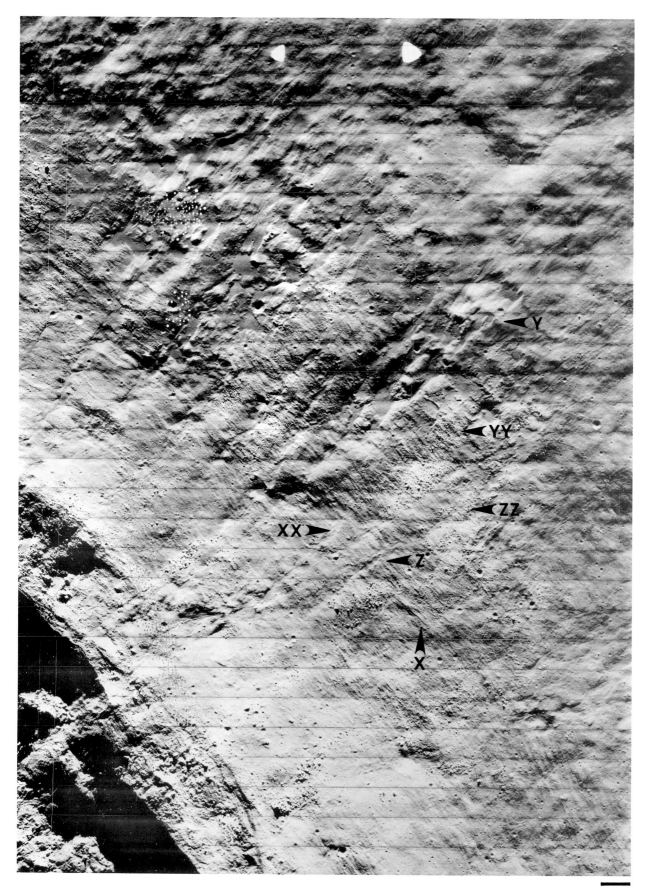

V-197-H1   0.53 km

Aristarchus

Plate 247a shows a flow on the southwestern rim that is indicated by arrow YX in Plate 236a. The pooled area (Plate 247a, arrow X) is not thought to be the source of this unit. Close examination in stereo suggests that the flow was derived from a larger pooled depression about 5 km to the north-northeast of the area shown in Plate 247a. Flow levees are clearly recognized (arrow Y) along the partly emptied channel. Near the end of this channel, transverse pressure ridges developed (arrow Z), and the lobate-fronted flow material extends across the rim surface to the area marked by arrow XX.

The flow seen in Plate 247a obviously postdates deposition of bulk ejecta and appears to postdate one of the larger craters (arrow YY). Note that its trend is perpendicular to the radially directed ejecta. If the flow represents molten material ejected from Aristarchus, then it must have had a high trajectory, and, during its flight, deposition of the bulk ejecta was complete. Moreover, it is remarkable that, after the secondary impact, there was enough unspattered material to form this flow. These constraints suggest that the flow was derived from an internal source, which may have been a peripheral vent or a trapped pocket of molten ejecta.

Plate 247b shows the western rim-wall contact of Aristarchus. Numerous fractures are revealed under this favorable solar illumination, and it is possible that future slumping will follow these weaknesses. Note the clusters of blocks on the wall that occur at approximately the same distance beneath the rim crest (arrows X, Y, and Z); they may represent layering.

PLATE 247

a. V-197-H3   0.53 km

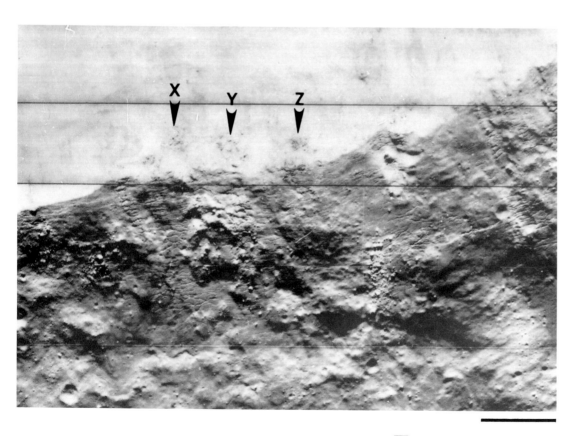

b. V-198-H3   0.53 km

Aristarchus

Plate 248 is a high-resolution photograph of the southwestern rim area (arrow YY; Plate 243, *a* and *b*). It shows the corrugated terrain (arrow X), numerous craters, and elongate depressions, which typically have flow units extending from them (arrow Y, Z, and XX). Some of the elongate depressions appear to be small relative to the flow material (arrows YY and ZZ) and commonly do not exhibit rims. They are interpreted as the results of limited flank eruptions or impacts by low-velocity molten ejecta.

PLATE 248

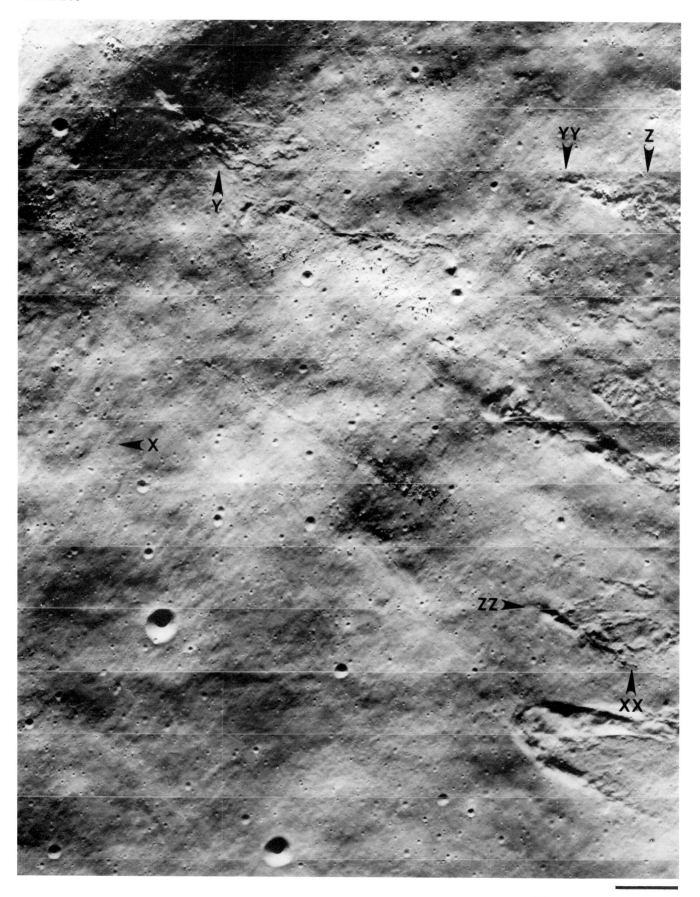

V-197-H3   0.53 km

Vitello

The crater Vitello, shown in Plate 249, is about 42 km in diameter and is on the southern margin of Mare Humorum (Plate 191a, arrow Y). The floor of Vitello is extensively fractured in a concentric pattern. The central peaks are within those fractures and are on a relatively unfractured central block. A hummocky zone encircles the fractures (arrow X) and is generally composed of poorly defined wall slumps. Several small craters (arrows Y and Z) have been severed along the top of the wall; therefore, mass wasting of the wall scarp is a continuing process. The rim of Vitello is raised (arrow XX), and its ejecta blanket is extremely subdued, although a few radial lineations can be identified (arrows YY and ZZ). Any ejecta blanket clearly predates emplacement of the mare units to the north and east.

The region around Vitello exhibits numerous small features that indicate an interesting history. Note the following forms:

1. A chain of subdued craters that extends across the peninsular feature (arrow XY) and abruptly terminates at the wall of Vitello.
2. Broad-rimmed crater (arrow XZ) with a hummocky floor that is interpreted as an endogenetic form (see Plate 87b).
3. Rectilinear rille whose profile changes across different units, probably as the result of a different competence of the surfaces (arrow YX; see Plates 178b and 87b).
4. Small crater on the mare (arrow YZ) that appears to be genetically related to a lobate flow extending to the south.
5. Irregular depressions that are interpreted as collapse features associated with mare emplacement (arrow ZX); also note the subdued craterform on this surface (see Plate 87a).
6. Narrow well-defined terrace that typically marks the contact between the highlands and mare (arrow ZY); at several locations, this contact is moatlike (arrow XXX). These contact features are also interpreted as results of mare emplacement, and their preservation indicates relatively slow degradational processes.
7. Numerous wrinkle ridges. One ridge (arrow YYY) appears to postdate a crater displaying an extensive ejecta blanket. Several small craters are interpreted to be related genetically to the ridges (arrow ZZZ). A ropy ridge can be identified on the highlands (arrow XXY), and other photographs show that it extends several hundred kilometers.

PLATE 249

V-168-M   5.1 km (*top*)   6.0 km (*bottom*)

Vitello

The floor of Vitello is shown in detail in Plate 250. The multiple central peaks are domelike, and five relatively large components of this complex surround a small central mound (arrow X). Such an arrangement suggests that radial tension fractures split the peaks and that the central portion was dropped (see discussion with Plate 49).

The rilles on the floor generally have U-shaped profiles (arrow Y). The floor-wall and wall-rim transitions are rounded. A coarse tree-bark-like pattern characterizes the rille floor, wall, and rim. Blocks occur in patches along the upper rille wall (see Plate 251b) and exhibit a higher reflectivity than do the surrounding surfaces. Few blocks occur outside these patches. The rille floors are notably devoid of blocks and exhibit lineations that mostly follow the rille plan. Note the moatlike contact between floor and wall within one rille (arrow Z).

Areas outside the rilles also exhibit extensive surface texturing, which resembles the tree-bark pattern but does not necessarily reflect surface topography. In at least one location (arrow XX), the pattern appears to be perpendicular to surface contours. Blocky patches are typically associated with low-relief mounds (arrows YY and ZZ).

Relatively few craters 30 m–300 m occur on the floor of Vitello, although there are small areas that do exhibit a higher density of small craters (arrow XY) than is usual. Pristine craters smaller than 30 m, however, appear to be abundant. Larger craters are typically rimless (arrow XZ) or broad rimmed (arrow YX). The latter crater has a ropelike form extending southward (arrow YZ; see Plate 251a).

These observations suggest that the floor of Vitello is covered with a blanket of material. The underlying surface is exposed in blocky patches along the rille walls and on top of mounds. The nature of this material may have been ashlike, but the ropelike ridge (arrow YZ) extending from the small crater may represent a flow unit. In addition, two domes (arrows ZX and ZY) are encircled by an apron and moat similar to domes on the mare (Plate 128) and within Tycho (Plate 33b). Consequently, the floor may have been covered with a unit that compacted (ashlike) or drained (lavalike). Alternatively, the elevation of the floor, as indicated by the floor fractures, was accompanied by local Moonquakes. Such quakes may have triggered movement of an overying regolith that produced surface textures, erased small craters, and exposed an underlying blocky surface along regions having the greatest slopes.

PLATE 250

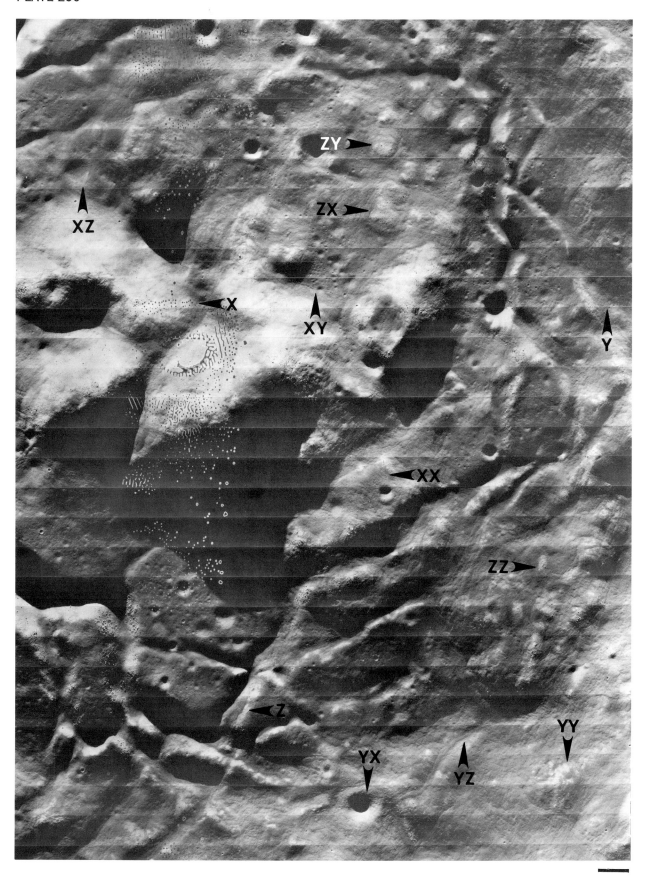

V-168-H2   0.71 km

Vitello

The broad-rimmed crater identified in Plate 250 by arrow YX is shown in greater detail in Plate 251a. Tree-bark-like textures on the rim are not concentric with, but are commonly radial to, the crater (arrow X); however, they do not appear to be part of an ejecta blanket, since this radial pattern is not uniform. Patterns from the adjacent terrains (arrow Y) commonly merge with the rim textures. Blocks are not widely distributed, but some can be seen on portions of the wall and rim. The smaller crater to the northeast (arrow Z) also displays prominent rim textures.

Both craters have flowlike structures extending from them. The ropelike form (arrow XX) from the larger crater is similar to the feature in Gassendi (Plate 86b) and to structures within floor fractures of Gassendi (Plate 38c, arrow Y). It appears to be controlled by a furrow and crosses the lineations marked by arrow YY. The flowlike structure from the small crater is identified by prominent NW-trending lineations (arrow ZZ). It is proposed that these craters were volcanically formed and that the flowlike features are associated extrusions.

The surface textures shown in Plate 251a are complex, and other possible flow units are easily identified. Confirmation of such identifications, however, requires stereo coverage, which is not available. The textures are thought to be produced by movement of material, although the nature of the material must remain speculative. Note the domelike feature at right (arrow XY) that is surrounded by a moatlike contact (arrow XZ) with the textured wall and floor material. Also note the highly reflective blocky patches (arrow YX) near the crest of the domelike feature.

Plate 251b shows an enlarged portion of floor fractures west of the central peaks. The E-trending fracture (arrow X) is oriented so that its floor and both walls are visible. The facing walls (arrows Y and Z) are strewn with blocks, whereas the lower wall and floor (arrow XX) have only a few. This particular fracture is deeper than the others and approximately **V** shaped in profile. Most of the floor fractures, however, resemble either the E-trending depression (arrow YY), which has only patches of blocks on its walls, or the subdued NE-trending depression (arrow ZZ), which has very few blocks.

The surface outside the rilles seen in Plate 251b exhibits a prominent NW-trending grain (arrow XY), presumably reflecting smaller scale fracturing. Note the small crater (arrow XZ) that is breached and has a flowlike feature extending onto the rille wall.

PLATE 251

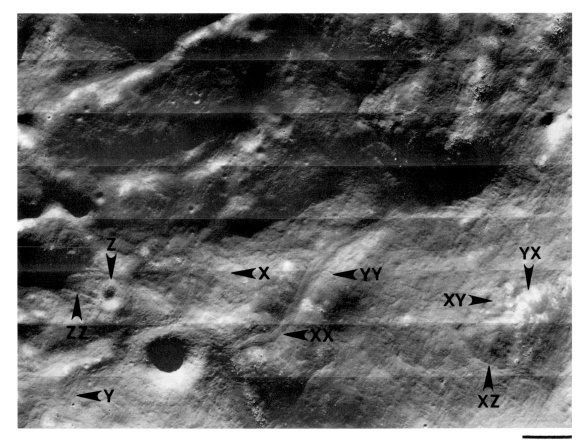

a. V-168-H2   0.71 km

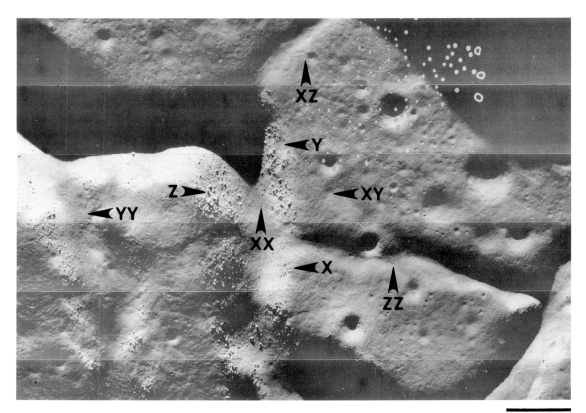

b. V-168-H2   0.71 km

Vitello

The blanket of material on the floor of Vitello, shown in Plate 252a, extends onto the western wall area. Little detail is recognized on the fully illuminated east-facing scarp, but the textured terrain characteristic of the floor is clearly identified on the hummocky region adjacent to the wall. This region is interpreted as remnants of slumps (see Plate 249). As on the floor, highly reflective patches occur near the crests of local topographic highs and generally correspond to blocky exposures (arrows X and Y). Relatively few exposures are recognized on the lower portions of the scarp (arrow Z). Ropelike features occur within linear depressions (arrow XX) and are analogous to features in Gassendi (Plate 38c, arrow Y). They do not appear to be associated with craters, as do those shown in Plate 251a, and are suggested to be the result of mass movement.

Plate 252b shows the eastern rim of Vitello. The textured terrain is characteristic of this region as well as of the wall and floor regions. Blocky outcrops also occur (arrow X) but are less common than those within Vitello. Numerous craters on the rim display narrow moatlike features along their rims (arrows Y and Z). Such a feature, marked by arrow XX, is connected to a fracturelike rille (arrow YY) that extends to the southeast and forms the wall of a small crater (arrow ZZ).

The moatlike ring in the craters may correspond to an underlying competent layer. The relatively shallow depth of such a layer seems inconsistent with the location of these craters on highland terrain and on a battered ejecta blanket. Perhaps they indicate a history different from the secular erosion of impact ejecta. Note that there are no pristine craters larger than 100 m.

PLATE 252

a. V-168-H3   0.74 km

b. V-168-H1   0.69 km

Messier

Plate 253 is an oblique view of the craters Messier (bottom) and Messier A (top) in Mare Fecunditatis (see Plate 7b for a vertical view). Messier has the approximate dimensions of 8 km by 15 km. The ejecta blanket of Messier is asymmetrically distributed. The hummocky Zone II region is well developed on either side of its major axis (for example, arrow X) but is essentially absent within sectors at both ends of the major axis (arrows Y and Z). The western rim of Messier (arrow Y), however, exhibits a smooth-surfaced plains-forming unit. The oblique view reveals the saddlelike rim, in which the lowest rim heights are at both ends of the major axis.

The outer ejecta blanket around Messier A is characterized by two long whiskerlike rays that extend to the west. It also has an asymmetric distribution of the hummocky interior ejecta blanket and has smooth-surfaced units on both ends of the major axis (arrows Y and XX; also see Plate 254a). The complex appearance of Messier A is the result of at least two overlapping craters. The rim of the pristine eastern crater is subdued on the floor of the western crater (arrow YY). Note the subdued craters around Messier A that have flat floors composed of units, perhaps similar to those seen near arrow Y (see arrows ZZ, XY, and XZ).

The rim-wall transition of Messier (arrow YX) is well defined and exhibits breakaway fractures. In contrast, this sharp transition on the southern rim of Messier A is well below the rim crest (arrow YZ; see Plate 254b). The wall and floor regions are shown in detail in Plate 254, c and d.

PLATE 253

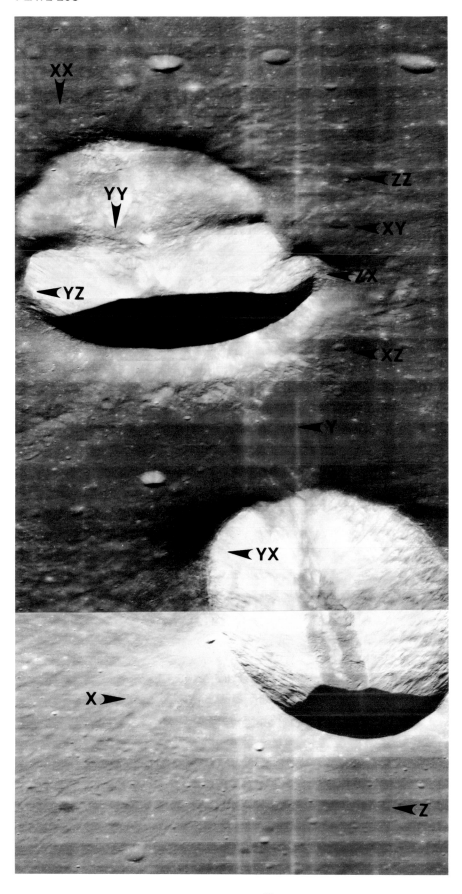

V-041-H2,H1   (oblique)

Messier

Plate 254 shows selected regions of Messier and Messier A. The western rim of the Messier A complex is included in Plate 254a (see Plate 253, arrow XX). Hummocky debris (arrow X) was ejected as a unit from the eastern member of the complex and across the floor and rim of the western member. This debris was heaved in line with the lowest elevation of the rim of the eastern crater (Plate 253, arrow YY). Perhaps it represents material that would have formed a rim if the parameters of the impact had been different or if the eastern crater had not overlapped the adjacent crater floor. Leveed furrows (Plate 254a, arrow Y) indicate possible channels for the nearby smooth units.

The southern rim and wall of Messier A are shown in Plate 254b (see Plate 253, arrow YZ). The hummocky ejecta debris is absent along the southwestern rim (arrow X) and occurs only near the area marked by arrow Y, where it is connected with the debris shown in Plate 254a. Breakaway scarps (arrow Z) suggest further crater expansion at the expense of the adjacent crater. The screelike wall makes an abrupt transition near the area marked by arrow Z, but farther to the east (arrow XX) this transition is below the rim crest. This occurs on the wall opposite a scallop on the northern wall (Plate 253, arrow ZX; also see Plate 7b), which may reflect a complex cratering event involving two or three projectiles.

Plate 254c shows the floor and wall of Messier. A hummocky high-albedo ridge (arrow X) is partly surrounded by a low-albedo floor unit. The ridge may represent a small rebound feature or the accumulation of debris from opposite walls, whereas the darker floor unit is interpreted as a solidified melt. Debris flows from the wall have encroached on the floor at several locations (arrow Y). Talus lines the northeastern floor-wall contact (arrow Z). The wall appears to be mottled, with relatively smooth patches (arrow XX) and outcrops with debris fans (arrow YY).

PLATE 254

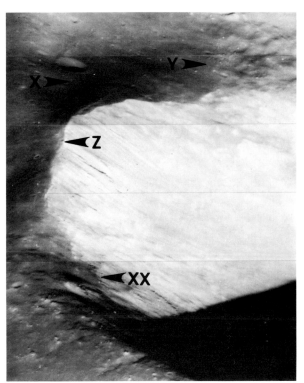

a. V-041-H2   (oblique)

b. V-041-H2   (oblique)

c. V-041-H1   (oblique)

Harbinger Mountains

The Harbinger Mountains (Plate 255) are about 100 km north-east of the crater Aristarchus, whose ejecta clearly blanket the region (see Plate 171), and are probably exposed parts of the outer mountainous ring of the Imbrium basin. In Lunar Orbiter photographs, these "mountains" are less impressive than are the numerous rilles that are members of the Rima Prinz system. Prinz is the large breached crater seen at right. Most of the continuous rilles have craters or irregular depressions at one end. Four of these craters or depressions occur at obvious topographic highs, being either on mounds (arrow X) or outside the mare-inundated regions (arrows Y, Z, and XX). Several discontinuous rilles, however, appear to lack such forms (arrows YY and ZZ), but this may be due to incomplete rille formation.

Three rilles (arrows ZZ, XY, and XZ) have meanderlike plans but are structurally controlled by NNE- and E-trends, which are also expressed by nonsinuous rilles (arrow YY). The branching rilles seen at bottom are less sinuous and can be traced to craters on the Aristarchus Plateau (see Plate 171). The rilles generally cross the mare plains but several either are on or cross portions of positive relief (arrows Y, YY, XY, XZ, YX, and YZ).

PLATE 255

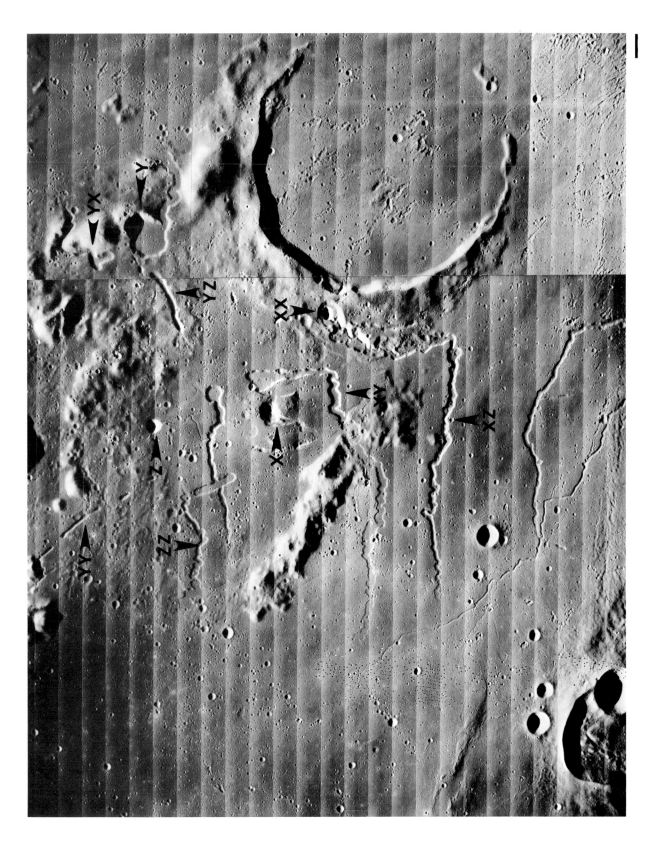

V-191-M; V-187-M   4.4 km

Harbinger Mountains

Plate 256 includes stereo views of the Harbinger Mountains region. Plate 256, *a* and *b*, shows the southern end of the rilles. Features of particular interest are noted:

1. The rille within a rille (arrows X and Y) has an interior "head" crater (arrow Z) on the rim of Prinz. This region is shown in detail in Plate 257, and the abrupt change in the rille path is shown in Plate 258. Note the loop (arrow XX; see Plate 261*b*).
2. One of the more sinuous rilles crosses a positive-relief form (arrow YY; see Plate 259).
3. An incipient rille (arrow ZZ; see Plate 260) has interior pits (arrow XY) that suggest multiphased formation.
4. The rille identified by arrow XZ has a relatively shallow crater at its southern end and a narrow interior rille (arrow YX). The rille is crossed by an elongate depression that parallels the E-trending structural grain. Also see Plate 261*a*.
5. A narrow rille shown near the top of these photographs (arrow YZ) is connected to the larger rille (arrow XZ) and to a rimless elongate crater (arrow ZX).

The structural origin of local nonsinuous rilles is indicated by incomplete formation on nonmarelike terrain (arrow ZY). It is also consistent with the sinuous rille that crosses the crested massif (arrow YY). The rille plans, however, are commonly meanderlike and suggest fluidlike flow. Such flows were guided by structural weaknesses that may have been expressed as pre-existing rilles. Where the rilles are widest, the facing walls do not always resemble the classical meander plan. This may be attributed to mass wasting, erosional processes associated with multiple stages of flowage, or collapses related to rille segments formed by structural processes prior to flowage.

Plate 256, *c* and *d*, shows the northern extensions of the rilles. The rilles typically become shallower and merge with the mare surface without an indication of flow deposits (arrows X and Y; see Plate 262, *a* and *b*, respectively). The narrow interior rille at top (arrow Z) appears to be linked with the continuation of the east-facing rille wall to the east (arrow XX).

Features indicated by arrows YY and ZZ are discussed with Plates 259 and 262, respectively.

PLATE 256

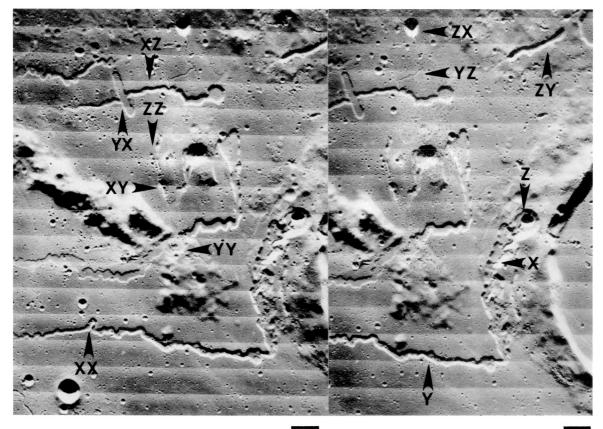

a. V-191-M   4.5 km   b. V-187-M   4.4 km

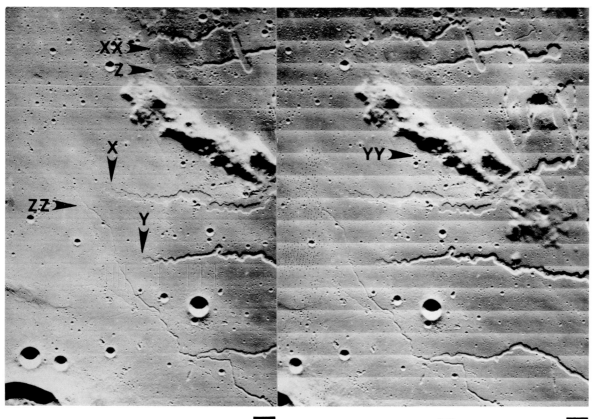

c. V-193-M   4.6 km   d. V-191-M   4.5 km

Harbinger Mountains

The rille on and adjacent to the rim of Prinz is shown in Plate 257 (see Plate 256, *a* and *b*, arrow Z). Secondary ejecta from Aristarchus (arrow X) have extensively blanketed the region, including the floor of the rille (arrow Y). The interior "head" crater (arrow Z) and interior rille (arrow XX) are less subdued than are the confining walls of the outer rille. The interior rille is adjacent to the northeast-facing outer rille wall and generally reflects its plan in this region. A similar configuration is displayed by portions of the interior rille of Schröter's Valley (Plate 171, *b* and *c*).

In addition, the following features are recognized:

1. The insular positive-relief features on the interior terrace (arrows YY and ZZ) that display basal aprons
2. Stippled surfaces (arrow XY) that probably represent myriads of small secondaries from Aristarchus

PLATE 257

V-188-H2; V-187-H2   0.58 km

Harbinger Mountains

Plate 258 shows the western extension of the rille included in Plate 257. Numerous flat-floored craters (arrows X and Y) suggest that the region has been blanketed. The rille floor also appears to be partly inundated by material (arrow Z). Lineations from Aristarchus (for example, area near arrows XX and YY) are generally ill defined in this region. This may indicate a different type of ejecta blanket composed of highly mobile material, such as molten debris. Cascadelike features are recognized on the leeward rille wall (arrows Z and ZZ) and are consistent with blanketing from Aristarchus.

The interior rille is not so prominent in this view as in the stereo pair in Plate 256 (a and b). It is recognized near the top of Plate 258, and an interior terrace is noted by arrow XY.

PLATE 258

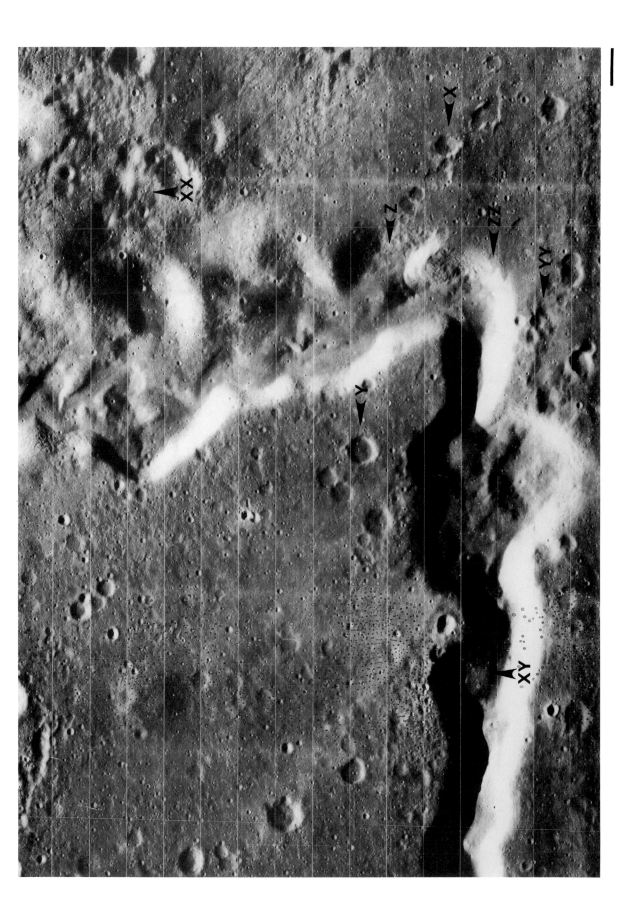

V-188-H3   0.58 km

Harbinger Mountains

Detail of the rille, Rima Prinz II, that crosses a crested massif (see Plate 256, *a* and *b*, arrow YY) is shown in Plate 259. The western rille wall on the southeastern side of the massif is connected to a shallow rille (arrow X) that makes a relatively large loop (arrow Y) around a portion of the massif. Along the main path of the rille, two spurlike features (arrows Z and XX) were produced by changes in the rille direction. Where the rille crosses the massif, wall scallops are not always mirrored by the facing wall (arrow YY); however, on either side of this positive relief, this is generally true (arrows ZZ and XY). Such mirroring in plan is not exact, for a concave element commonly does not face a convex element; rather, it faces a spurlike extension (arrow XZ). The rille is wide relative to the wavelength of the meanderlike plan. This is in contrast to terrestrial river meanders and to the narrow rille within Schröter's Valley.

Multiple phases of rille formation are suggested by the terrace at the base of the spurlike feature (arrow Z). Although other evidence is absent, it may have been buried beneath the material that produced the flat-bottomed profile. It is possible that continued use of the rille by later flows produced rille widening and unusual rille plans along certain segments.

Crossing of the massif by a surface flow possibly was accomplished by a pre-existing structurally formed rille. Alternatively, the level of the mare surfaces was significantly higher during rille formation. Lava flows, and subsequently the rille, passed through a low-relief area of the massif. Subsidence of the mare and a continued supply of flow material may have produced the apparent "carving" of the massif. Terraces adjacent to relief surrounding Mare Imbrium comprise evidence for this type of subsidence. Small terraces are shown in Plate 256, *c* and *d* (arrow YY).

PLATE 259

V-190-H2,H3; V-189-H2,H3   0.58 km

Harbinger Mountains

Plate 260 shows the eastern extension of the rille included in Plate 259. The broad flat-bottomed depression (arrow X) is not the craterform that typically is found at one end of a rille. The prominent south-facing wall (arrow Y) is within a larger depression whose southwest-facing wall (arrow Z) is approximately parallel to the direction of solar illumination. An extension of this wall is indicated by arrow XX. This wall is breached and connects with the adjacent pit (arrow YY), which is much deeper than the connecting rille floors. A shallow rille extends west-northwest from the pit (arrow ZZ), within which is an islandlike mass. Approximately 8 km west of the pit, this rille abruptly changes to an east-northeast direction (arrow XY) and exhibits numerous discontinuities of its plan (arrow XZ). Note that the rille and large pit display various levels (arrows YX and YZ, respectively), which suggests multiple stages of formation. Farther to the east, this rille dissolves into a crater chain.

The rilles shown in Plate 260 indicate a complex and multi-phased history. Further evidence is illustrated by the rimless pit (arrow ZX) and the crescent-shaped scarp (arrow ZY), which is interpreted as an incipient collapse feature or an inundated crater.

PLATE 260

V-189-H1,H2; V-188-H1,H2   0.58 km

Harbinger Mountains

Close-ups of two members of the Rima Prinz system are shown in Plate 261. The rille shown in Plate 261a is identified in Plate 256, a and b, by arrow YX. The following features are of interest:

1. The shallow flat-floored "head" crater (arrow X)
2. The "hanging" rille (arrow Y)
3. The narrow inner rille (arrow Z)
4. The apron at the base of the rille wall within the WNW-trending depression (arrow XX) as well as in the "head" crater
5. Portions of the rille floor that are more hummocky and debrislike (arrow YY) in contrast to the flat floor of the "head" crater and the transecting depression

Plate 261b (see Plate 256, a and b, arrow XX) includes a prominent meander loop (arrow X). Note that the spurlike feature (arrow Y) has a flat-topped surface similar to the surrounding terrain. The facing rille walls to the south (arrow Z) are not exact mirror images; in particular, the concave element is faced by a convex element having a smaller radius of curvature. Farther to the north (arrow XX) this departure of symmetry is greater. In the area near arrow XX the rille exhibits an interior terrace, and in that near arrow YY a narrow interior rille. Consequently, the rille plan was probably modified by processes associated with multiple phases of fluidlike transport.

Outside the rilles shown in Plate 261 the surfaces are heavily scarred by Aristarchus ejecta. One of the secondary complexes (Plate 261a, arrow ZZ) contains small interior depressions similar to those found on the floor of Hyginus (Plate 161b). Note the craters with well-defined interior terraces (Plate 261b, arrow ZZ).

PLATE 261

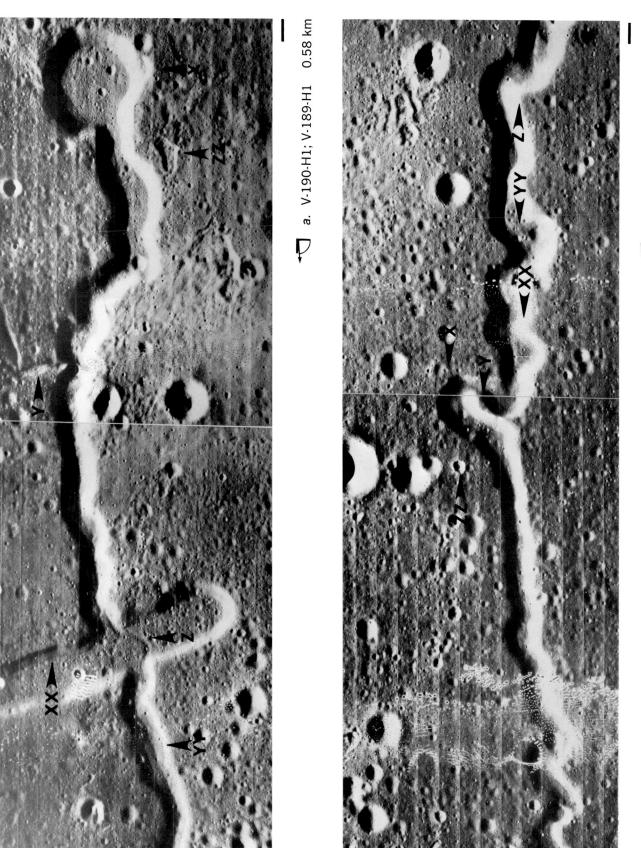

a. V-190-H1; V-189-H1   0.58 km

b. V-191-H3; V-190-H3   0.59 km

Harbinger Mountains

Plate 262, a and b, shows termini of two of the rilles shown in Plate 256, c and d, at arrows X and Y, respectively. The region seen in Plate 262a is heavily blanketed by Aristarchus ejecta, and traces of any possible terminal deposits cannot be clearly recognized. It is perhaps significant, however, that the narrow NE-trending rille (arrow X; also see Plate 256, c and d, arrow ZZ) disappears in this region, perhaps under terminal deposits as well as Aristarchus ejecta. The rille shown in Plate 262a also lacks obvious deposits where it merges with the surrounding mare surface (left).

If these rilles and the surrounding mare formed concurrently over a long period of time, then the absence of terminal deposits is not necessarily inconsistent with hypotheses of rille formation by transport of material. In particular, the rilles may have fed the mare plains with significant volumes of low-viscosity basalt. Eventually these flows and flows from other sources "backed up" and partly inundated the rilles. Therefore, the region shown in Plate 262 does not show the termini of flows from the rille channels, because they have been buried. In addition, these rilles may have been relatively early sources of mare basalts that have survived complete inundation by their relatively high elevation. Such a picture requires very low viscosity material and a large pool of molten lava represented by portions of the mare plains.

PLATE 262

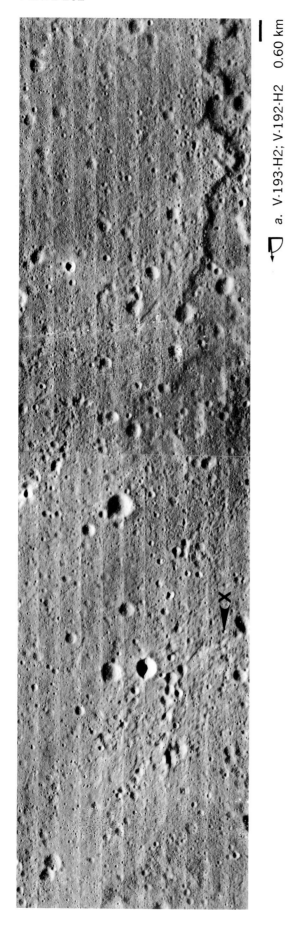

a. V-193-H2; V-192-H2   0.60 km

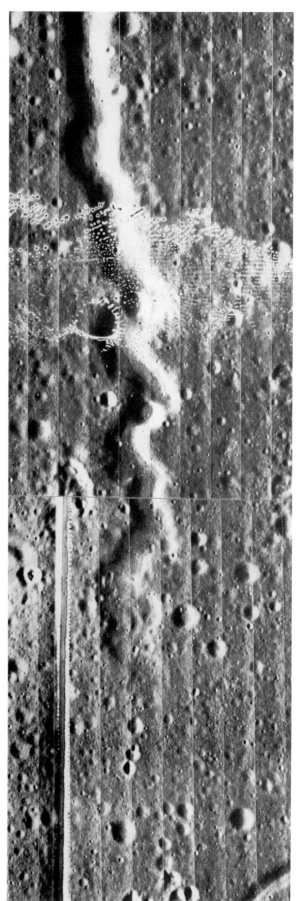

b. V-192-H3; V-191-H3   0.60 km

Marius Hills

Plate 263 is an oblique view of the Marius Hills region in Oceanus Procellarum and includes the crater Marius (arrow X; also see Plates 110*b* and 183*a*), whose marelike floor is noticeably below the surrounding mare. This region exhibits the results of a complex volcanic history that may be related to the intersection of two major structural trends expressed by NW- and NE-trending systems of wrinkle ridges. The northwest ridge system extends southeastward from the Marius Hills across Oceanus Procellarum (see Plate 149*b*) to Mare Humorum (see Plates 182*a* and 191*a*), and members of this system are shown in Plate 263 (arrow Y). The northeast ridge system extends between the Marius Hills to the Aristarchus Plateau. These ridges and the Marius Hills are associated with positive gravity anomalies (see Sjogren et al., 1971). This perhaps reflects locally thick layers of extruded lava, the sites of the vents and magma chambers, or a regionally thin crust overlying a denser mantle.

Plate 263 includes numerous structural and volcanic forms:

1. Sinuous rilles (arrow Z)
2. Calderalike depressions on low-relief mounds (arrow XX)
3. Rimless pits (arrow YY)
4. Isolated platforms (arrow ZZ)
5. Mounds with (arrow XY) and without (arrow XZ) summit pits

Further review of features found in this region can be found in the discussions by McCauley (1967), Guest (1971), and Greeley (1971*b*).

PLATE 263

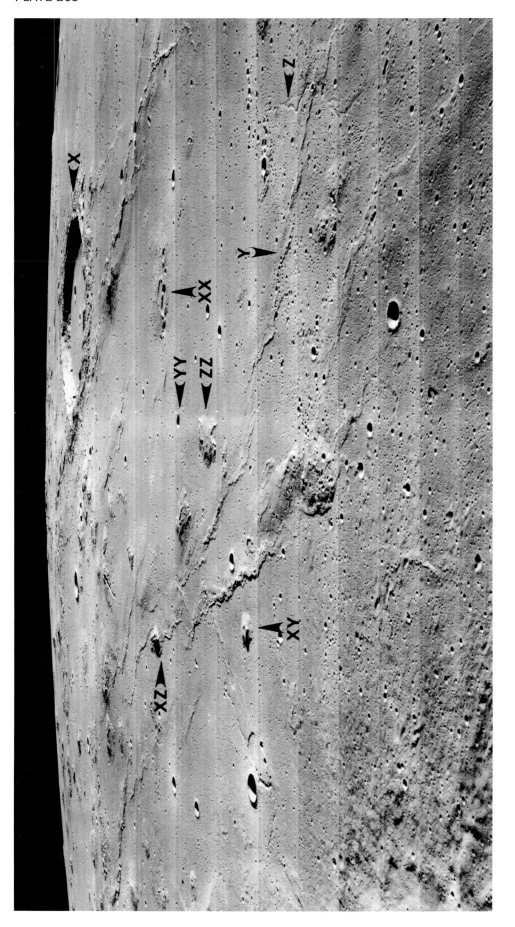

II-213-M   (oblique)

Marius Hills

Plate 264a shows a portion of the Marius Hills that is not included in the oblique view of Plate 263 but is visible in Plate 183a. The "hills" represent numerous isolated positive-relief features that are described as irregular platforms (arrow X), cratered and textured domes (arrow Y), crater-topped cones (arrow Z), and isolated peaks (arrow XX). Many of these features are surrounded by a low-relief apron (arrow X) that may indicate extrusive units, intrusive doming, or withdrawal of surrounding lava units. Several irregular platforms have rimmed borders (arrow YY) and may represent partly inundated calderas. Such platforms are not uncommon in other maria and occur completely isolated from such complexes as the Marius Hills. The positive-relief forms appear to surround relatively featureless plains (arrows ZZ and XY). This is more clearly illustrated in lower resolution views and diagrammetric maps that reveal a cell-like arrangement (see McCauley, 1967) that may reflect preferential volcanic constructions along ring fractures.

The region shown in Plate 264a also includes a variety of negative-relief features:

1. Sinuous rilles (arrow XZ)
2. Rectilinear rilles (arrow YX)
3. Incipient rilles (arrow YZ)
4. Elongate, flat-bottomed depressions (arrow ZX)

The region identified by arrow ZY in Plate 264a (see Plate 265, c and d, arrow XY) is shown in detail in Plate 264b and includes two elongate craterforms (arrows X and Y) with large rims that appear to be analogous with terrestrial tephra constructions. The feature marked by arrow Y is associated with a subdued rille extending to the south-southeast and north-northwest. Other breached tephra-conelike features are noted by arrows Z, XX, and YY.

The two major positive-relief forms shown in Plate 264b are bordered by relatively abrupt scarps (arrows ZZ and XY). The arclike scarp marked by arrow XY appears to be the remnants of a crater. This suggests that these forms were probably subsequently surrounded by mare units, although the tephra-conelike forms may have been contemporaneous with mare emplacement. The 1.3-km-diameter crater (arrow XZ) with a large interior peak is interpreted as a once-active volcanic vent with later tholoid construction. Consequently, these "hills" are believed to have been volcanic vents with relatively long eruptive histories spanning the period of mare emplacement.

Plate 264b also includes a ropy wrinkle ridge (arrow YX) that is shown in more detail in Plate 266, a and b.

PLATE 264

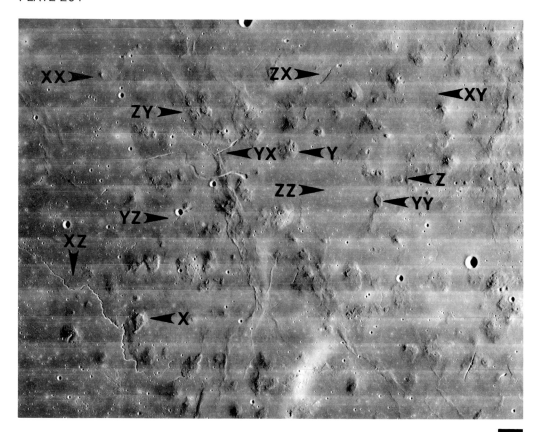

a. IV-157-H2   11 km

b. V-216-H3   0.49 km

Marius Hills

Plate 265 shows stereo views of the region seen in Plate 264a. Plate 265, a and b, includes the area shown near the bottom center of Plate 264a. Features of interest are listed below:

1. Platform with relatively well defined border scarps (arrow X)
2. Irregular platform with relatively small border scarps and indications of multiple stages of formation by multiple boundaries (arrow Y)
3. Small summit craterform with an irregular plan (arrow Z)
4. Hogbacklike ridge (arrow XX) that makes an abrupt change in plan and appears to be related to the crater chain on the textured dome (arrow YY)
5. Incipient rille overlapped by a crater (arrow ZZ)
6. Irregular depression (arrow XY)
7. Narrow ropy ridge that extends across the elevated hills (arrow XZ)

Plate 265, c and d (corresponds to the central area shown in Plate 264a) reveals additional evidence for several positive-relief features being postmare volcanic constructions. Lobate extensions from one of the highly textured domes (arrow X) indicate an extrusive unit that appears to overlie the mare. An elongate summit depression (arrow Y) appears to be genetically related to the rectilinear rille to the southwest (see Plate 268). The floor of this depression exhibits corrugations that may be pressure ridges from the last stages of eruptions. Note the irregular-bordered apron around this complex (arrow Z). The construction identified by arrow XX is more peaked and it possibly represents the buildup of late-stage viscous extrusions.

The two prominent rilles shown in Plate 265, c and d, are considered in more detail in Plates 267–268. Their termini are distinctly different. Arrow YY marks the abrupt end of the larger rille (Plate 267) that is similar to the rilles in the Harbinger Mountains. This termination appears to be related to a subdued west-facing scarp believed to be a boundary of a subsided lava lake that this rille once fed. In contrast, the terminus of the smaller rille (arrow ZZ) branches into narrower and shallower rilles that merge with the mare surface. These are reminiscent of the small rilles on the Imbrium flow front (Plate 192, b and c).

PLATE 265

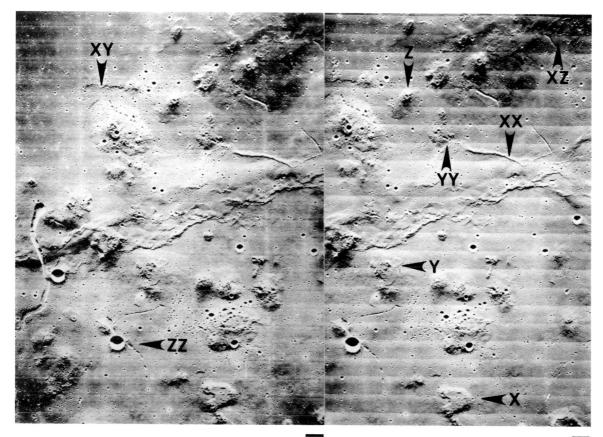

a. V-211-M 3.4 km (*top*) 3.8 km (*bottom*)   b. V-210-M 3.4 km (*top*) 3.8 km (*bottom*)

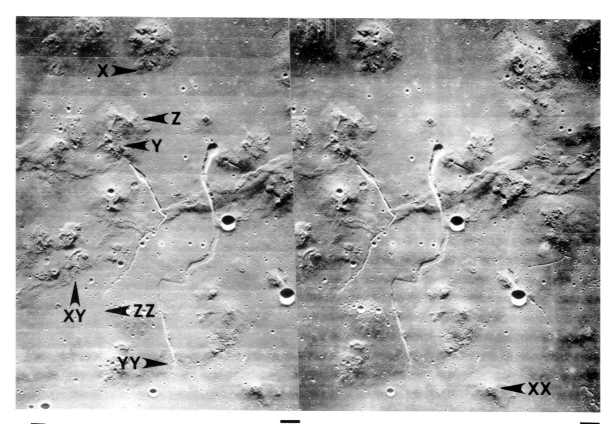

c. V-215-M 3.4 km (*top*) 3.9 km (*bottom*)   d. V-212-M 3.4 km (*top*) 3.9 km (*bottom*)

Marius Hills

Plate 266 shows close-ups of wrinkle ridges that cross the Marius Hills. Plate 266a is a high-resolution view of the narrow ropy ridges identified by arrow XY in Plate 265, c and d. The ridges generally disregard local topographic relief, such as the subdued craters (arrows X, Y, and Z). In addition, the ridges are commonly crossed by narrow fractures (arrows XX and YY).

Plate 266b is a closer view of the ridges shown in Plate 266a, and the fracture seen at arrow X corresponds to that seen at arrow YY in Plate 266a. The sinuous ridges exhibit abrupt displacements (arrow Y) and indicate sequential construction because one ridge overlaps another (arrow Z). Note the displacement of the ridge near the subdued crater (arrow XX). Several subdued craters are in distinct contrast to the preserved state of the ridges on which they occur and are suggested to be genetically related to or modified by ridge formation.

Plate 266c is a high-resolution view of the broad ridge identifiable near the left center of Plate 265, a and b. It is bordered by two ropy ridges (arrows X and Y). The mare-facing scarps of these ropy ridges have narrow aprons (arrow Z) or wrinkles (arrow XX) at their bases. Note the breached crater (arrow YY). The surface of these ropy ridges and the area between them are crossed by sets of small-scale parallel lineations (see the original photograph) that may have an origin similar to that of the lineations shown in Plate 142.

PLATE 266

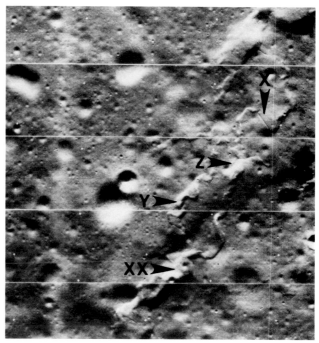

a. V-215-H3   0.49 km

b. V-216-H3   0.49 km

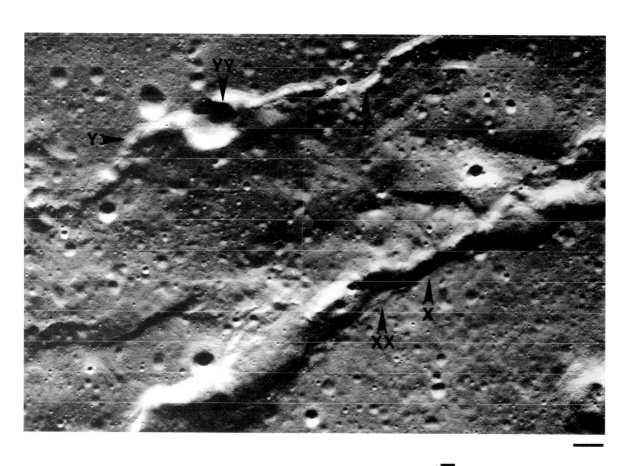

c. V-210-H2   0.48 km

Marius Hills

The rille shown in Plate 267 is included in Plate 265, c and d. It crosses a broad wrinkle ridge and therefore must be considered a later structure. Across this relief and up to the "head" crater (arrow X), it has a V- and U-shaped profile, whereas the rectilinear western extension (see Plate 265, c and d) is flat floored. Note the outcrops that appear to be associated with the ropy components of the wrinkle ridge (arrows Y and Z). In addition, note the elongate crater on the north-facing wall (arrow XX) that has no preserved ejecta blanket on the rille floor; it is interpreted as an endogenetic feature.

PLATE 267

V-213-H2,H3   0.48 km

Marius Hills

Plate 268 is a high-resolution view of the smaller prominent rille seen in Plate 265, c and d, that appears to be related to the textured relief to the northeast (arrow X). Arrow Y identifies two narrow ridges that extend from the rille to near the base of the textured relief. This region is interpreted as an uncollapsed segment of the rille. Farther to the southwest, the rille displays a well-defined raised rim (arrow Z) on the widest rille segment. Near the wrinkle ridge (arrow XX), the rille plan abruptly changes to the northwest, perhaps as the result of structural control expressed by the wrinkle ridge. The raised rims are pronounced at arrow YY and appear to be distinctly larger than rims at other portions of the rille. These may represent remnants of the wrinkle ridge rather than constructed levees or an uncollapsed roof. Rims are not pronounced farther to the northwest. Note the islandlike mass within the rille (arrow ZZ).

Plate 268 also includes narrow ropelike ridges (arrow XY) that either terminate at the scarp of the broader ridge (arrow XZ) or continue along its base. Outcrops associated with another ridge are recognized in the area near arrow YX.

The circular craters in this region may be endogenetic forms. The isolated crater (arrow YZ) displays a broad moundlike rim but lacks a structured ejecta blanket. In addition, the crater adjacent to the rille (arrow ZX) apparently has not blanketed the rille floor. This indicates either crater formation before the rille or an endogenetically produced collapse crater with only minor ejecta.

PLATE 268

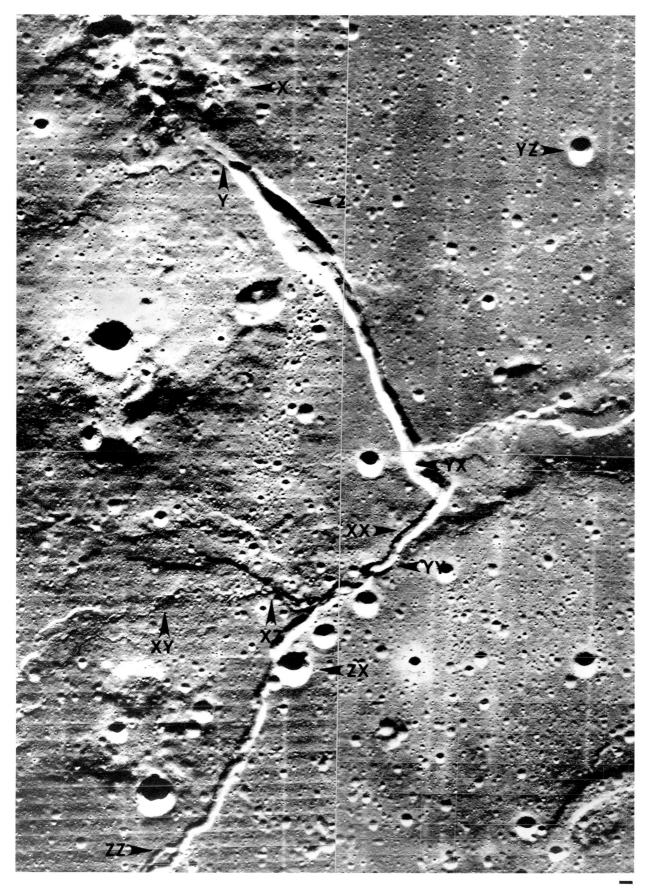

DATA FOR THE PLATES

The Emission Angle (EA), Incidence Angle (IA), and North Deviation Angle (NDA) are defined on page 5 and illustrated in Figure 1. They are listed below with the corresponding plates. Where more than one entry is included for a particular plate, the values are given in order of their identification with the Orbiter photographs as presented in the captions for that plate.

PLATE	EA	IA	NDA	PLATE	EA	IA	NDA
1a	3	74	6	16a	1	76	21
b	4	77	7	b	1	69	12
2a	10	81		c	7	75	85
b	6	66	19	d	4	69	12
c	8	70	21	17a	2	67	21
d	70	77	58	b	2	67	21
3a	2	73	6	c	2	67	21
b	7	75	5	18a	9	68	21
c	13	71	26	b	34	69	25
d	14	70	23	c	7	65	21
4a	1	70	6	19a	1	61	12
b	13	70	26	b	1	59	12
c	1	73	5	c	2	60	21
d	14	68	26	d	2	60	21
5a	36	71	115	20a	31	69	24
b	16	71	5	b	31	69	24
c	5	70	21	c	9	71	85
d	1	70	177	21a	5	70	21
6a	2	67	21	b	5	70	21
b	1	69	12	c	13	71	21
c	9	68	21	22a	3	65	13
d	9	68	21	b	5	71	5
7a	16	79	16	c	2	69	5
b	2	67	6	23a	13	70	85
c	1	67	6	b	13	70	85
d	10	71	86	c	13	70	85
8a	2	77	21	24a	1	70	177
b	1	76	21	b	1	70	177
c	1	70	12	c	6	70	20
d	1	71	12	d	6	70	20
9a	9	82	6	25a	7	75	85
b	4	66	19	b	6	75	85
c	6	66	19	c	6	75	85
10a	3	68	5		6	75	85
b	8	71	3	26a	15	69	85
c	2	66	6	b	15	69	85
d	12	68	25	27a	2	66	6
11a	24	70	181	b	13	70	4
b	12	65	23	c	13	70	4
c	5	68	18	28a	4	68	4
d	2	78	4	b	7	81	1
12a	1	70	177	c	10	69	20
b	9	81	6	d	5	72	84
c	6	70	20	e	3	71	84
d	2	60	21	f	2	73	5
13a	24	71	181	g	1	69	6
b	1	71	12	h	9	74	2
c	24	77	23	29a	45	79	4
d	4	68	12	b	12	70	173
14a	4	68	12	c	2	74	5
b	4	68	12	30a	4	71	84
c	2	60	12	b	5	72	22
d	2	60	12	c	5	72	22
15a	1	72	12	31a	6	81	84
b	0	75	21	b	6	81	84
c	0	75	21	c	6	81	84

PLATE	EA	IA	NDA	PLATE	EA	IA	NDA	PLATE	EA	IA	NDA	PLATE	EA	IA	NDA
32	6	81	84	c	2	65	6	c	5	65	19	86a	7	77	86
33a	6	81	84	d	3	66	5	d	3	66	5	b	7	77	86
b	6	81	84	49a	42	79	6	67	6	81	84	c	1	76	21
c	6	81	84	b	7	81	1	68	6	81	84	d	1	76	21
d	6	81	84	50a	18	79	23	69a	4	68	4	87a	11	80	83
34a	71	82	48	b	18	79	23	b	1	68	6	b	11	80	83
b	2	73	5	c	18	79	23	c	8	67	23	88a	24	71	181
c	13	80	2	d	18	79	23	d	8	67	23	b	24	71	181
d	25	82	9	51a	7	65	21	70a	3	73	85	89a	15	69	85
e	2	62	6	b	7	65	21	b	3	73	85	b	15	69	85
35	5	72	20	c	7	65	21	c	14	70	23	c	15	69	85
36a	6	77	86	d	7	64	21	d	13	70	26	90a	0	75	12
	7	77	86	52a	2	59	12	71a	2	68	6	b	0	75	12
	7	77	86	b	2	59	12	b	1	68	6	c	0	75	12
b	7	77	86	c	2	59	12	c	12	70	26	91a	3	70	5
c	7	77	86	d	2	59	12	d	14	68	26	b	1	70	6
37a	7	77	86	53a	13	72	21	72a	15	73	87	c	14	78	3
b	7	77	86	b	13	72	21	b	15	73	87	d	11	79	2
c	6	77	86	c	13	71	21	73a	1	74	86	92a	8	71	3
d	7	77	86	d	13	71	21	b	2	74	86	b	1	68	6
38a	6	77	86	54a	2	69	12	c	2	74	86	c	13	71	26
b	7	77	86	b	2	69	12	74a	3	66	5	d	1	69	6
c	7	77	86	c	2	69	12	b	17	79	173	e	2	68	6
39a	2	75	1	d	1	69	12	c	13	73	25	93	74	60	12
b	24	70	1	55a	34	69	25	d	1	69	6	94	3	70	5
c	3	69	5	b	9	68	21	75a	24	71	181	95	6	72	85
d	60	78	40	56a	0	75	12	b	24	70	181	96a	9	71	84
40a	6	68	19	b	0	75	12	76a	1	69	12	b	9	71	84
b	24	70	181	57a	55	72	86	b	1	69	12	c	9	71	85
41a	13	69	26	b	26	67	86	c	13	72	21	d	9	71	85
b	10	81	7	58a	2	60	21	d	2	59	12	97a	9	71	84
c	13	74	25	b	2	60	22	77a	13	71	21	b	9	71	85
d	14	73	25	59a	0	75	12	b	13	72	21	98a	4	69	17
42a	1	70	6	b	0	75	12	c	3	59	12		3	69	17
b	13	70	26	60a	15	69	85	d	2	60	12	b	3	77	6
c	1	64	5	b	31	69	24	78a	0	75	12	c	2	73	6
43a	1	69	6	c	2	62	13	b	0	75	12	99a	3	69	17
b	7	64	21	d	3	76	5	c	0	75	12	b	14	73	25
c	7	64	21	61a	3	75	6	d	0	68	12	c	13	73	25
d	2	60	21	b	1	76	5	79a	20	73	22	100a	1	74	85
44	1	69	6	c	5	72	20	b	1	77	11	b	1	73	85
45a	1	69	6	d	2	73	5	80a	36	71	115	c	1	73	85
b	2	68	6	62a	13	70	86	b	36	71	115	d	1	73	85
c	6	65	20	b	13	70	86	c	36	71	115	101a	13	70	26
d	2	65	6	63a	13	70	86	81a	31	67	86	b	6	81	84
e	1	69	6	b	12	70	86	b	31	67	86	c	5	81	84
f	1	70	3	c	12	70	86	82	31	67	86	102a	5	81	84
46a	12	65	23	d	12	70	86	83	31	67	86	b	6	81	84
b	2	67	6	64a	42	79	6	84a	4	69	12	c	5	81	84
c	1	67	6	b	oblique		16	b	4	69	12	d	6	81	84
d	2	62	6	65a	3	69	5	c	4	69	12	103a	4	81	84
47a	1	69	6	b	1	69	6	85a	10	71	86	b	6	81	84
b	14	68	26	c	5	72	22	b	13	71	21	c	4	81	84
c	2	66	6	d	5	72	22	c	9	73	81	d	6	81	84
48a	2	68	6	66a	6	66	19	d	15	73	87	104	5	81	84
b	13	68	26	b	5	66	19	e	1	71	6		4	81	84

PLATE	EA	IA	NDA	PLATE	EA	IA	NDA	PLATE	EA	IA	NDA	PLATE	EA	IA	NDA
105a	47	79	4	b	2	65	6	138a	3	69	85	156a	13	71	21
b	47	79	4	c	3	66	5	b	2	69	85	b	13	71	21
c	3	74	6	124a	2	65	6	c	2	69	85	c	13	71	21
106a	5	71	4	b	24	67	86	139a	3	69	85	d	13	71	21
b	12	72	86	c	24	67	85	b	3	69	85	157	13	71	21
c	12	72	86	d	24	68	86	140a	3	69	85		13	71	21
107a	4	71	84	125a	6	70	20	b	2	69	85	158a	13	71	21
b	4	71	84	b	6	70	20	c	2	69	85	b	13	71	21
108a	24	70	181	c	8	71	3	d	2	69	85	c	13	71	21
b	24	70	181	d	8	71	3	141	1	68	12	d	13	71	21
c	24	70	181	e	13	72	21		1	68	12	159a	55	83	37
109a	48	79	4	126a	46	79	12	142a	1	77	11	b	20	72	86
b	1	71	177	b	0	73	21	b	1	77	12	c	20	72	86
c	2	62	13	c	0	72	21	c	1	77	11	160	20	72	86
d	3	60	13	127	0	73	21		1	77	11		20	72	86
110a	2	73	6		70	77	58	143a	2	60	12	161a	20	72	86
b	3	72	6	128a	0	73	21	b	1	61	12	b	20	72	86
c	8	70	21	b	0	73	21	144a	10	69	20	162	20	72	86
d	1	70	6	c	15	83	2	b	6	68	19	163	2	74	5
111	3	74	6	d	15	82	2	c	2	69	85		2	73	5
112	3	69	5	e	15	77	1	d	2	69	85		1	73	5
113a	34	74	192	f	15	77	1	145a	72	78	32	164a	2	73	6
b	34	74	192	129a	4	74	1	b	1	70	177		3	74	6
114a	2	67	6	b	4	74	1	c	12	65	24	b	3	73	6
b	66	80	40	c	4	69	12	146a	34	69	25	c	2	73	5
c	66	80	40	130a	7	75	85	b	34	69	25	165a	2	72	5
115a	4	69	18	b	6	75	85	c	34	69	25	b	13	72	25
b	12	70	11	c	6	70	20	147a	13	71	21	c	4	68	5
c	13	73	22	d	4	70	18	b	13	71	21	d	3	71	84
d	1	73	12	e	1	70	6	c	3	69	85	166a	10	71	86
116a	1	83	11	f	6	70	20	d	2	69	85	b	10	71	86
b	8	73	81	131a	1	74	86	148a	3	78	5	c	10	71	86
c	8	73	81	b	2	74	86	b	3	73	6	167a	4	68	4
d	24	71	181	c	1	74	85	c	11	78	2	b	6	66	19
117a	26	67	86	132a	5	69	85	d	2	71	6	c	9	71	7
b	5	69	85	b	5	69	85	e	10	69	20	168	2	72	5
c	10	75	85	c	5	69	85	f	7	77	86	169a	2	72	5
d	1	68	6	d	19	75	23	149a	11	72	86	b	13	72	25
118a	3	69	85	e	8	71	3	b	6	70	20	c	1	73	5
b	2	69	85	133a	60	78	40	150a	3	70	5	d	12	73	24
c	1	69	6	b	4	68	4	b	26	67	86	170a	3	68	5
d	4	77	8	134a	2	73	85	c	6	66	19	b	14	76	12
119a	13	70	85	b	1	73	85	d	6	70	20	171a	3	73	15
b	13	70	85	c	8	71	3	151a	2	71	12	b	7	75	85
c	13	70	85	d	4	70	18	b	1	71	12	c	6	75	85
120a	1	70	6	e	10	69	20	152a	39	70	26	172	7	75	85
b	12	70	26	135a	5	66	174	b	9	68	21	173a	9	73	81
c	14	68	26	b	3	68	175	c	2	73	84	b	8	73	81
d	9	75	2	c	4	66	175	d	3	66	5	c	8	73	81
121a	10	81	7	136a	69	84	8	153a	1	74	11	174a	7	71	85
b	1	69	6	b	8	75	86	b	5	69	20		5	71	85
c	1	70	6	c	8	75	85	154a	2	72	6	b	5	71	85
122a	4	67	21	137a	19	74	86	b	20	73	22	c	5	71	85
b	20	65	22	b	8	75	86	c	20	73	22		5	71	85
c	26	67	86	c	8	75	86	155	20	73	22	175a	6	71	85
123a	1	73	85	d	19	74	86		20	73	22	b	5	71	85

610

PLATE	EA	IA	NDA	PLATE	EA	IA	NDA	PLATE	EA	IA	NDA	PLATE	EA	IA	NDA
c	5	71	85	c	2	73	84	215a	30	83	5		1	74	85
d	5	71	85	193	3	73	84	b	5	72	85	b	4	74	84
e	5	71	85		3	73	84	c	6	72	85	c	3	74	84
176	4	70	18	194a	1	77	11	216a	5	72	85	237	3	74	84
177a	5	66	19	b	1	77	11	b	5	72	85		2	74	84
b	7	75	85	c	0	75	12	c	5	72	85	238a	3	74	84
c	7	75	85	d	0	75	12	d	5	72	85	b	3	74	84
d	6	75	85	195a	1	71	6	217	6	72	85	c	3	74	84
178a	1	70	6	b	69	84	8	218a	74	65	3	d	3	74	84
b	11	80	83	c	18	79	23	b	74	65	3	239	3	74	84
c	55	72	86	d	18	79	23	219	5	72	85		3	74	84
d	3	69	5	196a	18	79	23	220a	74	65	3	240	3	74	84
179a	40	78	117	b	18	79	23	b	6	72	85	241a	3	74	84
b	9	73	81	c	18	79	23	c	5	72	85	b	3	74	84
c	8	73	81	d	18	79	23	221	5	72	85	c	3	74	84
180a	40	78	117	e	18	79	23		5	72	85	242a	3	74	84
b	40	78	117	197a	13	81	2	222a	74	65	3	b	3	74	84
c	40	78	117	b	7	81	1	b	5	72	85	243a	3	74	84
d	40	78	117	c	7	81	1		5	72	85	b	2	74	84
e	40	78	117	198a	3	75	6	223a	74	65	3	c	2	74	84
181a	7	75	85	b	3	76	6	b	5	72	85	d	1	74	85
b	6	75	85	c	3	76	6		5	72	85	244	2	74	84
c	6	75	85	d	3	76	6	224	5	72	85	245	4	74	84
182a	2	71	6	e	3	76	6	225	6	72	85	246	2	74	84
b	2	66	6	199a	11	83	0	226a	5	72	85	247a	2	74	84
c	3	66	5	b	11	83	3		6	72	85	b	3	74	84
d	10	71	86	c	2	74	5	b	6	72	85	248	2	74	84
e	3	72	6	200a	1	68	6	c	6	72	85	249	11	80	83
183a	66	80	29	b	5	67	21	227a	6	72	85	250	11	80	83
b	3	73	6	c	5	67	21	b	6	72	85	251a	11	80	83
	3	73	5	d	5	67	21	c	5	72	85	b	11	80	83
184a	24	71	181	201	1	76	5	228a	6	72	85	252a	11	80	83
b	34	75	188	202	3	76	5	b	6	72	85	b	11	80	83
185a	7	64	21	203	3	76	5	c	6	72	85	253	58	73	86
b	21	71	86	204	1	76	5	d	5	72	85	254a	58	73	86
c	10	71	86	205	2	75	5	229a	6	72	85	b	58	73	86
d	34	74	192	206a	1	76	5	b	5	72	85	c	58	73	86
e	6	70	20	b	1	75	5	c	6	72	85	255	3	74	85
186a	58	87	77	c	1	76	5	d	6	72	85		2	74	85
b	15	73	87	d	1	76	5	230a	5	72	85	256a	3	74	85
c	15	73	87	207	1	76	5	b	6	72	85	b	2	74	85
187	15	73	87	208a	2	75	5	c	6	72	85	c	4	74	84
188a	3	70	5	b	13	74	24	231a	74	65	3	d	3	74	85
b	43	81	18	c	1	76	5	b	6	72	85	257	2	74	85
189a	14	73	86	d	13	74	24	c	5	72	85		2	74	85
b	14	73	86	209	15	81	1	232	6	72	85	258	2	74	85
c	14	73	86	210a	3	75	6		6	72	85	259	3	74	85
d	14	73	86	b	1	75	5	233	6	72	85		2	74	85
190a	13	73	25	211	13	74	24	234a	6	72	85	260	2	74	85
b	oblique			212a	13	74	25	b	6	72	85		2	74	85
c	9	71	7	b	13	74	24	c	6	72	85	261a	3	74	85
d	2	72	5	c	1	75	5	d	6	72	85		2	74	85
191a	2	71	6	d	13	74	25	235a	6	72	85	b	3	74	85
b	14	70	26	213	1	75	5	b	6	72	85		3	74	85
192a	10	69	20	214a	1	75	5	c	6	72	85	262a	4	74	84
b	3	73	84	b	3	74	6	236a	4	74	84		3	74	85

PLATE	EA	IA	NDA
b	3	74	85
	3	74	85
263	73	79	37
264*a*	3	73	5
b	10	75	86
265*a*	10	75	85
b	10	75	85
c	10	75	86
d	10	75	85
266*a*	10	75	86
b	10	75	86
c	10	75	85
267	10	75	86
268	10	75	86
	10	75	86

REFERENCES CITED

Adams, J. B. 1967. Lunar surface composition and particle size: Implication from laboratory and lunar spectral reflectance data. *Jour. Geophys. Research* 72:5717–5720.

Adams, J. B., and McCord, T. B. 1971. Alteration of lunar optical properties: Age and composition effects. *Science* 171:567–571.

Adler, J. E., and Salisbury, J. W. 1969*a*. Circularity of lunar craters. *Icarus* 10:37–52.

———. 1969*b*. Behavior of water in a vacuum: Implications for "lunar rivers." *Science* 164:589.

Anderson, A. T., and Miller, E. R. 1971. Lunar Orbiter photographic supporting data. National Space Science Data Center, NSSDC 71–13.

Arthur, D. W. G. 1962. Some systematic visual lunar observations. *Univ. of Arizona Lunar and Planetary Lab. Commun.* 1(3):23–26.

Arthur, D. W. G.; Agnieray, A. P.; Horvath, R. A.; Wood, C. A.; and Chapman, C. R. 1963. The system of lunar craters, quadrant I. *Univ. of Arizona Lunar and Planetary Lab. Commun.* 2(30). 84 p.

———. 1964. The system of lunar craters, quadrant II. *Univ. of Arizona Lunar and Planetary Lab. Commun.* 3(40). 71 p.

Arthur, D. W. G.; Agnieray, A. P.; Pellicori, R. H.; Wood, C. A.; and Weller, T. 1965. The system of lunar craters, quadrant III. *Univ. of Arizona Lunar and Planetary Lab. Commun.* 3 (50). 155 p.

Arthur, D. W. G.; Pellicori, R. H.; and Wood, C. A. 1966. The system of lunar craters, quadrant IV. *Univ. of Arizona Lunar and Planetary Lab. Commun.* 5(70). 220 p.

Baldwin, R. P. 1949. *The face of the Moon.* Chicago: Univ. of Chicago Press. 273 p.

———. 1963. *The measure of the Moon.* Chicago: Univ. of Chicago Press. 488 p.

Beales, C. S. 1971. Crustal thickness and the forms of impact craters. *Jour. Geophys. Research* 76:5586–5594.

Blackwelder, E. 1933. The insolation hypothesis of rock weathering. *Am. Jour. Sci.* 26:97–113.

Borg, J.; Maurette, M.; Durrieu, L.; and Jouret, C. 1971. Ultramicroscopic features in micron-sized lunar dust grains and cosmophysics. In *Proceedings of the second lunar science conference,* ed. A. A. Levinson; vol. 3: *Physical properties,* pp. 2027–2040. Cambridge, Mass.: M.I.T. Press.

Bowker, D. E., and Hughes, J. K. 1971. *Lunar Orbiter photographic atlas of the Moon.* Washington, D.C.: U. S. Govt. Printing Office, N.A.S.A. SP-206.

Cameron, W. S. 1964. An interpretation of Schröter's Valley and other lunar sinuous rilles. *Jour. Geophys. Research* 69:2423–2430.

Cameron, W. S., and Coyle, G. J. 1971. An analysis of the distribution of boulders in the vicinity of small lunar craters. *The Moon* 3:159–188.

Carr, M. H. 1964. Impact-induced volcanism. In *Astrogeologic studies,* pp. 52–66. U.S. Geol. Survey open-file Ann. Prog. Rept., pt. A. 134 pp.

———. 1966. *Geologic map of the Mare Serenitatis region of the Moon.* U.S. Geol. Survey Misc. Inv. Map I-489.

Chadderton, L. T.; Krajenbrink, F. G.; Katz, R.; and Poveda, A. 1969. Standing waves on the Moon. *Nature* 223:259–263.

Chapman, C. R.; Mosher, J. A.; and Simmons, G. 1970. Lunar cratering and erosion from Orbiter V. *Jour. Geophys. Research* 75:1445–1466.

Cloud, P.; Margolis, S. V.; Moorman, M.; Barker, J. M.; Licari, G. R.; Krinsley, D.; and Barnes, V. E. 1970. Micromorphology and surface characteristics of lunar dust and breccia. In *Proceedings of the Apollo 11 lunar science conference,* ed. A. A. Levinson; vol. 2: *Chemical and isotope analysis,* pp. 1793–1798. New York: Pergamon Press.

Cohen, A. J., and Hapke, B. W. 1968. Radiation bleaching of thin lunar surface layer. *Science* 161:1237–1238.

Colton, G. W.; Howard, K. A.; and Moore, H. J. 1972. Mare ridges and arches in southern Oceanus Procellarum. In *Apollo 16 preliminary science report,* pp. 29:90–93. Washington, D.C.: U. S. Govt. Printing Office, N.A.S.A. SP-315.

Crittenden, M. D. 1967. Terrestrial analogues of lunar mass wasting. In *Preliminary geologic evaluation and Apollo landing analysis of areas photographed by Lunar Orbiter III,* pp. 125–127. Hampton, Va.: Langley working paper 407.

Cruikshank, D., and Wood, C. A. 1972. Lunar rilles and Hawaiian volcanic features: Possible analogues. *The Moon* 3:412–447.

De Hon, R. A. 1971. Cauldron subsidence in lunar craters Ritter and Sabine. *Jour. Geophys. Research* 76:5712–5718.

Dence, M. R. 1968. Shock zoning at Canadian craters: Petrography and structural implications. In *Shock metamorphism of natural materials,* ed. B. M. French and N. M. Short, pp. 169–183. Baltimore: Mono Book Corp. 664 p.

———. 1971. Impact melts. *Jour. Geophys. Research* 76: 5552–5565.

Dollfus, A.; Geake, J. E.; and Titulaer, C. 1971. Polarimetric properties of the lunar surface and its interpretation. Pt. 3: Apollo 11 and Apollo 12 lunar samples. In *Proceedings of the second lunar science conference,* ed. A. A. Levinson; vol. 3: *Physical properties,* pp. 2285–2300. Cambridge, Mass.: M.I.T. Press.

Donaldson, J. R. 1969. The lunar crater Dawes. *Photogramm. Eng.* 35:239–245.

Eggleton, R. E., and Schaber, G. G. 1972. Cayley Formation interpreted as basin ejecta. In *Apollo 16 preliminary science report,* pp. 29:7–16. Washington, D.C.: U.S. Govt. Printing Office, N.A.S.A. SP-315.

El-Baz, F. 1972*a*. The Alhazen to Abul Wafa swirl belt: An extensive field of light-colored, sinuous markings. In *Apollo 16 preliminary science report,* pp. 29:93–97. Washington, D.C.: U.S. Govt. Printing Office, N.A.S.A. SP-315.

———. 1972*b*. Discovery of two lunar features. In *Apollo 16* *preliminary science report*, pp. 29:33–38. Washington, D.C.: U.S. Govt. Printing Office, N.A.S.A. SP-315.

Fielder, G. 1965. *Lunar geology*. Chester Springs, Pa.: Dufour. 184 p.

Fielder, G., and Fielder, J. 1968. *Lava flows in Mare Imbrium*. Boeing Research Lab. Doc. D1-82-0749. 36 p.

Filice, A. L. 1967. Observations on the lunar surface disturbed by the footpads of Surveyor I. *Jour. Geophys. Research* 72: 5721–5728.

French, B. 1970. Possible relations between meteorite impact and igneous petrogenesis, as indicated by the Sudbury structure, Ontario, Canada. *Bull. Volc.* 34:466–517.

Gault, D. E. 1968. Personal communication.

———. 1973. Displaced mass, depth, diameter, and effects of oblique trajectories for impact craters formed in dense crystalline rocks. *The Moon* 6:32–44.

———. 1973. Personal communication.

Gault, D. E., and Heitowit, E. D. 1963. The partition of energy for hypervelocity impact craters formed in rock. *Proceedings of the 7th hypervelocity impact symposium* 2:419–456.

Gault, D. E.; Quaide, W. L.; and Oberbeck, V. R. 1968. Impact cratering mechanics and structures. In *Shock metamorphism of natural materials*, ed. B. M. French and N. M. Short, pp. 87–99. Baltimore: Mono Book Corp. 644 p.

Gilbert, G. K. 1893. The Moon's face—a study of the origin and its features. *Philos. Soc. Washington Bull.* 12:241–292.

Goodacre, W. 1931. Lunar statistics, craters with and without central peaks. *Brit. Astron. Assoc. Jour.* 42:24–25.

Greeley, R. 1970. Terrestrial analogs to lunar dimple (drainage) craters. *The Moon* 1:237–252.

———. 1971a. Lunar Hadley Rille: Considerations of its origin. *Science* 172:722–725.

———. 1971b. Lava tubes and channels in the lunar Marius Hills. *The Moon* 3:289–314.

———. 1971c. Observations of actively forming lava tubes and associated structures, Hawaii. *Modern Geol.* 2:207–223.

Greeley, R., and Gault, D. E. 1971. Endogenetic craters interpreted from crater counts on the inner wall of Copernicus. *Science* 171:477–479.

Green, J., and Short, N. M., eds. 1971. *Volcanic landforms and surface features*. New York: Springer-Verlag. 519 p.

Griggs, D. T. 1936. The factor of fatigue in rock exfoliation. *Jour. Geol.* 44:783–796.

Guest, J. E. 1971. Centers of igneous activity in the maria. In *Geology and Physics of the Moon*, ed. G. Fielder, pp. 41–53. Amsterdam: Elsevier Publishing Co.

Guest, J. E., and Fielder, G. 1968. Lunar ring structures and the nature of the maria. *Planet. Space Sci.* 16:665–673.

Guest, J. E., and Murray, J. B. 1969. Nature and origin of Tsiolkovsky crater, lunar farside. *Planetary Space Sci.* 17: 121–141.

———. 1971. A large scale surface pattern associated with the ejecta blanket and rays of Copernicus. *The Moon* 3:326–336.

Gutschewski, G. L.; Kinsler, D. C.; and Whitaker, E. 1971. *Atlas and gazetteer of the near side of the Moon*. Washington, D.C.: U.S. Govt. Printing Office, N.A.S.A. SP-241.

Hapke, B. W. 1963. A theoretical photometric function for the lunar surface. *Jour. Geophys. Research* 68:4571–4586.

———. 1965. Laboratory photometric studies relevant to the lunar surface. *Astronomical Jour.* 70:322.

Hapke, B. W., and Van Horn, H. 1963. Photometric studies of complex surfaces, with application to the Moon. *Jour. Geophys. Research* 68:4545–4570.

Hapke, B. W.; Cohen, A. J.; Cassidy, W. A.; and Wells, E. N. 1970. Solar radiation effects in lunar samples. *Science* 167: 745–747.

Harter, J. W., and Harter, R. G. 1971. Classification of lava tubes (submitted to *Bull. Speleo. Soc. America*), referenced in Greeley, R., 1971c. Observations of actively forming lava tubes and associated structures, Hawaii. *Modern Geol.* 2:207–223.

Hartmann, W. K. 1963. Radial structures surrounding lunar basins, I: The Imbrium system. *Univ. of Arizona Lunar and Planetary Lab. Commun.* 2:1–15.

———. 1964. Radial structures surrounding lunar basins, III: Orientale and other systems; conclusions. *Univ. of Arizona Lunar and Planetary Lab. Commun.* 2:175–191.

———. 1966. Early lunar cratering. *Icarus* 5:406–418.

———. 1967. Lunar crater counts, I: Alphonsus. *Univ. of Arizona Lunar and Planetary Lab. Commun.* 6:31–38.

———. 1970. Lunar cratering chronology. *Icarus* 13:299–301.

Hartmann, W. K., and Kuiper, G. P. 1962. Concentric structures surrounding lunar basins. *Univ. of Arizona Lunar and Planetary Lab. Commun.* 1:51–66.

Hartmann, W. K., and Wood, C. A. 1971. Moon: Origin and evolution of multi-ring basins. *The Moon* 3:3–76.

Head, J. W. 1972. Small-scale analogs of the Cayley Formation and Descartes in impact-associated deposits. In *Apollo 16 preliminary science report*, pp. 29:16–20. Washington, D.C.: U.S. Govt. Printing Office, N.A.S.A. SP-315.

Hinners, N. W., and El-Baz, F. 1972. Surface disturbances at the Apollo 15 landing site. In *Apollo 15 preliminary science report*, pp. 25:50–53. Washington, D.C.: U.S. Govt. Printing Office, N.A.S.A. SP-289.

Holcomb, R. 1971. Terraced depressions in lunar maria. *Jour. Geophys. Research* 76:5703–5711.

Howard, K. A., and Head, J. W. 1972. Regional geology of the Hadley Rille. In *Apollo 15 preliminary science report*, pp. 25: 53–58. Washington, D.C.: U.S. Govt. Printing Office, N.A.S.A. SP-289.

Hunt, G. R.; Salisbury, J. W.; and Vincent, R. K. 1968. Lunar eclipse: Infrared images and an anomaly of possible internal origin. *Science* 162:252–254.

614

Husain, L.; Sutter, J. F.; and Schaeffer, O. A. 1971. Ages of crystalline rocks from Fra Mauro. *Science* 173:1235–1236.

Jaggar, T. A. 1947. *Origin and development of craters.* Geol. Soc. America Mem. 21. 508 p.

Kosofsky, L. J., and El-Baz, F. 1970. *The Moon as viewed by Lunar Orbiter.* Washington, D.C.: U.S. Govt. Printing Office, N.A.S.A. SP-200.

Kuiper, G. P. 1965. Volcanic sublimates on the Earth and Moon. *Univ. of Arizona Lunar and Planetary Lab. Commun.* 3:36–60.

———. 1966a. Terrestrial and lunar collapse depressions. In *Ranger VIII and IX, part II. Experimenters' analyses and interpretations,* ed. R. L. Heacock, G. P. Kuiper, E. M. Shoemaker, H. C. Urey, and E. A. Whitaker, pp. 51–90. Pasadena: Cal. Inst. Tech., J.P.L., N.A.S.A. Tech. Rept. No. 32–800. 382 p.

———. 1966b. Ranger IX: The structure of the crater Alphonsus and surroundings. In *Ranger VIII and IX, part II. Experimenters' analyses and interpretations,* ed. R. L. Heacock, G. P. Kuiper, E. M. Shoemaker, H. C. Urey, and E. A. Whitaker, pp. 118–160. Pasadena: Cal. Inst. Tech., J.P.L., N.A.S.A. Tech. Rept. No. 32-800. 382 p.

———, ed. 1960. *Photographic lunar atlas.* Chicago: Univ. of Chicago Press.

Kuiper, G. P.; Whitaker, E. A.; Strom, R. G.; Fountain, J. W.; and Larson, S. M. 1967. Consolidated lunar atlas. *Univ. of Arizona Lunar and Planetary Lab. Contr.* no. 4.

Latham, G. V.; Ewing, M.; Press, F.; Sutton, G.; Dorman, J.; Nakamura, Y.; Toksoz, N.; Duennebier, F.; and Lammlein, D. 1971. Passive seismic experiment. In *Apollo 14 preliminary science report,* pp. 163–174. Washington, D.C.: U.S. Govt. Printing Office, N.A.S.A. SP-272.

Latham, G. V.; Ewing, M.; Press, F.; Sutton, G.; Dorman, J.; Nakamura, Y.; Toksoz, N.; Lammlein, D.; and Duennebier, F. 1972. Passive seismic experiment. In *Apollo 16 preliminary science report,* pp. 9:1–29. Washington, D.C.: U. S. Govt. Printing Office, N.A.S.A. SP-315.

Lingenfelter, R. E.; Peale, S. J.; and Schubert, G. 1968. Lunar rivers. *Science* 161:266–269.

Lipskiy, Yu N.; Pskovskiy, Y. P.; Gurshteyn, A. A.; Shevchenko, V. V.; and Pospergelis, M. M. 1966. Current problems of Moon's surface. *Kosmicheskiya Issledovaniya* 4:912–922 (in Russian); translated U.S. Goddard Space Flight Center, ST-LPS-10, 547.

Lowman, P. D., Jr. 1969. *Lunar panorama: A photographic guide to the geology of the Moon.* Feldmeilen/Zurich, Switzerland: Weltflugbild-Rheinhold A. Müler. 101 p.

Lunar Sample Preliminary Examination Team. 1971. Preliminary examination of lunar samples. In *Apollo 14 preliminary science report,* p. 129. Washington, D.C.: U.S. Govt. Printing Office, N.A.S.A. SP-272. 309 p.

McCall, G. J. H. 1965. The caldera analogy in selenology. In *Geological problems in lunar research,* ed. H. E. Whipple, pp. 843–873. *New York Acad. Sci. Annals* 123(2):367–1257.

McCauley, J. F. 1967. *Geologic map of the Hevelius region of the Moon.* U.S. Geol. Survey Misc. Geol. Inv. Map I-491.

McGetchin, T. R., and Head, J. 1973. Lunar cinder cones. *Science* 180:68–71.

McGill, G. E. 1971. Attitude of fractures bounding straight arcuate lunar rilles. *Icarus* 14:53–58.

Mackin, J. H. 1969. Origin of lunar maria. *Geol. Soc. America Bull.* 80:735–748.

Manley, W. D., Jr. 1969. Unpublished work.

Masursky, H. 1964. A preliminary report on the role of isostatic rebound in the geologic development of the lunar crater Ptolemaeus. In *Astrogeologic studies,* pp. 102–134. U.S. Geol. Survey open-file Ann. Prog. Rept., pt. A. 134 p.

———. 1968. Preliminary geologic interpretations of Lunar Orbiter photography. In *1969 N.A.S.A. authorization, pt. 3,* p. 664.

Milton, D. J. 1968. Geologic map of the Theophilus quadrangle of the Moon. U.S. Geol. Survey Map I-546.

Minneart, M. 1961. Photometry of the Moon. In *Planets and satellites,* ed. G. P. Kuiper and B. M. Middlehurst, pp. 213–248. Chicago: Univ. of Chicago Press. 601 p.

———. 1969. The effect of pulverization on the albedo of lunar rocks. *Icarus* 11:332–337.

Moore, J. A. 1967. Base surge in recent volcanic eruptions. *Bull. Volc.* 30:337–363.

Morris, E. C., and Wilhelms, D. E. 1967. *Geologic map of the Julius Caesar quadrangle of the Moon.* U.S. Geol. Survey Misc. Geol. Inv. Map I-510.

Muehlberger, W. R.; Baston, R. M.; Boudette, E. L.; Duke, C. M.; Eggleton, R. E.; Elston, D. P.; England, A. W.; Freeman, V. L.; Hait, M. H.; Hall, T. A.; Head, J. W.; Hodges, C. A.; Holt, H. E.; Jackson, E. D.; Jordan, J. A.; Larson, K. B.; Milton, D. J.; Reed, V. S.; Rennilson, J. J.; Schaber, G. G.; Schafer, J. P.; Silver, L. T.; Stuart-Alexander, D.; Sutton, R. L.; Swann, G. A.; Tyner, R. L.; Ulrich, G. E.; Wilshire, H. G.; Wolfe, E. W.; and Young, J. W. 1972. Preliminary geologic investigation of the Apollo 16 landing site. In *Apollo 16 preliminary science report,* pp. 6:1–81. Washington, D.C.: U.S. Govt. Printing Office, N.A.S.A. SP-315.

Murase, T., and McBirney, A. R. 1970. Viscosity of lunar lavas. *Science* 167:1491–1493.

Murray, J. B. 1971. Sinuous rilles. In *Geology and physics of the Moon,* ed. G. Fielder, pp. 27–39. Amsterdam: Elsevier Publishing Co.

Mutch, T. A. 1970. *Geology of the Moon: A stratigraphic view.* Princeton: Princeton Univ. Press. 324 p.

Nash, D. B., and Conel, J. E. 1971. Luminescence and reflectance of Apollo 12 samples. In *Proceedings of the second lunar science conference,* ed. A. A. Levinson; vol. 3: *Physical properties,* pp. 2235–2244. Cambridge, Mass.: M.I.T. Press.

Nichols, R. L. 1946. McCarthy's basalt flow, Valencia County, New Mexico. *Geol. Soc. America Bull.* 57:1049–1086.

REFERENCES CITED

Oberbeck, V. R. 1970. Lunar dimple craters. *Modern Geol.* 1: 161–171.

———. 1971. A mechanism for the production of lunar crater rays. *The Moon* 2:263–278.

Oberbeck, V. R., and Morrison, R. H. 1974. Laboratory simulation of the herringbone pattern associated with lunar secondary crater chains. *The Moon* (in press).

Oberbeck, V. R., and Quaide, W. L. 1967. Estimated thickness of a fragmental surface layer of Oceanus Procellarum. *Jour. Geophys. Research* 72:4697–4704.

———. 1968. Genetic implications of lunar regolith thickness. *Icarus* 9:446–465.

Oberbeck, V. R.; Quaide, W. L.; and Greeley, R. 1969. On the origin of lunar sinuous rilles. *Modern Geol.* 1:75–80.

Oberbeck, V. R.; Aoyagi, M.; Greeley, R.; and Lovas, M. 1972. Planimetric shapes of lunar rilles. In *Apollo 16 preliminary science report*, pp. 29:80–88. Washington, D.C.: U.S. Govt. Printing Office, N.A.S.A. SP-315.

O'Keefe, J.; Lowman, P.; and Cameron, W. S. 1966. Lunar ring dikes from Lunar Orbiter I. *Science* 155:77–79.

O'Keefe, J.; Cameron, W. S.; and Masursky, H. 1969. Hypersonic gas flow. In *Analysis of Apollo 8 photography and visual observations*, pp. 30–32. Washington, D.C.: U.S. Govt. Printing Office, N.A.S.A. SP-201. 337 p.

Öpik, E. J. 1969. The Moon's surface. In *Annual review of astronomy and astrophysics*, ed. L. Goldberg, pp. 473–526. Palo Alto, Cal.: Ann. Review Inc. 528 p.

Peacock, K. 1968. Multicolor photometry of the lunar surface. *Icarus* 9:16–66.

Pike, R. J. 1967. Schroeter's rule and the modification of lunar crater impact morphology. *Jour. Geophys. Research* 72:2099–2106.

———. 1971. Initial photographic analysis: Some preliminary interpretations of lunar mass-wasting processes from Apollo 10 photography. In *Analysis of Apollo 10 photography and visual observations*, pp. 14–20. Washington, D.C.: U.S. Govt. Printing Office, N.A.S.A. SP-232. 226 p.

Quaide, W. L. 1965. Rilles, ridges, and domes—clues to maria history. *Icarus* 4:374–389.

Quaide, W. L., and Oberbeck, V. R. 1968. Thickness determinations of the lunar surface layer from lunar impact craters. *Jour. Geophys. Research* 73:5247–5270.

Quaide, W. L.; Gault, D. E.; and Schmidt, R. A. 1965. Gravitative effects on lunar impact structures. In *Geological problems in lunar research*, ed. H. E. Whipple, pp. 563–572. *New York Acad. Sci. Annals* 123(2):367–1257.

Rehfuss, D. E. 1972. Lunar winds. *Jour. Geophys. Research* 77: 6303–6315.

Roberts, W. A. 1968. Shock crater ejecta characteristics. In *Shock metamorphism of natural materials*, ed. B. M. French and N. M. Short, pp. 101–114. Baltimore: Mono Book Corp. 644 p.

Roddy, D. J. 1968. The Flynn Creek crater, Tennessee. In *Shock metamorphism of natural materials*, ed. B. M. French and N. M. Short, pp. 291–322. Baltimore: Mono Book Corp. 644 p.

Ronca, L. B. 1965. A geologic model of Mare Humorum. *Icarus* 4:390–395.

Ross, H. P. 1968. A simplified mathematical model for lunar crater erosion. *Jour. Geophys. Research* 73:1343–1354.

Russell, I. C. 1889. Quaternary history of Mono Valley, California. *U.S. Geol. Survey Ann. Rept.* 8:261–394.

Ryan, J. A. 1962. The case against thermal fracturing at the lunar surface. *Jour. Geophys. Research* 67:2549–2558.

Salisbury, J. W. 1970. Albedo of lunar soil. *Icarus* 13:509–512.

Salisbury, J. W., and Smalley, V. G. 1964. The lunar surface layer. In *The lunar surface layer*, ed. J. W. Salisbury and P. E. Glaser, pp. 411–413. New York: Academic Press. 532 p.

Schubert, G.; Lingenfelter, R. E.; and Peale, S. J. 1970. The morphology, distribution, and origin of lunar sinuous rilles. *Rev. Geophys. Space Sci.* 8:199–224.

Schumm, S. A. 1970. Experimental studies on the formation of lunar surface features by fluidization. *Geol. Soc. America Bull.* 81:2539–2552.

Scott, D. H.; West, M. N.; Lucchitta, B. K.; and McCauley, J. F. 1971. Preliminary geologic results from orbital photography. In *Apollo 14 preliminary science report*, pp. 274–283. Washington, D.C.: U.S. Govt. Printing Office, N.A.S.A. SP-272. 309 p.

Scott, R. F. 1967. Viscous flow of craters. *Icarus* 7:139–148.

Sharpe, C. F. S. 1938. *Landslides and related phenomena.* New York: Columbia Univ. Press. 137 p.

Shoemaker, E. M. 1962. Interpretation of lunar craters. In *Physics and astronomy of the Moon*, ed. Z. Kopal, pp. 283–359. New York: Academic Press. 453 p.

Shoemaker, E. M., and Morris, E. C. 1968. Preliminary geologic map of the Surveyor VII landing site. In *Surveyor VII mission report, part II*, p. 29. Pasadena: Cal. Inst. Tech., J.P.L. Tech. Rept. 32-1264. 344 p.

———. 1969. Disturbances of the surface. In *Surveyor program results*, pp. 98–99. Washington, D.C.: U.S. Govt. Printing Office, N.A.S.A. SP-184. 425 p.

Shoemaker, E. M.; Batson, R. M.; Holt, H. E.; Morris, E. C.; Rennilson, J.J.; and Whitaker, E. A. 1968. Television observations from Surveyor VII. In *Surveyor VII mission report, part II*, p. 976. Pasadena: Cal. Inst. Tech., J.P.L. Tech. Rept. 32-1264. 344 p.

Shoemaker, E. M.; Hait, M. H.; Swann, G. A.; Schleicher, D. L.; Dahlem, D. H.; Schaber, G. G.; and Sutton, R. L. 1970. Lunar regolith at Tranquility Base. *Science* 167:452–455.

Silver, L. T. 1971. U-Th-Pb isotope systems in Apollo 11 and 12 regolithic materials and a possible age for the Copernicus impact event (abs.). Program, fifty-second annual meeting of the American Geophys. Union, Washington, D.C.

Sjogren, W.; Muller, P. M.; Gottlieb, P.; Wong, L.; Buechler, G.;

Downs, W; and Prislin, R. 1971. Lunar surface mass distribution from dynamical point-mass solution. *The Moon* 2:338–352.

Sjogren, W. L.; Muller, P. M.; and Wollenhaupt, W. R. 1972. S-band transponder experiment. In *Apollo 16 preliminary science report*, pp. 24:1–7. Washington, D.C.: U.S. Govt. Printing Office, N.A.S.A. SP-315.

Smalley, V. G. 1965. The lunar crater Dionysius. *Icarus* 4:433–436.

Smith, E. T. 1970. A comparison of selected lunar and terrestrial volcanic domes. Ph.D. dissertation, Univ. of New Mexico, Albuquerque.

Smith, J. V.; Anderson, A. T.; Newton, R. C.; Olsen, E. J.; Crewe, A. V.; Isaacson, M. S.; Johnson, D.; and Wyllie, P. J. 1970. Petrologic history of the Moon inferred from petrography, mineralogy, and petrogenesis of Apollo 11 rocks. In *Proceedings of the Apollo 11 lunar science conference*, ed. A. A. Levinson; vol. 1: *Mineralogy and petrology*, pp. 897–925. New York: Pergamon Press. 990 p.

Smith, R. L., and Bailey, R. 1968. Resurgent cauldrons. In *Studies in volcanology*, ed. R. R. Coats, R. L. Hay, and C. A. Anderson, pp. 613–662. Geol. Soc. America Mem. 116. 678 p.

Smith, R. L.; Bailey, R. A.; and Ross, C. S. 1961. Structural evolution of the Valles caldera, New Mexico, and its bearing on the emplacement of ring dikes. *U.S. Geol. Survey Prof. Paper 424D*, pp. 145–149.

Soderblom, L. A. 1970. A model for small-impact erosion applied to the lunar surface. *Jour. Geophys. Research* 75:2655–2661.

Spurr, J. E. 1944. *Geology applied to selenology: I. The Imbrium Plain region of the Moon.* Lancaster, Pa.: Science Press Pub. Co. 112 p.

———. 1945. *Geology applied to selenology: II. The features of the Moon.* Lancaster, Pa.: Science Press Pub. Co. 318 p.

———. 1948. *Geology applied to selenology: III. Lunar catastrophic history.* Concord, N.H.: Rumford Press. 253 p.

———. 1949. *Geology applied to selenology: IV. The shrunken Moon.* Concord, N.H.: Rumford Press. 207 p.

Steinberg, G. S. 1968. Comparative morphology of lunar craters and rings and some volcanic formations in Kamchatka. *Icarus* 8:387–403.

Strom, R. G. 1964. Analysis of lunar lineaments. I. Tectonic maps of the Moon. *Univ. of Arizona Lunar and Planetary Lab. Commun.* 2:205–216.

———. 1965. Map of flows photographed in Mare Imbrium. In *Ranger VII, part II. Experimenters' analyses and interpretations*, ed. R. L. Heacock, G. P. Kuiper, E. M. Shoemaker, H. C. Urey, and E. A. Whitaker, p. 32. Pasadena: Cal. Inst. Tech., J.P.L. Tech. Rept. No. 32-700. 154 p.

———. 1966. Sinuous rilles. In *Ranger VIII and IX, part II. Experimenters' analyses and interpretations*, ed. R. L. Heacock, G. P. Kuiper, E. M. Shoemaker, H. C. Urey, and E. A. Whitaker, pp. 199–210. Pasadena: Cal. Inst. Tech., J.P.L. Tech. Rept. No. 32-800. 382 p.

———. 1971. Lunar mare ridges, rings, and volcanic ring complexes. *Modern Geol.* 2:133–157.

Strom, R. G., and Fielder, G. 1970. Multiphase eruptions associated with the lunar crater Tycho and Aristarchus. *Univ. of Arizona Lunar and Planetary Lab. Commun.* 8:235–288.

Stuart-Alexander, D. E., and Howard, K. A. 1970. Lunar maria and circular basins—a review. *Icarus* 12:440–456.

Sutter, J. F.; Husain, L.; and Schaeffer, O. A. 1971. Argon 40/Argon 39 ages from Fra Mauro. *Earth Planet. Sci. Lett.* 11:249–253.

Titley, S. R. 1966. Seismic energy as an agent of morphologic modification on the Moon. In *Astrogeologic studies*, pp. 87–114. U.S. Geol. Survey open-file Ann. Prog. Rept., pt. A. 305 p.

Tjia, H. D. 1970. Lunar wrinkle ridges—indicative of strike-slip faulting. *Geol. Soc. America Bull.* 81:3095–3100.

Turner, G. 1971. Argon 40–Argon 39 dating: The optimization of irradiation parameters. *Earth Planet Sci. Lett.* 10:227–234.

Ulrich, G. 1966. Probable igneous relations in the floor of the crater J. Herschel. In *Astrogeologic studies*, pp. 123–132. U.S. Geol. Survey open-file Ann. Prog. Rept., pt. A. 305 p.

Urey, H. C. 1967. Water on the Moon. *Nature* 216:1094–1095.

Van Dorn, W. G. 1968. Tsunamis on the Moon? *Nature* 220:1102.

———. 1969. Lunar maria: Structure and evolution. *Science* 165:693–695.

de Vaucouleurs, G. 1964. Geometric and photometric parameters of the terrestrial planets. *Icarus* 3:187–235.

Warner, B. 1961. Holistic approach to selenology. *Nature* 191:586.

Whitaker, E. A. 1969. Sublimates. In *Analysis of Apollo 8 photography and visual observations*, pp. 34–35. Washington, D.C.: U.S. Govt. Printing Office, N.A.S.A. SP-201. 337 p.

Wilhelms, D. E. 1964. Fra Mauro and Cayley formations in Mare Vaporum and Julius Caesar. In *Astrogeologic studies*, pp. 13–28. U.S. Geol. Survey open-file Ann. Prog. Rept., pt. A. 124 p.

———. 1970. Summary of lunar stratigraphy—telescopic observations. *Contributions to astrogeology, U.S. Geol. Survey Prof. Paper 599-F.* Washington, D.C.: U.S. Govt. Printing Office. 47 p.

———. 1971. Initial photographic analysis: Terra volcanics of the near side of the Moon. In *Analysis of Apollo 10 photography and visual observations*, pp. 26–29. Washington, D.C.: U.S. Govt. Printing Office, N.A.S.A. SP-232. 226 p.

Wilhelms, D. E., and McCauley, J. F. 1969. Volcanic materials in the lunar terrae—Orbiter observations (abs.). *Am. Geophys. Union Trans.* 50:230.

Williams, H. 1941. Calderas and their origin. *Univ. of California Pub. Geol. Sci.* 25:239–346.

Williams, H., and McBirney, A. R. 1968. *An investigation of volcanic depressions, part I. Geological and geophysical features of calderas.* Univ. of Oregon, Center for Volcanology, progress report for N.A.S.A. Research Grant NGH-38-033-012. 87 p.

617

Wise, D. U., and Yates, M. T. 1970. Mascons as structural relief on a lunar "Moho." *Jour. Geophys. Research* 75:261–268.

Wood, C. A. 1968. Statistics of central peaks in lunar craters. *Univ. of Arizona Lunar and Planetary Lab. Commun.* 7:157–160.

Wood, J. A.; Dickey, J. S., Jr.; Marvin, U. B.; and Powell, B. N. 1970. Lunar anorthosites and a geophysical model of the Moon. In *Proceedings of the Apollo 11 lunar science conference*, ed. A. A. Levinson; vol. 1: *Mineralogy and petrology*, pp. 965–987. New York: Pergamon Press. 990 p.

Wright, F. E.; Wright, F. H.; and Wright, H. 1963. The lunar surface: Introduction. In *The solar system*, ed. G. P. Kuiper and B. M. Middlehurst; vol. 4: *Moon, meteorites, and Comets*, pp. 1–56. Univ. of Chicago Press.

Young, G. A. 1965. *The physics of the base surge*. White Oak, Md.: U.S. Naval Ordnance Lab., NOL-TR64-103.

Young, J. 1933. Preliminary report of a statistical investigation of the diameters of lunar craters. *Brit. Astron. Assoc. Jour.* 43:201–209.

―――. 1940. A statistical investigation of the diameters and distribution of lunar craters. *Brit. Astron. Assoc. Jour.* 50: 309–326.

Young, R. A. 1972. Lunar volcanism: Mare ridges and sinuous rilles. In *Apollo 16 preliminary science report*, pp. 29:79–80. Washington, D.C.: U.S. Govt. Printing Office, N.A.S.A. SP-315.

INDEX

Boldface numbers refer to plates.